Applications of Palaeontology

Techniques and Case Studies

Palaeontology, the scientific study of fossils, has developed from a descriptive science to an analytical science used to interpret relationships between Earth and life history. This book provides a comprehensive and thematic treatment of applied palaeontology, covering the use of fossils in the ordering of rocks in time and in space, in biostratigraphy, palaeobiology and sequence stratigraphy.

In this new book, Robert Wynn Jones presents a practical workflow for applied palaeontology, including sample acquisition, preparation and analysis, and interpretation and integration. He then presents numerous case studies that demonstrate the applicability and value of the subject to areas such as petroleum, mineral and coal exploration and exploitation, engineering geology and environmental science. Specialist applications outside the geosciences (including archaeology, forensic science, medical palynology, entomopalynology and melissopalynology) are also addressed.

Abundantly illustrated and referenced, *Applications of Palaeontology* provides a user-friendly reference for academic researchers and professionals across a range of disciplines and industry settings.

The author, struggling somewhat to come to terms with the equestrian demands of fieldwork in the Zagros mountains of Iran, April 2002. Photo courtesy of the author.

ROBERT WYNN JONES gained his B.Sc. in geological sciences at the University of Birmingham in 1979 and his Ph.D. at the University College of Wales, Aberystwyth, in 1982. Since then, he has worked as a micropalaeontologist and biostratigrapher in the petroleum industry, and has been a Principal Consultant Biostratigrapher at BG

Group since 2009. His industrial work involves analysis of micropalaeontological samples; interpretation of micro- and macropalaeontological data; and integration of palaeontological interpretations into geological models for petroleum exploration and reservoir exploitation. He has worked on petroliferous sedimentary basins from all around the world, and on rocks and fossils from wide ranges of ages and facies. Outside work, Dr Jones also maintains an active interest in academic research, especially in the study of foraminiferal taxonomy, palaeobiology, biostratigraphy and historical micropalaeontology. He has over 100 scientific publications to his name, including six books. He is a Scientific Associate in the Department of Palaeontology at the Natural History Museum in London.

Applications of Palaeontology

Techniques and Case Studies

Robert Wynn Jones
The Natural History Museum, London

CAMBRIDGE
UNIVERSITY PRESS

University Printing House, Cambridge CB2 8BS, United Kingdom

One Liberty Plaza, 20th Floor, New York, NY 10006, USA

477 Williamstown Road, Port Melbourne, VIC 3207, Australia

4843/24, 2nd Floor, Ansari Road, Daryaganj, Delhi - 110002, India

79 Anson Road, #06-04/06, Singapore 079906

Cambridge University Press is part of the University of Cambridge.

It furthers the University's mission by disseminating knowledge in the pursuit of education, learning and research at the highest international levels of excellence.

www.cambridge.org
Information on this title: www.cambridge.org/9781108446976

© Robert Wynn Jones 2011

First published 2011
First paperback edition 2017

A catalogue record for this publication is available from the British Library

Library of Congress Cataloging in Publication data
Jones, Robert Wynn.
  Case studies in applied palaeontology / Robert Wynn Jones.
    p.   cm.
  Includes bibliographical references and index.
  ISBN 978-1-107-00523-5
  1. Paleontology–Research–Case studies.   2. Earth sciences–Research–Case studies.
  3. Engineering geology–Research–Case studies.   I. Title.
  QE711.3.J665 2011
  560.72–dc22

                                                              2011001069

ISBN 978-1-107-00523-5 Hardback
ISBN 978-1-108-44697-6 Paperback

Dedicated to the memory of my late colleague and friend, Garry D. Jones, a great exponent and proponent of applied palaeontology.

Contents

Preface

Humankind has always been fascinated by fossils, by their beauty and their mystery, their charm and their strangeness, their mute testimony to lives and worlds lost unimaginably long ago. In prehistoric times, our forebears not only collected fossils, but evidently treated them as valued artefacts, as evidenced, for example, by the discovery of an ammonite at an Upper Palaeolithic burial site in Aveline's Hole in Burrington in the West Country in Britain, and numerous different types of fossil at Cro Magnon sites in the Vezere valley in the Perigord region of France, truly the birthplace of European civilisation (many of which are now displayed in the magnificent 'Museum of Prehistory' in Les Eyzies). The habit persisted both in so-called primitive and so-called advanced societies through historical times.

Palaeontology, that is, the scientific study of fossils, may be said to have originated at least as long ago as the sixteenth century, and, obviously, continues to be practised in the present day. The earliest written observations on fossils were made by the German Bauer, or Agricola, in his book '*De natura fossilum*', and the earliest illustrations by the Swiss Gesner in his book '*De rerum fossilium lapidum et gemmarum*', both of which date from the sixteenth century. The usage by these and other early observers of the term 'fossil', from the Latin *fodere*, meaning 'to dig', pertained to literally anything dug up from the ground or mined, including what we would now classify as minerals, crystals and gemstones. The earliest interpretations as to the nature of what we would now accept as fossils were made by the Danish anatomist Stensen, or Steno, working in the Medici court in Florence, in his publications dating from the latter part of the seventeenth century. Steno applied Descartes' 'method of doubt' and his own deductive logic to demonstrate that the so-called *glossopetrae* or 'tongue stones' much valued in medieval Europe for their supposed medicinal properties were in fact not the tongues of snakes turned to stone by St Paul, as was the superstition, but the fossilised equivalents of the sharks' teeth he was familiar with from his dissection work. Elsewhere in his writings, Steno established three important principles of stratigraphy, namely the 'principle of superposition', the 'principle of original horizontality' and the 'principle of lateral continuity', such that he is regarded by many as the true founder of that science. Incidentally, in later life, he renounced science for religion, and was recently made a saint by John Paul II.

There may be said to have been three, partially overlapping, areas or phases of subsequent palaeontological study: the descriptive; the synoptic; and the interpretive. The emphasis through the three phases has shifted from the documentation of fossils to their application in establishing the ordering of containing rocks in time and in space; from pure to applied. The pure descriptive phase began with the first descriptions of fossil species conforming to modern standards, made following the introduction of the binomial system for the naming of species by Linne, or Linnaeus, in the late eighteenth century. The synoptic phase has continued into the twenty-first century, with the establishment of higher-level taxonomic classification systems based on morphology and phylogeny, made following the publication of '*On the Origin of Species . . .*' by Darwin in the late nineteenth century, and the advances in cladistics and molecular biology in the twentieth. The applied interpretive phase, ultimately resulting in the development of, and advances in, the applied sub-disciplines of biostratigraphy and palaeobiology, began with the establishment of the 'Law of Superposition' and the 'Law of Strata identified by organised Fossils' by the Briton William ('Strata') Smith in the late eighteenth and early nineteenth centuries; and by the publication also by Smith of the first geological map, of Great Britain, 'the map that changed the world'. The first application of biostratigraphy in the petroleum industry was by the Pole Josef

Grzybowski in the late nineteenth and early twentieth centuries. At a time when (micro)palaeontology was essentially in a stage of synthesis, it was he who first used the discipline in an analytical fashion to solve geological problems encountered in the oil-fields of the eastern Carpathians, those around the village of Potok being the oldest still in production anywhere in the world. His contribution to biostratigraphy and also to palaeobiology has long been recognised and justly acclaimed in his own country, but is sadly seldom acknowledged in the west. In the future, applied palaeontology will continue to play a vital role in exploiting the world's discovered petroleum and other mineral resources, and in exploring for undiscovered reserves. In view of the growing concern about the environment, applications in environmental science, and outwith the exploitative industries, are also likely to come to the fore.

Unlike its immediate predecessor (Jones, 2006), this book deals essentially only with applied palaeontology, and contains many more case studies of applications. It will be of value to professionals in industry and elsewhere; including not only applied palaeontologists, palaeobiologists and biostratigraphers, but also petroleum, minerals, mining and engineering geologists, and environmental scientists, and, to a lesser extent, archaeologists, forensic scientists and others.

The first part, comprising Chapters 1–4, deals with general applications of palaeontology in the interpretation of Earth and life history and environments.

Chapter 1 deals with *work-flows in applied palaeontology*, and includes sections on project specification and management; on sample acquisition, processing and analysis; and on analytical data acquisition.

Chapter 2 deals with *biostratigraphy and allied disciplines, and stratigraphic time-scales*, and includes sections on the biostratigraphic significance and usefulness of the principal fossil groups; on biostratigraphy; on Proterozoic, on Palaeozoic, on Mesozoic and on Cenozoic biostratigraphy; on biostratigraphic technologies; on allied disciplines; and on stratigraphic time-scales. The section on biostratigraphy includes sub-sections on biostratigraphic zonation or biozonation, and on correlation. The section on biostratigraphic technologies includes sub-sections on graphic correlation; on constrained

optimisation (CONOP); on ranking and scaling (RASC); and on cluster analysis. The section on allied disciplines includes sub-sections on chemostratigraphy; on cyclostratigraphy; on heavy minerals; on magnetostratigraphy; on radiometric dating; and on Quaternary dating methods. The section on stratigraphic time-scales includes a sub-section on global stratotype sections and points (GSSPs).

Chapter 3 deals with *palaeobiology*, and includes sections on the palaeobiological significance and usefulness of the principal fossil groups; on palaeobiological, palaeoecological or palaeoenvironmental interpretation; on palaeobathymetry; on palaeobiogeography; on palaeoclimatology; on palaeo-oceanography; on quantitative and other interpretive techniques in palaeobiology; and on key biological events in Earth history. The section on palaeobiological, palaeoecological or palaeoenvironmental interpretation includes sub-sections on palaeoenvironmental interpretation on the basis of analogy; and palaeoenvironmental interpretation on the basis of functional morphology. The section on palaeobathymetry includes sub-sections on non-marine environments; on marine environments; on marginal marine environments; on shallow marine environments; and on deep marine environments. The sub-section on marginal marine environments covers deltas; and, in some detail, the palaeontological characterisation of marginal marine, peri-reefal sub-environments. The sub-section on shallow marine environments covers reefs, and covers, in some detail, the palaeontological characterisation of peri-deltaic sub-environments. The sub-section on deep marine environments covers oxygen minimum zones, submarine fans, and deep marine hydrothermal vents, 'nekton falls' and 'cold (hydrocarbon) seeps'; and, in some detail, deep-water agglutinated foraminifera, the palaeontological characterisation of submarine fan sub-environments, and benthic foraminifera and ostracods associated with cold (hydrocarbon) seeps. The section on palaeobiogeography includes sub-sections on Palaeozoic, on Mesozoic and on Cenozoic palaeobiogeography. The section on palaeoclimatology includes sub-sections on Palaeozoic, on Mesozoic and on Cenozoic palaeoclimatology. The section on quantitative and other interpretive techniques in palaeobiology includes sub-sections on palaeobathymetric interpretation techniques; palaeobiogeographic and palaeoclimatological interpretation techniques; cluster analysis and 'fuzzy C means' (FCM) cluster analysis; and 'fuzzy logic'. The section on key biological events in Earth history includes sub-sections on the Proterozoic, on the Palaeozoic, on the Mesozoic and on the Cenozoic. It also covers in detail foraminiferal diversity trends through time, new evidence for land mammal dispersal across the North Atlantic in the Early Eocene, and aspects of the palaeogeography and palaeoclimate of the Oligocene–Holocene of the Old World, and consequences for land mammal evolution and dispersal.

Chapter 4 deals with *sequence stratigraphy*, and includes sections on definitions; on general and clastic sequence stratigraphy; on carbonate sequence stratigraphy; on mixed sequence stratigraphy; on seismic facies analysis; on integration of palaeontological data; and on chronostratigraphic diagrams. It also covers, in some detail, palaeontological inputs into the characterisation of systems tracts.

The second part, comprising Chapters 5–10, deals with specific applications of palaeontology in industry and elsewhere, and with case studies of applications.

Chapter 5 deals with *petroleum geology*, and includes sections on petroleum source-rocks and systems, reservoir-rocks, and cap-rocks and traps; applications and case studies in petroleum exploration, in reservoir exploitation, and in well-site operations; and unconventional petroleum geology. It also covers, in some detail, palaeontology and health, safety and environmental issues in the petroleum industry. The section on petroleum source-rocks and systems, reservoir-rocks, and cap-rocks and traps covers, in some detail, palaeontological inputs into petroleum systems analysis, nummulite banks and reservoirs, rudist reefs and reservoirs, the palaeontological characterisation of cap-rocks, and stratigraphic and palaeobiological controls on source-, reservoir- and cap-rock distributions. The section on applications in petroleum exploration includes case studies on the Middle East; on the North Sea; on northern South America and the Caribbean; on the South Atlantic; and on the Indian subcontinent. The section on

applications in reservoir exploitation includes case studies on reservoir characterisation in shallow marine, peri-reefal carbonate reservoir (Al Huwaisah, Dhulaima, Lekhwair and Yibal fields in Oman, and Shaybah field in Saudi Arabia); in marginal to shallow marine, peri-deltaic clastic reservoirs (Gullfaks, Snorre and Statfjord fields in the Norwegian sector, and Ninian and Thistle fields in the UK sector, in the North Sea; and Pedernales field in Venezuela); and in a deep marine, submarine fan, clastic reservoir (Forties field, UK sector, North Sea). The section on applications in well-site operations includes case studies in 'biosteering' in a shallow marine carbonate reservoir (Sajaa field, United Arab Emirates); in a deep marine carbonate reservoir (Valhall field, Norwegian sector, North Sea); in a shallow marine clastic reservoir (Cusiana field, Colombia); and in a deep marine, submarine fan, clastic reservoir (Andrew field, UK sector, North Sea). The section on unconventional petroleum geology covers coal-bed methane (CBM) or coal-seam gas (CSG), shale gas, and gas or methane hydrate.

Chapter 6 deals with *mineral exploration and exploitation*, and includes sections on applications and case studies in mineral exploration and exploitation. The section on applications in mineral exploration includes a case study on La Troya mine, Spain. The section on applications in mineral exploitation includes case studies on Pitstone quarry, Hertfordshire, UK, and on East Grimstead quarry, Wiltshire, UK.

Chapter 7 deals with *coal geology and mining*, and includes sections on applications and case studies in coal geology and mining, in Great Britain, and in South Africa.

Chapter 8 deals with *engineering geology*, and includes sections on applications and case studies in site investigation, and in seismic hazard assessment. The section on applications in site investigation includes case studies on the Channel tunnel, UK; on the Thames barrier, UK; on 'Project Orwell', UK, and on site investigation in the petroleum industry, in Azerbaijan and Egypt. The section on applications in seismic hazard assessment includes case studies on the Strait of Juan de Fuca, Vancouver Island, Canada; on Iyo-nada Bay, Japan; on Hawke's Bay, New Zealand; on the Cabo de Gata lagoon, Almeria, Spain; on Lake Sapanca, Turkey; and on western Crete.

Chapter 9 deals with *environmental science*, and includes sections on applications and case studies in environmental impact assessment (EIA), in environmental monitoring, in bioremediation, and in anthropogenically mediated global change. The section on applications in EIA includes case studies on the North Slope of Alaska, and on Wytch Farm, Dorset, UK. The section on applications in environmental monitoring includes case studies on environmental monitoring of natural and anthropogenic effects on water quality, including domestic and industrial pollution, and environmental monitoring of coral reef vitality. The section on applications in bioremediation includes case studies on bioremediation of the *Exxon Valdez* oil spill, and bioremediation in the aftermath of Operation Desert Storm. The section on applications in anthropogenically mediated global change includes case studies on ocean acidification and carbon dioxide sequestration.

Chapter 10 deals with *other applications and case studies*, and includes sections on applications and case studies in archaeology, and in forensic science; and on miscellaneous other applications. The section on applications in archaeology includes case studies on Westbury Cave, Somerset, UK (Early Palaeolithic); on Boxgrove, Sussex, UK (Early Palaeolithic); on Massawa, Eritrea (Middle Palaeolithic); on Goat's Hole, Paviland, Gower, UK (Middle Palaeolithic); on 'Doggerland', North Sea (Mesolithic); on Mount Sandel, Coleraine, Co. Derry, Northern Ireland (Mesolithic); on Fayum, Egypt ('Epipalaeolithic'–Neolithic); on Littleton, Co. Tipperary, Ireland (Neolithic to Medieval); on Skara Brae, Orkney, UK (Neolithic); on the Tyrolean Alps (Chalcolithic or Copper Age); and on the City of London (Medieval). It also covers, in some detail, the palaeoenvironmental interpretation of the Pleistocene–Holocene of the British Isles, using proxy Recent benthic foraminiferal distribution data. The section on applications in forensic science includes case studies on the use of calcareous nannofossils, of diatoms, of spores and pollen, and of insects, in forensic science. The section on miscellaneous other applications contains sub-sections on medical palynology; on entomopalynology; and on melissopalynology.

Acknowledgements

Firstly, I would like to acknowledge the American Association for the Advancement of Science, Blackwell, Cambridge University Press, the Canadian Society of Petroleum Geologists, the Cushman Foundation for Foraminiferal Research, the Geological Society of London, the Geological Survey of South Africa, the Geologischen Bundesantalt (Wien), the Grzybowski Foundation, Gulf Petro-Link, Longman, the Micropalaeontological Society, Katharina von Salis Perch-Nielsen, Poyser, the Society of Economic Paleontologists and Mineralogists, the Smithsonian Institution Press, Springer and Wiley, for providing permission to reproduce copyright figures (and especially Wiley, for generously waiving their customary fee).

I would also like to acknowledge Susan Francis, Chris Hudson, Chris Miller and Lindsay Nightingale at Cambridge University Press for their help and support in seeing the project through to production; and, for their various contributions to the work, Tony Barwise, Paul Batey, Andrei Belopolsky, Jon Bennett, Jaewan Bhajan, Scott Carmichael, Bruno David, Alison Davies, Richard Dixon, Helen Doran, Dan Finucane, Karize Hosein, Jake Hossack, Nev Jones, Mike Larby, Steve Lowe, Steve Matthews, Joaquin Naar, Mark Osborne, Simon Payne, Julian Penge, Neil Piggott, Dave Pocknall, Pat Randell, Johan Sydow, Aubrey Thomas and the late Shakeel Akhter at BP; John Argent, Andrew Barnett, Jason Canning, Nigel Cross, James Derry, Laurent de Verteuil, Moumita Doubey, John Fisher, Dave Freemantle, Mark Houchen, Nick Lee, Mike Martin, Israel Polonio, Andy Racey, Sandip Roy, Shouvik Roy, Adriana del Pino Sanchez, Matt Wakefield and Paul Wright at BG Group; Paul Barrett, Andy Currant, Jerry Hooker, Noel Morris, Simon Parfitt, Brian Rosen, Andrew Smith, Jon Todd and John Whittaker at the Natural History Museum, London; and Elisabeth Alve, Christine Barras, Martin Bates, Erica Bicchi, Barry Carr-Brown, Eric Deville, Flavia Fiorini, John Frampton, David Haig, Wyn Hughes, Frans Jorissen, Sev Kender, Tony King, John Murray, Jan Pawlowski, Cesare Pappazoni, Jim Pindell, Pratul Saraswati, Dave Shaw, Mike Simmons and Brent Wilson.

Lastly, and on a more a more personal note, I would also like to acknowledge the usual suspects, plus Air, the Alabama 3, Allegri, the American Music Club, And You Will Know Us By The Trail Of Dead, Antony and the Johnsons, Any Trouble, Arcade Fire, Archer, the Arctic Monkeys, Richard Ashcroft, Asobi Seksu, Dan Auerbach, Babyshambles, Shirley Bassey, Bat for Lashes, Beck, Belle and Sebastian, Ben's Brother, the Blind Boys of Alabama, Bloc Party, the Bombay Bicycle Club, The Boy Who Trapped the Sun, British Sea Power, The Broken Family Band, Jeff Buckley, Byrd, The Chemical Brothers, Tony Christie, The Cinematic Orchestra, Jarvis Cocker ('I met her in the Museum of Palaeontology, and I make no bones about it . . .'), The Coral, Cousteau, Danger Mouse and Sparklehorse, Ray Davies, de Machault, Desmond Dekker, the Detroit Social Club, Dinosaur Jr, Nick Drake, Dufay, Duffy, Richard Durand, the Editors, the Eels, Ludovico Einaudi, Mark Eitzel, Elbow, Electrelane, The Fall, the Fleet Foxes, Renee Fleming, Florence and the Machine, Brandon Flowers, the Foo Fighters, The Fray, Charlotte Gainsbourg, Jan Garbarek, The Gaslight Anthem, Mary Gauthier, Gesualdo, God Help the Girl, Howard Goodall, Ellie Gouldring, Grinderman, Grizzly Bear, Charlotte Hatherley, Richard Hawley, Levon Helm, The Horrors, Hanne Hukkelberg, I Am Kloot, Tom Jones, Janis Joplin, Josquin, the Kenyan Boys Choir, Angelique Kidjo, The Killers, Diana Krall, Alison Krauss, Kremerata Baltica, Jon Legend and the Roots, the Libertines, Lord Tanamo, Los Lobos, Lotti, The Low Anthem, Laura Marling, Cecilia McDowall, Maximo Park, The Ministry of Sound, Janelle Monae, Matt Monro, the Morriston Orpheus Male Voice Choir, the Mull Historical Society, Mumford and Sons, The Mystery Jets, Nirvana, Palestrini, Paolo Nutini, Roy Orbison, Orbital, John Otway, Parry, Passion Pit, The Phantom Band, Placebo, Quatuor Ebene, Razorlight, Lou Reed, Jim Reeves, the Rhymney Silurian (!) Male Choir,

Alasdair Roberts, Rumer, Nitin Sawnhey, Seasick Steve, Schutz, Scroobious Pip, Bob Seger, Sigur Ros, Sleater-Kinney, Soft Cell, Sophie Solomon, Regina Spektor, Starsailor, Status Quo, the Stone Roses, Joe Strummer and the Mescaleros, The Tallest Man on Earth, They Might Be Giants ('I am a palaeontologist . . .'), Tindersticks, Toots and the Maytals, Vieux Farka Toure, the Undertones, Vangelis, The View, Rufus Wainwright, White Denim, Eric Whitacre, the White Stripes, The Who, Patrick Wolf, and Wolf People, for their musical inspiration; and my wife Heather, and my sons Wynn and Gethin, now grown-up, for their continuing love and support (and in Gethin's case also for continuing technological assistance). I will walk the dog now.

A generic work-flow in applied palaeontology is shown in Fig. 1.1.

It will be seen that the key constituent elements are: project specification and management; sample acquisition, processing and analysis; and analytical data acquisition. Each of these elements is discussed in turn below (and interpretation and integration in Chapters 2–10).

1.1 PROJECT SPECIFICATION AND MANAGEMENT

Project specification involves, firstly, the identification of the technical and business objectives of the project; and secondly, the formulation of a plan and budget appropriate to the timely delivery of these objectives to the customer, taking into account such factors as timing, resourcing and third-party involvement.

Project management involves assurance of adherence to the plan and budget. It also involves assurance of quality, and of compliance with Health, Safety and Environmental, or HSE, standards, as appropriate.

1.2 SAMPLE ACQUISITION

The sample acquisition strategy is determined by the technical or research objectives of the project, and by the available budget. In most cases, the principal factor to be considered is the number and spacing of samples, which will determine the ultimately achievable biostratigraphic resolution, and hence the value of the project, as well as the cost.

1.2.1 Surface sample acquisition

Acquisition of surface samples for their fossil content is required to constrain surface geological mapping and correlation, among other reasons.

Equipment. The (more-or-less) technical equipment required or useful for the palaeontologist in the field is as follows: a global positioning satellite (GPS) system; a topographic map or aerial photographs or satellite images of the area of interest; a compass/clinometer; an altimeter; a range-finder;

a pair of binoculars; a digital camera or video; a portable lap-top computer on which to upload digital images; a portable solar panel with which to recharge electronic equipment; a measuring tape; a $\times 10$ to $\times 20$ magnifying glass or a pocket microscope; a bottle of dilute hydrochloric acid to test for carbonates; a sledge or 4-lb (2-kg) lump hammer; a 2-lb (1-kg) hammer; a set of chisels; a set of dental tools; a pick-axe; an entrenching tool; an auger; a supply of sample bags; a supply of indelible pens for labelling them; a waterproof notebook and a supply of pencils for recording observations (Jones, 2006; Coe, in Coe, 2010).

Safety. Safety equipment should include clothing and footwear appropriate to the season and terrain; sun-cream; personal protection equipment, including a hard hat or climbing helmet, and goggles for use when hammering; sufficient food and water to see out an emergency; fire-lighting equipment; a survival blanket; a torch (flashlight); a whistle, for attracting attention; and a first-aid kit (Goldring, 1999; Oliveri & Bohacs, 2005; Jones, 2006; Coe, in Coe, 2010).

Recommended safety procedures are as follows (Goldring, 1999):

Listen to the daily weather forecast (including wind direction), which may determine where it is prudent to work. Take account of the time and height of tides when planning coastal work. Write down each day your approximate route, working area and time of return, and leave it for others to see. In worsening conditions, do not hesitate to turn back if it is still safe to do so. If you get lost, disabled, benighted, or cut off by the tide, … stay where you are until conditions improve or until you are found. Supposed short cuts can be lethal.

Distress codes are as follows (Goldring, 1999):

On mountains: 6 long blasts, flashes, shouts or waves in succession, repeat(ed) at minute intervals. At sea: 3 short then 3 long, then 3 short blasts or

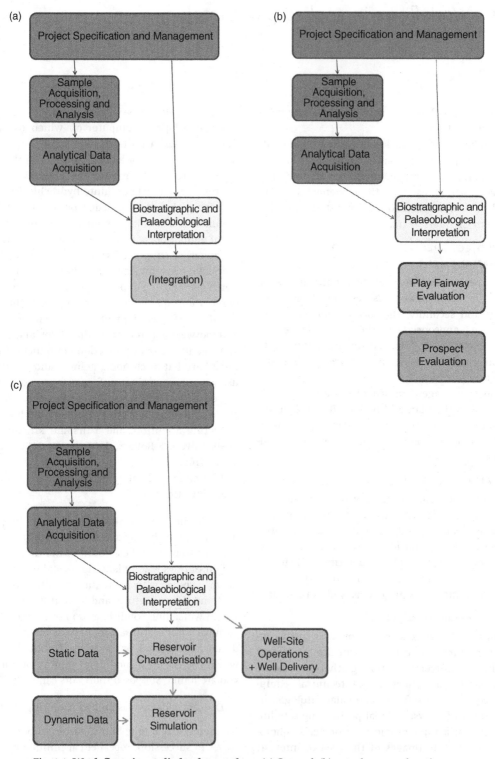

Fig. 1.1 **Work-flows in applied palaeontology**. (a) General; (b) petroleum exploration, as discussed in Section 5.2; (c) reservoir exploitation.

flashes [Morse code for SOS], ... repeat(ed). Rescuers reply with 3 blasts or flashes repeated at minute intervals.

Acquisition of surface samples for their macrofossil content

Acquisition of surface samples for their macrofossil content, and of macrofossils, has to be responsible and sustainable, so as to conserve or preserve what is a finite natural resource for future generations, preferably in place (Goldring, 1999; Jones, 2006; Spicer, in Coe, 2010). In the United Kingdom, acquisition in designated 'Sites of Special Scientific Interest' or SSSIs is restricted to that for genuine and justifiable scientific purposes only, otherwise it would constitute an 'operation likely to damage' (OLD) the resource. Elsewhere in the United Kingdom, it is restricted by recommendation or voluntary code of conduct. The Geologists' Association 'geological fieldwork code' of conduct recommends the following actions. Firstly, 'Observe and record, and *do not hammer indiscrimately.*' Secondly, 'Keep collecting to a minimum. Avoid removing *in situ* fossils, unless they are *genuinely* needed for serious study.' Thirdly, 'The collecting of actual specimens should be restricted to those localities where there is a plentiful supply, or to scree, fallen blocks and waste tips.' Fourthly, 'Never collect from walls or buildings. Take care not to undermine fences, walls, bridges or other structures.' In Germany, acquisition in so-called 'geotopes' ('parts of the geosphere ... clearly distinguishable from their surroundings in a geoscientific fashion') is restricted by nature conservation and by national monument protection legislation.

Macrofossils are generally large enough to be seen in surface outcrops or in float. However, careful observation may be required in order that they may actually be seen. The angle of the Sun is important in this regard. Early mornings and late afternoons, when the Sun is low and the shadows long, are often the best times for searching for fossils. (Similarly, tilting slabs can cast shadows that throw previously unseen and unsuspected fossils into unexpected relief.) Intensive searching can commence once extensive searching has revealed a fossiliferous horizon. Hard rocks can be broken open using a lump hammer, or split along bedding planes using a hammer and chisel, in both cases carefully, so as not to damage specimens. Contained fossils are typically harder than containing rocks, and can be readily extracted. In the event that the fossils are softer than the rock, thay can nonetheless still be extracted, carefully, using dental tools, a process often started in the field and finished in the laboratory. Collecting fossils from certain hard rocks, such as massive limestones, can be effectively impossible. Specimens are probably better photographed than removed from these rocks. Soft rocks can be trenched and samples removed for laboratory preparation.

Acquisition of surface samples for their microfossil content

Special care must be taken in the acquisition of surface samples for their microfossil content so as to avoid contamination, which can arise from, for example, the failure to clean hammers or other tools, or the use of cloth rather than plastic sample bags.

Sample spacing. The overall objectives of the fieldwork should be considered when determining the appropriate strategy for sampling. For example, if the objective is reconnaissance mapping, spot sampling might be all that is required, whereas if the objective is detailed logging, targeted or close systematic sampling would be required. As a general comment, the biostratigraphic or palaeoenvironmental resolution of the analytical results will depend as much on the sampling density as on the fossils themselves. Partly on account of this, and partly on account of the logistical effort and financial cost of mobilising field parties, it is always advisable to collect what might be thought of as too many rather than too few samples. However, any restrictions on access or sampling imposed by the land-owner should be respected, as should the code of conduct (see above). The particular microfossil groups to be expected in the ages and environments of the rocks expected to be encountered should also be considered, together with any sampling requirements specific to those groups (see below).

Sample size. The size of sample required depends to an extent on the group targeted (see also below). Samples for most micropalaeontological or palynological analysis should generally be at least 30–60 g, or 'One Standard British Handful', while those for nannopalaeontological analysis should be at least 5–10 g (and more in areas of high sedimentation rate and dilution of fossil content). Note, though, that samples for conodont analysis should be at least 500 g or 0.5 kg, and, in the case of the Devonian, which contains only rare specimens, 10 kg.

Sample lithology. The lithologies most likely to be productive for microfossils are fine-grained clastics such as shales and mudstones, especially where calcareous, and fine-grained limestones such as lime mudstones, wackestones and packstones. Those least likely to be productive are coarse-grained clastics such as sandstones and conglomerates, coarse-grained limestones such as grainstones, rudstones and framestones, altered dolomites, and evaporites. Note, though, that coarse-grained clastics can contain reworked microfossils that can provide useful information as to provenance.

Importantly, weathered rocks of any lithology are less likely than unweathered rocks to be productive for calcareous microfossils, on account of the likelihood of leaching; and are unlikely to be productive for organic-walled microfossils, on account of the likelihood of oxidation (which can also occur in the sample processing laboratory or storage facility, and which can be species-selective in its effects: Kodrans-Nsiah *et al.*, 2008). This is a particular problem in tropical climes, for example in the Kufra basin in southeast Libya, where surface weathering can affect up to 50–75 m of section, necessitating digging, trenching, auguring or even drilling, using appropriate tools, to obtain fresh, unweathered samples. (Note also that 'palaeo-weathering' affected up to 50 m of section below the Permo-Carboniferous unconformity in the North Sea.) Unweathered rocks can be recognised by their generally blocky rather than slabby, platy, fissile or earthy texture. Note that if it is simply not possible to access unweathered rocks, because the effects of weathering have pervaded so deep, it is nonetheless still worth sampling any calcareous concretions that might be present, since experience has shown that these can be productive for calcareous microfossils.

Thermally altered rocks of any lithology are less likely than unaltered rocks to be productive, particularly for organic-walled microfossils. The effects of thermal alteration can be either local or regional.

Specific groups of microfossils. Calcareous microfossils are locally so abundant in rocks of the appropriate age-range and facies as to be rock-forming, as in the case of '*Globigerina*' or planktonic foraminiferal oozes. They are common in essentially all fine-grained marine limestones and marls, and even in indurated ones, which cannot be easily disaggregated and which are therefore best studied in thin-section (although they may be difficult to identify in altered dolomites). Calcareous microfossils are also common in essentially all fine-grained marine calcareous mudstones, and, in the case of agglutinated foraminifera, in non-calcareous mudstones. Even non-marine, lacustrine calcareous mudstones can contain calcareous microfossils, in the form of ostracods and branchiopods, which may be sufficiently large to be discernible on bedding planes with the aid of a hand-lens. Samples are best collected by chiselling along bedding planes rather than hammering, so as to avoid damage to specimens. One large sample bag is generally sufficient to ensure recovery of calcareous microfossils, especially if the material is fresh and unweathered. It is invariably worth the effort ensuring that this is so.

Siliceous microfossils are locally so abundant in rocks of the appropriate age-range and facies as to be rock-forming, as in the case of diatomites, radiolarian cherts or radiolarites, and spiculites. Diatomites often resemble volcanic tuffs when weathered. Diatoms can be common not only in diatomites but also in siliceous mudstones, such as those of the Miocene of California, or in so-called 'opokas', such as those of the Miocene of Sakhalin. Radiolarians can be common not only in radiolarites but also in shales and in calcareous rocks of marine origin. Unfortunately, the silica of which diatoms is composed is an unstable variety (Opal-A), which converts to a more stable variety (cristobalite or Opal-CT) under the sort of pressure and temperature conditions encountered at burial depths of the order of 2 km, often resulting in the destruction of diagnostic morphological features. Even under these conditions, though, diatoms can be preserved, with their diagnostic morphological features intact, through recrystallisation, replacement – typically by pyrite or calcite – or entombment in concretions. Radiolarians are generally more robust, and more resistant to diagenetic alteration.

Phosphatic microfossils such as conodonts are at least locally common in most marine rocks of the appropriate age-range and facies. They are perhaps most common in limestones, especially bioclastic wackestones or packstones. The occurrence of macrofossils such as crinoids or brachiopods in a rock is an encouraging sign that it will be productive for conodonts. Cherts are also sometimes productive for conodonts on treatment with hydrofluoric acid. Conodonts are generally resistant to chemical attack, and also to diagenetic dolomitisation and thermal alteration. They can occasionally be seen on bedding planes with the aid of a hand-lens. They can be concentrated in lag deposits such as bone beds. The abundance of conodonts varies through time, such that sample sizes need to be adjusted accordingly (see above) The facies preference of conodonts also varies through time. Older conodonts are more common in shallower-water, younger ones in deeper-water, deposits.

Organic-walled microfossils or palynomorphs are present to common in most clastic rocks of the appropriate age and facies that contain clay-sized particles and that have not been subject to excessive oxidation or thermal alteration (carbonates are generally poorly productive). Organic-walled microfossils can be extremely abundant, with up to 100 000 grains per gram present in some carbonaceous deposits, such that small samples are generally sufficient. Even individual conglomerate clasts can be analysed, in order to provide an indication of provenance of reworking. Organic-walled microfossils are prone to reworking on account of their small size and resistance to chemical attack.

Calcareous nannofossils are locally so abundant in marine rocks of the appropriate age-range and facies as to be rock-forming, as in the case of calcareous nannofossil oozes and chalks. They are common in essentially all fine-grained marine limestones, marls and calcareous mudstones. Recrystallised limestones and dolomites should be avoided, though, as they are likely to have had their original calcareous components destroyed by diagenesis. Marine red beds deposited below the calcite compensation depth should also be avoided.

1.2.2 Subsurface sample acquisition

In the petroleum industry, acquisition of subsurface samples for their fossil content is required to constrain the biostratigraphic, or age, interpretation, and the palaeobiological, palaeoecological or palaeoenvironmental, or facies, interpretation, of subsurface wells, either during or after drilling, and to calibrate subsurface seismic interpretation, among other reasons (see Chapter 5 below).

Sample type. Conventional and side-wall core samples are generally preferred over cuttings samples (see Section 5.2). This is because cuttings samples are prone to contamination by material in the drilling mud and also by material sloughing off the walls of the bore and caving down-hole. For example, in my working experience, cuttings samples from the Pedernales field in the eastern Venezuelan basin were contaminated by mangrove pollen in drilling mud formulated from local river water that were indistinguishable from those in the reservoir (Jones, in Jones & Simmons, 1999; see also sub-section 5.3.3 below). Note, though, that contamination of cuttings samples appears to be much less of a problem with modern than with historical mud systems. Note also that in many ways cuttings are more representative and informative than core samples, as they provide continuous rather than point coverage. Wet cuttings are generally preferred over washed and dried, unless an oil-based drilling mud has been used.

Sample spacing. The generally preferred spacing of cuttings samples is every 10 m or 30′, with a contingency to close to every 3 m or 10′ over intervals of interest, such as the reservoir target (Fig. 1.2). The preferred spacing of conventional and side-wall core samples is every 1 m or 3′, with a contingency to close to every 0.3 m or 1′ over intervals of interest (Fig. 1.2).

Sample size. As noted above, the size of sample required depends to an extent on the group targeted. Cuttings and conventional core samples for micropalaeontological or palynological analysis should generally be at least 30–60 g, while those for nannopalaeontological analysis should be at least 5–10 g. Side-wall core samples for micropalaeontological or palynological analysis should be approximately half the size of the core, while those for nannopalaeontological analysis should be approximately a quarter of the size of the core.

Sample lithology. Again as noted above, the lithologies most likely to be productive for microfossils are fine-grained clastics and fine-grained limestones,

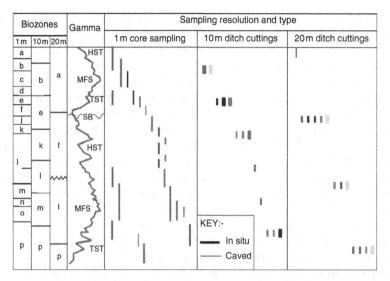

Fig. 1.2 **The effect of sampling on biostratigraphic resolution (and hence value).** Only fine sampling (down to 1 m) is sufficient to resolve fine detail such as thin rock units or systems tracts, or short-lived biozones. HST, high-stand systems tract; TST, transgressive systems tract, as defined in Section 4.1. Biozones a–p are invented categories, included for illustrative purposes only. Caving refers to down-hole contamination (see 1.2.2). Modified after Sturrock, in Emery and Myers (1993).

and those least likely to be productive are coarse-grained clastics, coarse-grained limestones, altered dolomites and evaporites.

Drilling environment. In the petroleum industry, the drilling environment also needs to be considered, as the drilling technology, bit type and mud system can all impact sample quality.

In terms of drilling technology, conventional drilling, controlled mud pressure drilling and riserless mud recovery (RMR) drilling have no or low impact on sample quality; coiled tubing drilling, drilling with casing and underbalanced drilling (UBD) can have a moderate impact; and managed pressure drilling (MPD) has a high impact, as it essentially does not permit the return of samples to the surface. Note, though, that coiled tubing, slim-hole and UBD technologies have recently been successfully employed in combination in the drilling of the Sajaa field in Sharjah in the United Arab Emirates, without impacting the quality of the samples used in micropalaeontological analysis and in 'biosteering' (Jones *et al.*, in Koutsoukos, 2005; see also 5.4.1 below).

In terms of bit type, roller cone bits have no or low impact; diamond bits, polycrystalline diamond

compact (PDC) bits, and PDC bits with down-hole motors can have a moderate impact; and PDC bits with turbines can have moderate to high impact, as they can effectively metamorphose samples and render them useless for analytical purposes.

In terms of mud systems, water-based and polymer-based systems can have a moderate impact; and oil-based systems can have a moderate to high impact.

1.3 SAMPLE PROCESSING

In the petroleum industry, in the case of micropalaeontological sample processing, for example for calcareous microfossils such as foraminifera and ostracods, samples should generally be simply disaggregated in water, with or without the addition of a solution of washing soda or of hydrogen peroxide, or of heat, to speed the process (Jones, 2006). The individual microfossils should then be picked out of the sieved residue with a moistened artist's brush, and sorted into numbered squares on a cardboard slide for identification under a reflected light microscope. Indurated limestone samples should be thin-sectioned for analysis under a transmitted light microscope.

In the case of nannopalaeontological processing, for calcareous nannofossils, samples should again be simply disaggregated in distilled water. The disaggregated sample should then be strewn onto a glass slide with a cover slip for identification under a powerful transmitted light microscope.

In the case of conventional palynological processing for organic-walled microfossils or palynomorphs, the non-palynomorph components of the sample should be dissolved in hydrochloric, hydrofluoric and fuming nitric acids. The sieved residue should then be strewn onto a glass slide with a cover slip for identification under a transmitted light microscope.

The use of hydrochloric, hydrofluoric and fuming nitric acids in conventional palynological sample processing raise some serious Health, Safety and Environmental (HSE) issues (see Box 5.6).

It is important that sample processing is undertaken by a best-in-class facility, both to ensure compliance with HSE standards, and also to ensure that

no palaeontological information is lost, because if it is lost, it is lost irretrievably.

1.4 SAMPLE ANALYSIS

In the petroleum industry, it is also important that sample analysis is undertaken by best-in-class analysts, so as to maximise the quality and value of the analytical data acquired. It is sometimes undertaken in-house in oil and gas companies, but more often externally by third party consultancies.

1.5 ANALYTICAL DATA ACQUISITION

In the petroleum industry, where, as noted above, analysis is often undertaken by third parties, it is important that the full suite of raw analytical data is acquired rather than simply an interpretation or summary thereof, so as to enable independent in-house quality assurance and interpretation.

It is preferable that the full suite of data is acquired in a digital format, for ease of manipulation, display and storage.

2 • Biostratigraphy and allied disciplines, and stratigraphic time-scales

Biostratigraphy involves the use of fossils in establishing the ordering of containing rocks in time and in relation to evolving Earth history (McGowran, 2005; Jones, 2006). It is one of the principal bases for chronostratigraphic subdivision and correlation of lithological units, thus providing a spatio-temporal context for their interpretation, and is a fundamental building block of Earth science.

I make no formal distinction between biostratigraphy (which essentially records relative age, or time) and lithostratigraphy (which records rock), in the characterisation of sequences, that is, intervals of time represented by rock, such as the Devonian Old Red Sandstone. I do so only in the characterisation of non-sequences, that is, intervals of time unrepresented by rock, such as that between the Late/Upper Devonian Old Red Sandstone and underlying Silurian greywacke observed at 'Hutton's Unconformity' at Siccar Point in East Lothian in Scotland. Note in this context that the absolute age, or extent in time, of the intervals of time either represented or unrepresented by rock can only be determined by absolute chronostratigraphy or geochronology, which actually measures time rather than simply recording or representing it (or by biostratigraphy calibrated against the absolute chronostratigraphic or geochronological time-scale).

I make no distinction at all between time- and rock-stratigraphic nomenclature. Note, though, that other authors do, and use the descriptors 'early', 'middle' and 'late' only when referring to time-stratigraphic units, and 'lower', 'middle' and 'upper' only when referring to rock-stratigraphic units.

2.1 SUMMARY OF BIOSTRATIGRAPHIC SIGNIFICANCE AND USEFULNESS OF PRINCIPAL FOSSIL GROUPS

The biostratigraphic significance and usefulness of the principal fossil groups is discussed in this section and in Sections 2.2–2.6 below, and summarised in Fig. 2.1 (from Jones, 2006; see also 'Chronos' website, www.chronos.org).

The ranges over which they are biostratigraphically significant are shown by broad bands. It is evident that most are only biostratigraphically significant over certain time intervals, and then only in the appropriate facies. Note also that the potential biostratigraphic significance of fossils can be impaired by natural factors, such as *post-mortem* transportation and diagenetic effects, and reworking. The biostratigraphic significance of fossils can also be impaired by artificial factors, such as sample acquisition and processing, and subjectivity in specific identification.

The particular usefulness and applicability of some groups is in not only biostratigraphy, but also palaeobiology, discussed in Chapter 3, and sequence stratigraphy, discussed in Chapter 4. Case studies of applications of these groups in industry and elsewhere are discussed in Chapters 5–10.

Characteristics of biostratigraphically significant and useful fossil groups

Biostratigraphically significant and useful fossil groups share two common characteristics: firstly, relatively rapid rates of evolutionary turnover, and hence restricted stratigraphic distributions, and/or essentially isochronous first and last appearances; and secondly, essentially unrestricted ecological distributions (for example throughout the marine realm, and across a range of biogeographic provinces, in the case of many planktonic or nektonic forms). The most useful groups for practical purposes are also, typically, abundant, well preserved, and easy to identify. These are referred to as 'marker fossils' or 'index fossils'. Conversely, the least stratigraphically useful groups characteristically exhibit relatively slow rates of evolutionary turnover, and/or diachronous or time-transgressive first and last appearances. Alternatively, they may exhibit restricted ecological distributions (for example to individual bathymetric zones, in the case of many benthic forms). Note, though, that the very ecological restriction exhibited by these groups renders them palaeobiologically useful 'facies fossils' (see Chapter 3 below).

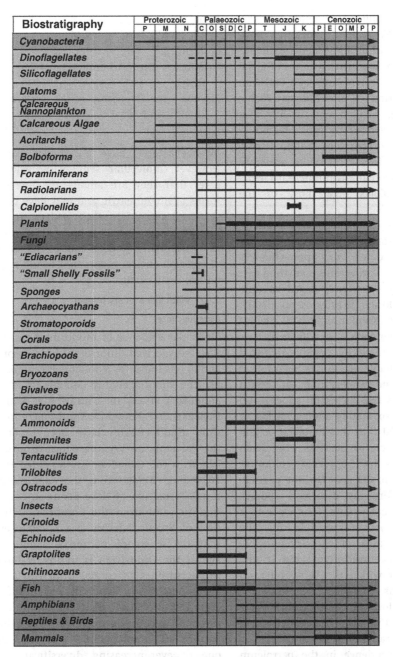

Fig. 2.1 **Stratigraphic distribution of selected fossil groups.** Modified from Jones (2006).

2.1.1 Bacteria

Cyanobacteria

Cyanobacteria evolved in the Archaean, approximately 3500 Ma, and have ranged through to the Recent (Batistuzzi & Hedges, in Hedges & Kumar, 2009). They diversified through the Palaeoproterozoic, Mesoproterozoic and Neoproterozoic, but underwent something of a decline in the Neoproterozoic, from around 1000 Ma. Some authors have speculated that this decline was due to excessive grazing of stromatolitic mats by early 'ediacarans'. However, other authors have hypothesised that it was brought about by environmental change associated with a series of glaciations, resulting in a so-called 'Snowball Earth' in the 'Cryogenian' period of the Neoproterozoic (Moczydlowka, 2008). Incidentally, it has also been hypothesised that diversity promotes environmental stability, and therefore that low-diversity biotas, like those of the Proterozoic, are more susceptible to such external environmental factors than high-diversity biotas, like those of the Phanerozoic.

The oldest known Cyanobacteria and/or microbialites are those from the Pilbara craton of Western Australia, dated to approximately 3500 Ma; and from the Barberton greenstone belt of South Africa and Swaziland, dated to approximately 3500–3300 Ma (Brasier *et al.*, 2005; Allwood *et al.*, 2007; Konhauser, 2007; Schopf *et al.*, 2007; van Kranendonk *et al.*, 2008; de Gregorio *et al.*, 2009). Incidentally, organic material of somewhat questionable origin has been found in the Witwatersrand supergroup of South Africa, which overlies the Barberton supergroup and is dated to approximately 2900–2700 Ma. It has been hypothesised that this material originated either from Bacteria or Algae, or from lichen-like organisms (the observed columnar form as representing *in situ* growths, the particulate or 'fly-speck' form as dispersed spores). Whether or not this is the case, it is clear that whatever organism was responsible for the organic material was also somehow responsible for the observed concentration of gold in the organic material, for which the Witwatersrand is rather more famous.

The exceptionally slow rate of evolutionary turnover exhibited by the Cyanobacteria renders them of limited use in biostratigraphy.

In my working experience in the petroleum industry, Cyanobacteria have proved of use in the following areas:

Proterozoic – the Neoproterozoic, Tonian of east Siberia, and the Neoproterozoic, Tonian–Cryogenian of the Jagbub High, Cyrenaica, northeast Libya, North Africa;

Proterozoic or Palaeozoic – the 'Infracambrian' of the Middle East, and of Mauritania in northwest Africa;

Palaeozoic – the Carboniferous of Libya in north Africa;

Mesozoic – the Cretaceous of the western margin of the British Isles.

'Archaebacteria'

'Archaebacteria' have no known fossil record. However, molecular sequencing evidence indicates that they are among the most primitive forms of life on earth, and may have been among the earliest (Batistuzzi & Hedges, in Hedges & Kumar, 2009). This is supported by the observation that many modern species inhabit the sorts of extreme environments that would have existed on the early earth.

2.1.2 Plant-like protists (Algae)

Dinoflagellates

Molecular evidence in the form of arguably dinoflagellate-derived dinosteranes in oil source-rocks and in oils indicates a possible Precambrian origin for the dinoflagellates (Moldowan *et al.*, 1996; Moldowan *et al.*, in Zhuravlev & Riding, 2001; Delwiche, in Falkowski & Knoll, 2007). Indeed, possible dinoflagellates have been recorded from the Precambrian Wynniatt formation of Victoria Island in the Canadian Arctic, dated to between 1081 and 721 Ma, from the Cambrian MacLean Brook formation of Nova Scotia in the Canadian Atlantic, and from the Cambrian Oville formation of Spain (Palacios *et al.*, 2009). (Note also that the existence of zooxanthellates as long ago as the Ordovician is indirectly indicated by the occurrence of corals, with which the group at present has a symbiotic relationship.) However, definite dinoflagellates do not appear in the rock record until the Triassic. The overall pattern of dinoflagellate evolution has been one of ever-increasing diversification through the Mesozoic and Cenozoic, with comparatively little loss of diversity other than that associated with the Late Cenomanian mass extinction, and, more especially, the end-Cretaceous mass extinction. Interestingly, dinoflagellates appear to have evolved (possibly iteratively) through the incorporation by a protist of a haptophyte, arising from the incorporation by a protozoan of an alga, arising in turn through the incorporation by a eukaryote of a cyanobacterium!

The rapid rate of evolutionary turnover exhibited by the dinoflagellates renders them of considerable use in biostratigraphy, at least in appropriate, marine, environments. Their usefulness is enhanced by their essentially unrestricted ecological distribution, and facies independence, a function of their pelagic habit. High-resolution biostratigraphic zonation schemes based on the dinoflagellates have been established that have at least local to regional applicability (Helby et al., in Jell, 1987; Riding et al., 2010). Biostratigraphic zonation schemes based on the 'calcispheres' have been established that have at least local to regional applicability within a given palaeobiogeographic realm or province (Rehanek & Cecca, 1993).

In my working experience, dinoflagellates have proved of use in a number of areas, and of particular use in the following areas:

Mesozoic – the Triassic–Cretaceous of Australasia, the Jurassic–Cretaceous of the North Sea, of the Arctic including west Siberia, of Qatar and Kuwait in the Middle East, and of the Indian subcontinent, and the Cretaceous of northern South America and the Caribbean;

Cenozoic – the Cenozoic of the North Sea, of the Arctic, of Paratethys and of the Indian subcontinent.

Silicoflagellates

Silicoflagellates range from Cretaceous to Recent. The moderate rate of evolutionary turnover exhibited by the silicoflagellates renders them of some use in biostratigraphy, at least in appropriate, marine, environments (Fig. 2.2). Their usefulness is enhanced by their essentially unrestricted ecological distribution, and facies independence, a function of their habit. Biostratigraphic zonation schemes based on the silicoflagellates have been established that have at least local to regional applicability.

In my working experience, silicoflagellates have proved of use in the Palaeogene of the North Sea and northwest Europe, and the Neogene of central Paratethys.

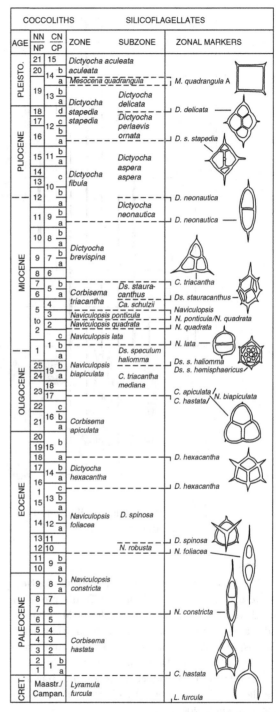

Fig. 2.2 **Stratigraphic zonation of the Cretaceous and Cenozoic by means of silicoflagellates.** © Cambridge University Press. Reproduced with the permission of the publisher and of the author from Perch-Nielsen, in Bolli et al. (1985).

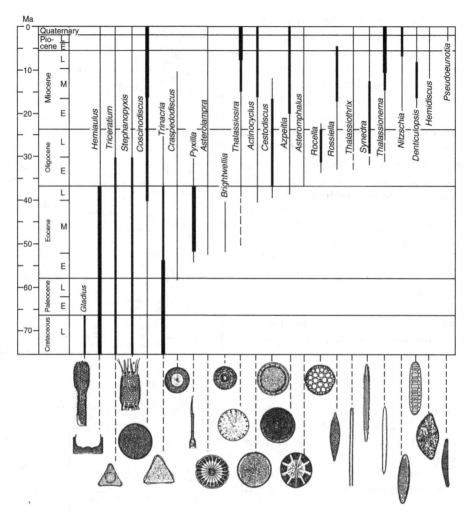

Fig. 2.3 **Stratigraphic distribution of selected diatoms.** © Wiley. Reproduced with permission from Lipps (1993).

Diatoms

Problematic isolated siliceous scales from the Neoproterozoic, specifically from the Tindir group of northwest Canada, dated to *c.* 650–600 Ma (that is, Ediacaran), may be derived from diatoms (Gaucher & Germs, in Gaucher *et al.*, 2010). Certain biomarkers that occur from the Late Palaeozoic to Recent, and in abundance from Cretaceous to Recent, may also be derived from diatoms (Kooistra *et al.*, in Falkowski & Knoll, 2007).

However, incontrovertible diatoms only appear in the rock record in the ?Jurassic (age attribution questionable) (Kooistra *et al.*, in Falkowski & Knoll,

2007; Medlin, in Hedges & Kumar, 2009). The oldest known non-marine forms (*Aulacoseira*) are from the latest Cretaceous intertrappean beds of India (Ambwani *et al.*, 2003). The moderate rate of evolutionary turnover exhibited by the diatoms renders them of some use in biostratigraphy, at least in appropriate, aquatic, environments (Fig. 2.3). Unfortunately, their biostratigraphic as well as palaeobiological usefulness can be compromised by their destruction in diagenesis. Nonetheless, biostratigraphic zonation schemes based on the diatoms have been established that have at least local to regional applicability. Important schemes applicable to the Neogene and Pleistogene of the North Pacific have been established by Motoyama and Maruyama (1998) and Yanagisawa and Akiba (1998).

In my working experience, diatoms have proved of use in the following areas:

Mesozoic – the 'Mid' Cretaceous of the Sergipe–Alagoas basin in Brazil in the South Atlantic;

Cenozoic – the Palaeogene of the North Sea and the Davis Strait, the Palaeogene–Neogene of Sakhalin in the former Soviet Union, and the Neogene of Angola in west Africa in the South Atlantic.

Calcareous nannoplankton

Problematic isolated siliceous scales from the Neoproterozoic, specifically from the Tindir group of northwest Canada, dated to *c.* 650–600 Ma (that is, Ediacaran), may be derived from haptophytes (Gaucher & Germs, in Gaucher *et al.*, 2010).

However, incontrovertible calcareous nannofossils only appear in the rock record in the Triassic (de Vargas *et al.*, in Falkowski & Knoll, 2007; Medlin, in Hedges & Kumar, 2009). They diversified through the Mesozoic, before sustaining severe losses in the end-Cretaceous mass extinction. Although they have staged some form of recovery from this event, they have not regained their pre-extinction diversity. Calcareous nannofossil diversity appears to exhibit cyclical change generally paralleling that of sea level, but perturbed by intermittent volcanic and impact events (Erba, 2006).

The rapid rate of evolutionary turnover exhibited by the calcareous nannofossils renders them of considerable use in biostratigraphy, at least in appropriate, marine, environments (Fig. 2.4). Their usefulness is enhanced by their essentially unrestricted ecological distribution, and facies independence, a function of their pelagic habit. High-resolution biostratigraphic zonation schemes based on the calcareous nannofossils have been established for the Jurassic–Cenozoic that have regional to essentially global applicability. The schemes for the later Cenozoic have been calibrated against the absolute chronostratigraphic time-scale by astronomical tuning. There is a recently published quantitative 'unitary association zonation' scheme for part of the Jurassic that has regional applicability within western Tethys (Mailliot *et al.*, 2006).

In my working experience, calcareous nannofossils have proved of use in a number of areas, and of particular use in the following areas:

Mesozoic – the Jurassic–Cretaceous of the Middle East, and the Cretaceous of the North Sea;

Cenozoic – the Cenozoic of Angola in west Africa, and of Parathethys, and the Neogene–Pleistogene of northern South America and the Caribbean, and the Gulf of Mexico.

Calcareous Algae

The oldest known rhodophytes are Mesoproterozoic; the oldest chlorophytes Neoproterozoic (Taylor *et al.*, 2009). The overall pattern of rhodophyte and chlorophyte evolution through the Phanerozoic has been one of diversification offset by loss of diversity associated with the various mass extinction events of that time. The oldest known charophytes are Silurian (McCourt *et al.*, 2004).

The moderate rate of evolutionary turnover exhibited by the calcareous Algae renders them of some use in biostratigraphy, at least in appropriate environments. Unfortunately, the usefulness of many forms is compromised by their restricted bathymetric and biogeographic distribution and facies dependence. Nonetheless, biostratigraphic zonation schemes based in part on the calcareous Algae have been established that have at least local to regional applicability. Charophytes are of at least some biostratigraphic use in, for example, the non-marine facies of the Mesozoic–Cenozoic of Europe. The ranges of some Eocene species from the Pyrenees have been calibrated against those of associated shallow marine larger benthic foraminifera (LBFs), and hence indirectly against the absolute time-scale (Martin-Closas *et al.*, 1999).

In my working experience, calcareous – and non-calcareous – Algae have proved of use in the following areas:

Palaeozoic – the Devonian–Permian of the former Soviet Union, the Carboniferous (Lisburne group) of Arctic Alaska, and the Permian of the Middle East;

Mesozoic – the Mesozoic of the Middle East;

Mesozoic–Cenozoic – the Late Cretaceous–Palaeocene of the Moroccan Atlas in North Africa;

Cenozoic – the Cenozoic of Papua New Guinea in the Far East, and the Neogene of Nigeria and of Azerbaijan.

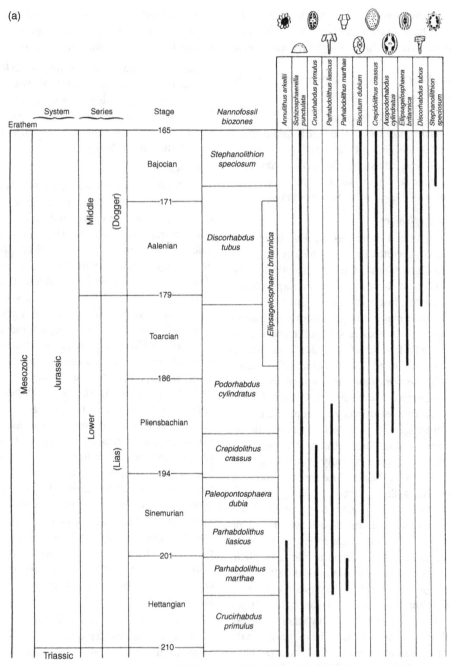

Fig. 2.4 **Stratigraphic distribution of calcareous nannofossils.** © Wiley. Reproduced with permission from Lipps (1993). (a) Early–Middle Jurassic; (b) Middle–Late Jurassic; (c) Early Cretaceous; (d) Late Cretaceous; (e) Palaeogene; (f) Neogene–Pleistogene.

(b)

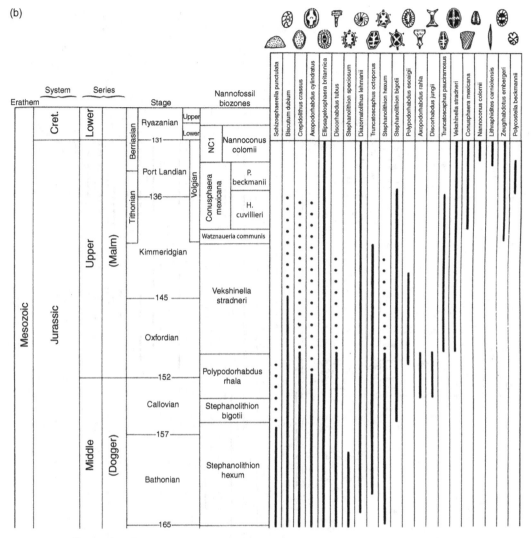

Fig. 2.4 (*cont.*)

Acritarchs

Acritarchs evolved in the Precambrian, Palaeoproterozoic (Javaux & Marshal, 2006; Yongbo Peng *et al.*, 2009; Gaucher & Sprechmann, in Gaucher *et al.*, 2010). The oldest known, sphaeromorph, forms are from the Palaeoproterozoic Gunflint chert of Western Ontario, dating to approximately 1900 Ma. The oldest known acanthomorphs are from the Mesoproterozoic Lakhanda formation of east Siberia.

Acritarchs diversified in the Precambrian, Neoproterozoic, although they also sustained

significant losses in the mid Cryogenian, 770–740 Ma, late Cryogenian, *c.* 700–635 Ma, and late Ediacaran, *c.* 560–542 Ma (Moczydlowka, 2008; Gaucher & Sprechmann, in Gaucher *et al.*, 2010). They underwent further adaptive radiations in the Cambrian, arguably in response to selection pressure exerted by the grazing activities of the newly evolved arthropods, and in the Ordovician. They then sustained further significant losses in the end-Ordovician mass extinction. However, they recovered from this event, too, and diversified again

(c)

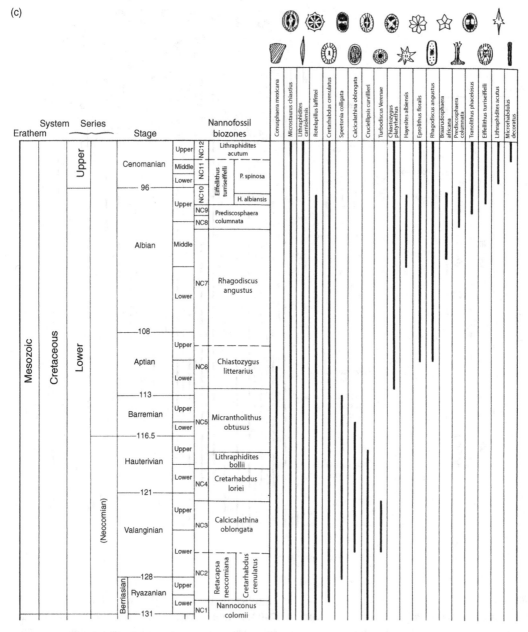

Fig. 2.4 (*cont.*)

in the Silurian. They sustained severe losses in the Late Devonian mass extinction, from which they may be regarded as never having really recovered.

The moderate rate of evolutionary turnover exhibited by the acritarchs renders them of some use in biostratigraphy, at least in appropriate, marine, environments. Their usefulness is enhanced by their essentially unrestricted ecological distribution, and facies independence, a function of their planktonic habit. Biostratigraphic zonation schemes

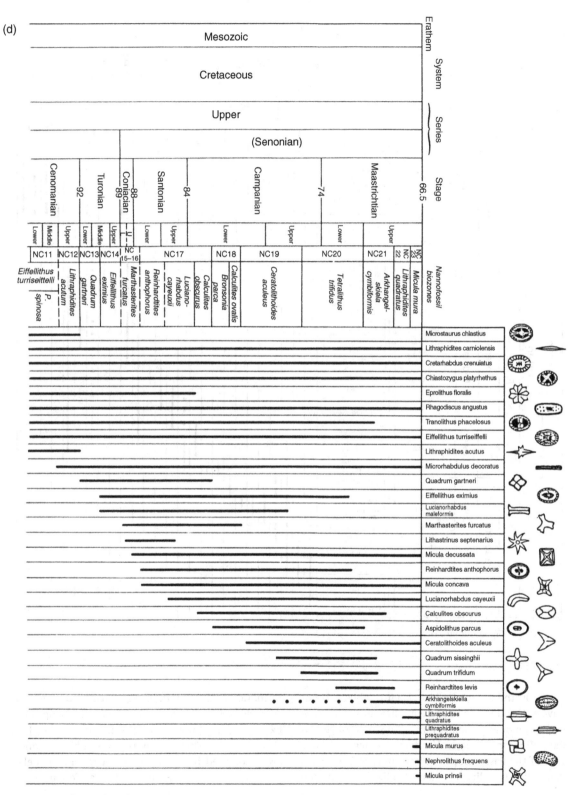

Fig. 2.4 (cont.)

(e)

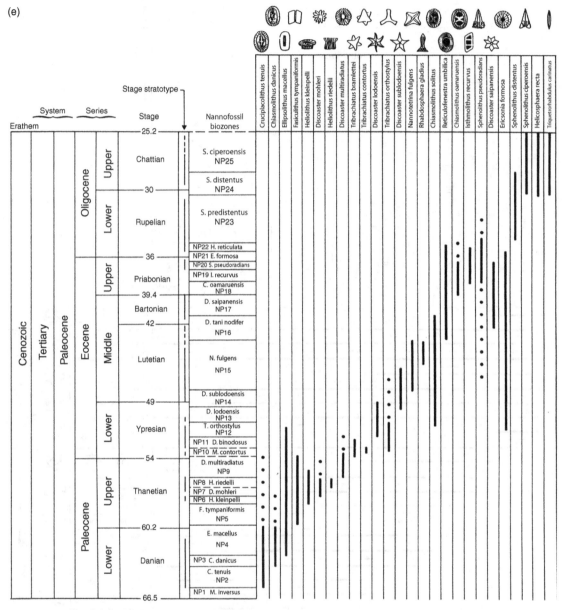

Fig. 2.4 (cont.)

based on the acritarchs have been established that have at least local to regional applicability. There are even provisional schemes for the Tonian–Cryogenian (late Riphean to early Vendian) Vychegda formation of the East European Platform (Vorobeva et al., 2009); for the Ediacaran (late Vendian) Doushantuo formation of the Yangtze Gorges area of south China (McFadden et al., 2009); and indeed for the Neoproterozoic generally (Gaucher & Sprechmann, in Gaucher et al., 2010). Three acritarch assemblage zones are recognised in the Tonian–Cryogenian, namely, in ascending stratigraphic order, a Tonian–early

(f)

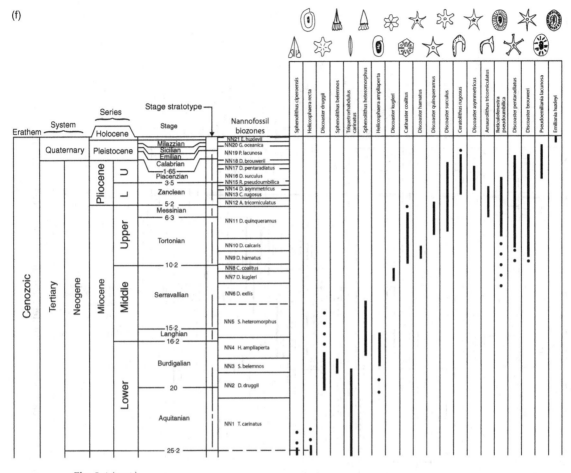

Fig. 2.4 (*cont.*)

Cryogenian assemblage, and late Cryogenian *Bavlinella* assemblages I and II; and a further three in the Ediacaran, namely, again in ascending stratigraphic order, an 'Early Ediacaran leiosphere palynoflora' or EELP, an 'Ediacaran complex acanthomorph palynoflora' or ECAP, and a 'Late Ediacaran leiosphere palynoflora' or LELP.

In my working experience, acritarchs have proved of use in the following areas:

Proterozoic – the Neoproterozoic, Tonian–Cryogenian of the Jagbub High, Cyrenaica, northeast Libya, north Africa, and late Cryogenian–early Ediacaran of the Middle East;

Proterozoic to Palaeozoic – the Late Precambrian to Early Cambrian of east Siberia, and the Late

Precambrian to Devonian of the Indian subcontinent;

Palaeozoic – the Palaeozoic of the Middle East and north Africa, and of Argentina and Bolivia in the 'southern cone' of South America.

Bolboforma

Bolboforma evolved in the Eocene and became extinct in the Pliocene.

The moderate rate of evolutionary turnover exhibited by *Bolboforma* renders it of some use in biostratigraphy, at least in appropriate, marine, environments (Fig. 2.5). Its usefulness is enhanced by its essentially unrestricted ecological distribution, and facies independence, a function of

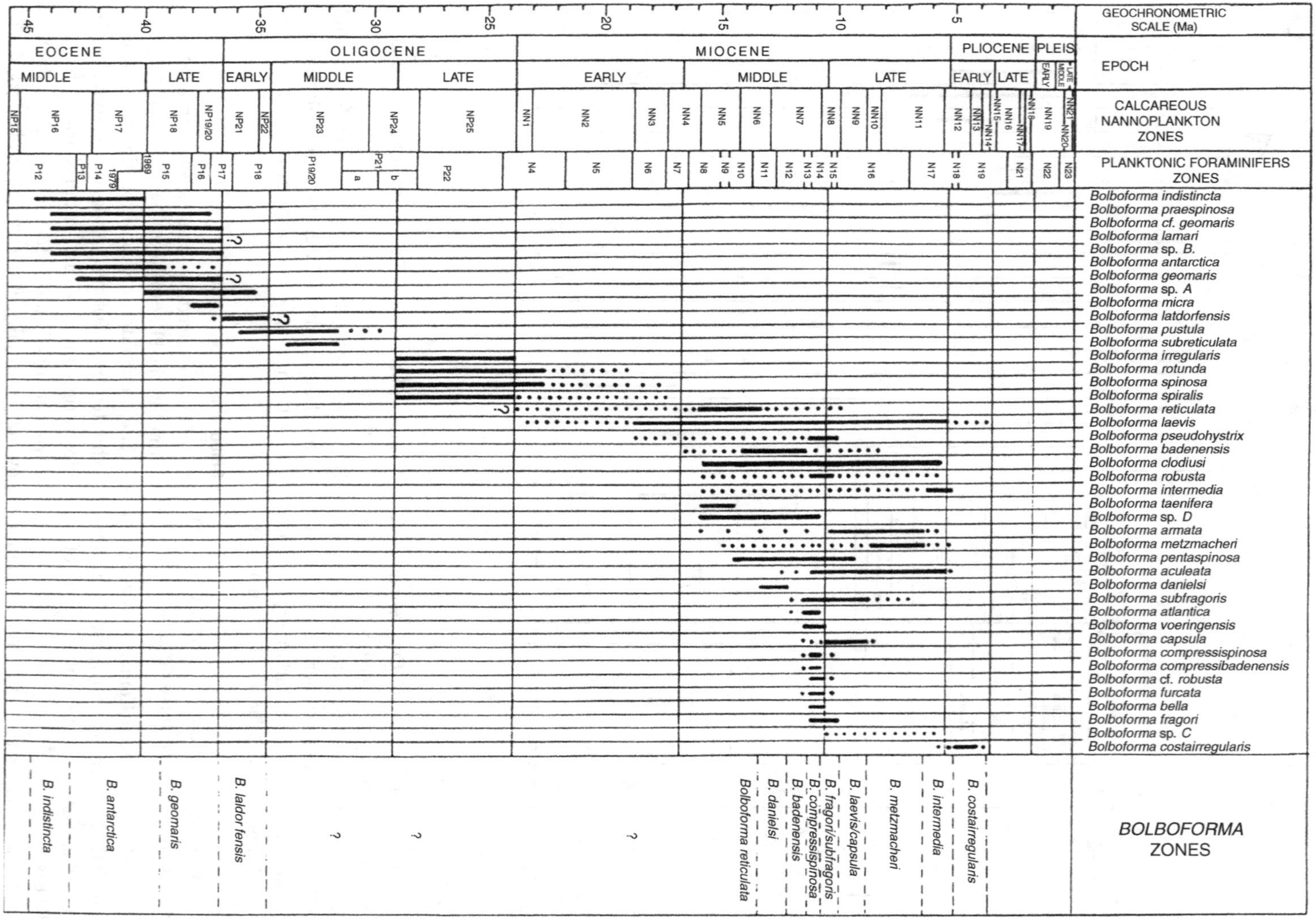

Fig. 2.5 **Stratigraphic distribution of *Bolboforma*.** © Cushman Foundation for Foraminiferal Research. Reproduced with permission from Spiegler and von Daniels (1991).

its planktonic habit. Biostratigraphic zonation schemes based on *Bolboforma* have been established that have at least local to regional applicability.

In my working experience, *Bolboforma* has proved of use in the Oligocene–Miocene of central Paratethys and the Miocene of the North Sea.

2.1.3 Animal-like protists (Protozoa)

Testate Amoebae

The pre-Pleistocene fossil record of the testate *Amoebae* is poor. They are only known from isolated records from the Palaeozoic, specifically from the Early Permian of the Himalayas; from the Mesozoic, specifically from the Early Cretaceous of the Scotian shelf, and from the Late Cretaceous, Cenomanian of Schliersee in Bavaria in southern Germany, in which latter case they are preserved in amber; and from the Cenozoic, specifically, from the Pliocene of the Pedernales field of eastern Venezuela, in which case they are preserved in pyrite (Jones *et al.*, in Jones & Simmons, 1999; Fiorini *et al.*, 2009; Farooqui *et al.*, 2010). Note, though, that some of the so-called 'vase-shaped microfossils' or VSMs from the Neoproterozoic/Precambrian, specifically from the Cryogenian Chuar group of the Grand Canyon, radiometrically dated to *c.* 780–742 Ma, and Ediacaran Jacadigo group of southwest Brazil and Tindir group of northwest Canada, dated to *c.* 650–600 Ma, may actually be testate *Amoebae* (Gaucher & Germs, in Gaucher *et al.*, 2010).

Foraminifera

The agglutinated *Titanotheca* from the Precambrian, late Ediacaran Arroyo del Soldado group of Uruguay, Corumba group of Brazil and Nolloth group of Namibia, all dated to 570–555 Ma, may be a foraminifer (Gaucher & Sprechman, 1999; Kaminski, in Bubik & Kaminski, 2004; Gaucher & Germs, in Gaucher *et al.*, 2010). However, the first incontrovertible agglutinated foraminifer, *Platysolenites*, has not been recorded from the Precambrian, only from the earliest Cambrian, 'Nemakit–Daldynian' to 'Tommotian' of Newfoundland, Great Britain, Scandinavia, Estonia and Russia. Taphonomic, teleological and ultrastructural studies suggest that ancient *Platysolenites* had a similar grade of organisation and mode of life to modern *Bathysiphon* (an epifaunal detritus feeder), although morphological evidence, specifically the presence of an agglutinated proloculus, indicates that it was probably not directly related. Microgranular foraminifera evolved in the Devonian, and diversified rapidly, before becoming extinct in the end-Permian mass extinction. Porcelaneous foraminifera evolved at the end of the Palaeozoic, and diversified through the Mesozoic and Cenozoic. Among the hyaline foraminifera, the nodosariides evolved in the ?Devonian, and the robertinides in the Triassic. Both diversified through the Jurassic and Early Cretaceous, and declined through the Late Cretaceous and Cenozoic, possibly owing to increasing competition from diversifying rotaliides, especially in shelfal environments. The buliminides evolved in the Jurassic, and diversified through the Cretaceous and Cenozoic. The rotaliides evolved in the Triassic, and diversified dramatically in the Cretaceous, before sustaining significant losses in the end-Cretaceous mass extinction. They recovered from this event, though, and in general have been radiating throughout the Cenozoic. The globigerinides, the planktonic foraminifera, evolved at the beginning of the Middle Jurassic, coincidentally or otherwise immediately after the end-Toarcian mass extinction. They diversified dramatically in the Cretaceous, before sustaining severe losses and indeed almost becoming extinct in the end-Cretaceous mass extinction. They staged a slow recovery from this event, and in general have been radiating through the Cenozoic.

The rapid rate of evolutionary turnover exhibited by at least the larger benthic foraminifera (LBFs), and the planktonic foraminifera renders them of considerable use in biostratigraphy, at least in appropriate, marine, environments. In practice, the usefulness of the LBFs is somewhat compromised by their restricted ecological, bathymetric and biogeographic distribution, and facies dependence, a function of their habit. Although a number of local to regional biostratigraphic zonation schemes based on LBFs have been established, no interregional or global schemes have been established that can be correlated across palaeobiogeographic realms (BouDagher-Fadel, 2008; BouDagher-Fadel & Price, 2010). The usefulness of the planktonic foraminifera is enhanced by their essentially unrestricted ecological distribution and facies independence, a function of their habit. A range of high-resolution biostratigraphic zonation schemes based on the planktonic foraminifera have been established that have inter-regional to global applicability (Fig. 2.6).

In my working experience, foraminifera have proved of use in a number of areas, and of particular use in the following areas:

Palaeozoic – the Cambrian of Senegal and Guinea in west Africa, the Devonian–Permian of the former Soviet Union and of the Arctic, the Carboniferous of China, and the Carboniferous–Permian of the Middle East and north Africa;

Mesozoic – the Triassic–Cretaceous of the western margin of the British Isles and of the Arctic, the Jurassic of Transcaucasia and the Jurassic–Cretaceous of West Siberia in the former Soviet Union, the Jurassic–Cretaceous of the Middle East and north Africa, and of the North Sea; and the Cretaceous of the Canadian interior, of the Indian subcontinent, and of northern South America and the Caribbean;

Cenozoic – the Cenozoic of the Arctic, of Sakhalin in the former Soviet Union, of Paratethys, of the Middle East and north Africa, of the North Sea and the West of Shetland basin, of the Indian subcontinent, of the South Atlantic, of northern South America and the Caribbean, and the Neogene–Pleistogene of the Gulf of Mexico and of the Niger Delta in Nigeria.

Radiolarians

Radiolarians evolved in the Cambrian, and underwent a dramatic diversification in the Ordovician, (Braun *et al.*, in Vickers-Rich & Komarower, 2007). They then sustained significant losses in the end-Permian mass extinction. They eventually recovered from this event, though, and underwent further diversifications in the Mesozoic and Cenozoic. The diversification in the Apto-Albian appears to have been associated with 'Oceanic Anoxic Event 1'. Molecular evidence has demonstrated that they did not evolve from siliceous sponges, as had been hypothesised (Danielian & Moreira, 2004).

The moderate rate of evolutionary turnover exhibited by the radiolarians renders them of some use in biostratigraphy, at least in appropriate,

marine, environments (Fig. 2.7). Their usefulness is enhanced by their essentially unrestricted ecological distribution, and facies independence, a function of their planktonic habit. Biostratigraphic zonation schemes based on the radiolarians have been established that have at least regional applicability, for parts of the Mesozoic (O'Dogherty *et al.*, in Lucas, 2010), and for the Cenozoic (Motoyama & Maruyama, 1998; Sanfilippo & Nigrini, 1998; Haslett, 2004).

In my working experience, radiolarians have proved of use in the following areas:

Palaeozoic – the Carboniferous (Lisburne group) of Arctic Alaska;

Mesozoic – the Middle Jurassic to earliest Cretaceous of the Middle East and north Africa, the Late Jurassic–Early Cretaceous of the Boreal and Arctic realms, and the 'mid' Cretaceous of the Sergipe–Alagoas basin in Brazil in the South Atlantic;

Cenozoic – the Cenozoic of the European platform and central Paratethys, the Palaeogene of the Boreal and Arctic realms, and of the Caribbean, and the Neogene of the North Pacific, including California and Sakhalin in the former Soviet Union.

Calpionellids

Calpionellids evolved in the latest Jurassic, Tithonian and became extinct in the Early Cretaceous, Neocomian (?Barremo–Aptian). (The morphologically similar problematicum *Vautrinella lapparenti* occurs in the Late Devonian, for example in north Africa: Sacal & Cuvillier, 1963.)

The rapid rate of evolutionary turnover exhibited by the calpionellids renders them of considerable use in biostratigraphy, at least in appropriate, deep marine, environments, in low latitudes. Unfortunately, their usefulness is somewhat compromised by their restricted bathymetric and biogeographic distribution. Nonetheless, high-resolution biostratigraphic zonation schemes based on the calpionellids have been established that have at least regional applicability.

In my working experience, calpionellids have proved of use in the latest Jurassic to earliest Cretaceous of the Middle East.

(a)

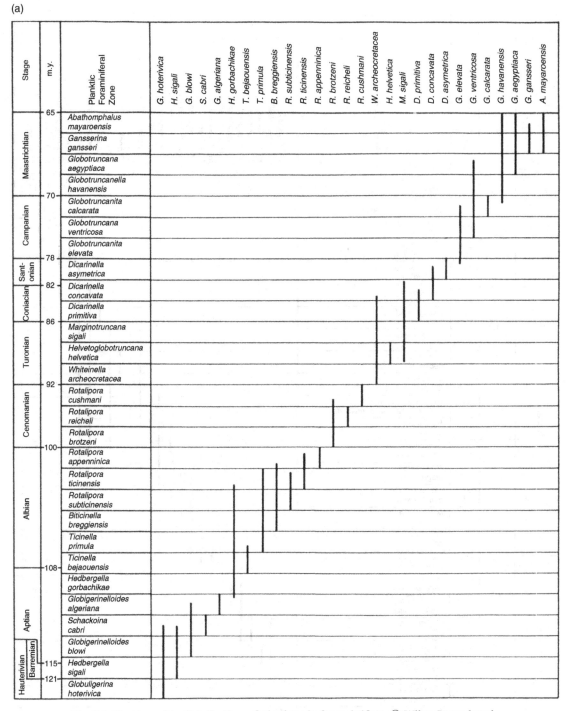

Fig. 2.6 **Stratigraphic distribution of planktonic foraminifera**. © Wiley. Reproduced with permission from Lipps (1993). (a) Cretaceous; (b) Palaeogene; (c) Neogene–Pleistogene.

(b)

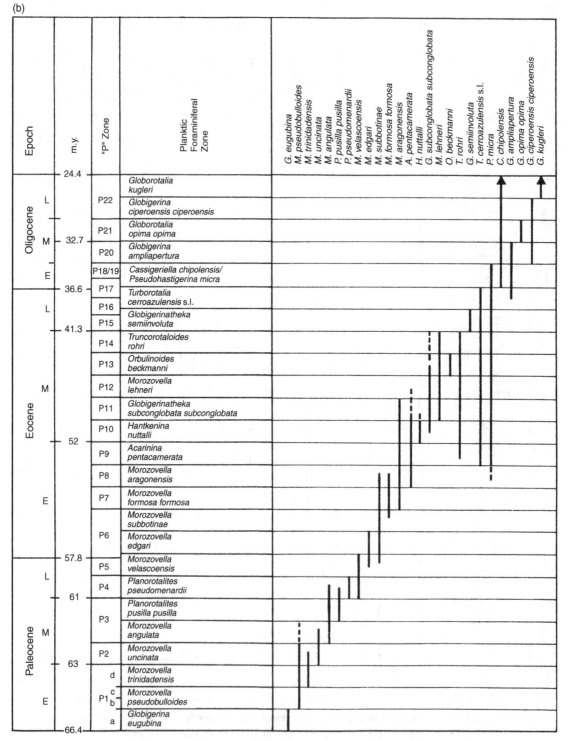

Fig. 2.6 (cont.)

(c)

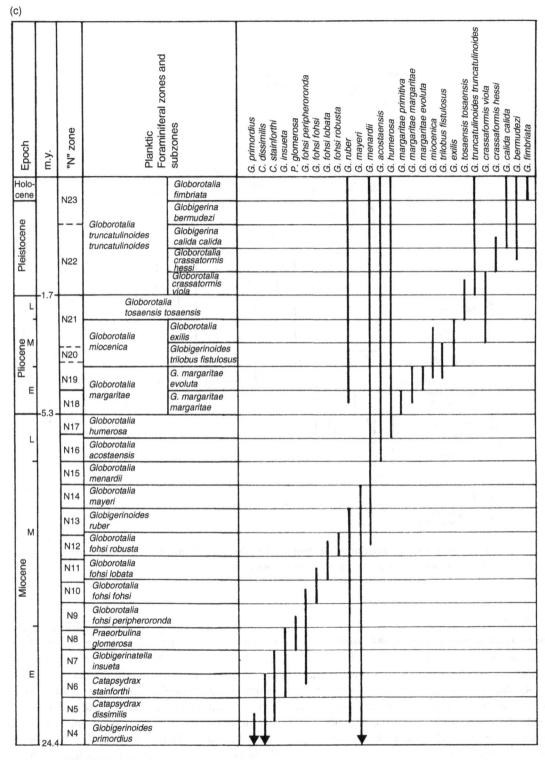

Fig. 2.6 (*cont.*)

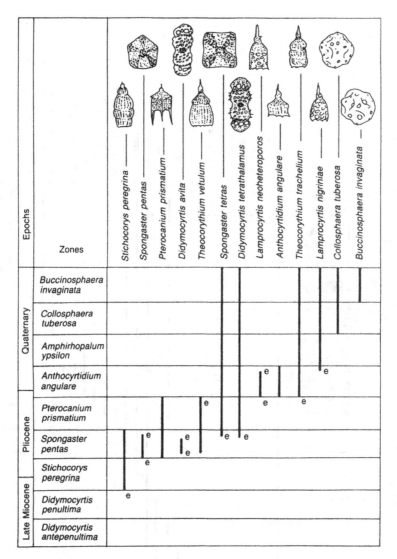

Fig. 2.7 **Stratigraphic zonation of the tropical late Cenozoic by means of radiolarians.**
© Wiley. Reproduced with permission from Lipps (1993).

2.1.4 Plants

Plant macrofossils, spores and pollen, and phytoliths

Molecular biological data indicates that plants evolved somewhere between the Precambrian, ~700–600 Ma and Silurian–Devonian, ~440–350 Ma (Magallon & Hilu, in Hedges & Kumar, 2009). Note, though, that there are issues surrounding the calibration of the 'molecular clock' (Benton *et al.*, in Hedges & Kumar, 2009).

Fossil spores and pollen first appeared in the Ordovician, with the oldest known forms from the Middle Ordovician, Llanvirn (=Darriwilian) of the Middle East and Late Ordovician, Caradoc?–Ashgill (=Sandbian?–Hirnantian) of North Africa, and range through to the Cenozoic (Gray *et al.*, 1982; Strother *et al.*, 1996; Edwards & Wellman, in Gensel & Edwards, 2001; Wellman *et al.*, 2003; Steemans & Wellman, in Webby *et al.*, 2004; Labandeira, 2005;

Wellman, 2005; le Herisse *et al.*, 2007; Vecoli *et al.*, 2010). ('Enigmatic, spore-like organic-walled microfossils' have recently been recorded in the Cambrian of Algeria: Vecoli *et al.*, 2007.) Different sub-groups made their appearances at different times.

Bryophytes first appeared in the Ordovician, with the oldest known forms from the Late Ordovician, Sandbian of the Appalachians; pteridophytes in the Silurian; gymnosperms in the Devonian; and angiosperms in the ?Triassic (Klitzsch *et al.*, 1973; Boureau *et al.*, 1978; Douglas & Lejal-Nicol, 1981; Krassilov, 1997; Beerling & Woodward, 2001; Hochuli & Feist-Burkhardt, 2004; Soltis *et al.*, 2005; Friis *et al.*, 2006; Beerling, 2007; Tomescu *et al.*, 2009). Pteridophytes diversified and dominated the plant kingdom in the Carboniferous, before sustaining significant losses at the time of the glaciation in the Late Carboniferous to Early Permian. Gymnosperms dominated in the later part of the Palaeozoic and Mesozoic. Angiosperms rose to dominance in the Cenozoic.

The moderate rate of evolutionary turnover exhibited by the plants renders them of some use in biostratigraphy, at least in appropriate, non-marine and associated marine, environments (in all other than the highest (palaeo)latitudes). Biostratigraphic zonation schemes based on plant macrofossils and, especially, microfossils have been established that have at least local to regional applicability, for example in the Silurian–Devonian of the 'Old Red Sandstone' continent, in the Carboniferous of Euramerica, in the Permian of Pangaea, specifically Gondwana, in the Triassic of Pangaea, specifically Euramerica and Gondwana, in the Jurassic–Cretaceous of Australasia, in the Cretaceous and Cenozoic of northern South America, and in the Cenozoic of southeast Asia (Helby *et al.*, in Jell, 1987; Morley, 1991; see also Section 5.2.3).

In the Triassic of Euramerica, numerous pollen zones and 'stages' have been recognised in the non-marine facies of North America that have been correlated with ammonoid zones in associated marine facies and thus calibrated against the standard stratigraphic time-scale (Fraser, 2006; Kurschner & Herngreen, in Lucas, 2010; Cirilli, in Lucas, 2010). The Langobardian 'stage' has been correlated with the Middle Triassic, Ladinian; the Cordevolian, Julian and Tuvalian with the Late Triassic, Carnian;

the Lacian and Alaunian with the Norian; and the Devatian with the Rhaetian.

In my working experience, plant macrofossils and microfossils have proved of use in the following areas:

Palaeozoic – the Ordovician–Permian of the Middle East and north Africa, the Devonian of western Europe, the Devonian–Permian of the Arctic and Former Soviet Union, and of eastern South America, and the Late Carboniferous–Permian of the Indian subcontinent;

Mesozoic – the Triassic–Jurassic of the North Sea, the Triassic–Cretaceous of the western margin of the British Isles, of the Canadian Arctic, of the Indian subcontinent, and of the Middle East and north Africa, the Early Cretaceous of the South Atlantic, and the Cretaceous of northern South America and the Caribbean, of the Melut and Muglad basins in Sudan, and of western Canada;

Cenozoic – the Cenozoic of the Indian subcontinent, of northern South America and the Caribbean, and of the Melut and Muglad basins, Sudan, the Palaeogene of the North Sea and the American Arctic, the 'Mid'–Late Cenozoic of the Black Sea and Caspian in eastern Parathethys, and the Neogene–Pleistogene of Nigeria.

2.1.5 Fungi

Fungal spores and hyphae

Molecular biological data indicates that fungi evolved in the Precambrian, much earlier than their first appearance as fossils in the Ordovician (Magallon & Hilu, in Hedges & Kumar, 2009). Note, though, that there are issues surrounding the calibration of the 'molecular clock' (Benton *et al.*, in Hedges & Kumar, 2009).

The slow rate of evolutionary turnover exhibited by the fungi renders them of limited use in biostratigraphy.

In my working experience, fungal spores and fruiting bodies have proved of use in the following areas:

Mesozoic – the Late Cretaceous of Alberta in Canada;

Cenozoic – the Oligo-Miocene of offshore Angola in the South Atlantic, and the Pliocene of offshore Azerbaijan.

2.1.6 Invertebrate animals

'Ediacarans'

Ediacarans first appeared in the Precambrian, Ediacaran (late Vendian), immediately above late Cryogenian, Varangerian (early Vendian), tillites dated to approximately 650–620 Ma. The abundance and diversity of the ediacarans already in the Ediacaran is arguably indicative of a still older origin. One indication that this was the case comes from the observation that microbialites began to decline in the Tonian–Cryogenian, from around 1000 Ma, which some authors have suggested was due to excessive grazing by ediacarans. Ediacarans appear to disappear at the end of the Precambrian, although morphologically similar forms are known from the Cambrian. The extinction may be more apparent than real, through, and an artefact of generally poor preservation in the Cambrian, in turn resulting from an increase in atmospheric oxygen and/or in scavengers and predators. Perhaps significantly, one of the better-known 'ediacaran-like' organisms of the Cambrian, *Xenusion*, is characterised by modification to the basic body-plan to incorporate spines interpreted as a defence against predation.

In my working experience, 'ediacarans' have proved of use in the Precambrian, late Cryogenian-early Ediacaran of the Middle East.

'Small shelly fossils' (SSFs)

Calcareous SSFs first appeared in the Precambrian, late Ediacaran, approximately 560 Ma (Gaucher & Germs, in Gaucher *et al.*, 2010). However, SSFs are generally regarded as characterising the earliest Cambrian, 'Nemakit–Daldynian' to 'Botoman'. They did not appear until the base of the 'Nemakit–Daldynian' in Siberia, which must now be accepted as defining the base of the Cambrian, on the basis of correlation, by means of the marker trace fossil *Phycodes pedum*, with the Precambrian/Cambrian boundary stratotype section in southeast Newfoundland. They disappeared in the 'Botoman', at around the level of with the so-called 'Sinsk [anoxic] event' at the *Bergeroniellus micmacciformis–Erbiella/B. gurarii* (Trilobite) Zone boundary.

Phosphatic SSFs also first appeared in the Precambrian, late Ediacaran, approximately 560 Ma (Gaucher & Germs, in Gaucher *et al.*, 2010). Hyolithans first appeared in the early Cambrian, 'Tommotian',

and disappeared in the 'Botoman'; in Britain, they have been recorded in the Hartshill formation of Warwickshire in the West Midlands. *Halkieria* and *Microdictyon* first appeared in the 'Atdabanian', and may more accurately be regarded to be representative of the evolutionary diversification of that, rather than of any earlier, time. Non-SSF phosphatic fossils such as bradoriide and phosphatocopide arthropods also first appeared in the 'Atdabanian'.

The rapid rate of evolutionary turnover and short range exhibited by SSFs renders them of use in biostratigraphy, at least in appropriate, marine, environments. Two SSF zones mark the Cambrian, 'Nemakit–Daldynian', three the 'Tommotian', and four the 'Atdabanian', in Siberia. Five SSF zones mark the 'Meishucunian' (the equivalent of the 'Nemakit–Daldynian' to early 'Atdabanian'), one zone with three sub-zones the 'Qiongzhusian' (the equivalent of the late 'Atdabanian'), and four zones the 'Canglangpuian' (the equivalent of the 'Botoman'), in south China (Steiner *et al.*, 2007).

In my working experience, SSFs have proved of use in the following areas:

Proterozoic – the Precambrian of the Middle East;
Palaeozoic – the Early Cambrian of east Siberia, and of Senegal and Guinea in west Africa.

Sponges

The problematic fossil *Otavia* and associated isolated spicules from the Precambrian, Cryogenian Otavi and Nama groups of Namibia, dated to c. 760 Ma, may be, or be derived from, sponges (Gaucher & Germs, in Gaucher *et al.*, 2010). Certain biomarkers and certain textures that occur in the even older rocks may also be attributable to sponges (Love *et al.*, 2009; Neuweiler *et al.*, 2009; Gaucher *et al.*, in Gaucher *et al.*, 2010).

However, incontrovertible (calci)sponges only appear in the rock record in the latest Precambrian, Ediacaran (Serezhnikova & Ivantsov, 2007; Gaucher *et al.*, in Gaucher *et al.*, 2010). Sponges rose to prominence in the Cambrian, 'Atdabanian', diversified in the Ordovician, and remained important through the remaining part of Palaeozoic, but became progressively less so through the Mesozoic and Cenozoic. There are at least some 5000 living species.

In my working experience, sponges have proved of use in the following areas:

Palaeozoic – the Late Palaeozoic of the former Soviet Union, and the Carboniferous (Lisburne group) of Arctic Alaska;
Mesozoic – the Late Jurassic of the North Sea;
Cenozoic – the Neogene of central Paratethys.

Archaeocyathans

Archaeocyathans first appeared in the Cambrian, 'Tommotian', and diversified and dispersed in the 'Botoman'. They underwent a major extinction in the 'Botoman–Toyonian', and declined thereafter, with only a few genera recorded from the Middle Cambrian, only one from the Late Cambrian, and none after the end of the Cambrian.

The rapid rate of evolutionary turnover and short range exhibited by the archaeocyathans renders them of considerable use in biostratigraphy, at least in appropriate, marine, environments. Unfortunately, their usefulness is somewhat compromised by their restricted bathymetric and biogeographic distribution and facies dependence. Nonetheless, high-resolution biostratigraphic zonation schemes based on the archaeocyathans have been established that have at least local to regional applicability. For example, three regional archaeocyathan zones are recognisable in the 'Atdabanian' of Australia, all with an average duration or resolution of less than 1 Ma. Moreover, some stratigraphic significance can be assigned simply on the basis of observation of archaeocyathan morphology, since older forms tend to be solitary, and younger ones modular.

In my working experience, archaeocyathans have proved of use in the Early Cambrian of east Siberia.

Stromatoporoids

Stromatoporoids first appeared in the Cambrian, 'Botoman', almost certainly evolving from a soft-bodied and/or sponge-like ancestor. They diversified in the Ordovician, before sustaining losses in the end-Ordovician mass extinction, taking much of the Silurian to recover. They then diversified again in the Devonian, before being almost eliminated in the Late Devonian mass extinction, which caused the Timan Pechora basin reef ecosystem alluded to above essentially to collapse, and the Canning basin

reef ecosystem of Western Australia to undergo a major reorganisation (Stock, in Over *et al.*, 2005). After the Late Devonian mass extinction, stromatoporoids took until the Jurassic to stage a partial recovery. They appear to have finally disappeared during the end-Cretaceous mass extinction. Note, though, that some living so-called calcified sponges exhibit stromatoporoid grades of organisation.

In my working experience, stromatoporoids have proved of use in the Devonian of Arctic Canada, and the Jurassic of the Middle East.

Corals

Cnidarians first appeared in the Precambrian, and coral-like cnidarians or coralimorphs in the Cambrian (Rogers, in Hedges & Kumar, 2009; Gaucher *et al.*, in Gaucher *et al.*, 2010). However, incontrovertible tabulate and rugose corals did not appear until the Ordovician. These sub-groups diversified through the Ordovician, before sustaining losses in the end-Ordovician mass extinction. They recovered from this event, though, and diversified again in the middle part of the Palaeozoic, before sustaining severe losses in the Late Devonian mass extinction. The tabulate corals never really recovered from this event, and finally disappeared in the end-Permian mass extinction. As noted above, the rugose corals staged some form of recovery from the Late Devonian mass extinction in the Carboniferous, but they, too, disappeared in the end-Permian mass extinction. Recognisable scleractinian corals first appeared in the Triassic, although arguably ancestral corals that exhibit scleractinian patterns of septal insertion have been described from the Ordovician of the Southern Uplands of Scotland. Scleractinian corals diversified in the Mesozoic and again in the Cenozoic, and are the dominant sub-group, and the only stony sub-group, at present. Interestingly, their rise through the Mesozoic–Cenozoic mirrors the decline of the essentially Palaeozoic stromatoporoids. There are some 3000 living species.

The moderate rate of evolutionary turnover exhibited by the corals renders them of some use in biostratigraphy, at least in appropriate, marine, environments (Fig. 2.8). Unfortunately, their usefulness is compromised by their restricted bathymetric and biogeographic distribution and facies

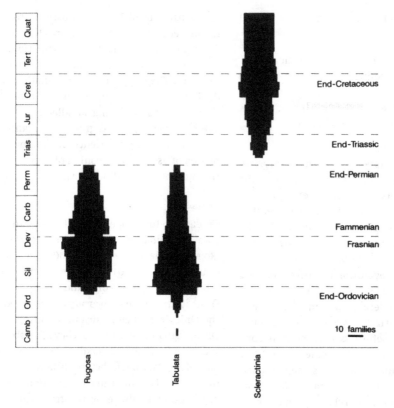

Fig. 2.8 **Stratigraphic distribution of corals**. © Wiley. Reproduced with permission from Doyle (1996).

dependence. Nonetheless, they are of some use in stratigraphic subdivision and correlation, for example in the British Isles, where they form part of the basis of the biozonation of the Carboniferous Limestone.

In my working experience, corals have proved of use in the following areas:

Palaeozoic – the Ordovician to Permian of the former Soviet Union, and the Carboniferous of the Middle East and north Africa;

Mesozoic – the Jurassic–Cretaceous of the Middle East and north Africa;

Cenozoic – the Oligo-Miocene of the Middle East.

Brachiopods

Brachiopods first appeared in the Cambrian; lingulates at the base of, and calciates within, the 'Tommotian' (Ruban, 2009). They diversified in the Ordovician, before sustaining significant losses in the end-Ordovician mass extinction. They recovered

from this event, though, and diversified again in the Silurian and Devonian, before sustaining significant losses again in the Late Devonian mass extinction. They recovered from this event, too, and diversified yet again in the Carboniferous and Permian, before sustaining severe losses in the end-Permian mass extinction. They may be said never to have really recovered from this event, and to have been in general decline ever since. There are some 12 000 fossil but only 350 living species.

The observed post-Palaeozoic decline of the brachiopods could be due in part to competition from the bivalve molluscs, whose rise appears to mirror it (brachiopods would probably have been less able to disperse, and to colonise infaunal niches, than bivalves). Some authors have suggested that it could be due to predation pressure (Leighton, in Kelley *et al.*, 2003). However, there are few records of predation on living brachiopods, either in the wild, for example in South Georgia, or in the laboratory,

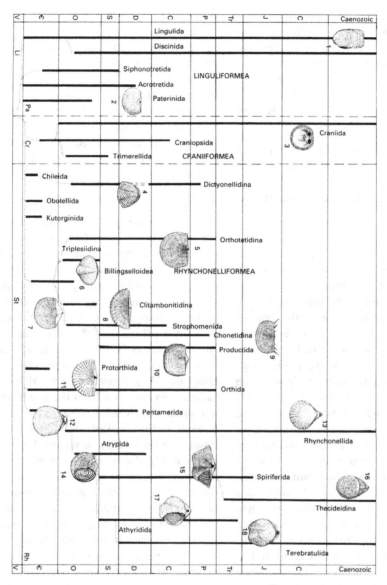

Fig. 2.9 **Stratigraphic distribution of brachiopods.** © Wiley. Reproduced with permission from Clarkson (1998).

where they are invariably ignored by potential predators such as fish and asteroids. This could be because they make use of chemical defences to deter predators, as do bryozoans.

The moderate rate of evolutionary turnover exhibited by the brachiopods renders them of some use in biostratigraphy, at least in appropriate, marine, environments (Fig. 2.9). Unfortunately, their usefulness is compromised by their restricted ecological distribution and facies dependence, a function of their benthic habit. Nonetheless, they are of some use in stratigraphic subdivision and correlation. For example, *Eocoelia* is extremely useful in the Silurian, where it forms the basis of a biozonation of the Llandovery to Wenlock that has inter-regional applicability throughout the 'Silurian cosmopolitan province'. Brachiopods are also useful in the Carboniferous of Britain, where they form part of the basis of the biozonation of the Carboniferous Limestone, and are locally important in the identification of marine bands in the 'Coal Measures', as in Northumberland and Durham. Also in Britain, *Terebratula lata* forms part of the basis of the biozonation of the Late Cretaceous Chalk.

In my working experience, brachiopods have proved of use in the following areas:

Palaeozoic – the Palaeozoic of the Middle East and north Africa, and the Late Palaeozoic of Arctic Alaska and of the former Soviet Union;
Mesozoic – the Jurassic–Cretaceous of the Middle East and north Africa.

Bryozoans

Cryptostomate bryozoans first appeared in the Late Cambrian, Series 4 (Furongian), Stage 10, stenolaemates in the Early Ordovician, Tremadocian, and gymnolaemates in the Late Ordovician, 'Ashgillian' (Feng-Sheng Xia *et al.*, 2007; Sen-Gui Zhang *et al.*, 2009; Landing *et al.*, 2010). Bryozoans sustained significant losses in the end-Ordovician mass extinction. They recovered from this event, though, and diversified again, even through the Late Devonian mass extinction event (Gutak *et al.*, 2008). They then sustained severe losses in the end-Permian mass extinction. The stenolaemates may be said to have been in general decline ever since this event, and their only living representatives are the cyclostomates. The gymnolaemates, in contrast, recovered from the end-Permian mass extinction, and have been diversifying throughout the Mesozoic to Cenozoic. In fact, in detail, the diversity of one constituent group, the ctenostomates, appears to have remained more or less constant through the Cretaceous to Recent, while that of the other, the cheilostomates, has expanded almost exponentially over that same time interval. Cheilostomates are especially abundant in the shallow marine facies of the Late Cretaceous to Early Palaeocene Danskekalk or Danish Chalk of the Danish North Sea. There are some 16 000 fossil and 5000 living species.

The moderate rate of evolutionary turnover exhibited by the bryozoans renders them of some use in biostratigraphy, at least in appropriate, aquatic, environments. Unfortunately, their usefulness is compromised by their restricted ecological distribution and facies dependence, a function of their benthic habit.

In my working experience, bryozoans have proved of use in the following areas:

Palaeozoic – the Ordovician–Permian of the former Soviet Union, and the Carboniferous (Lisburne group) of Arctic Alaska;
Mesozoic – the Late Cretaceous of the Middle East;
Cenozoic – the Miocene of central and eastern Paratethys.

Bivalves

Ancestral molluscs first appeared in the Precambrian (Gaucher *et al.*, in Gaucher *et al.*, 2010). However, incontrovertible bivalves did not appear until the Cambrian. They underwent significant diversification in the Ordovician, by the end of which all of the known sub-groups and life strategies had become established. Indeed, they continued to diversify until the end of the Permian, when they sustained losses in the end-Permian mass extinction, although not to the same extent as the brachiopods. They recovered from this event, and radiated, to take over from the brachiopods as the dominant filter-feeding group in the marine environment, in the so-called 'Mesozoic Revolution'. The group as a whole may be regarded as having been continuously diversifying through the Cenozoic, although locally, as in the US Atlantic and Gulf coastal plains, sustaining severe losses over the Eocene/Oligocene transition, characterised by falling sea level and cooling climate. The evident shift toward efficiency in burrowing through the Cenozoic has been interpreted as an evolutionary response to an increasing diversity of durophagous predators or an accelerating rate of sediment reworking. There are some 20 000 living species of bivalves.

The moderate rate of evolutionary turnover exhibited by the bivalves renders them of some use in biostratigraphy, at least in appropriate environments (Fig. 2.10). Unfortunately, their usefulness is somewhat compromised by their facies dependence. Nonetheless, they are of local use, especially in the absence of more suitable stratigraphic index fossils. In the Palaeozoic, non-marine bivalves form the basis of biozonation of the Carboniferous Coal

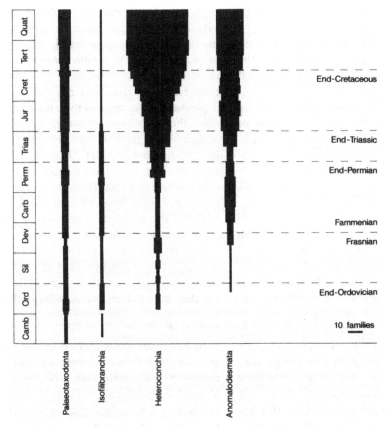

Fig. 2.10 **Stratigraphic distribution of bivalves**. © Wiley. Reproduced with permission from Doyle (1996).

Measures of Great Britain. In the Mesozoic, *Inoceramus labiatus* and *'Ostrea' lunata* form part of the basis of biozonation of the Late Cretaceous Chalk of Great Britain. Incidentally, inoceramid debris is locally sufficiently abundant in the Late Cretaceous of the adjoining North Sea basin as to be rock-forming (and even, in the case of the Turonian of Bruce field, to contribute to the formation of a petroleum reservoir). Inoceramids and other bivalves form part of the basis of biozonation of the Late Cretaceous of the 'Western Interior Seaway' of the United States. Species of *Lopha*, *Pycnodonte*, *Exogyra* and *Ostrea* form part of the basis of the biozonation of the Late Cretaceous Austin Chalk of Texas. I have on my desk before me as I write a specimen of the splendidly named *Exogyra ponderosa* from the Austin Chalk. In the Cenozoic,

bivalves are useful in the stratigraphic subdivision and correlation, and in the palaeoenvironmental interpretation, of the Eocene London clay and associated deposits, and Bracklesham and Barton beds. They are especially useful in the stratigraphic subdivision and correlation, and in the palaeoenvironmental interpretation, of the essentially Pleistocene 'Crags' and raised-beach deposits of the coasts of the British Isles, and indeed of the Pleistocene throughout northwest Europe. Interestingly, the classical subdivisions of the Cenozoic (Palaeocene, Eocene, Oligocene, Miocene, Pliocene, Pleistocene, Holocene) are based on Lyell's estimates of the proportions of still-extant bivalves in successive stages in the marginal basins of northwest Europe.

In my working experience, bivalves have proved of use in the following areas:

Palaeozoic – the Carboniferous of the former Soviet Union;

Mesozoic – the Triassic–Cretaceous of the Arctic, and of the Middle East and north Africa, and the Cretaceous of the western interior of the United States;

Cenozoic – the Cenozoic of Sakhalin in the former Soviet Union, and of the Arctic, the Eocene of north Africa, the Oligocene–Pleistocene of Paratethys, the Miocene of the Middle East, and the Miocene–Pleistocene of northern South America and the Caribbean.

Gastropods

Gastropods evolved in the Cambrian, 'Nemakit–Daldynian', probably from ancestors with a monoplacophoran grade of organisation (Parkhaev, in Vickers-Rich & Komarower, 2007). They diversified in the Ordovician, and by the Carboniferous had colonised freshwater and terrestrial as well as marine environments (Stworziewicz *et al.*, 2009). They sustained losses in the end-Ordovician, Late Devonian and, especially, end-Permian mass extinctions, but recovered from all of these events. They have been diversifying from the Triassic, Jurassic or Cretaceous to the present day, that is, through much of the Mesozoic and, especially, Cenozoic, sustaining only comparatively minor losses in the end-Cretaceous and end-Eocene mass extinctions. Terrestrial forms diversified most dramatically during the sea-level low-stands at the end of the Jurassic and at the end of the Cretaceous. Marine forms diversified during the mid-Cretaceous, coincident with a prolonged sea-level high-stand, and also, significantly, with a diversification of predators within and outwith the group. (Interestingly, it was a study of the land-snail *Poecilizontes* from the Pleistocene of Bermuda that first led to the development of the 'punctuated equilibrium' model of evolution, involving long periods of stasis interrupted by short periods of rapid evolutionary change, interpreted as the result of allopatric speciation at times of isolation of populations by glacioeustatically mediated sea-level rise. A similar study of snails from the Pliocene of Lake Turkana in the east African rift valley in Kenya later led to similar interpretations.) There are some 15 000 fossil and 30–40 000 living species of gastropods.

The moderate rate of evolutionary turnover exhibited by the gastropods renders them of some use in biostratigraphy, at least in appropriate environments (Fig. 2.11). Unfortunately, the potential usefulness of most is somewhat compromised by their restricted ecological distribution and facies dependence, a function of their benthic habit. However, the usefulness of some, the pteropods, is enhanced by their essentially unrestricted ecological distribution, and facies independence, a function of their nektonic habit. Biostratigraphic zonation schemes based on pteropods have been established that have at least local to regional applicability, as in the Cenozoic of the onshore parts of the North Sea basin in northwest Europe.

In my working experience, gastropods have proved of use in the Devonian of the Middle East, the Jurassic and Cretaceous of the Middle East, and the Oligocene to Pleistocene of Paratethys.

Ammonoids

Goniatite ammonoids evolved in the Early Devonian. The group then underwent a dramatic diversification in the Middle and Late Devonian before almost becoming extinct in the Late Devonian mass extinction. Nonetheless, they recovered from this event, and underwent renewed dramatic radiation in the Carboniferous, before finally becoming extinct in the end-Permian mass extinction. The rapid rate of evolutionary turnover exhibited by the goniatites renders them of considerable use in biostratigraphy, at least in appropriate, marine, environments (Fig. 2.12). Their usefulness is enhanced by their essentially unrestricted ecological distribution, and facies independence, a function of their nektonic habit. High-resolution biostratigraphic zonation schemes based on the goniatites have been established that have been calibrated against standard conodont schemes, or by orbital tuning, and that have regional to essentially global applicability. Anarcestids and clymeniids form the bases of the biozonations of the Middle and Late Devonian respectively of the so-called Rhenish facies of western Europe, including the British Isles. Goniatitids form part of the basis of the biozonation of the Late Devonian and Carboniferous of the British Isles, including the Carboniferous Coal Measures (Fig. 2.13).

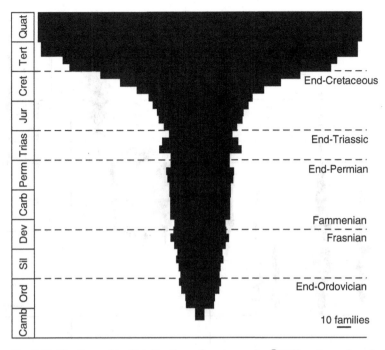

Fig. 2.11 **Stratigraphic distribution of gastropods**. © Wiley. Reproduced with permission from Doyle (1996).

Ceratites are essentially restricted to the Triassic. The rapid rate of evolutionary turnover exhibited by ceratites renders them of considerable use in biostratigraphy, at least in appropriate, marine, environments (Balini et al., in Lucas, 2010; Fig. 2.12). Their usefulness is enhanced by their essentially unrestricted ecological distribution, and facies independence, a function of their nektonic habit.

Ammonites evolved in the Triassic. The group then underwent rapid diversification through the Jurassic and Cretaceous, before becoming extinct in the end-Cretaceous mass extinction. The rapid rate of evolutionary turnover exhibited by the ammonites renders them of considerable use in biostratigraphy, at least in appropriate, marine, environments (Fig. 2.12). Unfortunately, their usefulness is somewhat compromised by their restricted biogeographic distribution. However, high-resolution biostratigraphic zonation schemes based on the ammonites have been established, that have regional to inter-regional applicability within a given palaeobiogeographic realm or province; and, moreover, the independent schemes

established for the various realms or provinces have been calibrated against one another, at least for the most part. Individual ammonite zones or sub-zones can have a duration or resolution of as little as 200 000 years. This level of temporal resolution is sufficient to enable the identification of even relatively minor unconformities or sequence boundaries. Significantly, it is also sufficient to provide a meaningful framework within which to assess at least some of the effects of climatic change, as inferred principally from isotope geochemistry. As long ago as the nineteenth century, ammonites were first used by the German palaeontologists Oppel and Quenstedt for stratigraphic subdivision and correlation in the Jurassic of the Swabian and Franconian Alb of southern Germany, the extension of the Jura Mountains of France and Switzerland from which the Jurassic takes its name. By the middle of the twentieth century, they were being used for stratigraphic subdivision and correlation of the Jurassic throughout the world (Arkell, 1956). Ammonites form the bulk of the basis of the biostratigraphic zonation of the marine facies of the Jurassic and, to a slightly lesser extent, the

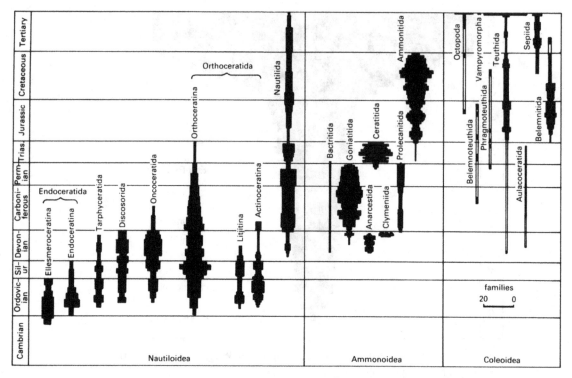

Fig. 2.12 **Stratigraphic distribution of ammonoids.** © Wiley. Reproduced with permission from Clarkson (1998).

Cretaceous of the British Isles, especially in the classical areas on the south and east coasts of England (Figs. 2.14–2.17).

In my working experience, goniatites, ceratites and ammonites have proved of use in the following areas:

Palaeozoic – the Late Palaeozoic, Devonian–Permian of the former Soviet Union and Arctic, and of the Middle East and north Africa;

Mesozoic – the Triassic–Cretaceous of the former Soviet Union and Arctic, of the Middle East and north and east Africa, and of the Indian subcontinent, the latest Jurassic to Cretaceous of northern South America, the Cretaceous of the Gulf Coast and 'Western Interior Seaway' of North America, and the 'Middle'–Late Cretaceous of west Africa in the South Atlantic.

Belemnites

True belemnites evolved in the Jurassic, and became extinct during the end-Cretaceous mass extinction. The arguably ancestral aulacocerids ranged from ?Devonian to Jurassic.

The moderate rate of evolutionary turnover exhibited by the belemnites, and their generally good preservation, even in diagenetically altered or even metamorphic rocks, renders them of some use in biostratigraphy, at least in appropriate, marine, environments. Their usefulness is somewhat compromised by the restricted, regional biogeographic distribution they exhibited in the earliest and again in the latest stages of their evolution. However, not only high-resolution stratigraphic subdivision but also inter-regional correlation is possible in intervening stages (Baraboshkin & Mutterlose, 2004). In Great Britain and Northern Ireland, belemnites form part of the basis of the biozonation of the Late Cretaceous Chalk, with *Gonioteuthis quadrata* and *Belemnitella mucronata* marking the middle and late parts of the Campanian, respectively, and *Belemnella lanceolata* the early part of the Maastrichtian (Fig. 2.18).

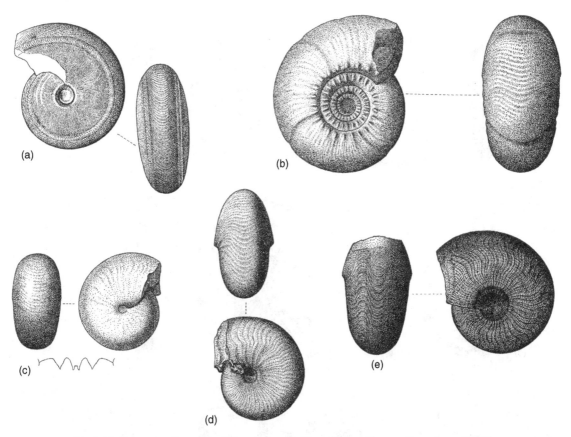

Fig. 2.13 **Some stratigraphically useful goniatites, Carboniferous**. From Jones (2006).
(a) *Reticuloceras bilingue*, Late Carboniferous, Namurian (Marsdenian), *Reticuloceras* Zone R2,
Hebden Bridge, North Yorkshire; (b) *Gastrioceras carbonarium*, Late Carboniferous,
Westphalian A (Langsettian), *Gastrioceras* Zone G2, near Leek, Staffordshire; (c)
Beyrichoceras obtusum, Early Carboniferous, Visean, *Beyrichoceras* Zone B, near Clitheroe,
Lancashire; (d) *Homoceras diadema*, Late Carboniferous, Namurian (Chokierian–Alportian),
Homoceras Zone H, West Yorkshire; (e) *Reticuloceras reticulatum*, Late Carboniferous,
Namurian (Kinderscoutian), *Reticuloceras* Zone R1, Hebden Bridge, North Yorkshire.

In my working experience, belemnites have proved of use in the Late Cretaceous of the Crimea, north Caucasus, peri-Caspian, Russian platform, Urals and west Siberia in the former Soviet Union.

Tentaculitids

Tentaculitids evolved in the ?Ordovician or Silurian. They became extinct in the Late Devonian mass extinction, sustaining severe losses in the late *rhenana* Conodont Zone of the Frasnian, and dying out in the *triangularis* Conodont Zone of the Famennian.

The rapid rate of evolutionary turnover exhibited by the tentaculitids renders them of considerable use in biostratigraphy, at least in appropriate, marine, environments, and in low to moderate latitudes. High-resolution biostratigraphic zonation schemes based on the tentaculitids have been established for the Devonian, that have widespread applicability, especially in central Europe, and that have been calibrated against conodont standards. In the Barrandian area of the Czech Republic, the Lochkovian is represented by two zones; the Pragian by three; the Zlichovian by four; the Daleian by three; the Eifelian by three; and the Givetian by one.

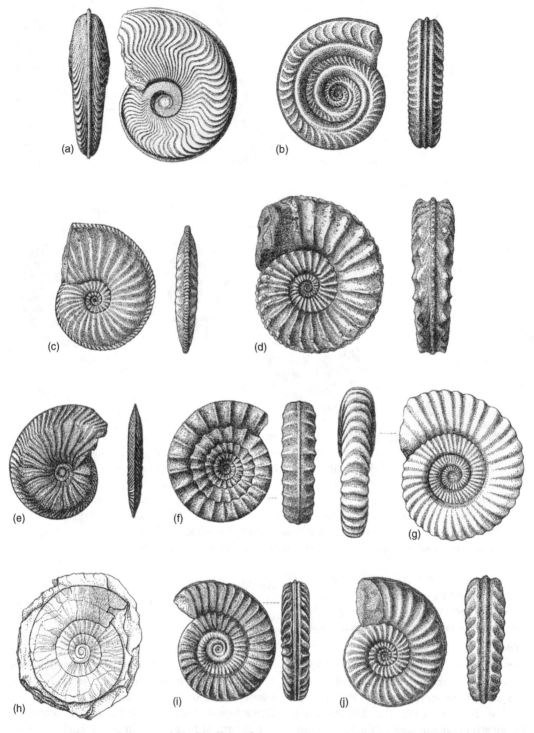

Fig. 2.14 **Some stratigraphically useful ammonites, Early Jurassic.** From Jones (2006).
(a) *Harpoceras falciferum*, nominate taxon of the Toarcian *Harpoceras falciferum* Zone, Upper
Lias, Lincoln; (b) *Hildoceras bifrons*, nominate taxon of the Toarcian *Hildoceras bifrons* Zone,

In my working experience, tentaculitids have proved on use in the Devonian of north Africa, for example in the Ahnet basin, Ougarta Hills and elsewhere in Algeria, in the Anti-Atlas in Morocco, and in Libya.

Trilobites

Ancestral arthropods first appeared in the Precambrian (Gaucher *et al.*, in Gaucher *et al.*, 2010). However, incontrovertible trilobites did not appear until the Early Cambrian, 'Atdabanian'. They underwent a dramatic diversification through the remainder of the Cambrian and Ordovician, before sustaining severe losses in the end-Ordovician mass extinction. Nonetheless, they recovered from this event, only to sustain severe losses again in the Late Devonian mass extinction. They finally became extinct in the end-Permian mass extinction. (Interestingly, trilobites, for example those of the Ordovician *Teretiusculus* Shale of Wales, played a pivotal role in the development of the 'phyletic gradualism' model of evolution.)

The rapid rate of evolutionary turnover exhibited by the trilobites renders them of considerable use in biostratigraphy, at least in appropriate, marine, environments (Fig. 2.19). Unfortunately, their usefulness is somewhat compromised by their restricted biogeographic distribution. However, high-resolution biostratigraphic zonation schemes based on the trilobites have been established that have regional to interregional applicability within a given palaeobiogeographic realm or province. Moreover, the independent schemes established for the various realms or provinces have been calibrated against one another,

at least for the most part. It can thus now be demonstrated that the conventional placement of the Early/Middle Cambrian boundary on trilobite evidence has been stratigraphically lower in parts of Gondwana than elsewhere in the world. In Morocco in western Gondwana, the boundary has been placed at the base of the *Hupeolenus* Zone at the base of the 'Tissafinian' (Geyer & Landing, 2004). In contrast, in Iberia, it has been placed at the base of the *Acadoparadoxides mureroensis* Zone at the base of the 'Leonian', and in Australia, it has been placed at the base of the *Xystridura templetonensis/Redlichia chinensis* Zone at the base of the 'Templetonian', both of which are in stratigraphically higher positions (Geyer & Landing, 2004; Gradstein *et al.*, 2004). The resolution of certain local trilobite zonation schemes is capable of enhancement through the use of graphic correlation technology, as in the case of the Late Cambrian of Pennsylvania.

In England and Wales, olenellids, paradoxidids and agnostids are useful for stratigraphic subdivision and correlation in the Cambrian (Fig. 2.20). *Merlinia* is locally useful in the Ordovician, and *Calymene* in the Silurian. Disarticulated pygidia of *Merlinia* are so conspicuous around Carmarthen in southwest Wales as to have gone down in local legend as butterflies petrified by a spell cast by the magician Merlin. (Incidentally, the Welsh for Merlin is Myrddin, as, in mutated form, in Caer-Fyrddin, meaning 'Merlin's Fort', and, in anglicised form, in Carmarthen.) *Calymene blumenbachi* is so conspicuous in the Wenlock limestone around the town of Dudley in the West Midlands as to have acquired the local name of the 'Dudley bug', and even to have had its image incorporated into the town's

---

Caption for fig. 2.14 (*cont.*) Upper Lias, Whitby, Yorkshire; (c) *Amaltheus margaritatus*, nominate taxon of the late Pliensbachian (Domerian) *Amaltheus margaritatus* Zone, Middle Lias, Ilminster, Somerset; (d) *Pleuroceras spinatum*, nominate taxon of the late Pliensbachian (Domerian) *Pleuroceras spinatum* Zone, Down Cliff, near Bridport, Dorset; (e) *Oxynoticeras oxynotum*, nominate taxon of the Sinemurian *Oxynoticeras oxynotum* Zone, Lower Lias, Cheltenham, Gloucestershire; (f) *Echioceras raricostatum*, nominate taxon of the Sinemurian *Echioceras raricostatum* Zone, Lower Lias, Radstock, Somerset; (g) *Uptonia jamesoni*, nominate taxon of the early Pliensbachian (Carixian) *Uptonia jamesoni* Zone, Lower Lias, Munger Quarry, Paulton, Somerset; (h) *Psiloceras planorbis*, nominate taxon of the Hettangian *Psiloceras planorbis* Zone, Lower Lias, Watchet, Somerset; (i) *Arnioceras semicostatum*, nominate taxon of the Sinemurian *Arnioceras semicostatum* Zone, Lower Lias, Robin Hood's Bay, Yorkshire; (j) *Asteroceras obtusum*, nominate taxon of the Sinemurian *Asteroceras obtusum* Zone, Lower Lias, Lyme Regis, Dorset.

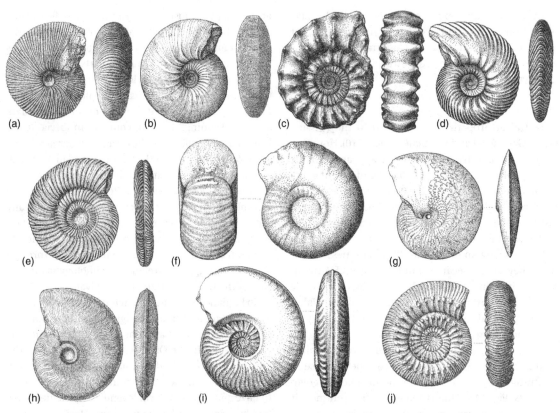

Fig. 2.15 **Some stratigraphically useful ammonites, Middle Jurassic.** From Jones (2006). (a) *Macrocephalites macrocephalus*, nominate taxon of the Callovian *Macrocephalites macrocephalus* Zone, Upper Cornbrash, Scarborough, Yorkshire; (b) *Sigaloceras calloviense*, nominate taxon of the Callovian *Sigaloceras calloviense* Zone, Kellaways Rock, Kellaways, Wiltshire; (c) *Peltoceras athleta*, nominate taxon of the Callovian *Peltoceras athleta* Zone, Oxford Clay, Eye, Cambridgeshire; (d) *Quenstedtoceras lamberti*, nominate taxon of the Callovian *Quenstedtoceras lamberti* Zone, Oxford Clay, Ashton Keynes, Wiltshire; (e) *Parkinsonia parkinsoni*, nominate taxon of the Bathonian *Parkinsonia parkinsoni* Zone, Upper Inferior Oolite, Sherborne, Dorset; (f) *Tulites subcontractus*, nominate taxon of the Bathonian *Tulites subcontractus* Zone, Great Oolite, Fuller's Earth Rock, Somerset; (g) *Clydoniceras discus*, nominate taxon of the Bathonian *Clydoniceras discus* Zone, Lower Cornbrash, Closeworth, Somerset; (h) *Leioceras opalinum*, nominate taxon of the Aalenian *Leioceras opalinum* Zone, Lower Inferior Oolite, Bridport, Dorset; (i) *Ludwigia murchisonae*, nominate taxon of the Aalenian *Ludwigia murchisonae* Zone, Lower Inferior Oolite, Beaminster, Dorset; (j) *Stephanoceras humphriesanum*, nominate taxon of the Bajocian *Stephanoceras humphriesanum* Zone, Middle Inferior Oolite, Dundry, Somerset.

coat-of-arms. *Calymene [Neseuretus] ramseysnsis* is sufficiently abundant as to lend its name to the *Calymene* ashes of the Arenig of the type area in north Wales.

In my working experience, trilobites have proved of use in the Cambrian of the Irkutsk amphitheatre area of east Siberia, the Cambrian–Silurian of the Middle East and north Africa, and the Ordovician–Silurian of Argentina and Bolivia in the 'southern cone' of South America.

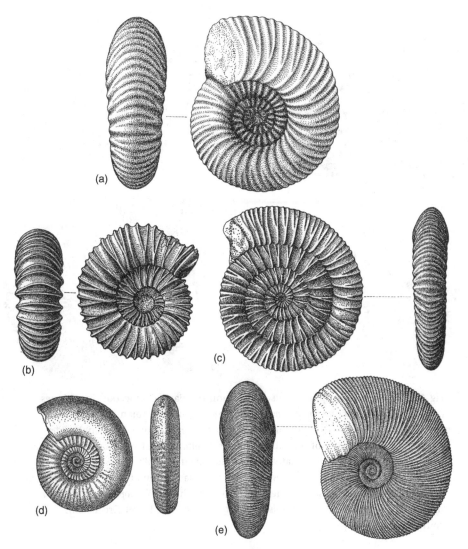

Fig. 2.16 **Some stratigraphically useful ammonites, Late Jurassic.** From Jones (2006).
(a) *Titanites giganteus*, nominate taxon of the 'Portlandian' *Titanites giganteus* Zone, Upper
Portland Beds (Portland Stone), Tisbury, Wiltshire; (b) *Pavlovia pallasoides*, nominate taxon
of the Kimmeridgian *Pavlovia pallasoides* Zone, Hartwell Clay, Hartwell, Buckinghamshire;
(c) *Crendonites (Glaucolithes) gorei*, nominate taxon of the 'Portlandian' *Glaucolithes gorei*
Zone, Lower Portland Beds, Portland, Dorset; (d) *Pictonia baylei*, nominate taxon of the
Kimmeridgian *Pictonia baylei* Zone, Lower Kimmeridge Clay, Wootton Bassett, Wiltshire;
(e) *Pectinatites pectinatus*, nominate taxon of the Kimmeridgian *Pectinatites
pectinatus* Zone, Upper Kimmeridge Clay, Swindon, Wiltshire.

Ostracods

Ostracods first appeared in the Cambrian, 'Atdabanian', or Ordovician if the bradoriides are excluded.
They diversified in the Ordovician, by the end of
which all of the major sub-groups had appeared.
They sustained significant losses in the end-Ordovician mass extinction, when the bradoriides disappeared. The leperditicopides became extinct at the

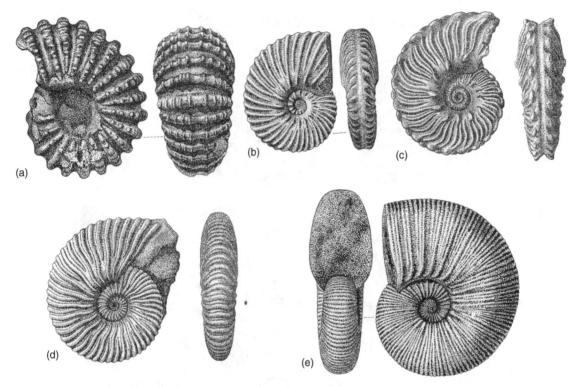

Fig. 2.17 **Some stratigraphically useful ammonites, Early Cretaceous**. From Jones (2006). (a) *Douvilleiceras mammillatum*, nominate taxon of the Albian *Douvilleiceras mammillatum* Zone, Lower Greensand/Gault junction, Folkestone, Kent; (b) *Hoplites dentatus*, nominate taxon of the Albian *Hoplites dentatus* Zone, Lower Gault, Folkestone, Kent; (c) *Euhoplites lautus*, nominate taxon of the Albian *Euhoplites lautus* Zone, Lower Gault, Folkestone, Kent; (d) *Deshayesites forbesi*, nominate taxon of the Aptian *Deshayesites forbesi* Zone, Lower Greensand, Atherfield, Isle of Wight; (e) *Parahoplites nutfieldensis*, nominate taxon of the Aptian *Parahoplites nutfieldensis* Zone, Lower Greensand, Nutfield, Surrey.

time of the Late Devonian mass extinction; the palaeocopides either in the Permian, at the time of the end-Permian mass extinction, or in the Triassic, at the time of the end-Triassic mass extinction. The remaining sub-groups diversified during the Mesozoic and Cenozoic, sustaining some losses in the Late Triassic, Pliensbachian, Bathonian, 'Purbeckian', Barremian, Turonian, Maastrichtian, Late Eocene and Late Miocene. There are some 6000 living species of ostracods.

The moderate rate of evolutionary turnover exhibited by the ostracods renders them of some use in biostratigraphy, at least in appropriate environments. Unfortunately, their usefulness is compromised by their restricted ecological distribution and facies dependence, a function of their benthic habit. Nonetheless, biostratigraphic zonation schemes based on the ostracods have been established that have at least local to regional applicability.

In my working experience, ostracods have proved of use in the following areas:

Palaeozoic – the Palaeozoic of the former Soviet Union, the Devonian of Canada, the Devonian–Permian of the Middle East and north Africa, the Carboniferous (Lisburne group) of Arctic Alaska, and the Carboniferous–Permian of Texas;

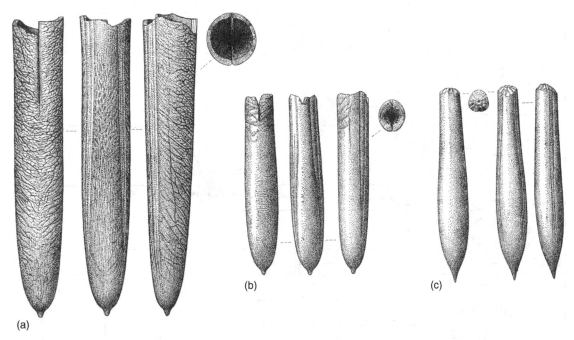

Fig. 2.18 **Some stratigraphically useful belemnites.** From Jones (2006). (a) *Actinocamax plenus*, Late Cretaceous, late Cenomanian–early Turonian *Plenus* Marl, West Cliffs, Dover, Kent; (b) *Gonioteuthis quadrata*, nominate taxon of the Late Cretaceous, Campanian *Gonioteuthis quadrata* Zone, Upper Chalk, near Salisbury, Wiltshire; (c) *Belemnitella mucronata*, nominate taxon of the Late Cretaceous, Campanian *Belemnitella mucronata* Zone, Upper Chalk, Whitlingham, Norfolk.

Mesozoic – the Triassic–Cretaceous of the United Kingdom continental shelf, including the North Sea, the Jurassic of the Arctic, the non-marine Early Cretaceous of the South Atlantic, and the 'Middle'–Late Cretaceous of the Middle East and north Africa;

Cenozoic – the Eocene of north Africa, and the Oligocene–Pleistocene of eastern Paratethys, of northern South America and the Caribbean, and of the Gulf of Mexico.

Branchiopods

Branchiopods first appeared in the Silurian, and have ranged through to the Recent. There are some 800 living species.

The moderate rate of evolutionary turnover exhibited by the branchiopods renders them of some use in biostratigraphy, at least in appropriate environments. Unfortunately, their usefulness is compromised by their restricted ecological distribution and facies dependence, a function of their benthic habit. Nonetheless, biostratigraphic zonation schemes based on the branchiopods have been established, that have at least regional applicability, for example in the Triassic of the northern hemisphere (Kozur & Weems, in Lucas, 2010).

In my working experience, branchiopods have proved of biostratigraphic and/or palaeobiological use in the Permo-Triassic of the Gondwanan basins of east and South Africa and of the Indian subcontinent, and in the Early Cretaceous 'Gondwana Wealden' facies of the South Atlantic salt basins.

Insects

Wingless insects first appeared in the Devonian, as exemplified by the Rhynie chert of Aberdeenshire in Scotland. Primitive winged, dragonfly-like, insects, unable to fold their wings against their bodies, appeared next, in the Early Carboniferous; and many more advanced groups in the Late

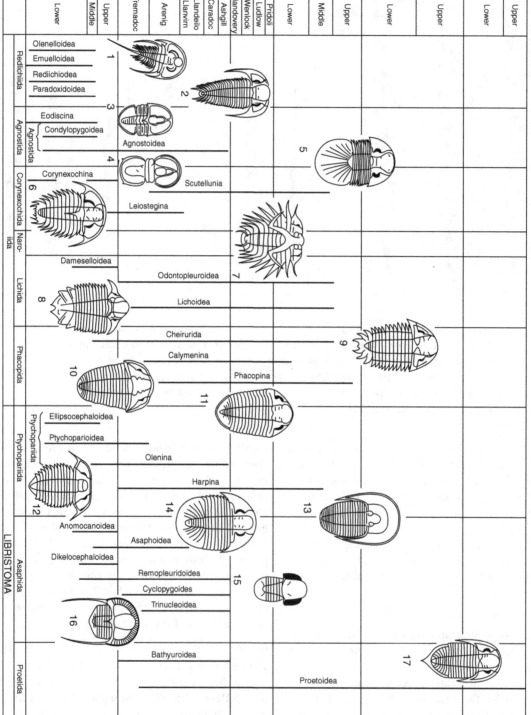

Fig. 2.19 **Stratigraphic distribution of trilobites.** © Wiley. Reproduced with permission after Clarkson (1998).

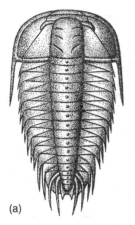

(a)

(b)

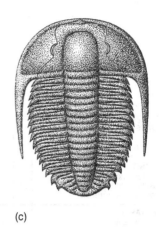

(c)

Fig. 2.20 **Some stratigraphically useful trilobites**. From Jones (2006). (a) *Parabolina spinulosa*, nominate taxon of the Cambrian, Merioneth *Parabolina spinulosa* Zone, Dolgellau, Gwynedd; (b) *Shumardia pusilla*, nominate taxon of the Ordovician, Tremadoc *Shumardia pusilla* Zone, Sheinton, Shropshire; (c) *Angelina sedgwickii*, nominate taxon of the Ordovician, Tremadoc *Angelina sedgwickii* Zone, Tremadoc, Gwynedd.

Carboniferous through Triassic (Bethoux *et al.*, 2005). Most of the remaining groups, including the lepidopterans or butterflies and allied forms, and hymenopterans or bees and allied forms, that are now important plant pollinators appeared in the Jurassic through Cretaceous, more or less coincident with the radiation of the flowering plants. The last remaining group, the siphonapterans or fleas, appeared in the Cenozoic, coincident with the radiation of the mammals that they parasitise. The overall pattern of insect evolution has been one of ever-increasing diversification, with little or no evident loss of diversity associated with any of the main mass extinction events, other than, arguably, the end-Permian mass extinction. All nine subgroups of pterygotes are found as fossils in the Late Jurassic Solnhofen Limestone of Bavaria in southern Germany, the Early Cretaceous Crato formation of the Araripe basin of northeastern Brazil, and the Eocene lacustrine deposits of Florissant in Colorado; and one sub-group of apterygotes and six of pterygotes in the Miocene amber deposits of the Dominican Republic, together with collembolans and myriapods (Martill *et al.*, 2007; Nudds & Selden, 2008). In Britain, diverse insects are found in the Carboniferous Coal Measures, and in Early Cretaceous and Oligocene continental deposits on the Isle of Wight. There are over a million living species of insect described, and probably millions more undescribed. It has been estimated that there are 10 quintillion or 10 000 000 000 000 000 000 living individuals!

The rapid rate of evolutionary turnover exhibited by the insects renders them useful in biostratigraphy (Fig. 2.21).

They form the basis of the biozonation of the Late Carboniferous–Early Permian of Euramerica, with a mean temporal resolution of 1.5–2 Ma (Schneider & Werneburg, in Lucas *et al.*, 2006).

In my working experience, insects have not proved of use biostratigraphically, although they have proved of use palaeobiologically, in the Middle Miocene, Karaganian of Vishnevaya Balka in the Caucasus in the former Soviet Union.

Crinoids

Crinoids first appeared in the earliest Ordovician, Tremadoc, if the Cambrian echmatocrinids are excluded (Guensburg & Sprinkle, 2009). They diversified through the Ordovician, but sustained significant losses in the end-Ordovician mass extinction. They recovered from this event, though, and diversified again through the Silurian and Devonian, but sustained further significant

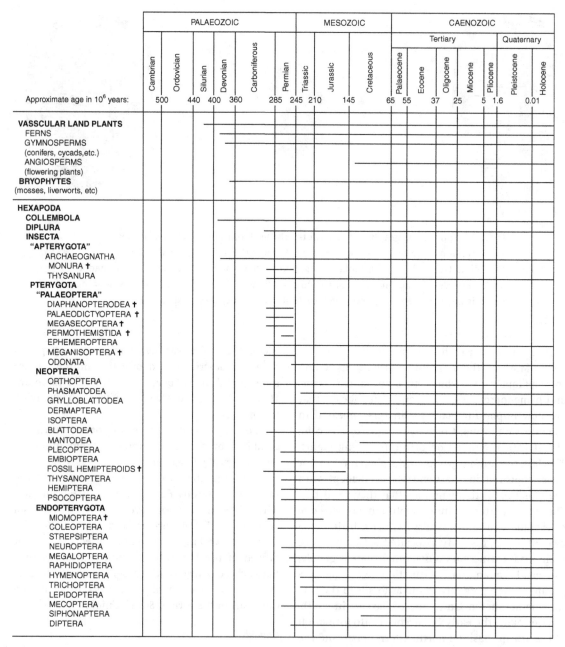

Fig. 2.21 **Stratigraphic distribution of insects**. From Jones, 2006 (reproduced with the permission of Blackwell from Gullan & Cranston, 2000).

losses in the Late Devonian mass extinction. They recovered from this event, too, and underwent renewed radiation in the Carboniferous and Permian, before almost becoming extinct in the end-Permian mass extinction. They did stage some

form of recovery from this event, and diversified through the Triassic and Jurassic, but never attained the prominence in the Mesozoic that they had had in the Palaeozoic. They may be said to have been in decline since the Mesozoic, possibly

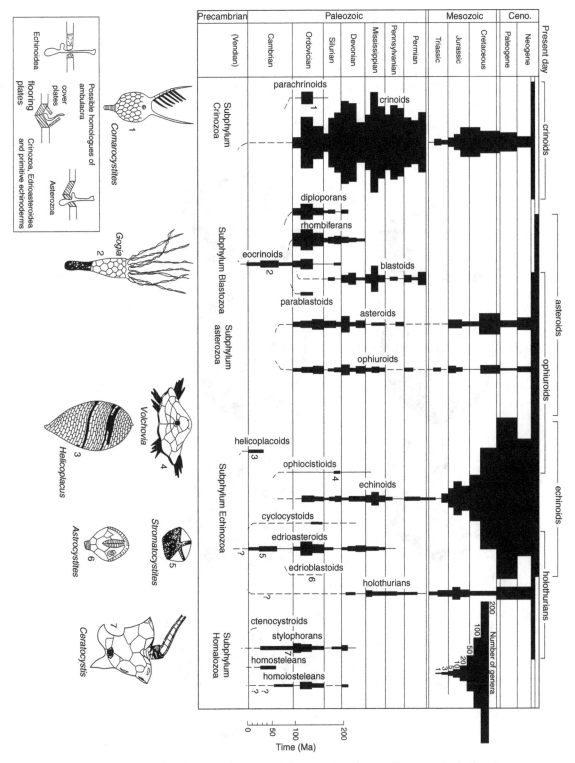

Fig. 2.22 **Stratigraphic distribution of crinoids and echinoids**. © Wiley. Reproduced with permission from Clarkson (1998).

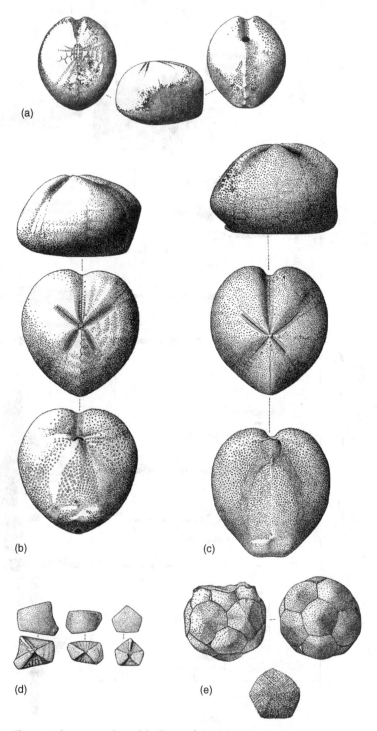

Fig. 2.23 **Some stratigraphically useful echinoids (a–c) and crinoids (d, e).** From Jones (2006). (a) *Sternotaxis (Holaster) planus*, nominate taxon of the Late Cretaceous, Coniacian *Holaster planus* Zone, Upper Chalk, east of Dover, Kent; (b) *Micraster cortestudinarium*,

owing either to increased predation by fish, especially in shallow marine environments, or to increased competition from echinoids. The perceived decline may be more apparent than real, though, and an artefact of preservation. There are some 600 living species of crinoids.

The moderate rate of evolutionary turnover exhibited by the crinoids renders them of some use in biostratigraphy, at least in appropriate, marine, environments (Fig. 2.22).

Crinoids are of use in, for example, the Ordovician and Cretaceous of the United Kingdom. Here, crinoid columnals are of use in the Ordovician and Cretaceous; crinoids in the Late Cretaceous, where they form part of the basis of the biozonation of the Chalk, *Uintacrinus socialis* and *Marsupites testudinarius* marking the middle and late parts of the Santonian, respectively (Fig. 2.23). Note also, incidentally, that exquisitely preserved specimens of *Pentacrinites fossilis* occur in the Early Jurassic, lower Lias of Charmouth in Dorset.

Crinoids are also of use in, for example, the Early Carboniferous of North America, the Permian of Australia, and the Triassic of Europe.

In my working experience, crinoids have proved of use in the following areas:

Palaeozoic – the Ordovician–Permian of the former Soviet Union;
Mesozoic – the latest Jurassic to earliest Cretaceous of the Middle East.

Echinoids

Echinoids first appeared in the 'Middle' Cambrian, Series 3, Stage 5, and diversified through the Ordovician, Silurian and Devonian, before sustaining significant losses in the Late Devonian mass extinction (Zamora, 2010). They recovered from this event, though, and underwent renewed radiation in the Permian, before sustaining further significant losses in the end-Permian mass extinction. They recovered from this event, too, and underwent a dramatic diversification in the Mesozoic, when the euechinoids evolved, at which time they overtook the crinoids as the dominant group of echinoderms. They were little if at all affected by the end-Cretaceous mass extinction, and indeed underwent renewed radiation in the Cenozoic. There are some 900 living species of echinoids.

In theory, the moderate rate of evolutionary turnover exhibited by the echinoids renders them of some use in biostratigraphy, at least in appropriate, marine, environments (Fig. 2.22). Unfortunately, in practice, their usefulness is compromised by their restricted ecological distribution and facies dependence, a function of their benthic habit. Moreover, their potential usefulness in the Palaeozoic is further compromised by their poor preservation potential at this time, prior to the innovation of the fusion of the plates in the Mesozoic. Nonetheless, echinoids are of use in, for example, the Late Cretaceous of Great Britain and Northern Ireland, where they form part of the basis of the biozonation of the Chalk (Fig. 2.23).

Here, *Holaster planus* marks the late part of the Turonian, *Micraster cortestudinarium* the early part of the Coniacian, *M. coranguinum* the late part of the Coniacian and the early part of the Santonian, and *Offaster pilula* the early part of the Campanian.

In my working experience, echinoids have proved of use in the Cretaceous of the Middle East and north Africa, and in the Oligocene–Miocene of the Middle East.

Graptolites

Dendroid graptolites first appeared in the Cambrian, and ranged through to the Carboniferous. Graptoloids evolved, from dendroids, in the Ordovician. They diversified through the Ordovician,

Caption for Fig. 2.23 (*cont.*) nominate taxon of the Late Cretaceous, Coniacian *Micraster cortestudinarium* Zone, Upper Chalk, Chatham, Kent; (c) *Micraster coranguinum*, nominate taxon of the Late Cretaceous, Santonian *Micraster coranguinum* Zone, Upper Chalk, Gravesend, Kent; (d) *Unitacrinus socialis*, nominate taxon of the Late Cretaceous, Santonian *Unitacrinus socialis* Zone, Upper Chalk, Kent; (e) *Marsupites testudinarius*, nominate taxon of the Late Cretaceous, Santonian *Marsupites testudinarius* Zone, Upper Chalk, southern England.

but then sustained significant losses in the end-Ordovician mass extinction. They recovered from this event, though, and diversified again in the Silurian, especially in the Llandovery and Ludlow. However, they then appear to have gone into a decline, and to have become extinct during the Devonian. (Note, though, that it has recently been argued that the pterobranch hemichordate *Rhabdopleura* is an extant graptolite: Mitchell *et al.*, 2010.)

The rapid rate of evolutionary turnover exhibited by the graptolites renders them of considerable use in biostratigraphy, at least in appropriate, deep marine environments (Fig. 2.24).

Indeed, high-resolution biostratigraphic zonation schemes based on the graptolites have been established that have essentially global applicability. The zonal resolution is locally capable of enhancement through the use of graphic correlation technology, for example in the Ordovician of the Appalachians, and the Silurian of central Nevada and elsewhere.

In Great Britain, 60 graptolite zones and subzones can be recognised over the Ordovician-Silurian, with a mean duration or resolution of the order of 1 Ma, sufficient to facilitate the construction of detailed chronostratigraphic diagrams (Zalasiewicz *et al.*, 2009). Here, *Dictyonema* is useful for stratigraphic subdivision and correlation in the Ordovician, Tremadoc; *Didymograptus* in the former Arenig–Llanvirn (current Floian–Darriwilian, according to Cocks *et al.*, 2010); *Glyptograptus* in the Llandeilo (Darriwilian); *Nemagraptus* in the Llandeilo–Caradoc (Darriwilian–Sandbian); *Climacograptus*, *Dicranograptus* and *Pleurograptus* in the Caradoc (Katian); and *Dicellograptus* and *Glyptograptus* again in the Ashgill (Katian–Hirnantian); *Akidograptus*, *Monograptus* and *Monoclimacis* in the Silurian, Llandovery; *Cyrtograptus*, *Monograptus* again and *Pristiograptus* in the Wenlock; and *Pristiograptus* again, *Monograptus* yet again and *Bohemograptus* in the Ludlow (Figs. 2.25–2.26).

In north Wales, graptolites are locally sufficiently abundant as to lend their name to the containing rocks, as in the case of the Tremadoc *Dictyonema* Band of the west of Portmadoc, and the Llanvirn (Darriwilian) *Didymograptus bifidus* [= *artus*] Beds of the area around Cader Idris in southern Snowdonia.

In my working experience, graptolites have proved of use in the Palaeozoic of the Middle East and north Africa, of Argentina and Bolivia in the 'southern cone' of South America, and of the American Arctic.

Chitinozoans

Chitinozoans evolved in the Cambrian, diversified in the Ordovician and Silurian, declined in the Devonian and became extinct in the Carboniferous.

The rapid rate of evolutionary turnover exhibited by the chitinozoans renders them of considerable use in biostratigraphy, at least in appropriate, deep marine, environments. Unfortunately, their usefulness is somewhat compromised by their restricted bathymetric distribution. Nonetheless, high-resolution biostratigraphic zonation schemes based on the chitinozoans have been established that have widespread applicability.

In my working experience, chitinozoans, like graptolites, have proved of use in the Palaeozoic of the Middle East and north Africa, of Argentina and Bolivia in the 'southern cone' of South America, and of the American Arctic.

2.1.7   Vertebrates

Fish macrofossils, conodonts, ichthyoliths and otoliths

The first fish, the jawless agnathans, evolved in the Early Cambrian (Grande, in Coggin *et al.*, 2003; Jastrow & Rampino, 2008; Kuraku *et al.*, in Hedges & Kumar, 2009). The group diversified through the Ordovician, Silurian and Devonian, when the placoderms, acanthodians, chondrichthyans and osteichthyans appeared, before sustaining significant losses in the Late Devonian mass extinction, when the placoderms disappeared, and in the end-Permian mass extinction, when the acanthodians disappeared (the acanthodians were also affected by a Late Silurian extinction event: Erikkson *et al.*, 2009). The overall pattern throughout the Mesozoic and Cenozoic has essentially been one of ever-increasing diversification, with little or no evident loss of diversity associated with any of the main mass extinction events of the time. There are some 28 000 living species of fish. Conodonts first appeared in the Late Cambrian. They diversified in the Ordovician, before sustaining significant losses in the end-Ordovician mass extinction. They recovered from this event, though, and diversified again in the Silurian and Devonian, when ramiform and platform types took over from conical types.

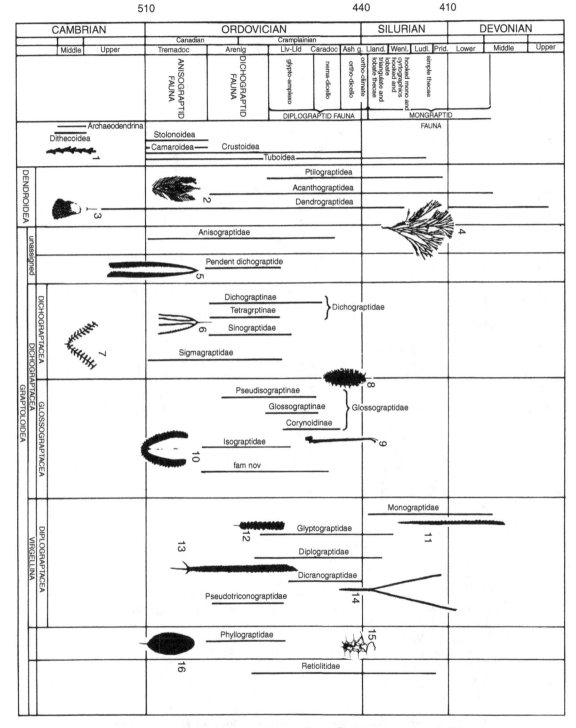

Fig. 2.24 **Stratigraphic distribution of graptolites.** © Wiley. Reproduced with permission from Clarkson (1998).

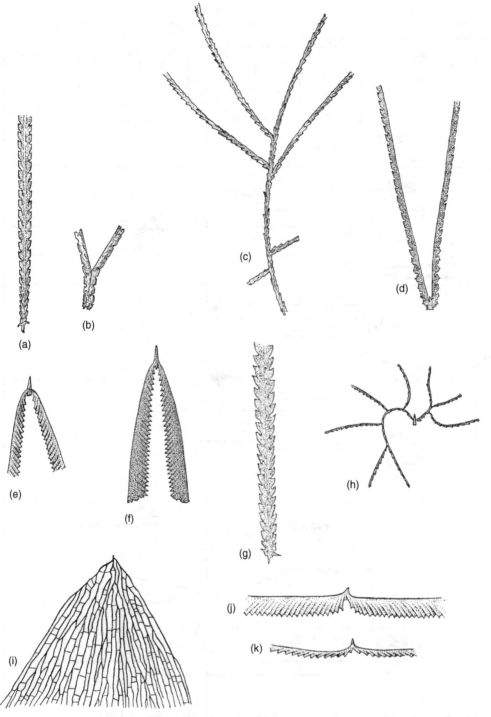

Fig. 2.25 **Some stratigraphically useful graptolites, Ordovician.** From Jones (2006).
(a) *Climacograptus wilsoni*, nominate taxon of the Caradoc (Soudleyan) *Climacograptus wilsoni*
Zone, Moffat, Dumfriesshire; (b) *Dicranograptus clingani*, nominate taxon of the Caradoc

They declined in the Permian, and disappeared in the Triassic (Orchard, in Lucas, 2010). Ichthyoliths have hitherto only been described from the Late Cretaceous to Cenozoic.

The locally rapid rate of evolutionary turnover exhibited by the fish macrofossils and fish-derived microfossils renders them of considerable use in biostratigraphy, at least in appropriate, aquatic, environments. Their usefulness is enhanced by their essentially unrestricted ecological distribution, and facies independence, at least within the aquatic realm. Locally, the usefulness of the conodonts has been further enhanced through the use of graphic correlation. High-resolution biostratigraphic zonation schemes based on the conodonts have been established that have essentially global applicability, across a range of palaeobiogeographic realms or provinces, for the Late Cambrian to Triassic. Local schemes are also available for, for example, the Culm facies of the Devonian and Carboniferous of the British Isles. Biostratigraphic zonation schemes based on ichthyoliths and otoliths have also been established that have at least local to regional applicability. Fish macrofossils are of use in, for example, the Silurian and Devonian of China. Here, seven biochrons have been recognised on the basis of their successive faunas: an Early Silurian *Dayongaspis* Biochron; an early Middle Silurian *Hanyangaspis* Biochron; a late Middle Silurian *Sinogaleaspis* Biochron; an unnamed Late Silurian Biochron; an Early Devonian *Yunnanolepis* Biochron; a Middle Devonian *Bothriolepis* Biochron; and a Late Devonian *Remigolepis* Biochron. Fish macrofossils are also of use in the non-marine latest Silurian-Devonian, 'Downtonian–Breconian' of the 'Old Red Sandstone' of Wales and the Welsh Borderlands.

In my working experience, fish macrofossils and fish-derived microfossils have proved of use in the following areas (see also Murray, 2000 (Palaeozoic, Mesozoic and Cenozoic of Africa); and Capetta *et al.*, 2000 (Upper Cretaceous and Palaeogene phosphate- and chert-bearing sequences of Jordan)):

Palaeozoic – the Palaeozoic of the Middle East and north Africa, the Early Palaeozoic of the Carnarvon Basin in Western Australia and of Argentina in South America, and the Late Palaeozoic of the former Soviet Union, the American and Canadian Arctic, and the Canadian Interior;

Mesozoic – the Triassic to Cretaceous of the former Soviet Union; the Triassic of the Carnarvon Basin, Western Australia, of the Canadian Interior, and of east Africa, and the Cretaceous of Brazil in the South Atlantic;

Cenozoic – the Cenozoic of the North Sea, of the Indian subcontinent, and of Brazil in the South Atlantic, and the Oligocene–Miocene of central and eastern Paratethys.

Amphibians

Amphibians evolved from fish in the Devonian (trace fossils first appear in the Middle Devonian, body fossils in the Late), and diversified through the Carboniferous and Permian (Shear & Selden,

---

Caption for fig. 2.25 (*cont.*) (Longvillian–Actonian) *Dicranograptus clingani* Zone, Moffat, Dumfriesshire; (c) *Pleurograptus linearis*, nominate taxon of the Caradoc (Onnian) *Pleurograptus linearis* Zone, Moffat, Dumfriesshire; (d) *Dicellograptus anceps*, nominate taxon of the Ashgill *Dicellograptus anceps* Zone, Moffat, Dumfriesshire; (e) *Didymograptus artus* [*D. bifidus*], nominate taxon of the Llanvirn *Didymograptus bifidus* Zone, Aberdaron, Gwynedd; (f) *Didymograptus murchisoni*, nominate taxon of the Llanvirn *Didymograptus murchisoni* Zone, Abereiddy Bay, Dyfed; (g) *Glyptograptus teretiusculus*, nominate taxon of the Llandeilo *Glyptograptus teretiusculus* Zone, Pwllheli, Gwynedd; (h) *Nemagraptus gracilis*, nominate taxon of the Llandeilo–Caradoc (Costonian) *Nemagraptus gracilis* Zone, Pwllheli, Gwynedd; (i) *Dictyonema flabelliforme*, nominate taxon of the Tremadoc *Dictyonema flabelliforme* Zone, Ffestiniog, Gwynedd; (j) *Didymograptus hirundo*, nominate taxon of the Arenig *Didymograptus hirundo* Zone, Skiddaw, Cumbria; (k) *Didymograptus extensus*, nominate taxon of the Arenig *Didymograptus extensus* Zone, Lleyn, Gwynedd.

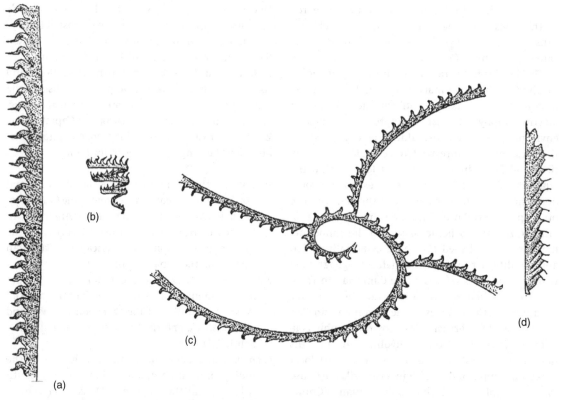

Fig. 2.26 **Some stratigraphically useful graptolites, Silurian**. From Jones (2006).
(a) *Monograptus sedgwickii*, nominate taxon of the Llandovery (Fronian) *Monograptus sedgwickii* Zone, Stockdale, Cumbria; (b) *Monograptus turriculatus*, nominate taxon of the Llandovery (Telychian) *Monograptus turriculatus* Zone, Stockdale, Cumbria; (c) *Cyrtograptus murchisoni*, nominate taxon of the Wenlock (Sheinwoodian) *Cyrtograptus murchisoni* Zone, Builth Wells, Powys; (d) *Monograptus leintwarinensis*, nominate taxon of the Ludlow (Leintwardinian) *Monograptus leintwarinensis* Zone, Leintwardine, Herefordshire.

in Gensel & Edwards, 2001; Rieppel, in Coggin *et al.*, 2003; Boisvert, 2005; Clack, in Briggs, 2005; Clack, 2006; Jones, 2006; Long, 2006; Long *et al.*, 2006; Blieck *et al.*, in Becker & Kirchgasser, 2007; Clack, 2007; Hall, 2007; Coates *et al.*, 2008; Jastrow & Rampino, 2008; Nudds & Selden, 2008; Cannatella *et al.*, in Hedges & Kumar, 2009; Carroll, 2009; Clement & Letenneur, 2009; Taylor *et al.*, 2009; Bernard *et al.*, 2010; Blieck *et al.*, in Vecoli *et al.*, 2010; Niedzwiedzki *et al.*, 2010). They then sustained severe losses in the end-Permian mass extinction, when the nectrideans, anthracosaurs and seymouriamorphs all disappeared (the microsaurs became extinct slightly earlier). They staged

some form of recovery from this event, though, and underwent renewed radiation in the Triassic, when the frogs, toads, salamanders, newts and allied forms appeared. They were essentially unaffected by the end-Jurassic and end-Cretaceous mass extinctions. There are some 5000 living species of amphibians. Many are threatened or endangered species.

The slow rate of evolutionary turnover exhibited by the amphibians renders them of limited use in biostratigraphy. Note, though, that certain species are of considerable biostratigraphic use in the Late Carboniferous–Early Permian of Europe, and indeed form the basis of a biozonation with a mean temporal resolution of 1.5–3 Ma (Werneburg & Schneider, in Lucas et al., 2006).

In my working experience, amphibians have not proved of use biostratigraphically, although they have proved of use palaeobiologically, in the Palaeocene of the Indian subcontinent.

Reptiles and birds

Reptiles evolved from amphibians in the Carboniferous, and diversified through the Carboniferous and Permian (Rieppel, in Coggin et al., 2003; Falcon-Long et al., 2008; Jastrow & Rampino, 2008; Hedges & Vidal, in Hedges & Kumar, 2009). They then sustained severe losses in the end-Permian mass extinction, when many synapsids disappeared (the pelycosaurs became extinct slightly earlier). They recovered from this event, though, and diversified again in the Triassic, when the diapsids, ichthyosaurs and lepidosaurs appeared, but sustained significant losses in the end-Triassic mass extinction, when the anapsids and many early archosaurs disappeared. They recovered from this event, too, and diversified yet again in the Jurassic, when the birds and the plesiosaurs appeared, but sustained severe losses in the end-Cretaceous mass extinction, when the pterosaurs, dinosaurs, ichthyosaurs and plesiosaurs became extinct. They recovered from this event, too. There are some 6000 living species of reptiles, and 9000 of birds.

The moderate rate of evolutionary turnover exhibited by the reptiles and birds renders them of some use in biostratigraphy, at least in appropriate environments. They are of use, for example, in the Permian–Jurassic of Pangaea, specifically in Argentina, Brazil, South Africa, Zimbabwe, Madagascar and India in Gondwana, in the western and eastern United States, Nova Scotia, Greenland, Scotland, Germany and Italy in Euramerica, in Siberia, and in China. Regional land vertebrate faunachrons (LVFs) have been defined for the Permian–Triassic, namely the Kapteinskraalian (Middle Permian), Gamkan, Hoedemakeran, Stellkransian, Platbergian (Late Permian), Lootsbergian (Late Permian–Early Triassic, Induan), Nonesian (Early Triassic, Olenekian), Perovkan (Middle Triassic, Anisian), Berdyankian (Ladinian), Otischalkian (Late Triassic, early Carnian), Adamanian (late Carnian), Revueltian (Norian), and Apachean (Rhaetian)

(Lucas & Morales, 1993; Veevers et al., in Veevers & Powell, 1994; Fraser, 2006; Lucas, in Lucas et al., 2006; Lucas, in Lucas, 2010); on the bases of the evolutionary appearances of Eodicynodon, Tapinocephalus, Tropidostoma, Ourenodon, Dicynodon, Lystrosaurus, Cynognathus, Shansiodon, Mastodonsaurus, Palaeorhinus, Rutiodon, Pseudopalatus and Redondasaurus, respectively, many of which are of use in correlation and field mapping, for example in the Karoo basins of southern Africa (Rubidge, 1995; Schluter, 1997; Gay & Cruickshank, 1999; MacRae, 1999; Hancox & Rubidge, 2001; Fig. 2.27).

Regional LVFs have also been defined for the Jurassic, namely the Wassonian (Early Jurassic, Hettangian–Sinemurian), Dawan (Early–Middle Jurassic, Sinemurian–Aalenian), Dashanpuan (Middle Jurassic, Bajocian), Tuojingian (Bathonian–Callovian), and Comobluffian and Ningliagouan (Late Jurassic) (Lucas, 2009).

In China, reptiles and birds are palaeobiologically important in the Late Jurassic–Early Cretaceous as regards elucidation of evolution, especially of feathered dinosaurs and of primitive birds. The feathered dinosaurs or primitive birds Sinornis, Cathayornis, Liaoningornis and Confuciornis were all first described from the Chaomidianzi, Yixian and Jiufotang formations of Liaoning province in northeast China, variously interpreted as Late Jurassic, that is, around the same age as the beds that yielded Archaeopteryx, or, more likely, Early Cretaceous, that is, slightly younger.

In Britain, marine reptiles such as ichthyosaurs and plesiosaurs are locally abundant in the Jurassic, as in the Lias around Lyme Regis in Dorset, where many exceptional specimens now housed in museums were found as long ago as the early nineteenth century by the pioneering collector Mary Anning (for accounts of whose extraordinary life the reader is referred to Torrens, 1995, Tickell, 1996 (the eminent environmentalist Sir Crispin Tickell is Anning's great-great-great-nephew), McGowan, 2001 and Pierce, 2006). Dinosaurs are locally abundant in the Early Cretaceous, as in the marginal- to non-marine 'Wealden' facies of the Weald and Wessex basins: indeed, they are locally sufficiently abundant as to lend their name to the containing rocks, as in the case of the Hypsiliphodon Bed of the Isle of Wight. Famously, an Iguanodon,

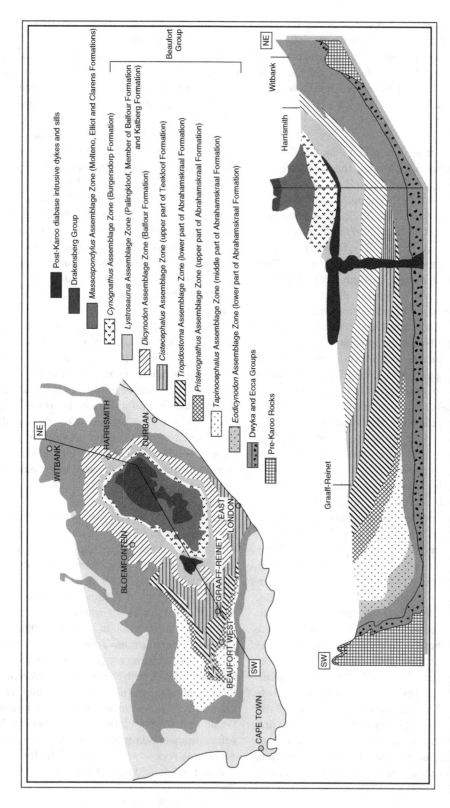

Fig. 2.27 **Stratigraphic zonation and correlation of the Permian and Triassic of the Karoo basin of South Africa by means of reptiles.** © Geological Survey of South Africa. Modified, with permission, after MacRae (1999).

Post-Karoo diabase intrusive dykes and sills

Drakensberg Group

Massospondylus Assemblage Zone (Molteno, Elliot and Clarens Formations)

Cynognathus Assemblage Zone (Burgersdorp Formation)

Lystrosaurus Assemblage Zone (Palingkloof, Member of Balfour Formation and Katberg Formation)

Dicynodon Assemblage Zone (Balfour Formation)

Cistecephalus Assemblage Zone (upper part of Teekloof Formation)

Tropidostoma Assemblage Zone (lower part of Abrahamskraal Formation)

Pristerognathus Assemblage Zone (upper part of Abrahamskraal Formation)

Tapinocephalus Assemblage Zone (middle part of Abrahamskraal Formation)

Eodicynodon Assemblage Zone (lower part of Abrahamskraal Formation)

Dwyka and Ecca Groups

Pre-Karoo Rocks

Beaufort Group

NE

Witbank

Harrismith

Graaff-Reinet

Graaff-Reinet

SW

NE

Harrismith

Witbank

CAPE TOWN

BEAUFORT WEST

GRAAFF-REINET

BLOEMFONTEIN

EAST LONDON

DURBAN

HARRISMITH

WITBANK

NE

SW

from the 'Wealden' of Sussex, was among the first dinosaurs ever to be described, by Gideon Mantell, in 1825, although the actual name 'dinosaur' was not coined until later, by Richard Owen, in 1842 (Dean, 1993, 1999; McGowan, 2001). (Incidentally, a *Megalosaurus*, from the Middle Jurassic of Stonesfield in Oxfordshire, was the first ever to be described, by the Reverend William Buckland, in 1824: Benson *et al.*, 2008.)

In my working experience, reptiles have proved of use biostratigraphically only in the Triassic of Madagascar and in the Cretaceous of Nigeria. Also in my working experience, reptiles and birds have proved of use palaeobiologically in the Late Jurassic–Early Cretaceous of the Mediterranean, in the Palaeocene of the Indian subcontinent, and in the Late Miocene of eastern Paratethys and the Middle East.

Mammals

Mammals evolved, from mammal-like reptiles, in the Late Triassic, and diversified during the Jurassic and Cretaceous (Jastrow & Rampino, 2008; Madsen, in Hedges & Kumar, 2009). They then sustained significant losses in the end-Cretaceous mass extinction. They recovered from this event, though, and diversified dramatically through the Cenozoic, filling the niches left vacant by the extinctions of other groups. The overall pattern throughout the Cenozoic has essentially been one of ever-increasing diversification, although some losses were sustained during the end-Eocene and Pleistocene mass extinctions. There are some 4500 living species of mammals.

The rapid rate of evolutionary turnover exhibited by the mammals renders them of considerable use in biostratigraphy, at least in appropriate, typically non-marine, environments. High-resolution biostratigraphic zonation schemes based on the terrestrial mammals have been established that have at least local to regional applicability. Calibration is typically by means of magnetostratigraphy.

For example, in Europe, the Cenozoic is divided into a total of 47 zones on the basis of mammals (Agusti *et al.*, 2001). It is also divided into 15 'European Land Mammal Ages' or ELMAs (Fig. 2.28).

Importantly, the Quaternary is divided into zones on the basis of voles and allied forms, providing a high-resolution temporal framework for assessing climatic and environmental change. Many Quaternary species range through to the Recent, such that their present day distributions can be used to infer the past climate. In North America, the Late Cretaceous is divided into five 'pre-Aquilan faunas' and four 'North American land mammal ages' or NALMAs, namely, the Aquilan, Judithian, 'Edmontonian' and Lancian. The Cenozoic is divided into a total of 19 NALMAs, namely the Palaeocene Puercan, Torrejonian, Tiffanian and Clarkforkian, the Eocene Wasatchian, Bridgerian, Uintan, Duchesnean and Chadronian, the Oligocene – to earliest Miocene – Orellan, Whitneyan and Arikareean, the Miocene – to earliest Pliocene – Hemingfordian, Barstovian, Clarendonian and Hemphillian, the Pliocene – to earliest Pleistocene – Blancan, and the Pleistocene Irvingtonian and Rancholabrean. (The issue of the time-transgressiveness of the NALMAs, and its implications for geochronology, is discussed by Alroy, 1998.) In South America, the Cenozoic is divided into 21 'South American land mammal ages' or SALMAs. In China, the Cenozoic is divided into 17 land mammal ages. In Australia, the Cenozoic is divided into six land mammal ages, namely the Oligocene Etadunnan, the Oligo-Miocene Wipajarian, the Miocene Canfieldian and Waitean, the Pliocene Tirarian, and the Plio-Pleistocene Naracoortean (Piper *et al.*, 2006; Megirian *et al.*, 2010). In South Africa, the Neogene–Pleistogene is divided into six land mammal ages, namely, the Namibian (18–14 Ma), Langebaanian (5.0–4.5 Ma), Makapanian (3.0–1.6 Ma), Cornelian (1.6–0.5 Ma), Florisian (200 000–12 000 years BP), and Recent (12 000 years BP to present). Incidentally, mammals are of palaeobiological importance as well as biostratigraphic use here, and, to paraphrase Darwin's famous phrase, shed much light on human origins.

In my working experience, mammals have proved of use in the Cenozoic of the Indian subcontinent, and in the Miocene of the Middle East and Paratethys.

Trace fossils

Trace fossils first appeared in the Late Precambrian, Ediacaran, approximately 565 Ma, and have ranged

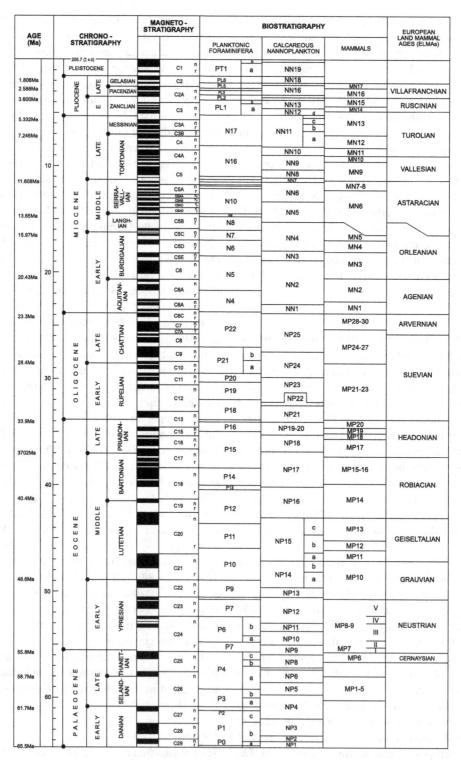

Fig. 2.28 **Stratigraphic zonation of the Cenozoic of Europe by means of mammals.**
From Jones (2006).

through to the present day (Liu *et al.*, 2009). The complete suite of interpreted behavioural types became developed by the Cambrian or Ordovician. Deep-water forms only became common in the Ordovician, probably through migration from shallow water (Buatois *et al.*, 2009). Terrestrial forms only became common in the Jurassic, coincident with the rise of the insects. Previously, such environments would not have been colonised and would not have had an ichnological signature, only a sedimentological one. Palaeoenvironmental interpretations should take observations such as this into account.

In my working experience, trace fossils have not proved of use biostratigraphically, although they have proved of use palaeobiologically. Note, though, that the first appearance of the arthrophycid trace fossil *Phycodes* or *Treptichnus pedum* has now been accepted as marking the base of the Cambrian, worldwide, despite potential issues of diachronism (Seilacher, 2007); and that *Phycodes* marks Cambrian Series 1 Stage 1, and *Teichichnus* Series 1 Stage 2 and Series 2 Stages 3–4, in the Cambrian basin (Loughlin & Hillier, 2010). Note also that other ichnospecies of *Phycodes*, and ichnospecies of the other arthrophycids *Arthrophycus* and *Daedalus*, are of some biostratigraphic use in the Palaeozoic, as are ichnospecies of the trilobite trace fossils *Cruziana* and *Dimorphichnus*, and ichnospecies of *Dictyodora* (Crimes, in Holland, 1981; Turner & Benton, 1983; Seilacher, in Said, 1990; Seilacher, in Sola & Worsley, 2000; Seilacher *et al.*, 2002; Seilacher, 2007). In the terrestrial realm, certain ichnospecies of tetrapod are of considerable biostratigraphic use in the Permian–Triassic, and indeed form the basis of a biozonation with inter-regional applicability (Lucas & Hunt, in Lucas *et al.*, 2006; Klein & Lucas, in Lucas, 2010).

2.2 BIOSTRATIGRAPHY

As intimated above, the moderate to rapid rates of evolutionary turnover exhibited by the dinoflagellates, calcareous nannoplankton, foraminifera, plants, ammonoids, trilobites, insects, graptolites, chitinozoans, conodonts, reptiles and mammals renders them of use in biostratigraphy, at least in appropriate environments or facies (calpionellids, 'small shelly fossils', archaeocyathans and tentaculitids are also

of local use). The usefulness of the dinoflagellates, calcareous nannoplankton, planktonic foraminifera, ammonoids, trilobites, graptolites, chitinozoans and conodonts is enhanced by their essentially unrestricted ecological distribution, and facies independence, in marine environments. That of the mammals is enhanced by their essentially unrestricted ecological distribution, and facies independence, in non-marine environments. High-resolution biozonation schemes based on the dinoflagellates, calcareous nannofossils, planktonic foraminifera, ammonoids, trilobites, graptolites, chitinozoans and conodonts have been established that have inter-regional to global applicability, across palaeobiogeographic realms or provinces, in marine facies (high-resolution biozonations based on larger benthic foraminifera have been established that have local to regional, but not inter-regional, applicability). Biozonation schemes based on the plants, insects, reptiles and mammals have been established that have local to regional applicability, within palaeobiogeographic realms or provinces, in non-marine facies.

In my working experience in the petroleum industry, dinoflagellates, calcareous nannoplankton and foraminifera have proved of particular use.

2.2.1 Biostratigraphic zonation or biozonation

The fundamental unit in biostratigraphy is the biozone (Jones, 2006; Fig. 2.29).

The main types of biozone are defined on a combination of: evolutionary or first appearance datums or FADs, also known as first occurrences, last down-hole occurrences or LDOs, or 'bases', of marker species; extinction or last appearance datums or LADs, also known as last occurrences, first down-hole occurrences or FDOs, or 'tops'; total, partial or concurrent ranges; and abundances or acmes. In the petroleum industry, the main types of biozone are defined on FDOs or 'tops', or first consistent or common down-hole occurrences or FCDOs, because 'bases' can appear low in cuttings samples owing to down-hole contamination or 'caving'. Note, though, that 'tops' can appear high owing to reworking. Note also that 'tops' can appear low in the event that rock record is incomplete, or contains unfavourable facies, a phenomenon known as range offset. Integration of bio- and

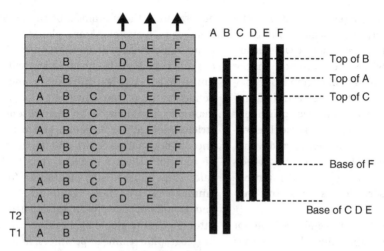

Fig. 2.29 **The basis of biostratigraphic zonation**. T1, T2 etc. are successive time-slices represented by rock units. A–F are fossils occurring in the various rock units. Their overall ranges, between their evolutionary inceptions, first appearances or bases and their extinctions, last appearances or tops, are indicated on the right. Biostratigraphic zones or biozones have been defined: between the bases of A and B, and the bases of C, D and E; between the bases of C, D and E and the base of F; between the base of F and the top of C; between the top of C and the top of A; and between the top of A and the top of B. From Jones (2006).

seismic sequence-stratigraphy enables the discrimination of real from apparent, or ecologically depressed 'tops'.

As noted above, biozonation schemes have been established that have inter-regional to global applicability, across palaeobiogeographic realms or provinces, in appropriate facies. As noted below, these – high-resolution – biozonations are applicable over part the whole of the Phanerozoic, that is, the Palaeozoic, Mesozoic and Cenozoic. Also as noted below, there is even a – low-resolution – biozonation applicable over part of the Proterozoic.

Resolution

Temporal resolution is essentially a function of evolution and extinction rates, rock accumulation rate, and sample spacing.

The mean resolution of the biozones in many of the published global standard biozonation schemes, such as the ammonite schemes for the Mesozoic, or the calcareous nannoplankton or planktonic foraminiferal schemes for the Cenozoic, is typically of the order of 1 Ma or less. This level of temporal

resolution is sufficient to enable the identification of even relatively minor unconformities or sequence boundaries. Significantly, it is also sufficient to provide a meaningful framework within which to assess at least some of the effects of climatic change, as inferred principally from isotope geochemistry.

The mean resolution of the biozones in the proprietary BP calcareous nannoplankton biozonation scheme for the Neogene–Pleistogene of the Gulf of Mexico is of the order of 0.1 Ma, 100 ka, or 100 000 years.

2.2.2 Correlation

The 'Law of Strata identified by organised Fossils', established by William ('Strata') Smith, states that particular ages of rock can be correlated by means of stratigraphic index fossils, irrespective of facies (Fig. 2.30).

Smith applied this principle not only in correlation but also in field mapping, and in the production of the first geological map of Great Britain, 'the map that changed the world'. (Incidentally, what might be thought of the first geological map of anywhere

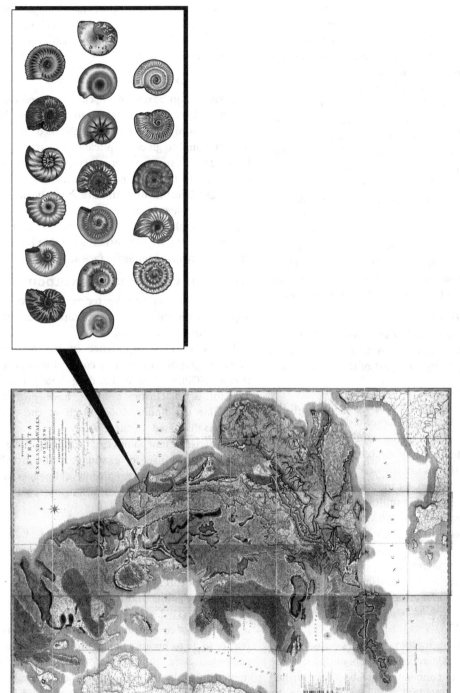

Fig. 2.30 An illustration of the basis of biostratigraphic correlation (William Smith's 'Law of Strata identified by organised Fossils'). The Jurassic strata of England, shown here on Smith's geological map cropping out from Yorkshire in the northeast to Dorset in the southwest, have been correlated by means of ammonites (inset). From Jones (2006).

in the world was produced in preparation for a quarrying expedition to Wadi Hammamat by Rameses IV's 'Scribe-of-the-Tombs', Amennakhte, in approximately 1160 BC. This map eventually came into the possession of an Italian attache to Napoleon's army in Egypt, Bernardino Drovetti, in approximately AD 1824. It currently resides in Turin's Museo Egizio, and is referred to as the 'Turin papyrus'.)

Biostratigraphic correlation is superior to lithostratigraphic correlation in that it allows the identification of rock units that are correlatable in time and space, or isochronous, and the discrimination of those that are not, or are diachronous.

This capability is critically important in the construction of accurate and meaningful time-slice palaeogeographic and facies maps, and risk maps, in play fairway analysis in petroleum exploration. It is also important in the characterisation and modelling of permeability distribution and flow behaviour in the exploitation phase, not least in order to ensure the optimal location of production and injection wells.

2.3 PROTEROZOIC

Low-resolution biozonation schemes based on acritarchs have been established that have local to regional applicability over part of the Proterozoic.

2.4 PALAEOZOIC

High-resolution biozonation schemes based on the ammonoids, trilobites, graptolites, chitinozoans and conodonts have been established that have inter-regional to global applicability, across palaeobiogeographic realms or provinces, in marine facies (see Section 2.9 below). Biozonation schemes based on larger benthic foraminifera, 'small shelly fossils', archaeocyathans, tentaculitids and trace fossils have been established that have local to regional, but not inter-regional, applicability in marine facies. Schemes based on plants, insects and reptiles have been established that have local applicability in non-marine facies.

2.5 MESOZOIC

High-resolution biozonation schemes based on the dinoflagellates, calcareous nannofossils, planktonic foraminifera and conodonts have been established that have inter-regional to global applicability, across palaeobiogeographic realms or provinces, in marine facies (see Section 2.9 below). Biozonation

schemes based on larger benthic foraminifera and calpionellids have been established that have local to regional, but not inter-regional, applicability. Schemes based on plants, reptiles and mammals have been established that have local applicability in non-marine facies.

2.6 CENOZOIC

High-resolution biozonation schemes based on the dinoflagellates, calcareous nannofossils and planktonic foraminifera have been established that have inter-regional to global applicability, across palaeobiogeographic realms or provinces, in marine facies (see Section 2.9 below). Biozonation schemes based on larger benthic foraminifera have been established that have local to regional, but not inter-regional, applicability. Schemes based on plants and mammals have been established that have local applicability in non-marine facies.

2.7 BIOSTRATIGRAPHIC TECHNOLOGIES

A number of quantitative techniques have been applied in the field of biostratigraphy (Hammer & Harper, 2006; Jones, 2006).

In my working experience in the petroleum industry, graphic correlation, constrained optimisation (CONOP), ranking and scaling (RASC), and various forms of cluster analysis have proved of particular use (Jones, 2006). Each of these techniques is discussed, in turn, below.

Biometric or morphometric techniques have also proved of some use, for example, in the establishment of larger benthic foraminiferal lineages in the Cenozoic of the Tethyan realm.

2.7.1 Graphic correlation

Graphic correlation involves correlation of an outcrop or well section against a global composite standard section by treating the two as coordinate axes and plotting the fossil events common to both as a series of points (Hammer & Harper, 2006; Jones, 2006; Figs. 2.31–2.32).

The 'line of correlation' (LOC), fitted between the points so as to honour palaeontological and other geological data, provides information as to sediment accumulation rates and stratigraphic breaks that is of use in sequence stratigraphy and in petroleum systems analysis. Rates are indicated by, and indeed can be calculated from, the gradient of

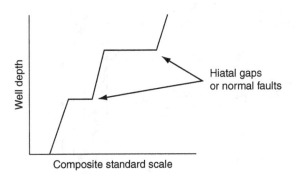

Fig. 2.31 **The basis of graphic correlation**. Fossil events in a well – or outcrop – section are plotted on the vertical axis, and the equivalent events in the established 'composite standard' on the horizontal axis, and a line of correlation or LOC is fitted to them. Plateaux or terraces in the LOC indicate sections represented in the 'composite standard' and unrepresented in the well or outcrop, and hence the presence of hiatuses (or normal faults) in the well or outcrop. It may be difficult to distinguish such hiatuses from condensed sections or CSs if sampling in the well or outcrop section is too coarse. From Jones (2006).

the LOC; breaks, and/or condensed sections, from plateaux (breaks and condensed sections can be difficult to distinguish using graphic correlation alone). Projecting composite standard units through the LOC into the section provides a precise and accurate indication of which units are represented and which unrepresented in the section, and of the degree of expansion or condensation.

Examples of applications of graphic correlation technology in petroleum exploration have been published by, among others: Neal et al., in Mann and Lane (1995) (Palaeogene, North Sea); Martin and Fletcher, in Mann and Lane (1995) (Plio-Pleistocene, Gulf of Mexico); Stein et al., in Steel et al. (1995) (Palaeogene, North Sea); Groves and Brenckle (1997) (Permo-Carboniferous, Tarim basin, China); Highton et al. (1997) (Neogene, Gulf of Thailand); Simmons, in Simmons (1994) (Cretaceous, Oman); Wakefield et al. (2001) (Palaeogene, Fleming field, North Sea); Wakefield and Monteil (2002) (Cretaceous and Cenozoic, Indus basin, Pakistan); and Jaramillo et al., in Powell and Riding (2005) (Palaeogene, Colombia) (see Fig. 2.32; also also Section 5.2).

Examples of applications in reservoir exploitation have been published by Wakefield et al.

(2001) (Fleming field, North Sea); and Gary et al., in Demchuk and Gary (2009) (Hawkins field, North Sea) (see also Section 5.3).

In the petroleum industry, graphic correlation has become fully automated, and capable of being imported into the work-station environment for integration with well log and seismic data (Wescott et al., 1998; Jones et al., in Jones & Simmons, 1999; Jones, 2006; Gary et al., in Demchuk & Gary, 2009; Fig. 4.3). Alternative LOCs are fitted using path-finding or slotting algorithms (Gary et al., in Powell & Riding, 2005; Gary et al., in Demchuk & Gary, 2009). The 'best fit' is selected by appropriate weighting of defining events. Confidence limits can be assigned.

2.7.2 Constrained optimisation (CONOP)

CONOP provides a deterministic sequence of biostratigraphic events in sections, using a methodology similar to that of automated graphic correlation, but with the best-fit LOC found with a 'simulated annealing heuristic search' (Kemple et al., in Mann & Lane, 1995; Hammer & Harper, 2006; Foote & Miller, 2007; Sadler & Cooper, in Harries, 2008; Sadler et al., in Harries, 2008).

Examples of applications of CONOP technology in the petroleum industry have been published by Cooper et al. (2001) and Sadler and Cooper, in Harries (2008) (Cenozoic, Taranaki basin, New Zealand).

A comparison between CONOP and UAgraph is given by Galster et al. (2010).

2.7.3 Ranking and scaling (RASC)

Ranking (RA) provides a probabilistic sequence of biostratigraphic events in sections, determined by manual scoring or computer-driven matrix permutation; scaling (SC), a measure of the inter-event distance determined by cross-over frequency (partly a function of sample spacing) (Hammer & Harper, 2006; Jones, 2006; Foote & Miller, 2007; Sadler & Cooper, in Harries, 2008; Fig. 2.33).

Examples of applications of RASC technology in petroleum exploration have been published by, among others: Gradstein et al. (1988) (Cenozoic, North Sea); Highton et al. (1997) (Neogene, Gulf of Thailand); and Cooper et al. (2001) (Cenozoic, Taranaki basin, New Zealand) (Fig. 2.33; see also Section 5.2 below). An example of the application of RASC,

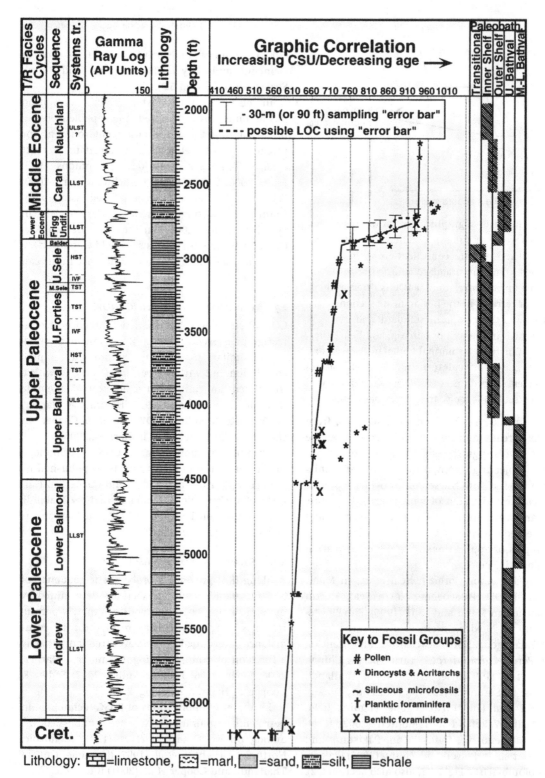

Lithology: ▦=limestone, ≈≈=marl, ░=sand, ▤=silt, ▬=shale

Fig. 2.32 A graphic correlation plot of a Palaeogene section in a North Sea well. From Jones (2006) (reproduced with the permission of the Society of Economic Paleontologists and Mineralogists from Neal *et al.*, in Mann & Lane, 1995). CSU, composite standard unit.

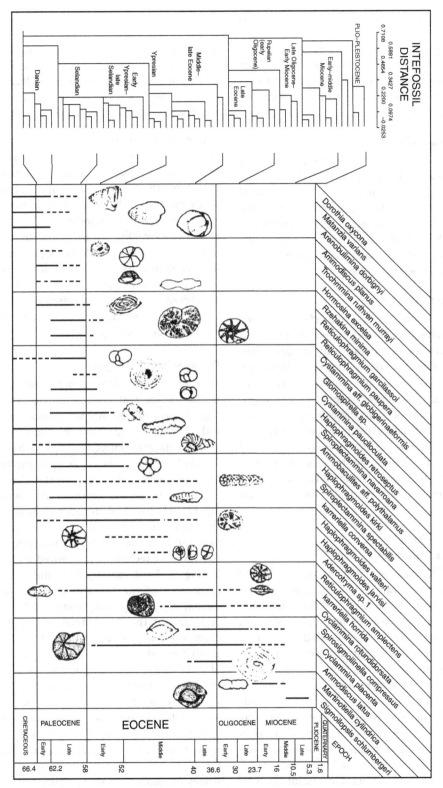

Fig. 2.33 **A North Sea Cenozoic micropalaeontological zonation generated by ranking and scaling.** From Jones (2006) (reproduced with the permission of der Geologischen Bundesanstalt, Wien, from Gradstein *et al.*, 1988).

and of the related technique of correlation and scaling or CASC, in reservoir exploitation has been published by Bowman *et al.*, in Demchuk and Gary (2009) (Miocene, Mad Dog field, Gulf of Mexico) (see also Section 5.3 below).

2.7.4   Cluster analysis

Various forms of cluster analysis have proved of use in biostratigraphy, for example, hierarchical cluster analysis in the establishment of a mollusc-based stratigraphy for the Miocene, Eggenburgian of Austria (Jones, 2006); and correspondence analysis in the establishment of a foraminiferal assemblage zonation for the Oligocene of Angola (Kender *et al.*, in Kaminski & Coccioni, 2008).

2.8   ALLIED DISCIPLINES

This section contains sub-sections on chemostratigraphy, cyclostratigraphy, heavy minerals, magnetostratigraphy, radiometric dating and miscellaneous Quaternary dating methods (Dunay & Hailwood, 1995; Jones, 2006). It is important to note that, like biostratigraphy, chemostratigraphy, cyclostratigraphy, heavy minerals, magnetostratigraphy and most Quaternary dating methods provide relative rather than absolute age-dating, unless calibrated against the absolute chronostratigraphic or geochronological time-scale. Radiometric dating and some Quaternary dating methods provide absolute age-dating.

2.8.1   Chemostratigraphy

Carbon isotope stratigraphy

Del^{13}C, or δ^{13}C, has fluctuated through geological time, in response to various biological and physico-chemical controls. A standard δ^{13}C curve has been constructed on the basis of measured values in pelagic carbonates in Tethys. Measured values from elsewhere have been calibrated against this standard in order to establish a local carbon isotope stratigraphy. In my working experience in the petroleum industry, the carbon isotope technique has proved useful in establishing a local stratigraphy for the platform carbonates of the Early Cretaceous Kharaib and Shu'aiba formations of Oman and the United Arab Emirates in the Middle East (Jones, 2006). In this case, stratigraphic resolution is of the order of 1 Ma, comparable with that attainable

using conventional biostratigraphic techniques. Importantly, this is sufficient to resolve the timing of the formation of the intra-shelf Bab basin. The carbon isotope technique is only applicable to sediments essentially unaltered by diagenesis.

Oxygen isotope stratigraphy

High-resolution palaeoclimate curves based on the oxygen isotopic record of changing temperature or ice volume that some fossil groups preserve in their shells, bones or teeth serve, when calibrated against biostratigraphy, magnetostratigraphy or absolute chronostratigraphy, as the basis of a workable marine isotope stage (MIS) or oxygen isotope stage (OIS) climatostratigraphy, at least for the Tertiary and Quaternary (Jones, 2006). A total of 63 numbered stages are recognised over the Quaternary. Odd-numbered stages represent interglacials; even-numbered ones, glacials.

Strontium isotope stratigraphy

The ratio between strontium-87 and strontium-86 in the marine realm has fluctuated through geological time, in response to various physico-chemical controls (MacArthur, in Whittaker & Hart, 2010). A standard curve of the ratio has been constructed on the basis of biostratigraphically constrained measured values worldwide. Measured values from specific localities can be calibrated against this standard in order to establish a local strontium isotope stratigraphy. In my working experience in the petroleum industry, the strontium isotope technique has proved useful in establishing a local stratigraphy for the platform carbonates of the Oligo-Miocene of the Kujung, Tuban, Ngryayong and Wonocolo formations of Java, and of the Darai limestone of Papua New Guinea, in the Far East (Sharaf *et al.*, 2005; Jones, 2006). Here, stratigraphic resolution is of the order of 1 Ma (note, though, that the resolution varies according to stratigraphic age, being highest in the Oligo-Miocene). The strontium isotope technique is only applicable in marine environments of normal salinity, and is not applicable in either hypo- or hyper-saline environments. Analysis can be performed on macrofossils, microfossils, marine mudstones or marine cements. Epifaunal rather than infaunal species should be selected on account of chemical differences between marine and interstitial pore waters.

In isolated basins, the strontium isotope ratio diverges from that of the open ocean, and the degree of divergence can be used as a measure of the degree of isolation, as in the case of the eastern Mediterranean in the Cenozoic (Cagatay *et al.*, 2006; Flecker & Ellam, 2006; Jones, 2006). In this case, the isolation inferred from the strontium isotope ratio is significantly earlier than that inferred from faunal evidence. This is interpreted to be because the fauna tracks salinity rather than isolation, and because salinity is partly controlled by evaporation, which is independent of isolation.

Trace element stratigraphy

The abundance or ratio of certain trace elements can be related to provenance, and used for provenance-based stratigraphic correlation. Data are acquired using mass spectrometry, and processed using various statistical techniques. The technique is useful in biostratigraphically barren strata, such as the continental red beds of the Triassic of the North Sea (Racey *et al.*, in Dunay & Hailwood, 1995; Jones, 2006).

2.8.2 Cyclostratigraphy

Various cyclostratigraphic techniques and tools are available, for example spectral analysis of gamma log signatures, which can be used to identify changes in rock accumulation rate through time (Yang & Kouwe, in Dunay & Hailwood, 1995; Weedon, 2003; d'Argenio *et al.*, 2004; Hammer & Harper, 2006; Jones, 2006).

In my working experience in the petroleum industry, cyclostratigraphy has proved of use in the Phanerozoic of the Middle East, in revising the ages of the regional 'maximum flooding surfaces' of Sharland *et al.* (2001) using 'astronomical tuning' or 'orbital forcing' (Matthews & Frohlich, 2002; Mattner & Al-Husseini, 2002; Al-Husseini & Matthews, 2005a, b, c, 2006a, b; Al-Husseini *et al.*, 2006).

2.8.3 Heavy minerals

The abundance or ratio of certain heavy minerals in sandstones can be related to provenance, and used for provenance-based stratigraphic correlation. The technique has proved useful in the continental

Devonian red beds west of Shetland and in the continental red beds of the Triassic of the North Sea, which are essentially biostratigraphically barren (Mange-Rajetsky, in Dunay & Hailwood, 1995; Morton & Hurst, in Dunay & Hailwood, 1995; Jones, 2006). Provenance-sensitive mineral pairs and apatite grain roundness have been used as the basis for 'geosteering' in the Devonian reservoir in Clair field.

2.8.4 Magnetostratigraphy

Magnetic polarity has reversed repeatedly through geological time, on various time-frames, as evidenced by bands of normal and reversed polarity oceanic crust on either side of the active sea-floor-spreading centre at the mid-oceanic ridge. A global standard geomagnetic time-scale has been constructed using independently dated magnetic reversals. Continuous measured – outcrop or oriented core – sections can be calibrated against this standard in order to establish a local magnetostratigraphy. Resolution is typically high, but varies according to stratigraphic age, and is low in the Cretaceous 'long normal zone' or 'quiet zone'. The technique has proved useful in the continental, fluvial to aeolian Triassic Sherwood sandstone of Devon, which is also essentially biostratigraphically barren, although locally containing tetrapods of Perovkan aspect (Hounslow & McIntosh, 2003). (Note that the Sherwood sandstone constitutes one of the oil reservoirs in BP's Wytch Farm field in neighbouring Dorset.)

In my working experience in the petroleum industry, magnetostratigraphy has proved of use in the following areas:

Palaeocene–Eocene – of the area west of the Shetland Islands;

Oligocene–Pliocene – of Paratethys. Here, specifically in the Pliocene of the south Caspian, the resolution of the magnetostratigraphy is sufficient to detect short-lived unconformities generated by structuration, with important implications for the timing of growth of prospective structures in relation to petroleum charge;

Miocene–Pliocene – of the Middle East. Here, magnetostratigraphy – and vertebrate palaeontology – has been used to date the thick 'molasse' sequence of the Zagros Mountains in southwest Iran, with important consequences for the timing of maturation of underlying Mesozoic and Palaeogene source-rocks.

2.8.5 Radiometric dating

Various radiometric dating techniques are available, for example Ar–Ar, Pb–Pb, and 'sensitive high-resolution ion microprobe' or SHRIMP dating of zircons (Roberts *et al.*, in Dunay & Hailwood, 1995; Russell, in Dunay & Hailwood, 1995; Jones, 2006). Note that K–Ar dating of authigenic glauconites is inherently inaccurate because of degassing.

In my working experience in the petroleum industry, radiometric dating has proved of use in constraining the age of the 'productive series' reservoir in Azerbaijan, which is bracketed by datable volcanic ash horizons (Jones, 2006). It has also proved of use in determining the provenance of Mesozoic and Cenozoic sediments surrounding the Caucasus, with important implications for reservoir quality. Sediments of local Caucasus provenance and regional Russian platform provenance are characterised by zircon suites of different ages.

2.8.6 Quaternary dating methods

A number of methods are employed in the dating of the Quaternary (Walker, 2005; Jones, 2006; Penkman *et al.*, 2010; Weiner, 2010). These include not only conventional biostratigraphy and oxygen isotope stratigraphy, but also accelerator mass spectometry (AMS), amino-acid racemisation, cosmogenic chlorine-36 rock exposure, dendrochronology, electron spin resonance, magnetic susceptibility, molecular stratigraphy, optically stimulated luminescence (OSL), radiocarbon dating, thermoluminescence (TL) and uranium-series dating.

In my working experience, only conventional biostratigraphy, AMS and OSL have proved of particular use, in dating and correlating superficial sediments in the Black Sea off Turkey, in the Caspian off Azerbaijan and in the Mediterranean off Egypt in the course of site investigation work (Jones, 2006; author's unpublished observations).

Conventional palynostratigraphy provides relative rather than absolute age-dating, unless calibrated against the absolute chronostratigraphic or geochronological time-scale. It can be calibrated against the absolute chronostratigraphic time-scale through the similarly essentially climatostratigraphic deep-sea oxygen isotope record.

AMS and OSL provide absolute age-dating. AMS is a form of radiocarbon dating of organic material that requires a smaller sample, and is therefore less likely to be affected by contamination, and that provides precise and accurate age determinations to 40 000 years BP (before present), that is, the last glacial. OSL is a form of dating of mineral grains that provides age determinations to at least 130 000 years BP, that is, the last interglacial, and under certain circumstances to 500 000 years BP (Rendell, in Dunay & Hailwood, 1995).

2.9 STRATIGRAPHIC TIME-SCALES

The principal published absolute chronostratigraphic or geochronological time-scales incorporating biostratigraphic data and covering substantial periods of time are those of Harland *et al.* (1982), Haq *et al.* (1987), Harland *et al.* (1990), Berggren *et al.* (1995), de Graciansky *et al.* (1998) and Gradstein *et al.* (2004) (see also Ogg *et al.*, 2008).

The Haq *et al.*, de Graciansky *et al.* and Gradstein *et al.* time-scales are widely used, and indeed have become unofficial standards, in the petroleum industry, undoubtedly because they also incorporate sequence stratigraphic data (Figs. 2.34–2.44).

Unfortunately, the Haq *et al.* and de Graciansky *et al.* time-scales do not deal in detail with Palaeozoic stratigraphy; the Gradstein *et al.* time-scale is not up-to-date with recent developments in Palaeozoic stratigraphy (Kaufmann, 2006; Babcock & Shanchi Peng, 2007); and none is up-to-date with recent developments in Triassic stratigraphy (Brack *et al.*, 2005; Mundil *et al.*, in Lucas, 2010; Sues & Fraser, 2010).

Fortunately, though, comprehensive and up-to-date stratigraphic information on the entire stratigraphic column, including the Palaeozoic, is available through the 'International Commission on Stratigraphy' or ICS website (www.stratigraphy.org).

2.9.1 Global stratotype sections and points (GSSPs)

Up-to-date stratigraphic information regarding global stratotype sections and points (GSSPs) is available through the ICS website. The philosophical basis of the GSSP concept is discussed by Walsh *et al.* (2004).

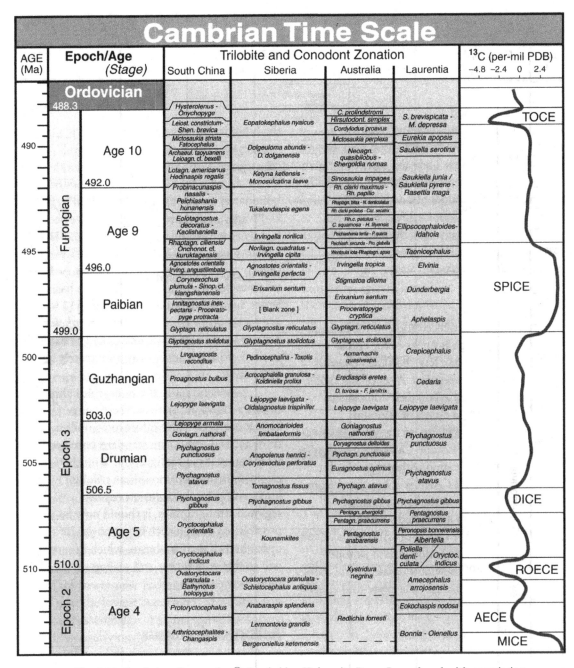

Fig. 2.34 Cambrian time-scale. © Cambridge University Press. Reproduced with permission from Ogg *et al.* (2008).

Palaeozoic

The position of the GSSP for the Precambrian–Cambrian (P–€) boundary has been fixed at the first appearance of the trace fossil *Phycodes* or *Treptichnus* *pedum* near the base of the Fortune Head section in southeastern Newfoundland on the Avalon plate, dated to approximately 542 Ma (Jones, 2006; Landing *et al.*, 2007; Ogg *et al.*, 2008).

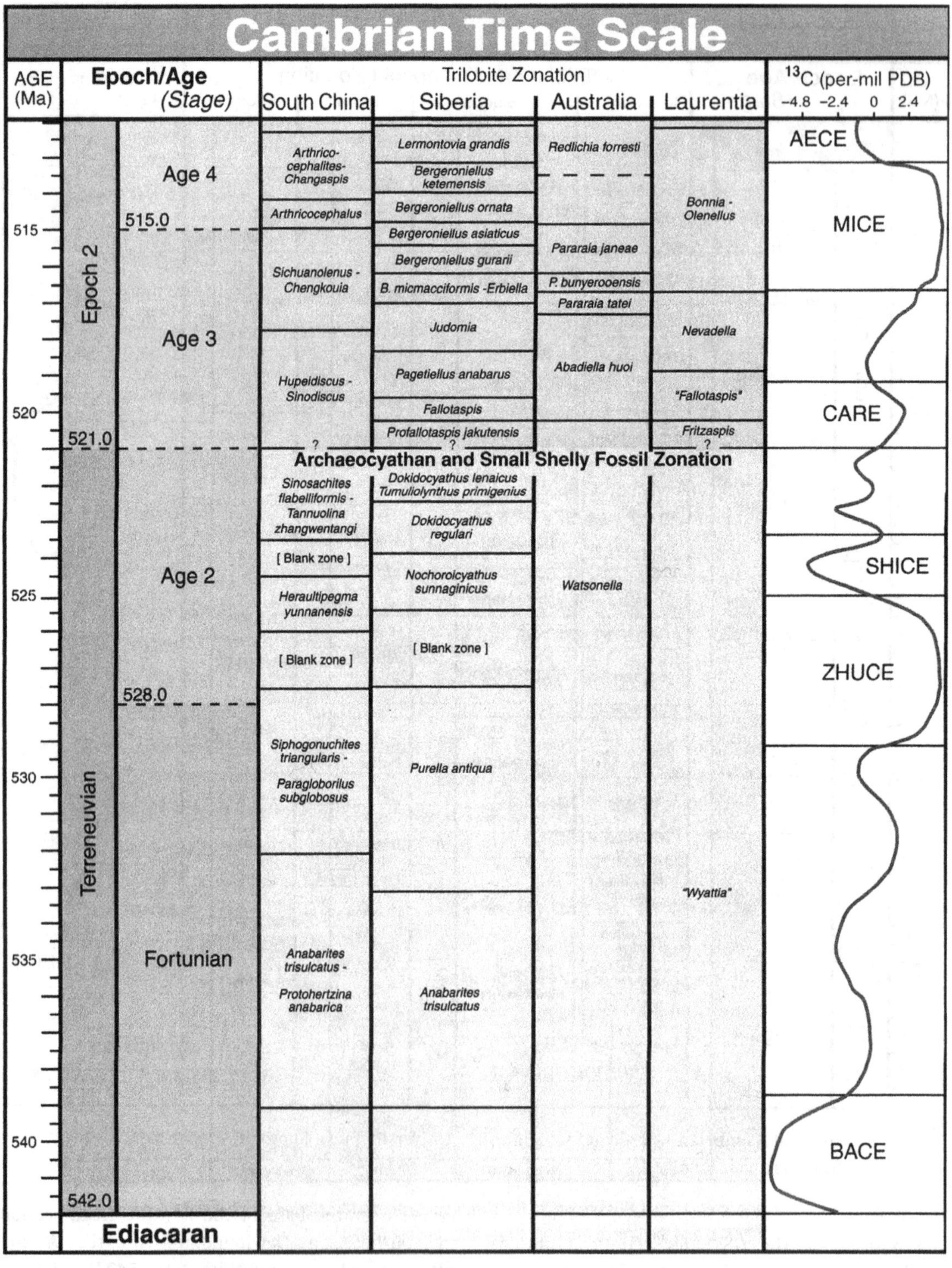

Cambrian Time Scale

AGE (Ma)	Epoch/Age (Stage)		Trilobite Zonation				^{13}C (per-mil PDB) −4.8 −2.4 0 2.4
		South China	Siberia	Australia	Laurentia		
	Age 4	Arthrico-cephalites-Changaspis	Lermontovia grandis	Redlichia forresti		AECE	
			Bergeroniellus ketemensis				
515 —	515.0	Arthricocephalus	Bergeroniellus ornata		Bonnia - Olenellus	MICE	
			Bergeroniellus asiaticus	Pararaia janeae			
	Age 3	Sichuanolenus - Chengkouia	Bergeroniellus gurarii	P. bunyerooensis			
			B. micmacciformis -Erbiella	Pararaia tatei			
			Judomia		Nevadella		
		Hupeidiscus - Sinodiscus	Pagetiellus anabarus	Abadiella huoi			
520 —			Fallotaspis		"Fallotaspis"	CARE	
	521.0	?	Profallotaspis jakutensis ?		Fritzaspis ?		

Archaeocyathan and Small Shelly Fossil Zonation

		Sinosachites flabelliformis - Tannuolina zhangwentangi	Dokidocyathus lenaicus Tumuliolynthus primigenius				
	Age 2		Dokidocyathus regulari				
		[Blank zone]				SHICE	
525 —		Heraultipegma yunnanensis	Nochoroicyathus sunnaginicus	Watsonella			
		[Blank zone]	[Blank zone]			ZHUCE	
	528.0						
530 —		Siphogonuchites triangularis - Paragloborilus subglobosus	Purella antiqua				
					"Wyattia"		
535 —	Fortunian	Anabarites trisulcatus - Protohertzina anabarica	Anabarites trisulcatus				
540 —						BACE	
	542.0						
	Ediacaran						

(Left margin labels: Epoch 2, Terreneuvian)

Fig. 2.34 (*cont.*)

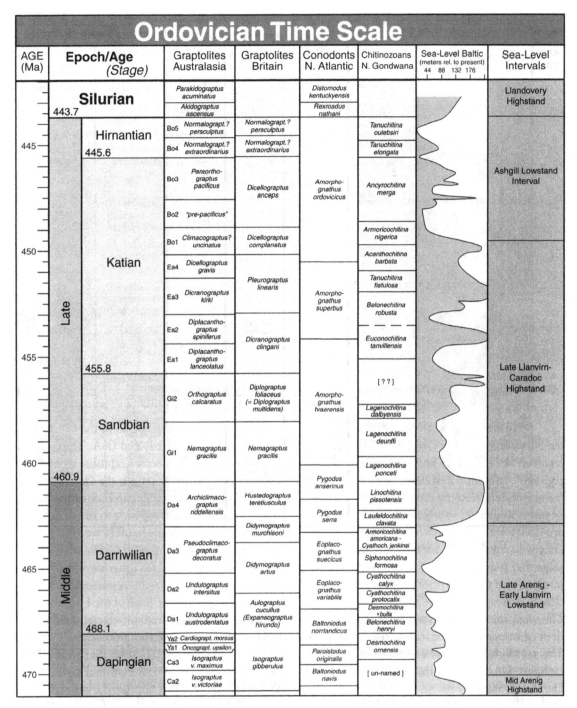

Fig. 2.35 **Ordovician time-scale.** © Cambridge University Press. Reproduced with permission from Ogg *et al.* (2008).

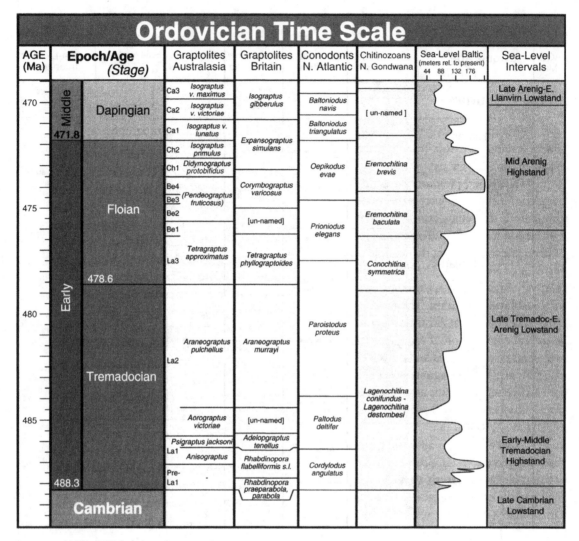

Fig. 2.35 (*cont.*)

Unfortunately, this datum has been demonstrated to be diachronous when calibrated against the locally established isotopic stratigraphic framework for the P–C boundary interval. The GSSP for the Cambrian–Ordovician boundary has been fixed at the first appearance of the conodont *Iapetognathus fluctivagus* in Bed 23 of the Green Point section in western Newfoundland, dated to 488.3 Ma (Jones, 2006). The GSSP for the Ordovician–Silurian boundary has been fixed at the first

appearance of the graptolite *Parakidograptus acuminatus* near the base of the Birkhill shale in the Dob's Linn section, near Moffat in the Southern Uplands of Scotland, dated to 443.7 Ma. The GSSP for the Silurian–Devonian boundary has been fixed at the first appearance of the graptolite *Monograptus uniformis* in Bed 20 of the Klonk section, near Prague in the Czech Republic, dated to 416 Ma. The GSSP for the Devonian–Carboniferous boundary has been fixed at the first appearance of the conodont

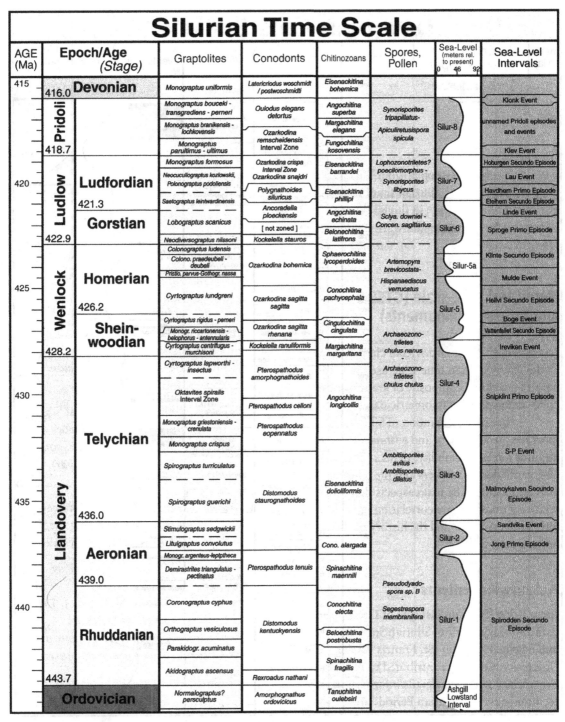

Fig. 2.36 **Silurian time-scale.** © Cambridge University Press. Reproduced with permission from Ogg *et al.* (2008).

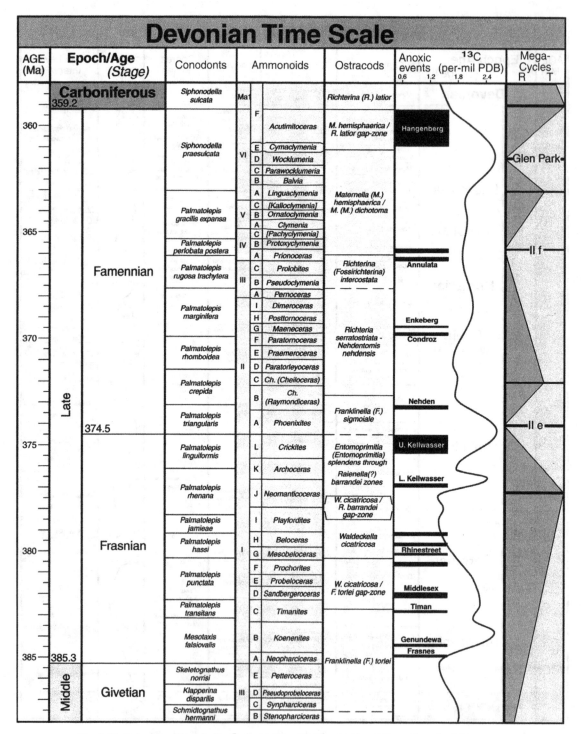

Fig. 2.37 **Devonian time-scale.** © Cambridge University Press. Reproduced with permission from Ogg *et al.* (2008).

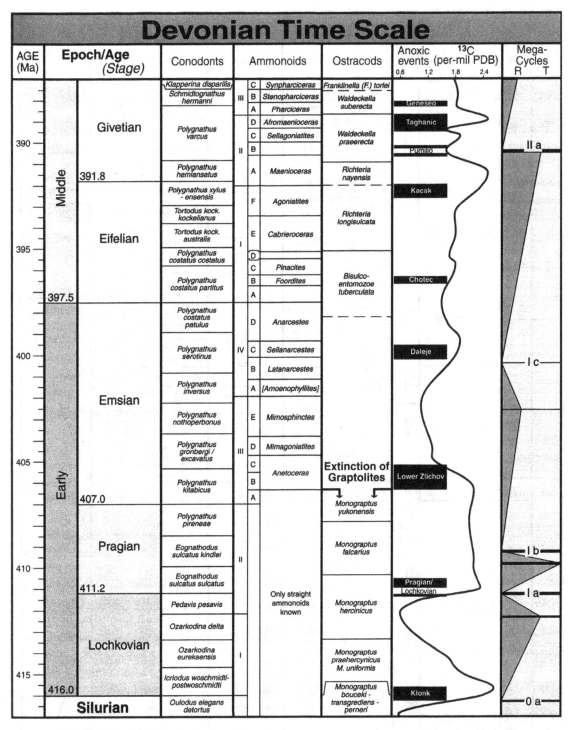

Fig. 2.37 (cont.)

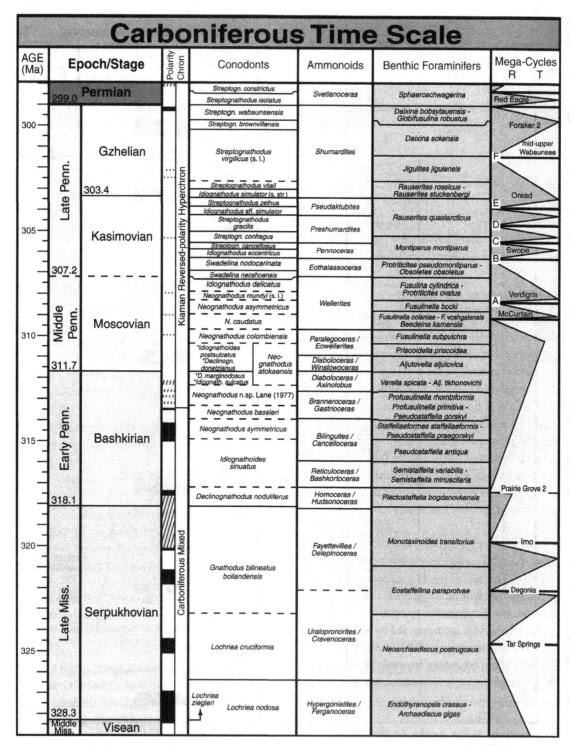

Fig. 2.38 **Carboniferous time-scale.** © Cambridge University Press. Reproduced with permission from Ogg *et al.* (2008).

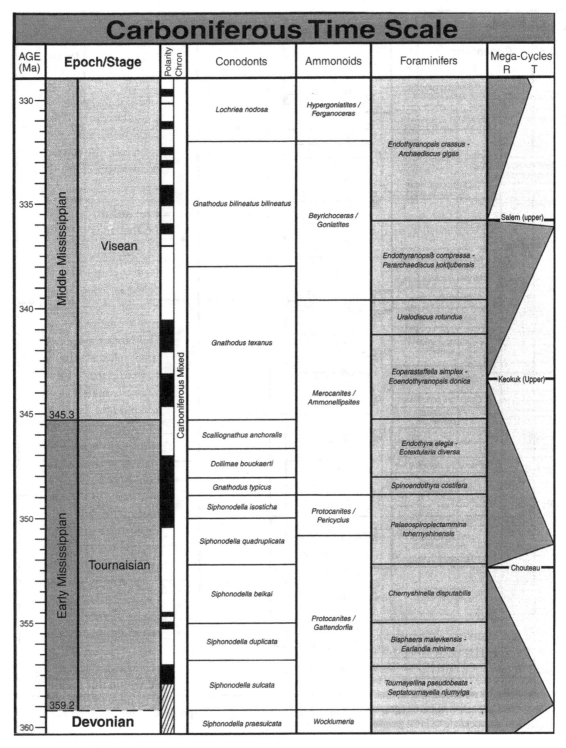

Fig. 2.38 (*cont.*)

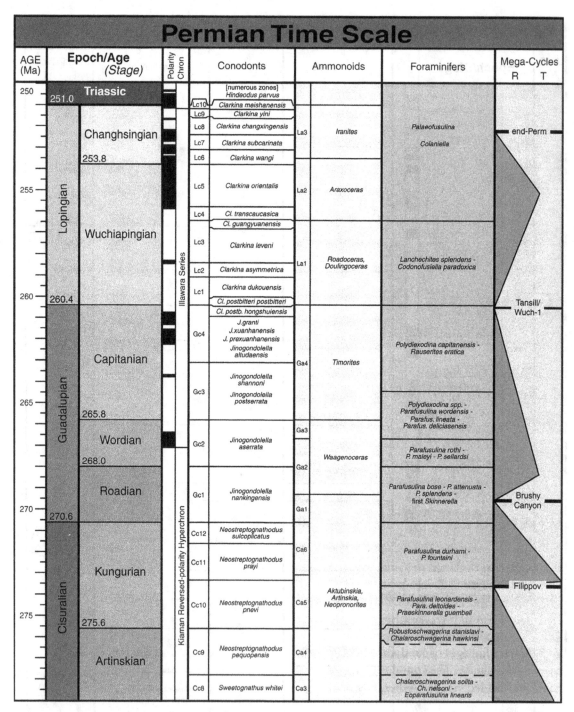

Fig. 2.39 **Permian time-scale.** © Cambridge University Press. Reproduced with permission from Ogg *et al.* (2008).

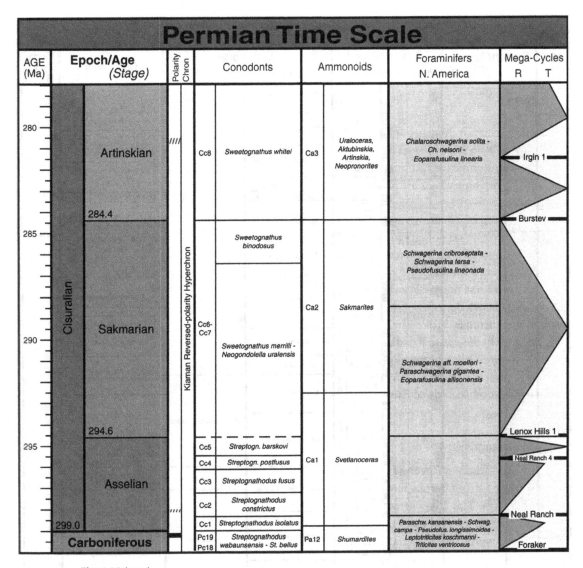

Fig. 2.39 (cont.)

Siphonodella sulcata in Bed 89 of the La Serre section near Cabrieres in the Montagne Noire region of southern France, dated to 359.2 Ma.

Incidentally, geochemical work has recently been undertaken on brachiopods from the boundary stratotype section. It is evident that the strontium isotopic composition of unaltered shells is a potentially powerful tool in precisely and accurately delineating the boundary (and also,

incidentally, in inter-regional correlation). Systematic variations in values seem to represent responses to changes in continental weathering patterns and riverine fluxes in turn associated with a glacial event in the late Middle *Praesulcata* Subzone. Variations in oxygen and carbon isotopic compositions of shells from the boundary section also appear to represent responses to this event. The GSSP for the Carboniferous–Permian

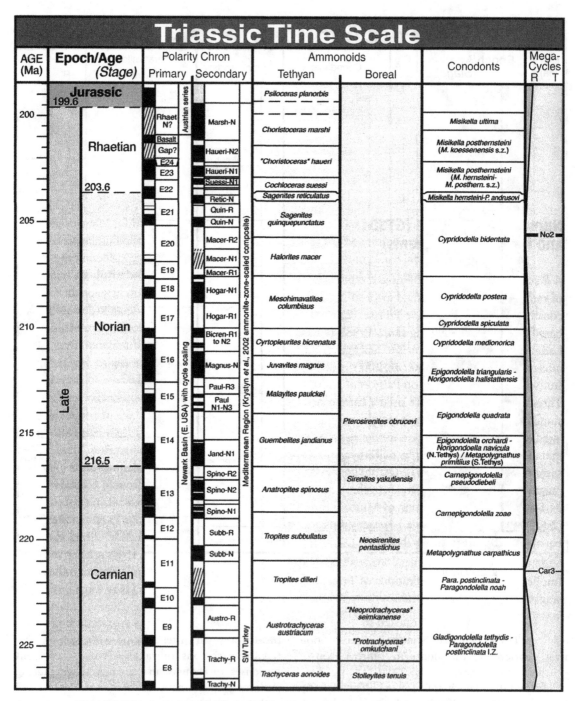

Fig. 2.40 **Triassic time-scale.** © Cambridge University Press. Reproduced with permission from Ogg *et al.* (2008).

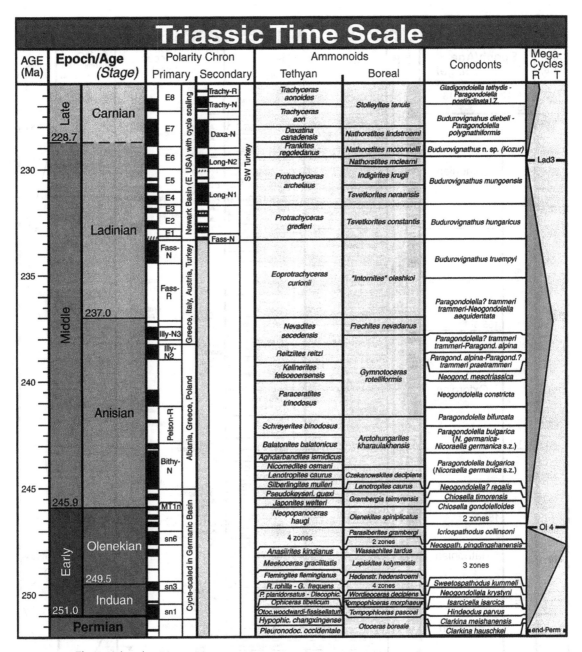

Fig. 2.40 (*cont.*)

boundary has been fixed at the first appearance of the conodont *Streptognathus isolatus* in Bed 19 of the Aidaralash River section near Aktobe in the southern Urals region of northern Kazakhstan, dated to 299 Ma.

Mesozoic

The GSSP for the Permian–Triassic boundary has been fixed at the first appearance of the conodont *Hindeodus parvus* in Bed 27c in Meishan section D in Changxing County, Zhejiang province, China,

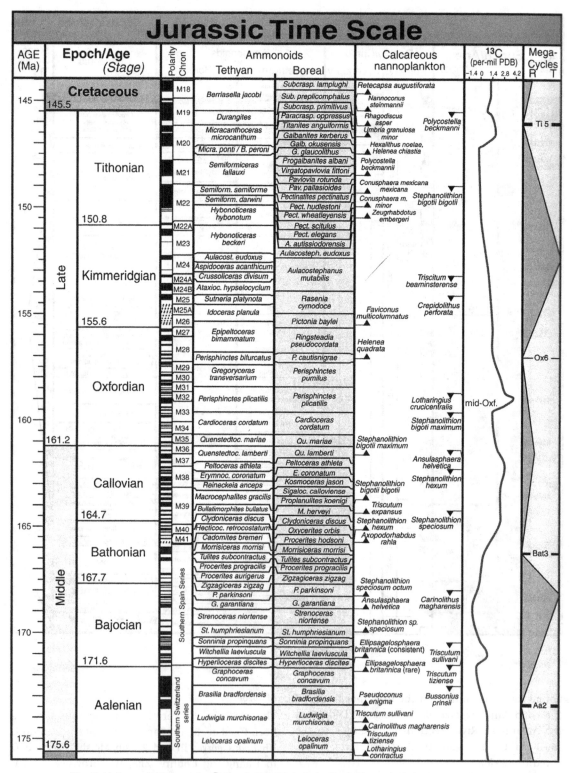

Fig. 2.41 **Jurassic time-scale.** © Cambridge University Press. Reproduced with permission from Ogg *et al.* (2008).

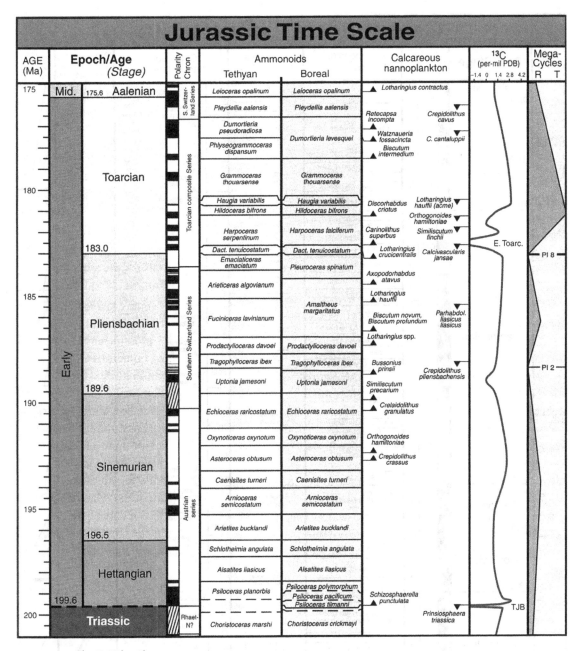

Fig. 2.41 (*cont.*)

dated to 251 Ma. The GSSP for the Triassic–Jurassic boundary is to be fixed at the first appearance of the ammonite *Psiloceras spelae* in the Kuhjoch section in the Karwendel mountains in the Tyrol in Austria, dated to 199.6 Ma (von Hillebrandt *et al.*, 2007; Lucas, 2010). The GSSP for the Jurassic–Cretaceous boundary, dated to 145.5 Ma, has not yet been selected.

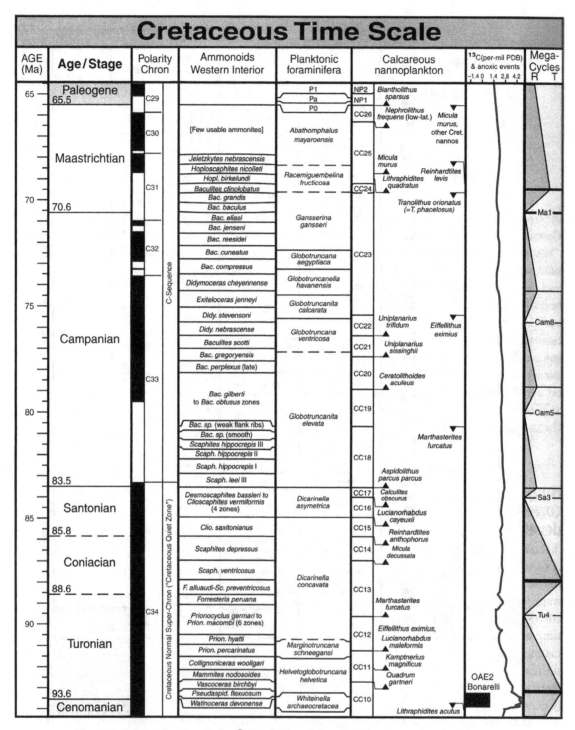

Fig. 2.42 **Cretaceous time-scale.** © Cambridge University Press. Reproduced with permission from Ogg *et al.* (2008).

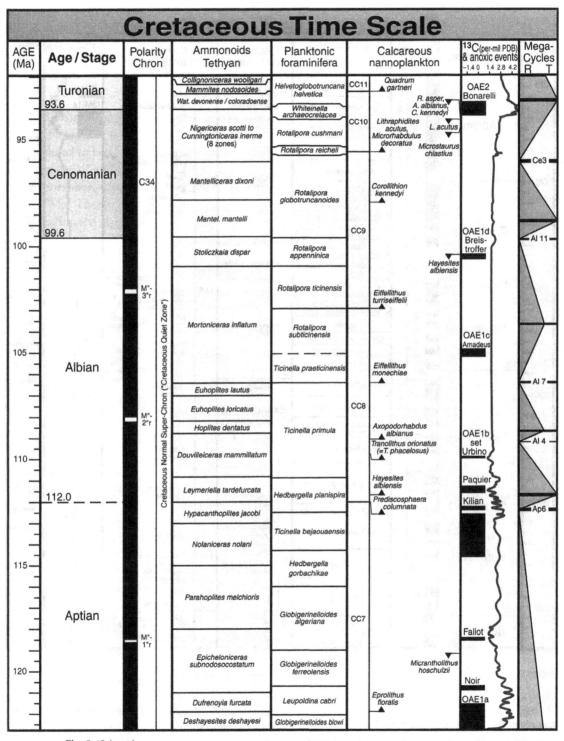

Fig. 2.42 *(cont.)*

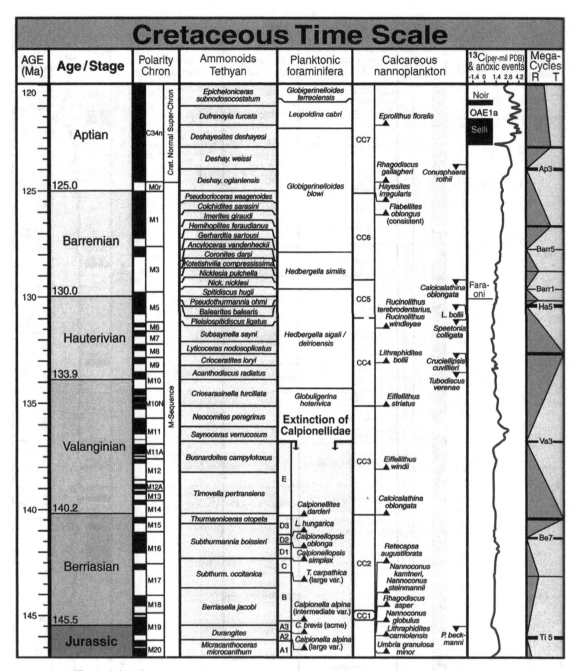

Fig. 2.42 (*cont.*)

Cenozoic

The GSSP for the Cretaceous–Tertiary boundary has been fixed at the iridium anomaly in the Boundary Clay in the El Kef section in Tunisia, dated to 65.5 Ma (Jones, 2006; Molina *et al.*, 2006). The GSSP for the Palaeocene–Eocene boundary has fixed at the base of the carbon isotope excursion in the Dababiya section near Luxor in Egypt, dated to 55.8 Ma (Jones,

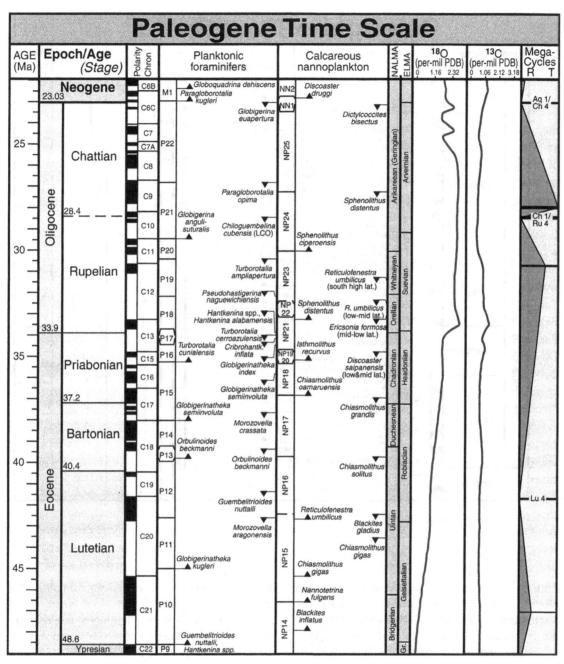

Fig. 2.43 **Palaeogene time-scale.** © Cambridge University Press. Reproduced with permission from Ogg *et al.* (2008).

2006; Aubry *et al.*, 2007; Ogg *et al.*, 2008). The GSSP for the Eocene–Oligocene boundary has been fixed at the last appearance of the planktonic foraminifer *Hantkenina* in the marl bed at the base of the exposed section in the Massigano section near Ancona in Italy, dated to 33.9 Ma (Jones, 2006). The GSSP for the Oligocene–Miocene boundary has been fixed at the base of magnetic polarity chronozone C6Cn.2n, 35 m from

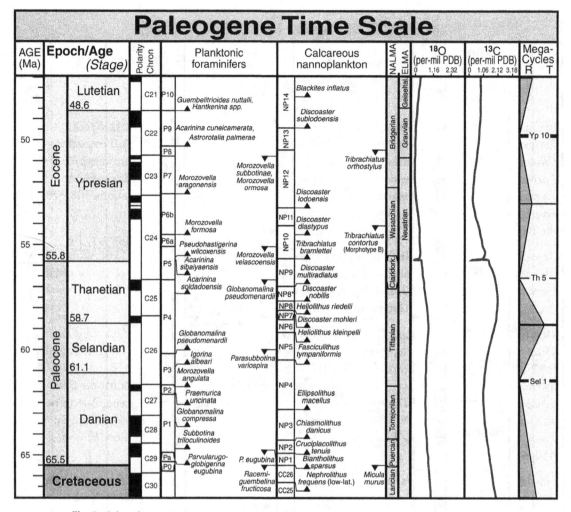

Fig. 2.43 (*cont.*)

the top of the Lemme–Carrosio section in Carrosio village near Genoa in Italy, dated to 23.03 Ma. The best biostratigraphic proxies for the GSSP are the first appearance of the planktonic foraminifer *Paragloborotalia kugleri* and the last appearance of the calcareous nannofossil *Reticulofenestra bisecta*. The GSSP for the Miocene–Pliocene boundary has been fixed at the base of Insolation Cycle 510 in the Trubi formation in the Eraclea Minoa section in Sicily, astrochronologically dated to 5.332 Ma. The best biostratigraphic proxies for the GSSP are the first appearance of the calcareous nannofossil *Ceratolithus acutus* and

the last appearance of the calcareous nannofossil *Triquetrorhabdulus rugosus*. The Pliocene–Pleistocene – and Tertiary–Quaternary – boundary has recently been moved from the base of the Calabrian stage to the base of the Gelasian stage, and the GSSP from immediately above Mediterranean precession related sapropel or MPRS 176 in the Vrica section in Calabria in Italy, astrochronologically dated to 1.806 Ma, to immediately above MPRS 250 in the Monte San Nicola section in Sicily, astrochronologically dated to 2.588 Ma (Mascarelli, 2009). The best biostratigraphic proxy for the GSSP is the last

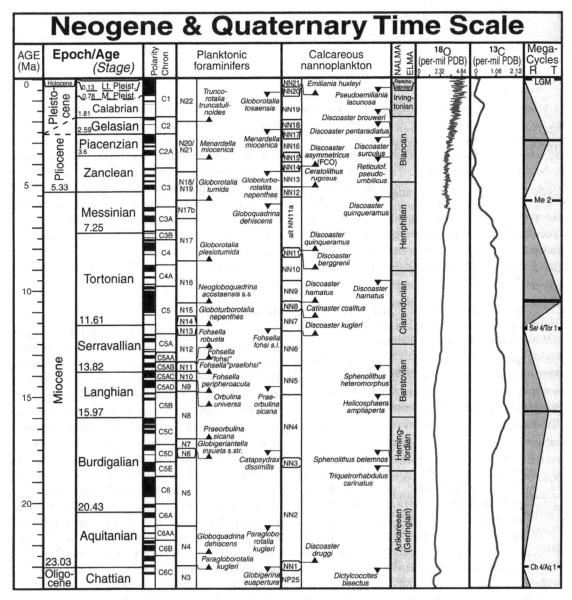

Fig. 2.44 **Neogene time-scale.** © Cambridge University Press. Reproduced with permission from Ogg *et al.* (2008).

appearance of the calcareous nannofossil *Discoaster pentaradiatus*. The GSSP for the Pleistocene–Holocene boundary is in the process of being fixed at the shift in deuterium excess marking the first indication of global warming at the end of the Younger Dryas/Greenland Stadial 1 in the NorthGRIP ice core from Greenland, dated to 11.5 ka (Walker *et al.*, in press, 2010).

3 • Palaeobiology

Palaeobiology involves the use of fossils in establishing the ordering of containing rocks in space and in relation to evolving earth environments. Readers interested in further details of the principles and practice of palaeobiology are referred to McKerrow (1978), Briggs and Crowther (1990), Behrensmeyer *et al.* (1992), Bosence and Allison (1995), Brenchley and Harper (1998), Briggs and Crowther (2001), Cohen (2003), Krassilov (2003), Jackson and Erwin (2006), Jones (2006), Cockell (2007), Foote and Miller (2007), Benton and Harper (2009) and Lieberman and Kaesler (2010).

3.1 SUMMARY OF PALAEOBIOLOGICAL SIGNIFICANCE AND USEFULNESS OF PRINCIPAL FOSSIL GROUPS

The palaeobiological significance and usefulness of the principal fossil groups is discussed below, and in Sections 3.2–3.5 below, and summarised in Fig. 3.1 (modified after Jones, 2006; see also 'Paleobiology Database' or PBDB website, www.paleodb.org).

The ranges over which they are palaeobiologically significant are shown by broad bands. Note that the potential palaeobiological significance of fossils can be impaired by natural factors, such as *post-mortem* transportation and diagenetic effects, and reworking. It can also be impaired by artificial factors, such as sample acquisition and processing, and subjectivity in specific identification.

The particular usefulness and applicability of some fossil groups is in not only palaeobiology, but also biostratigraphy, discussed in Chapter 2 above, and sequence stratigraphy, discussed in Chapter 4 below. Case studies of applications of these groups in industry and elsewhere are discussed in Chapters 5–10 below.

3.1.1 Bacteria

Cyanobacteria

Fossil Cyanobacteria are interpreted, essentially on the basis of analogy with their living counterparts, as having lived, and microbialites as having formed, in a range of marine and lacustrine environments; and as having thrived under adverse and even extreme environmental conditions. Microbialites are characteristic of the recovery intervals following mass extinctions, during which many other groups of organisms, including reef-building organisms, were effectively absent from marine ecosystems (Riding, 2005; Pruss *et al.*, 2006). They are also characteristic of fossil cold seeps, for example in the Early Jurassic of the Neuquen basin of Argentina.

'Archaebacteria'

'Archaebacteria' are typically small, morphologically simple to comparatively complex and differentiated, unicellular organisms (Garrett & Klenk, 2007; Batistuzzi & Hedges, in Hedges & Kumar, 2009). The only ones that are important in the field of applied palaeontology are the crenarchaeotes, derivatives of which are useful in palaeotemperature interpretation (Schouten *et al.*, 2002; Powers *et al.*, 2004; Weijers *et al.*, 2006; Rosell-Mele & McClymont, in Hillaire-Marcel & de Vernal, 2007; Jung-Hyun Kim *et al.*, 2008); and, more especially, the methanogenic euryarchaeotes or methanogens, which produce ('bacterial' or 'biogenic') methane either from biogenic or immature thermogenic source-rock, through the process known as 'methanogenesis', or from oil, through 'biodegradation' (Konig, in Vially, 1992; Martens, in Vially, 1992; Jorgensen, in Schulz & Zabel, 2000; Huang & Larter, in Ollivier & Magot, 2005; Jeanthon *et al.*, in Ollivier & Magot, 2005; Rabus, in Ollivier & Magot, 2005; Burdige, 2006; Konhauser, 2007; Fuhrman & Hagstrom, in Kirchman, 2008; Kirchman, in Kirchman, 2008; Moran, in Kirchman, 2008; Batistuzzi & Hedges, in Hedges & Kumar, 2009). It has been calculated that methanogens are responsible for the release of 300–700 million tonnes of methane into the atmosphere every year (and an unknown amount into traps). In terms of ecology, they live in

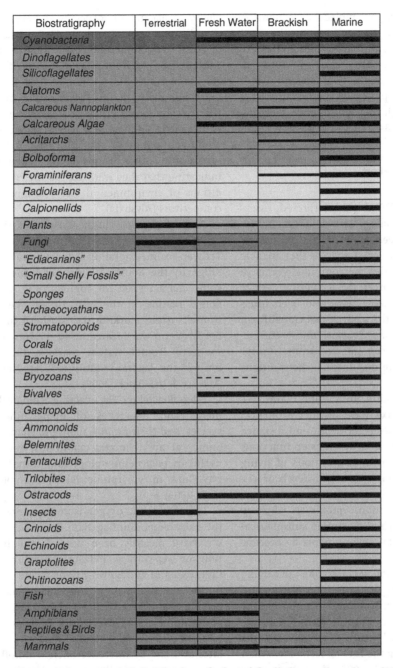

Biostratigraphy	Terrestrial	Fresh Water	Brackish	Marine
Cyanobacteria				
Dinoflagellates				
Silicoflagellates				
Diatoms				
Calcareous Nannoplankton				
Calcareous Algae				
Acritarchs				
Bolboforma				
Foraminiferans				
Radiolarians				
Calpionellids				
Plants				
Fungi				
"Ediacarians"				
"Small Shelly Fossils"				
Sponges				
Archaeocyathans				
Stromatoporoids				
Corals				
Brachiopods				
Bryozoans				
Bivalves				
Gastropods				
Ammonoids				
Belemnites				
Tentaculitids				
Trilobites				
Ostracods				
Insects				
Crinoids				
Echinoids				
Graptolites				
Chitinozoans				
Fish				
Amphibians				
Reptiles & Birds				
Mammals				

Fig. 3.1 Palaeoecological distribution of selected fossil groups. From Jones (2006).

anaerobic aquatic marine, brackish and freshwater environments, in soils, in manure, and in the intestines of animals, especially herbivores such as termites and ruminants. Many of the approximately 200 modern species are extremophiles, living in environments characterised by extremes of temperature and, to a lesser extent, acidity, such as would have existed on the early earth (they have no known fossil record, but molecular sequencing evidence indicates that they are among the most primitive forms of life on earth). Mesophiles are optimally developed at temperatures of 30–50 °C; thermophiles at temperatures of around 65 °C; and extreme thermophiles at temperatures of around 98 °C, although they will tolerate temperatures in the range 84–110 °C or possibly even higher.

3.1.2   Plant-like protists (Algae)

Dinoflagellates

Most fossil dinoflagellates are interpreted, essentially on the basis of analogy with their living counterparts, as having been more or less exclusively marine, and planktonic, or, strictly, as they are able to achieve some motility, nektonic (de Vernal & Marret, in Hillaire-Marcel & de Vernal, 2007). Importantly, though, some have also been interpreted, essentially on the basis of associated fossil and sedimentary facies, as brackish and even nonmarine. Salinity has been inferred from transfer functions based on the documented distributions of living species.

Palaeobathymetry. Within the marine realm, distinct dinoflagellate taxa have distinct ecological preferences. Some measure of depth and distance from shoreline is provided by transfer functions based on the documented distributions of living species. Moreover, some measure of depth or distance from shoreline is provided by assemblage morphogroup composition, proximal environments being characterised by simple subspherical morphotypes, and distal environments by process-bearing morphotypes.

Fossil calcareous dinoflagellates, like their living counterparts, appear to indicate either continental shelf or open oceanic environments. Those from the Early Pliocene of Cyprus in the eastern Mediterranean record the phased return to an open oceanic environment following the Late Miocene Messinian Salinity Crisis (Bison *et al.*, 2009). The first phase is characterised by the dominance of *Leonella granifera*, and is interpreted as indicating a continental shelf environment influenced by freshwater and nutrient run-off from the continent (and reduced salinity). The second and third phases are characterised by the dominance of *Caracomia stella, Calciodinellum albatrosianum* and other species, and are interpreted as indicating open oceanic environments.

Fossil 'calcispheres' also appear to have exhibited a preference for outer shelf to upper slope environments, although they also appear more abundant in the 'Waulsortian mounds' of the Carboniferous than in their level-bottom equivalents. Significantly, where they occur, they tend to do so in flood abundance and to the virtual exclusion of other groups of calcareous plankton, indicating that they were able to thrive under conditions of environmental stress inimical to their competitors (?dysoxia).

Palaeobiogeography. In the Cretaceous, the 'Malloy suite' characterised the low- to moderate-latitude Tethyan realm, extending as far south as the Walvis Ridge in the South Atlantic, and the 'Williams suite' characterised the high-latitude Austral and Boreal realms. In the Late Albian, the northern frontal system marking the southern limit of the Boreal realm was located at 40–45° N, whereas the southern frontal system marking the northern limit of the Austral realm was located at 50–70° S (Masure & Vrielynck, 2009).

Palaeoclimatology. The exacting ecological requirements and tolerances of many dinoflagellate species, and their rapid response to changing environmental and climatic conditions, render them useful in palaeoclimatology and climatostratigraphy, and in environmental archaeology (Sluijs *et al.*, 2005).

The carbon and oxygen isotopic compositions of the calcareous dinoflagellate *Pirumella krasheninnikovi* have recently been used in the interpretation of sea-surface palaeotemperatures in the Cretaceous of the South Atlantic, and specifically in the confirmation of a Campanian–Maastrichtian cooling phase identified on planktonic foraminiferal isotopic evidence (Friedrich & Meier, 2006).

Palaeo-oceanography. Some modern dinoflagellates are, and by inference ancient dinoflagellates were, characteristically associated with upwelling (Pospelova *et al.*, 2008). The ratio of autotrophic gonyaulacoid to heterotrophic peridinioid dinoflagellates has been used as an indicator of palaeoproductivity, as, for example in the Late Cretaceous, Cenomanian-Turonian of the 'Western Interior Seaway' of the United States.

Silicoflagellates

Fossil silicoflagellates are interpreted, essentially on the basis of analogy with their living counterparts, as having been marine and planktonic.

Palaeoclimatology. The ratio between the comparatively warm-water *Dictyocha* and the cool-water *Distephanus* provides some measure of palaeotemperature and palaeoclimate.

Diatoms

Fossil diatoms are interpreted, essentially on the basis of analogy with their living counterparts, as having occupied a range of aquatic, freshwater, brackish (oligohalobous, mesohalobous or polyhalobous) and marine, benthic and planktonic environments (Hustedt, 1953, 1957; Denys, 1992; Horton *et al.*, 2007; Krosta & Koc, in Hillaire-Marcel & de Vernal, 2007; Yongqiang Zong *et al.*, 2010). To (over) simplify, centric diatoms are typically marine and planktonic, and pennates typically freshwater and benthic.

A worked example of the application of diatoms to the palaeoenvironmental interpretation of the Middle–Late Miocene, Sarmatian of Romania has been provided by Saint-Martin & Saint-Martin (2005). Worked examples of the application of diatoms to the palaeoenvironmental interpretation of the Late Miocene, Messinian of the northeastern Rif in Morocco have been provided by Saint-Martin *et al.* (2003) and El-Ouahibi *et al.* (2007).

Note that the palaeobiological usefulness of diatoms can be compromised by their destruction in diagenesis, through the transformation of opaline silica into cristobalite and of cristobalite into quartz, often at quite shallow burial depths (note, though, that the transformation is controlled by temperature rather than by depth *per se*) (Ryves *et al.*, 2009).

Palaeobathymetry. Transfer functions based on the environmental distributions and tolerances of diatoms are of considerable use in sea-level reconstructions in temperate and tropical environments (Horton *et al.*, 2007).

Palaeobiogeography. In the Cenozoic, diatom evidence points toward the existence of a marine connection between Tethys and the Arctic Ocean in the Eocene, by way of the so-called Obik Sea or Turgai Strait.

Palaeoclimatology. The exacting ecological requirements and tolerances of many diatom species, and their rapid response to changing environmental, and especially climatic, conditions, render them extremely useful in palaeoclimatology and climatostratigraphy, and in environmental archaeology (Hustedt, 1953, 1957; Denys, 1992).

Palaeo-oceanography. Modern diatoms are, and by inference ancient diatoms were, characteristically associated with upwelling.

Calcareous nannoplankton

Calcareous nannofossils are interpreted, essentially on the basis of analogy with living coccolithophores, as having been essentially marine and planktonic (Giraudeau & Luc, in Hillaire-Marcel & de Vernal, 2007). Importantly, though, some forms are interpreted as having had some brackish tolerance, for example *Braarudosphaera* and *Emiliania*. Low-salinity 'events', such as heavy rainfall or increased meltwater, or upwelling, have had to be invoked to account for the ancient '*Braarudosphaera* chalks' observed in the Oligocene of the various DSDP and ODP Sites of the South Atlantic, which are clearly of open oceanic and not nearshore aspect: downslope transportation from a nearshore setting is another possible explanation.

Abundance and diversity peaks of calcareous nannofossils have been widely used to characterise consensed sections in sequence stratigraphic interpretation. Note, though, that certain of these peaks have been observed to be associated with what benthic foraminiferal and sedimentological data demonstrate to be allochthonous debris flows, such that they are recording allochthonous as well as autochthonous calcareous nannofossil species. Thus, nannofossil abundance and diversity peaks identified in the absence of such benthic foraminiferal and sedimentological evidence should be regarded as of questionable reliability as indicators of anything meaningful!

Palaeobathymetry. Different calcareous nannofossil sub-groups are interpreted as having had different vertical distributions within the water column. For example, ancient nannoconids are interpreted as characteristic of epipelagic environments. In the Pliensbachian of the Lusitanian basin in Portugal, *Schizosphaerella* – possibly a calcareous dinoflagellate – is interpreted as characteristic of epipelagic environments, and *Crepidolithus crassus* and *Mitrolithus jansae* of deeper water environments, although still within the photic zone (Reggiani et al., 2010).

Different calcareous nannofossil sub-groups are also interpreted as having had different horizontal distributions. In the late Pliensbachian–early Toarcian of the Umbria-Marche basin in central Italy, *Schizosphaerella* and *Crepidolithus crassus* are interpreted as characteristic of proximal facies, as represented by the limestones of the Latium–Abruzzi platform. *Mitrolithus jansae* has been interpreted as characteristic of distal facies, as represented by marls. *Lotharingius*, *Biscutum* and *Calyculus* are also interpreted as characteristic of distal facies.

Palaeoclimatology. Calcareous nannofossils recently been used in the interpretation of sea-surface palaeotemperatures in the Jurassic of eastern France, and specifically in the confirmation of a Callovian–early Oxfordian cooling phase identified on carbon and oxygen isotopic evidence (Tremolada et al., 2006). *Biscutum* spp. and *Triscutum* spp., together with representatives of the 'A-Group', comprising *Axopodorhabdus* spp. and other Axopodorhabdaceae, were positively correlated with cooling (and/or eutrophication associated with increased upwelling). *Schizosphaerella punctulata* and *Watznaueria manivitae* appeared correlated with warming (and/or oligotrophication).

Palaeo-oceanography. Some modern coccithophores are, and by inference ancient calcareous nannofossils were, characteristically associated with upwelling. The abundance and diversity of calcareous nannofossils has been used as an indicator of palaeoproductivity (or carbonate dilution), as, for example, in the Late Jurassic Kimmeridge Clay formation of the Wessex basin in the south of England or the Early Cretaceous of the Vocontian basin in southeast France (Lees et al., 2006). In the Kimmeridge Clay formation, diversity has been interpreted as inversely proportional to nutrient level. Here, low diversity assemblages, characterised, and indeed locally dominated, by the opportunistic species *Watznaueria britannica* and/or *Cyclagelosphaera margarelii*, are associated with oil shales and interpreted high nutrient levels. Higher, but still comparatively low, diversity assemblages, characterised by *Watznaueria barnesiae/fossacincta*, are associated with

'stone bands' and interpreted lower, but still high, nutrient levels.

It has been postulated that the evident 'blooms' resulting in the '*Braarudosphaera* chalks' observed in the Oligocene of the various DSDP and ODP Sites of the South Atlantic reflected rhythmic variations in the vigour of the South Atlantic gyre, and that their cessation towards the end of the Early Oligocene represented a response to an important shift in circulation, related to the opening of the Drake Passage. Interestingly, foraminiferal evidence indicates that benthic to thermocline oxygen and carbon isotope gradients were reduced during the blooms, a hallmark signature for strengthened upwelling. *Braarudosphaera* may have lived in the lower part of the euphotic zone, a favourable location for exploiting upwelling nutrients.

Calcareous Algae

Fossil calcareous Algae are interpreted, essentially on the basis of analogy with their living counterparts, as having occupied a range of aquatic, freshwater and shallow marine, habitats. Charophytes and non-calcareous green Algae are interpreted as freshwater to brackish; calcareous green Algae freshwater to shallow marine; and calcareous red Algae shallow marine. Those charophytes that have been observed in apparently fully marine environments, for example in the Devonian of the Canning basin in Australia, have generally been interpreted as allochthonous (Edgell, 2003). Note, though, that the umbellinids, which have been observed in deep but not in shallow marine environments, could be interpreted as autochthonous. Charophytes are of considerable use in the palaeoenvironmental interpretation of freshwater environments. Distinct deltaic, lagoonal, estuarine, riverine and lacustrine sub-environments have been identified in the Oligocene 'molasse' of the Penzberg syncline in south Germany. Distinct bathymetric zones and phases of high and low lake level have been identified in the Holocene lacustrine sediments of Lake Tigalmamine in the Moroccan Atlas (Soulie-Marsche et al., 2008).

Palaeobathymetry. Living calcareous Algae photosynthesise, and by analogy fossil ones photosynthesised, within the photic zone in order to manufacture food. They are especially useful in

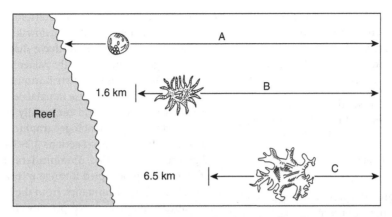

Fig. 3.2 **Palaeoecological distribution of acritarchs**. © Wiley. Reproduced with permission from Lipps (1993). (A) Simple, spherical 'morphogroup'; (B) thin-spined 'morphogroup'; (C) thick-spined 'morphogroup'.

palaeobathymetric interpretation, as specific taxa have specific light requirements and depth distributions even within the photic zone (light penetration being dictated primarily by water depth).

Palaeobiogeography. Many calcareous Algae appear to have had restricted or endemic biogeographic distributions, rendering them of use in the characterisation of palaeobiogeographic realms or provinces, and in turn in the constraint of plate tectonic reconstructions, in the Palaeozoic.

Palaeoclimatology. Charophytes are of some use in palaeoclimatic interpretation, although the isotopic records that they reveal have to be interpreted with caution on account of possible disequilibrium effects.

Acritarchs

Fossil acritarchs are interpreted, on the bases of analogy with dinoflagellates, and associated fossils and sedimentary facies, as having been essentially marine, although arguably with some brackish tolerance, and probably planktonic. In the Silurian of Gotland and of the Welsh borderlands, and in the Devonian of Alberta in Canada, proximal, intermediate and distal assemblages are distinguishable on the basis of abundance, diversity, dominance and dominant 'morphogroup' (Staplin, 1961; Williams, in Haq & Boersma, 1978; Dorning, in Neale & Brasier, 1981; Mendelson, in Lipps, 1993; Stricanne *et al.*, 2004; Fig. 3.2). Proximal assemblages are characterised by simple subspherical

'morphotypes', intermediate assemblages by thin-spined 'morphotypes', and distal assemblages by thick-spined 'morphotypes'. Some measure of the favourability or otherwise of the environment for reproduction is provided by the proportion of specimens possessing excystment apertures.

Palaeobathymetry. At least in the Ordovician of southern Europe, the Middle East, north Africa and China in Gondwana, and in the Silurian of Gotland in Baltica, acritarch diversity appears to be highest at times of sea-level high-stand and lowest at times of sea-level low-stand, and can therefore be used as a proxy for sea level (Vecoli, 2000; Jun Li *et al.*, 2004; Stricanne *et al.*, 2004; Vecoli & le Herisse, 2004). In the Ordovician of northern England in Avalonia, though, acritarch diversity appears to be highest at times of sea-level low-stand, and lowest at times of sea-level high-stand (Molyneux, 2009). One explanation for this observation might be the occurrence of reworked as well as *in situ* specimens in low-stand fans.

Palaeobiogeography. Many acritarchs appear to have had restricted or endemic biogeographic distributions, rendering them of use in the characterisation of palaeobiogeographic realms or provinces, and in turn in the constraint of plate tectonic reconstructions, in the Palaeozoic. An American palynological unit and an African palynological unit (or *Coryphidium bohemicum* province) are recognisable in the Early Ordovician, separated by an Appalachian oceanic gap; an African palynological unit or *Multiplicisphaeridum pilaris* province in the Arenig–Llanvirn; and a Baltic acritarch province and a 'Gondwanan' *Neoveryhachium carminae* facies in the Middle Silurian, late Llandovery–early Wenlock.

Bolboforma

Bolboforma is interpreted, essentially on the basis of associated fossils and sedimentary facies, as having been exclusively marine and planktonic.

3.1.3   Animal-like protists (Protozoa)

Testate Amoebae

Fossil testate *Amoebae* are interpreted on the basis of analogy with their living counterparts as having occupied a range of freshwater and brackish environments (Ogden & Hedley, 1980; Charman *et al.*, 2010). They have recently proved of use in the reconstruction of Holocene sea level in tidal salt-marsh environments in Maine and Nova Scotia on the Atlantic seaboard of North America (Charman *et al.*, 2010). Unfortunately from the point of view of their more widespread use in palaeobiological interpretation, their preservation potential and hence pre-Pleistocene fossil record is poor.

Foraminifera

Fossil foraminifera are interpreted, on the bases of analogy with their living counterparts, functional morphology, and associated fossils and sedimentary facies, as having occupied a range of brackish and marine, benthic and planktonic, environments, and as having assumed a variety of life positions and practised a variety of feeding strategies (Jorissen *et al.*, in Hillaire-Marcel & de Vernal, 2007; Kucera, in Hillaire-Marcel & de Vernal, 2007). Note in this context, though, that recent empirical evidence indicates that at least one modern species, *Cribroelphidium gunteri*, is apparently able to live in essentially freshwater environments, in salinities locally as low as 1–2 parts per thousand, in Lake Winnipegosis in Manitoba in Canada, into which it may have been introduced by migratory birds in the Holocene (Patterson *et al.*, 1997; Boudreau *et al.*, 2001; see also Pint & Frenzel, 2010). Note also that recent empirical and molecular biological evidence indicates that some species of 'allogromiids' are able to live in entirely freshwater environments in Lake Geneva in Austria, and even in damp essentially terrestrial environments

in Queensland in Australia (Meisterfeld *et al.*, 2001; Holzmann & Pawlowski, 2002; Pawlowski *et al.*, 2010).

Different taxa appear to have had different palaeoecological, and especially palaeobathymetric and palaeobiogeographic, distributions, rendering the group of considerable use in palaeoecological or palaeoenvironmental, and especially in palaeobathymetric and palaeobiogeographic, interpretation (see below; see also Sections 3.2–3.4).

Agglutinated benthic foraminifera are interpreted as having occupied a range of brackish and marine benthic environments, from the high marsh to the abyssal plain, and as especially characteristic of marginal marine, peri-deltaic and associated, and deep marine, submarine fan clastic environments. At least some of those from marginal marine, peri-deltaic and associated environments are tolerant of hyposalinity and hypoxia or dysoxia (Nagy *et al.*, 2010). In terms of functional morphology, living, and by inference fossil, epifaunal filter-feeding agglutinated benthic foraminifera – representing 'morphogroup' A – are characterised by tubular to branching morphologies. Epifaunal detritus-feeding (including, incidentally, phytodetritus-feeding) agglutinated benthic foraminifera – representing 'morphogroup' B – are characterised by globular to flattened, spiral morphologies. Infaunal deposit-feeding agglutinated benthic foraminifera – representing 'morphogroup' C – are characterised by elongate, spiral to serial morphologies with ratios of low surface area to volume (Fig. 3.3).

Triangular cross-plots of agglutinated benthic 'morphogroups' provide a means of distinguishing depth environments such as marginal marine, shelf to upper slope, middle to lower slope, and abyssal; or sedimentary sub-environments such as delta top, delta front and prodelta, in marginal to shallow marine, peri-deltaic environments (see 3.3.1 and Box 3.1 below; see also 5.3.2), or channel or turbiditic, channel levee or interturbiditic, and overbank or hemipelagic, in deep marine, submarine fan environments (see 3.3.3, Boxes 3.3–3.4 and Fig. 3.13 below; see also 5.3.4). 'Morphogroup' analysis has also been used in reservoir characterisation, specifically in the identification of barriers and baffles to fluid flow (see 5.4.4).

Smaller calcareous benthic foraminifera are interpreted as having occupied a range of brackish

Fig. 3.3 **'Morphogroups' of agglutinating foraminifera.** From Jones (2006). 'Morphogroup' A: (a) *Halyphysema tumanowiczii*; (b) *Saccodendron* sp; (c) *Pelosina* sp.; (d) *Dendrophrya erecta*; (e) *Halyphysema* sp.; (f) Pelosina *arborescens*; (g) *Jaculella obtusa*; (h) *Notodendrodes antarctikos*; (i) *Bathysiphon* sp.; (j) *Marsipella arenaria*; 'Morphogroup' B: (k) *Saccammina alba*; (l) *Saccammina sphaerica*; (m) *Hippocrepina* sp.; (n) *Marsipella sphaerica anglica.* (o) *Ammodiscus* sp.; (p) *Psammosphaera parva*; (q) *Glomospira* sp.; (r) *Labrospira [Veleroninoides] jeffreysi.* 'Morphogroup' C: (s) *Reophax subfusiformis*; (t) *Miliammina fusca*; (u) *Ammobaculites exiguus*; (v) textulariid. 'Morphogroup' D: (w) *Trochammina pacifica*; (x) *Trochammina inflata*; (y) *Jadammina [Entzia] macrescens.*

and marine benthic environments. In terms of functional morphology, epifaunal detritus-feeding smaller calcareous benthic foraminifera are characterised, like their agglutinated counterparts, by globular to flattened, biconvex or plano-convex morphologies. Infaunal smaller calcareous benthic foraminifera are characterised, again like their agglutinated counterparts, by elongate morphologies.

Infaunal buliminides are tolerant of hypoxia or dysoxia, and characteristic of oxygen minimum zones (see below).

Larger benthic foraminifera interpreted on the basis of analogy with their living counterparts as having hosted photosymbionts are further interpreted as diagnostic of the photic zone, and thus typically platformal and peri-reefal environments (BouDagher-Fadel, 2008; see 3.3.2 and Box 3.2 below). Those interpreted as having hosted green or red algal or dinoflagellate photosymbionts, such as the porcelaneous miliolides and microgranular fusulinides, are further interpreted as diagnostic of the euphotic zone, and thus typically shallow, back-reef and reef, sub-environments; those interpreted as having hosted diatom photosymbionts, such as the hyaline rotaliides, as diagnostic of the sub-euphotic zone, and thus typically deeper, reef and fore-reef, sub-environments. (Boron isotope composition and high-resolution liquid chromatography can be used as indicators of photosymbiosis.)

Planktonic foraminifera are interpreted as having been exclusively marine.

Palaeobathymetry. The ratio of planktonic to benthic foraminifera tends to increase with increasing depth, so serves as some measure of depth (see 3.7.1 below). Transfer functions based on the environmental distributions and tolerances of foraminifera are of considerable use in sea-level reconstructions. Abundance and diversity trends, and bathymetric trends, have been used in the characterisation of condensed sections (CSs) and systems tracts (STs) in sequence stratigraphic interpretation (see Chapter 4 and Box 4.1). They have also been used in the identification of barriers and baffles in reservoir characterisation and in seal capacity analysis (see 5.4.4 and Box 5.4).

Palaeobiogeography. In the Palaeozoic, fusuline LBFs were characteristic of the Palaeo-Tethyan realm.

In the Mesozoic, smaller benthic foraminiferal, LBF and planktonic foraminiferal distributions indicate the existence of generally distinct Austral, Boreal and Tethyan realms (smaller benthic foraminifera such as *Stensioeina*, LBFs, and keeled planktonic foraminifera such as globotruncanid globigerinides characterising the Tethyan realm). Note, though, that there was probably at least local or intermittent communication between these realms, as indicated, for instance, by the occurrence of the essentially Tethyan to sub-Boreal planktonic foraminifer *Lilliputianella globulifera [Hedbergella maslakovae]* in the Austral realm in the Early Cretaceous, Barremian to Aptian. Smaller benthic foraminiferal distributions indicate the existence of three distinct provinces within the Boreal realm in the Cretaceous, namely the Boreal–Atlantic province, characterised by epistominids, the Arctic province, characterised by ammodiscids, haplophragmoidids, trochamminids and ataxophragmiids, and the Boreal–Pacific province, characterised by endemic rzehakinids. Norway and Greenland would appear to have been in the Boreal–Atlantic province, Alaska and Canada in the Arctic province, and Kamchatka and Sakhalin in the Far East of the former Soviet Union in the Boreal–Pacific province.

In the Cenozoic, Palaeocene–Eocene, smaller foraminiferal distributions point toward the existence of distinct shallow (Midway) and deep (Velasco) waters in the Tethyan and low-latitude Atlantic realms. The Indo-Pacific and Mediterranean realms are characterised by LBFs (Renema, in Renema, 2007; Renema et al., 2008). The commonality of LBFs between the Indo-Pacific and Mediterranean is a key indicator of the palaeo(bio)geographic and palaeoclimatic evolution of Eurasia over the critical Oligocene–Miocene time interval (Jones, in Agusti et al., 1999; Box 3.8).

Palaeoclimatology. The exacting ecological requirements and tolerances of many foraminiferal species, and their rapid response to changing environmental and climatic conditions, render them useful in palaeoclimatology and climatostratigraphy, and in environmental archaeology. Importantly, foraminifera have been used in quantitative palaeotemperature estimation in the Cenozoic of the North Sea basin (de Man & van Simaeys, 2004). Interestingly, and counter-intuitively, while the maximum sizes attained by modern planktonic foraminiferal assemblages are typically in the tropics or in warm water, the maximum sizes attained by ancient, Cenozoic assemblages were typically during times of climatic cooling (Schmidt et al., 2004).

Palaeo-oceanography. Foraminifera are of considerable use in palaeo-oceanography, for example as tracers of water-masses with particular temperature and salinity characteristics, and as indicators of palaeoproductivity and organic carbon flux to the sea floor, and of oxygenation levels in the bottom sediment.

Certain benthic foraminifera have been used as indicators of high palaeoproductivity or organic carbon flux, or of low oxygenation levels in the sediment, or both. The incidence of infaunal agglutinated and especially calcareous buliminide 'morphotypes' appears to be more or less proportional to the organic content of the substrate, or inversely proportional to the oxygen content, and is as a key input in the calculation of the benthic foraminiferal 'dissolved oxygen index', or TRophic condition and OXygen concentration, TROX (Poag, 1985; Quinterno & Gardner, 1987; Kaiho, 1994; Jorissen et al., 1995; Fontanier et al., 2005; Jorissen et al., in Hillaire-Marcel & de Vernal, 2007; Koho et al., 2007; Koho, 2008a, b; Ashckenazi-Polivoda et al., 2010). Buliminides are especially characteristic of dysoxic to anoxic environments such as deep marine oxygen minimum zones or OMZ (see also 3.3.3).

Certain planktonic foraminifera have been used as indicators of high palaeoproductivity associated with upwelling.

Radiolarians

Fossil radiolarians are interpreted, essentially on the basis of analogy with their living counterparts, as having been exclusively marine and planktonic. They bloomed under particularly favourable environmental conditions, resulting in the accumulation of cherts and siliceous limestones.

Palaeobathymetry. Fossil spumellarians and nassellarians are interpreted, essentially on the basis of analogy with their living counterparts, as having been epipelagic to mesopelagic, and bathypelagic, respectively.

Palaeobiogeography. In the Mesozoic, radiolarian distributions point toward the development of three palaeobiogeographic realms, namely: a southern or Austral realm characterised by *Praeparvicingula* and *Parvicingula*; a central, Tethyan realm, characterised by high abundances and diversities of pantanellids and by *Mirifusus, Andromeda, Acanthocircus dicranocanthos* and *Ristola*; and a northern or Boreal realm, characterised by *Praeparvicingula, Parvicingula, Stichocapsa* and, in its southern part only, by *Ristola*.

In the Cenozoic, radiolarian evidence points toward the existence of a marine connection between Tethys and the Arctic Ocean in the Eocene, by way of the so-called Obik Sea or Turgai Strait.

Palaeoclimatology. Radiolarians are of use in palaeoclimatic as well as in palaeobiogeographic interpretation. For example, in the Middle to Late Miocene of the equatorial Pacific, collosphaerids have been interpreted, on the basis of analogy with their living counterparts, as having been essentially tropical to temperate, and a specific variant of the theoperid *Stichocorys delmontensis* as an indicator of cooler conditions characterised by enhanced upwelling. Data on the distributions of the respective cool- and warm-water markers, taken together, indicate a significant cooling within the Late Miocene. The Late Pleistocene–Holocene sea-surface temperature history of the southwest Pacific has been inferred using radiolarian distribution datasets (Luer et al., 2009). The Late Pleistocene–Holocene palaeoclimatic history of the Norwegian Sea has been inferred using artificial neural networks trained on radiolarian distribution datasets (Cortese et al., 2005).

Palaeo-oceanography. Ancient radiolarians appear to have been strongly associated with upwelling, as in the Early Carboniferous of the Culm basin of Germany, in the Miocene of the Pungo River formation of North Carolina, and in the Miocene to Recent of the Indo-Pacific and of the Benguela system in the eastern South Atlantic (Nigrini & Caulet, 1992; Lazarus et al., 2006; Kamikuri et al., 2009). In the Miocene to Recent of the Indo-Pacific and eastern South Atlantic, upwelling favoured high radiolarian productivity, measurable through the use of the Upwelling Radiolarian Index or URI. 'URI taxa' common to the Indo-Pacific and Atlantic include *Plectacantha cremastoplegma, Acrosphaera murrayana, Pterocanium auritum, Dictyophimus infabricatus, Lamprocyrtis migrinae* and *Pterocorys minythorax*.

Calpionellids

Fossil calpionellids are interpreted, essentially on the basis of associated fossils and sedimentary facies, as having been marine and pelagic. They appear to have exhibited a preference for open oceanic environments.

Palaeobiogeography. Calpionellids exhibited a wide geographic distribution in low to mid latitudes.

3.1.4 Plants

Plant macrofossils, spores and pollen, and phytoliths

Fossil plants are interpreted, essentially on the bases of analogy with their living counterparts, and functional morphology, as having been essentially terrestrial to aquatic, and as having occupied a range of habitats in all other than the highest latitudes, and at all other than the highest altitudes (Beerling & Woodward, 2001; Niklas, in Briggs, 2005; Olivarez *et al.*, 2005; Beerling, 2007). Hardiness, moisture needs and salinity tolerance have been inferred from transfer functions based on the documented distributions of living species. C3 or Calvin–Benson and C4 or Hatch–Slack plants utilise different photosynthetic carbon reduction pathways, initially forming compounds containing three and four carbon molecules respectively. In consequence, C3 plants have a lower photosynthetic efficiency, and have to use more water in the process, so are characteristic of comparatively wet, typically woodland, environments; while C4 plants have a higher efficiency, and use less water, so are characteristic of drier, grassland, conditions (Osborne & Beerling, 2006).

Palaeobiogeography. In the Palaeozoic, the floras of the Late Silurian, Devonian, Carboniferous and Permian were characterised by greater or lesser degrees of endemism (Raymond *et al.*, 2006). Plant distributions over this time interval point toward the existence of generally separate Angara, Euramerica, Cathaysia and Gondwana realms. In the Carboniferous–Permian, Gondwana was characterised by *Glossopteris* (Anderson *et al.*, 1999). Note, though, that in the Permian, part of Euramerica was also characterised by *Glossopteris*, specifically *Glossopteris anatolica* (Berthelin *et al.*, 2006).

In the Mesozoic, plant distributions point toward the existence of separate Angara, Euramerica and Gondwana realms in the Triassic. Plant and pollen distributions point toward the existence of a distinct West Africa/South America or WASA province in the Gondwana realm in the Early Cretaceous; and distinct *Aquilapollenites* and *Normapolles* provinces in the Euramerican realm, and a distinct *Nothofagidites* province in the Gondwana realm, in the Late Cretaceous (Traverse, 2008). The WASA province in the Gondwana realm in the Early Cretaceous is characterised

by *Classopollis* pollen and its parent plant, an extinct cheirolepidacean (Alvin, 1983; Seward, in McCabe & Parrish, 1992; Xiao-Yu Jang, 2008).

In the Cenozoic, Palaeocene–Eocene plant distributions point toward the palaeogeographic and palaeoclimatic evolution of the North Atlantic. In the Oligocene–Holocene, plant distributions point toward the existence, and the palaeogeographic and palaeoclimatic evolution, of Tethyan and Paratethyan realms.

Palaeoclimatology. As intimated above, plant macro- and microfossils are important in palaeoclimatic as well as in palaeobiogeographic interpretation (Jaramillo *et al.*, 2006; Peters-Kottig *et al.*, 2006; Uhl *et al.*, 2007; Punyasena, 2008; Traverse, 2008; Spicer *et al.*, 2009; Tao Su *et al.*, 2010). Climate has been inferred from plant macrofossil distribution, abundance and diversity, and also from functional morphology and leaf physiognomy (for example through the use of leaf margin analysis or LMA, and of the Climate Leaf Analysis Multivariate Program or CLAMP); tropical rainforest from a high incidence of plants with 'drip tips' at the ends of their leaves, etc. Climate has also been inferred from plant microfossil (spore and pollen) distribution, abundance and diversity. Atmospheric carbon dioxide concentration has been inferred from stomatal density (McElwain *et al.*, 1999; Beerling & Woodward, 2001; Beerling, 2007; Cheng Quan *et al.*, 2009).

In the Palaeozoic, the leaf physiognomy of Cathaysian gigantopterids has been used in palaeoclimatic interpretation. Specifically, LMA and CLAMP have been used.

In the Mesozoic, in the (Late Permian–) Triassic, plant macrofossil distributions point toward the development of forests at high palaeolatitudes, even in Antarctica, at 70–75° S (Taylor & Ryberg, 2007). The forests appear to have been associated with a time of at least local if not global warming, resulting in a prolonged growing season, as indicated by the occurrence of the frost-sensitive cycad *Antarcticycas*, and by tree-ring analysis, respectively. Interestingly, during this evident climatic amelioration, the *Glossopteris* flora of the Carboniferous–Permian came to be replaced by a – seasonally deciduous – *Dicroidium* flora. In the Late Jurassic–Early Cretaceous, the distribution of the evergreen? conifer *Metapodocarpoxylon libanoticum*, interpreted as a canopy forest species, points toward the development of a seasonally (summer) wet tropical forest over much of northern Gondwana, from Colombia and Peru in the west through north Africa in the centre to Lebanon in the east (Philippe *et al.*, 2003). In the latest Cretaceous, plant macrofossils distributions point toward the development of forests apparently at extraordinarily high palaeolatitudes, even in Arctic Alaska and Canada, at 75–85° N. These ecosystems evidently thrived under light and temperatures regimes that do not exist in the present world, such that they cannot be interpreted solely on the basis of the principle of uniformitarianism. Also in the Cretaceous, CLAMP has been used in palaeoclimatic interpretation. The analytical results are generally in good agreement with those derived from computer modelling. However, serious discrepancies exist in continental interiors, where the computer models predict lower mean annual temperatures. The cause of this mismatch is not understood, although it has been hypothesised that the flora of the time was somehow capable of buffering the extreme effects of the continental climate.

In the Cenozoic, in the Palaeogene, plant microfossils have been used in the palaeoclimatic interpretation of the Eocene of northern Tanzania. The vegetation is interpreted as having been dominated not by tropical rainforest but rather by caesalpinioid legume woodland, similar to the modern miombo. The mean annual precipitation (MAP) is interpreted as having been in the range 600–800 mm, similar to that of today. Most of the precipitation is interpreted as having occurred in the wet season, although a small but significant proportion is interpreted as having occurred in the dry season, indicating a more equable distribution than that of today. Plant microfossils have also been used in the palaeoclimatic interpretation of the Eocene/Oligocene transition of the Gulf coast of the United States, and plant macrofossils in the palaeoclimatic interpretation of the Eocene/Oligocene transition of the Pacific northwest. In the Neogene, plant microfossils have also used in the palaeoclimatic interpretation of the (Late Oligocene–) Miocene of Bulgaria (Ivanov *et al.*, 2002, 2007). The coexistence approach (CoA) has been used to infer a mean annual temperature (MAT) of 16–18 °C and a MAP

of 1100–1300 mm for the Middle Miocene, Badenian to middle Sarmatian (Bessarabian), and a significantly lower MAT of 13.3–17 °C and a MAP of 652–759 mm for the Late Miocene, upper Sarmatian (Chersonian). Plant macrofossils have been used in the palaeoclimatic interpretation of the Miocene Shanwang, Luhe and Xiaolongtan floras of China (Jing-Xian Xu *et al.*, 2008; Ke Xia *et al.*, 2009). Specifically, LMA, CLAMP and CoA have been used. Plant macrofossils have also been used in the palaeoclimatic interpretation of the Miocene–Pliocene of Australasia. In New Zealand, the Early Miocene, characterised by the southern beech, *Nothofagus*, has been interpreted as having been ever-wet and cool, but frost-free. The Middle Miocene, characterised initially by eucalypts and palms, and then by casuarinaceans, chenopods and asteraceans, has been interpreted as having been initially seasonally dry, and subject to bush fires, and then dry. The Late Miocene, characterised by *Nothofagus* and *Sphagnum*, has been interpreted as having been wet and cool again. In southeast Australia, the Pliocene has been interpreted as having been dry. The climatic changes have been interpreted as having been caused by oscillations in the positions of the subtropical high pressure system or subtropical front.

The exacting ecological requirements and tolerances of many plant species and associated pollen grains and phytoliths, and their rapid response to changing environmental, and especially climatic, conditions, render them useful in palaeoclimatology and climatostratigraphy, and in environmental archaeology (Court-Picon *et al.*, 2005; Piperno, 2006; Marsh & Cohen, 2008; Lopez-Saez & Dominguez-Rodrigo, 2009; Piperno, 2009; Leroy *et al.*, 2010; Messager *et al.*, 2010). Note in this context, though, that there are limitations on the climatic information provided by phytoliths (Webb & Longstaffe, 2010).

3.1.5 Fungi

Fungal spores and hyphae

Fossil fungal remains are interpreted, on the basis of analogy with living fungi, as having been terrestrial (Taylor & Krings, 2010). The occurrence of fungal remains is often indicative of the development of a palaeosol. It is worth noting in this context that outcrop samples can be contaminated by fungi living in the soil at the surface. The super-abundant occurrence of fungal remains may be indicative of the existence of significant amounts of dead vegetation, and thus of an ecological catastrophe. Note in this context that there was a pronounced 'fungal spike' following the end-Permian mass extinction.

3.1.6   Invertebrate animals
'Ediacarans'

'Ediacarans' are interpreted, essentially on the basis of associated fossils and sedimentary facies, as having been marine. They are also interpreted as having lived in shallow marine environments, some as part of the sessile (?or motile) epifaunal benthos, as 'mat stickers', 'flat recliners' or 'erect elevators', like living sea-pens; others as part of the infaunal benthos; still others, the medusioids, as part of the pelagos, like living jellyfish (Seilacher, 2007). 'Ediacarans' are interpreted as having adopted a number of feeding strategies, including, in the case of flattened, segmented 'bilaterians' such as *Dickinsonia* (often observed in association with serial resting traces), absorbing nutrients directly from a decomposing substrate through a pedal tegument or 'digestive foot' (Gehling *et al.*, in Briggs, 2005). It has been postulated, and widely accepted, that they were similarly able to obtain oxygen directly from sea-water through diffusion (which process would have been facilitated by their high surface area to volume ratio), thereby obviating any need for complex respiratory as well as digestive organs. It has also been postulated, although not widely accepted, that they hosted photosynthetic Algae, maintaining an autotrophic existence in a so-called 'garden of Ediacara'. There do not appear to have been many, if any, scavenging or predatory 'ediacarans' (although many modern turbellarian flatworms, to which smooth bilaterians exhibit certain similarities, do exhibit these feeding behaviours).

Palaeobiogeography. As a group, 'ediacarans' appear to have had an essentially cosmopolitan palaeobiogeographic distribution, but the bilaterally symmetrical sub-group appears to have had a provincial or endemic distribution, in equatorial palaeolatitudes. Parsimony analysis of endemism (PAE) points toward separate 'ediacaran' realms centred on Avalonia, the White Sea and Namibia.

'Small shelly fossils' (SSFs)

SSFs are interpreted, essentially on the basis of comparison with associated fossils and sedimentary facies, as having been shallow marine. They are further interpreted on the basis of hard- and soft- part functional morphology as having lived as part of the sessile (?or motile) epifaunal or semi-infaunal benthos, and, at least in the case of the tubular morphotypes such as *Cloudina*, *Anabarites* and hyolithans, as having fed by passive filter-feeding. Orthothecimorph hyolithans are interpreted on the basis of empirical observation as having lived in an upright position; hyolithomorph hyolithans on the basis of experimental observation as having lived in a prone position, orienting themselves into currents by means of keels in order to feed. *Cloudina* has been observed to have been attacked by boring parasites or predators, leading to the suggestion that its mineralised skeleton evolved as a measure to counter such attack.

Palaeobiogeography. As a group, SSFs appear to have had an essentially cosmopolitan palaeobiogeographic distribution, although *Cloudina* appears to have been essentially endemic to proto-Gondwana, and *Anabarites* to proto-Laurasia.

Sponges

Fossil sponges are interpreted, essentially on the basis of analogy with their living counterparts, as having been mostly shallow marine, and as having been active filter-feeders, although fossil freshwater sponges have also been described, from the Late Miocene Quiallaga formation of Chile (Pisera & Saez, 2003). Cup-shaped and discoidal morphotypes colonised soft substrates, and often built substantial structures. Flattened morphotypes encrusted hard substrates.

Palaeobiogeography. Commonality between assemblages in Spain and in North America in the Carboniferous indicates a marine connection between the western Palaeo-Tethys Sea and the eastern Panthalassan Ocean in the Moscovian (Garcia-Bellido & Rodriguez, 2005). The marine connection, the so-called 'Iberian-Midcontinent Seaway', was closed in the Kasimovian.

Archaeocyathans

Archaeocyathans are interpreted, essentially on the basis of associated fossils and sedimentary facies, as having been shallow marine. They are further interpreted, essentially on the basis of functional morphology, as having lived embedded in soft substrates, in the case of the ajacicyathids, or held fast on hard substrates, in the case of the archaeocyathids, thus forming part of the sessile benthos, and as having fed by filter-feeding.

Palaeogeography. Archaeocyathans as a group appear to have had an essentially cosmopolitan palaeobiogeographic distribution, although only in low palaeolatitudes.

Stromatoporoids

Stromatoporoids are interpreted, on the basis of associated fossils and sedimentary facies, as having been mostly shallow marine. They are also interpreted as having been active filter-feeders, creating their own currents, the better to utilise dispersed resources.

Corals

Fossil corals are interpreted, essentially on the bases of analogy with their living counterparts, functional morphology, and associated fossils and sedimentary facies, generally as having harboured algal photosymbionts and as having lived, as part of the sessile epifaunal benthos, in the photic zone, in shallow, and warm, seas (Fig. 3.4).

Fossil rugose corals are interpreted, on the basis of functional morphology, as having lived in a variety of positions. The most prevalent appears to have been ambitopic, or recumbent on the seafloor, concave side upward, with growth essentially vertical, giving rise to the stepped or roughened appearance that gives the group its name. Interestingly, the rugose coral *Hamarophyllum belkai* occurs in monospecific assemblages associated with interpreted ancient hydrothermal vents in the Devonian, Emsian of Morocco (Berkowski, 2004).

Palaeobathymetry. Fossil corals are interpreted on the basis of analogy with their living counterparts, as having been generally shallow marine. They are further interpreted on the basis of functional morphology, specifically on the degree of integration, as either having housed algal photosymbionts, or not, and thus as having been restricted to the photic zone, or not (Tapanila, 2004). Many fossil corals appear to have had restricted bathymetric distributions within the photic zone, rendering

them of use in the characterisation of palaeobathymetric zones (Hongo & Kayanne, 2010). Depth zonation and/or depth-related morphological trends may also be evident.

Practically all modern coral reefs are found in depths of less than 100–150 m, and temperatures of more than 16–17 °C (Sheppard *et al.*, 2009). This is essentially on account of the partial reliance of reef-constructing, or hermatypic, corals on food produced by photosymbionts, which can only live in the photic zone. Note in this context that, strictly speaking, the distributions of photosymbiont-bearing corals are controlled primarily by light availability, and hence turbidity and turbulence, and only secondarily and indirectly by water depth.

Some coral reefs, chiefly constructed by *Lophelia*, are found in cold, deep marine environments (Freiwald & Roberts, 2005; Dorschel *et al.*, 2007; Sherwood & Risk, in Hillaire-Marcel & de Vernal, 2007; Mortensen *et al.*, 2008; Mienis *et al.*, 2009; Roberts *et al.*, 2009; Wienberg *et al.*, 2009). Interestingly, there is some evidence to indicate that *Lophelia* is associated with cold seeps (Coles *et al.*, 1996; Hovland & Thomsen, 1997). In the Porcupine area off the west coast of Ireland in the North Atlantic, geophysical or seismic profiles have revealed the existence of a number of reefs up to 200 m high and 2000 m across that have been demonstrated by dredging, gravity coring and underwater photography to be chiefly constructed by *Lophelia pertusa*. The *Lophelia* reefs here have in turn been demonstrated by the results of geochemical analysis to be characterised by higher than average interstitial hydrocarbon concentrations, interpreted to be related to seepage associated with underlying geological faults. However, it has not yet been demonstrated that *Lophelia*, or the associated fauna, including other corals such as *Desmophyllum cristagalli*, is actually metabolising the seeping hydrocarbon.

Palaeobiogeography. In the Palaeozoic, coral distributions point toward the palaeogeographic evolution of Palaeo-Tethys and Panthalassa (Garcia-Bellido & Rodriguez, 2005). Commonality data indicates that the open connection between the marine areas, through the 'Iberian–Midcontinent Seaway', was closed by the Kasimovian. In the Cenozoic, coral distributions point toward the palaeogeographic and palaeoclimatic evolution of the

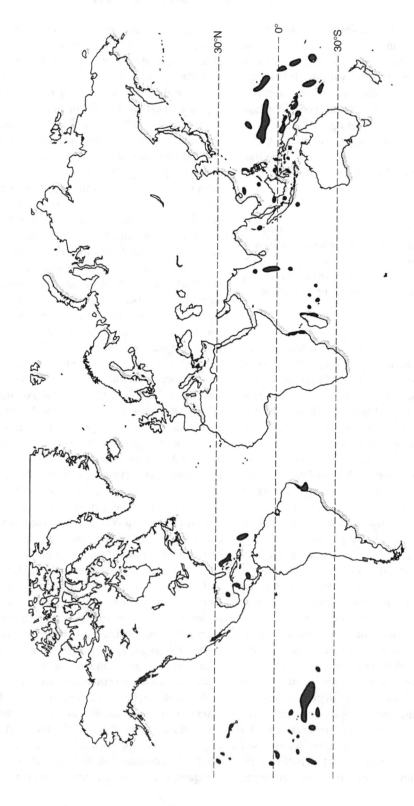

Fig. 3.4 **Biogeographic distribution of living corals.** © Wiley. Reproduced with permission from Milsom and Rigby (2004).

(Neo-)Tethyan and Paratethyan Realms (Chaix & Cahuzac, 2005).

Palaeoclimatology. Corals and/or coral isotopic records are of use in palaeoclimatic as well as in palaeobiogeographic interpretation (Pfeiffer *et al.*, 2004; Chaix & Cahuzac, 2005; Yu *et al.*, 2005; Correge, 2006; Bosellini & Perrin, 2008). For example, *Cladophora caespitosa* has been used in the palaeoclimatic interpretation of the Pleistocene of the Mediterranean. The $\delta^{18}O$, Sr/Ca and Mg/Ca isotopic records of *Porites* spp., including *P. lutea*, have been used in the palaeoclimatic interpretation of the Holocene of the South China Sea (Yu *et al.*, 2005), and of the last 120 years of the Chagos Archipelago in the Indian Ocean, the latter evidently driven by variability in the location of the Intertropical Convergence Zone and El Niño–Southern Oscillation monsoon system (Pfeiffer *et al.*, 2004). Scherochronology shows highest mean annual growth rates in warm phases. Coral diversity has been used as a measure of temperature, and, by inference, palaeotemperature (Fraser & Currie, 1996; Rosen, in Agusti *et al.*, 1999). However, it is also influenced by other factors, including depth and also biogeography, and distance to, and dispersal from, loci of evolution (see above). Coral diversity is higher in the Indo-Pacific than in the Atlantic/Caribbean, and is highest in, and decreases in proportion to increased distance from, the interpreted locus of evolution in southeast Asia.

Brachiopods

Fossil brachiopods are interpreted, essentially on the basis of analogy with their living counterparts, as having been marine, and on the bases of associated fossils and sedimentary structures, as having been more typically shallow marine than their living counterparts. Some smooth forms from the Jurassic of the Alpine-Mediterranean region – *Lingulithyris*, *Securithyris*, *Apringia*, *Bakonyithyris* and *Pisirhyncia* – are interpreted on the basis of analogy with their living counterparts – Basiliolidae, Dysocoliidae and Dallinidae – as having been deep marine, or 'thalassobathyal', in the Early Jurassic, although they also appear to have colonised shallow marine environments by the Middle Jurassic (Voros, 2005).

Fossil brachiopods are further interpreted, essentially on the basis of functional morphology, as having lived principally as part of the sessile epifaunal or semi-infaunal filter-feeding benthos (Rudwick, 1961, 1964; Alexander, in Savazzi, 1999; McGhee, in Savazzi, 1999). (Rudwick's work elegantly demonstrated the function of the often-observed zigzag morphology of the commissure, which effectively increases the intake area, while maintaining the same gape and particle-size restriction.) In the Late Carboniferous, Pennsylvanian of the Great Basin of Nevada and Utah in the United States, productides tended to dominate brachiopod associations, and attain large size, in deep-water, low-nutrient settings, on account of their lophophore geometry and interpreted ability to generate multi-directional inhalant currents (Perez-Huerta & Sheldon, 2006). In contrast, athyridides and spiriferides tended to dominate, and attain large size, in shallow-water, high-nutrient settings.

Interestingly, certain brachiopods are associated with interpreted ancient hydrothermal vents and cold seeps (Campbell & Bottjer, 1995; Gischler *et al.*, 2003; Kaim *et al.*, 2010).

Palaeobathymetry. Many fossil brachiopods appear to have had restricted bathymetric distributions, rendering them of use in the characterisation of palaeobathymetric zones. For example, in the Early Silurian, Llandoverian of the margins of the Anglo-Welsh basin, *Lingula*, *Eocoelia*, *Pentamerus*, *Costistricklandia* and *Clorinda* characterise successive benthic assemblage zones from shallowest to deepest. Here, the shallowest, *Lingula*, zone is interpreted as intertidal, and the deepest, '*Stricklandia*', zone as subtidal, below the lower limit for photosynthesis and reef-building activity. In the Carboniferous, Chesterian of the Antler foreland basin of Idaho, spiriferides and athyridides characterise environments between fair-weather- and storm-wave-base; productoids and orthotetoids, and *Orbiculoidea wyomingensis*, environments below storm-wave-base (Butts, 2005). Also here, bathymetric interpretations based on brachiopod assemblages indicate the initiation of high amplitude and frequency sea-level changes at the base of the *Adentognathus unicornis* Conodont Zone (319 Ma). These sea-level changes are in turn interpreted as associated with the onset of Gondwana glaciation.

Palaeobiogeography. In the Palaeozoic, the cold-adapted *Hirnantia* brachiopod fauna appears to have had an unrestricted, pandemic or cosmopolitan distribution during the Late Ordovician, late Ashgillian, Hirnantian glaciation; and warm-adapted taxa essentially cosmopolitan distributions during the succeeding Early Silurian, Llandoverian deglaciation (Sutcliffe *et al.*, 2001; Rong Jia-Yu *et al.*, 2002). The cold-adapted opportunistic r-strategist *Pachycyrtella omanensis* was the dominant taxon in the early stages of the Permian deglaciation in Oman. The *Pachycyrtella omanensis* community was succeeded by a more diverse and stable warm-adapted *Reedoconcha permixta–Punctocyrtella spinosa* community, also known from India, Afghanistan, the Himalayas and Thailand (the so-called 'Westralian province', possibly part of a still larger 'Indoralian province', also including eastern Australia), in the later stages of the deglaciation. It is possible that this apparent ecological succession may actually have been prompted by a 'catastrophic physical perturbation' that wiped out the *Pachycyrtella omanensis* community.

Bryozoans

Fossil bryozoans are interpreted, on the bases of analogy with their living counterparts, functional morphology and associated fossils and sedimentary facies, as having been mostly shallow marine (Taylor, in Savazzi, 1999). However, some appear to have been associated with deep marine vents, as in the Carboniferous of Newfoundland.

Palaeoclimatology. The inverse relationship between cheilostome zooid size and ambient water temperature has been used to infer palaeotemperature and seasonality in the Pliocene of the North Atlantic (Knowles *et al.*, 2009).

Bivalves

Fossil bivalves are interpreted, essentially on the bases of analogy with their living counterparts, and associated fossils and sedimentary facies, as having been freshwater, brackish or marine. Freshwater examples include *Unio* and allied forms from the Carboniferous Coal Measures of Great Britain, and *Pseudunio* and allied forms from the Early Cretaceous 'Wealden' facies of the Weald and Wessex basins in the southeast of England. Marine examples associated with interpreted fossil seeps include lucinids from the Late Jurassic of southeastern France and Late Cretaceous of Colorado in the

Infaunal shallow burrowers *Glycimeris*	equivaled, adductor muscles of equal sizes and commonly with strong external ornament.
Infaunal deep burrowers *Mya*	elongated valves, often lacking teeth and with permanent gape and a marked pallial sinus.
Epifaunal with byssus *Mytilus*	elongate valves with flat ventral surface and reduction of both the anterior part of the valve and the anterior muscle scar. Attached by thread-like byssus.
Epifaunal with cementation *Ostrea*	markedly differently shaped valves, sometimes with crenulated commissures; large single adductor muscle.
Unattached recumbents *Gryphaea*	markedly differently shaped valves sometimes with spines for anchorage or to prevent submergence in soft sediment.
Swimmers *Pecten*	valves dissimilar in shape and size with very large, single adductor muscle and commonly with hinge line extended as ears.
Borers and cavity dwellers *Teredo*	elongate, cylindrical shells with strong, sharp external ornament; cavity dwellers commonly grow in dimly lit conditions following the contours of the cavity.

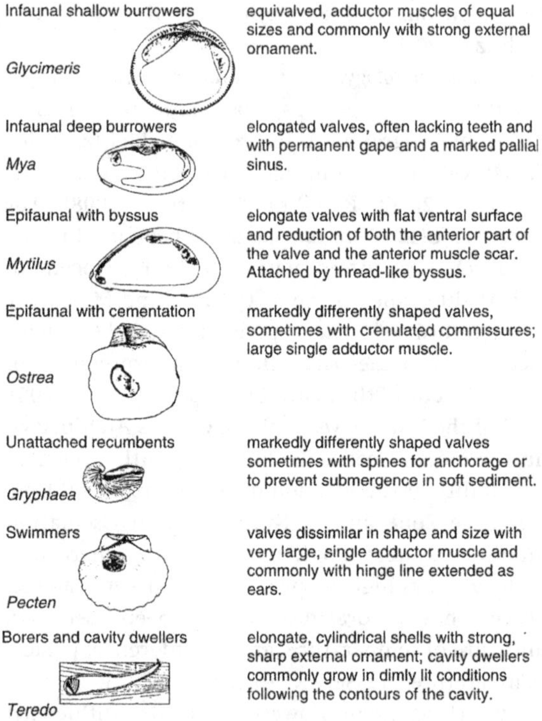

Fig. 3.5 **Functional morphology of bivalves**. From Jones (2006) (reproduced with the permission of Longman from Benton & Harper, 1997).

United States, *Cryptolucina* and allied forms from the Early Cretaceous of northeast Greenland, and '*Bathymodiolus*-like' seep-mussels, vesicomyids, lucinids, thyasirids, solemyids and nuculanids from the Palaeogene and Neogene of Barbados, Trinidad and Venezuela (Gill *et al.*, 2005; Jones, 2009).

Fossil bivalves are further interpreted, essentially on the basis of functional morphology, as having been either benthic or nekto-benthic; benthic forms as either free-living or motile, or byssally attached or cemented, and sessile. In detail, seven main life strategies are adopted by living or are inferred to have been adopted by ancient bivalves, namely: shallow infaunal benthic; deep infaunal benthic; byssally attached sessile epifaunal benthic; cemented sessile epifaunal benthic; motile epifaunal benthic; nekto-benthic; and boring or cryptic (Stanley, 1970; Seilacher, 1984; Fig. 3.5).

Each life strategy is represented by a particular functional morphological adaptation.

Palaeobathymetry. Fossil bivalves can be used as indicators of palaeobathymetry, or proximity to shoreline, in marine environments. It is important to note, though, that certain living bivalves appear to have their distributions controlled less by bathymetry than by other factors, such as temperature. Thus, *Venericardia borealis* occurs in shallow water in high latitudes and in deep water in lower latitudes (in the northwest Atlantic), indicating that it is a stenothermal, eurybathyal species.

Palaeobiogeography. In the Palaeozoic, specifically in the Early Ordovician, bivalves appear to have been endemic to Gondwana. In the Middle Ordovician, bivalves as a whole were more widespread, although sub-groups continued to exhibit endemic or provincial distributions, for example pteriomorphs and nuculoids in low latitudes, and heteroconchs to mid to high latitudes. In the Late Ordovician, bivalves as a whole exhibited cosmopolitan distributions, encompassing for the first time the carbonate shelf facies of Laurentia and Baltica, although the pteriomorphs continued to remain restricted to low latitudes.

In the Mesozoic, bivalve distributions can be used to characterise Tethyan and Boreal realms, Boreal–Atlantic, Boreal–Pacific, and Arctic provinces within the Boreal realm, and Chukotka–Canada and Greenland–North Siberia sub-provinces within the Arctic province. The Triassic bivalve *Claraia* and the large, aberrant Early Jurassic, Pliensbachian–Toarcian bivalve *Lithiotis* and the allied forms *Cochlearites*, *Lithioperna*, *Gervilleioperna* and *Mytiloperna* exhibited essentially Pan-Tethyan distributions, from Oregon and California states on the west coast of North America in the west, through southern Europe, north Africa and the Middle East, to East Timor in the east.

In the Cenozoic, marine bivalve distributions indicate the development of three palaeobiogeographic provinces in the northeast Pacific, namely the Gulf of Alaska province to the north, the Pacific Northwest province centred on British Columbia and Vancouver in Canada and Washington State and Oregon in the conterminous United States, and the California province to the south. The dispersal of the bivalve *Astarte* from the Arctic into the North Pacific between 4.8–5.5 Ma implies that the Bering Strait had opened by this time.

Marginal- to non-marine bivalve distributions point toward the palaeogeographic evolution of northern South America (Wesselingh & Macsotay, 2006). The marginal- to non-marine bivalves of the Middle Miocene Chaguaramas formation of eastern Venezuela, including *Pachydon hettneri*, are of Orinoco basin affinity, indicating that the palaeo-Orinoco had begun to discharge into the Atlantic by this time.

Palaeoclimatology. The exacting ecological requirements and tolerances of many bivalve species, and their rapid response to changing environmental and climatic conditions, render them useful in palaeoclimatology and climatostratigraphy, and in environmental archaeology.

In the Palaeozoic, the *Dickinsartella* bivalve fauna is associated with the cold-adapted *Pachycyrtella* brachiopod fauna of the Saiwan formation of Oman, interpreted as representing an early stage of the deglaciation following the Permian glaciation of Gondwana (Larghi, 2005).

In the Cenozoic, bivalve distributions point toward the palaeogeographic and palaeoclimatic evolution of the Tethyan and Paratethyan realms.

Transfer functions based on the environmental distributions and tolerances of terrestrial bivalves are of considerable use in palaeoclimatic interpretation (Moine & Rousseau, 2002).

Gastropods

Fossil gastropods are interpreted, essentially on the bases of analogy with their living counterparts, and associated fossils and sedimentary facies, as having occupied a range of aquatic and terrestrial environments (Pickford, 1995; Kerney, 1999; Pickford, 2004; Davies, 2008). Thusly interpreted freshwater gastropods include *Viviparus* (the 'river snail') from the Pleistocene of the Swanscombe archaeological site in Kent, and elsewhere. Terrestrial gastropods or land-snails, which, importantly, provide evidence as to substrate, ground cover and disturbance, include *Spermodea*, *Acanthinula*, *Aegopinella*, *Pupilla* and *Vallonia* from the Pleistocene of the Boxgrove archaeological site in Sussex, and elsewhere. Incidentally, calcitic plates attributed to the slug *Deroceras/Limax* also occur at the Boxgrove archaeological site. Fossil gastropods with complete apertural margins are interpreted on the basis of analogy as having been herbivorous; those with siphonal canals as either actively herbivorous or carnivorous; those with wide apertures as having been carnivorous, and as having swallowed their prey whole.

Thick-shelled, patelliform or low-spired, fossil morphotypes are interpreted, on the basis of functional morphology, as indicating high-energy epifaunal environments. Thin-shelled morphotypes are similarly interpreted as indicating low-energy, often freshwater, epifaunal environments; high-spired morphotypes, infaunal environments.

Some fossil gastropods are associated with interpreted ancient cold seeps, for example in the Jurassic and Cretaceous of California, the Cretaceous of Japan, and the Palaeogene and Neogene of Barbados, Trinidad and Venezuela (Gill *et al.*, 2005; Kiel & Campbell, 2005; Kiel *et al.*, 2008; Kaim *et al.*, 2009). These include trochids, turbinids, provannids, abyssochrysids, fissurellids, acmaeids, neomphaloids, seguenzioids and hokkaidoconchids.

Palaeobiogeography. In the Palaeozoic, three palaeobiogeographic realms are characterised by gastropods, namely the Old World, Eastern Americas and Malvinokaffric.

In the Mesozoic, the Tethyan realm is characterised by nerineoidean, acteonellid, pseudomelaniid, cassiopid, campanilid, strombid, cypraeid and chilodont gastropods.

In the Cenozoic, gastropod distributions point toward the opening of the Bering Strait in the Pliocene, and its closure in the Pleistocene, resulting in the evolution of separate populations in the North Atlantic and North Pacific; and also the closure of the Strait of Panama in the mid-Pliocene, resulting in the evolution of separate populations in the eastern Atlantic and western Pacific.

Palaeoclimatology. In the Cenozoic, gastropod distributions point toward the palaeogeographic and palaeoclimatic evolution of the Tethyan and Paratethyan realms. Gastropod distributions and turnovers also point toward the palaeoclimatic evolution of the North Pacific over the critical transition from Eocene to Oligocene. Palaeocene to Middle Eocene, unnamed to Tejon stage gastropod faunas of the west coast of the United States are of cosmopolitan, warm-water aspect, especially in the late Early to early Middle Eocene, Capay and Domengine stages, coincident with the global climatic optimum; Late Eocene, Galvinian stage faunas, including, interestingly, a number of species indicative of 'cold-seep' and 'whale-fall' habitats, of cooler- and/or deeper-water aspect; and Early Oligocene, Matlockian stage, and Late Oligocene, Juanian stage, faunas of cool-water aspect. Moreover, Middle Eocene gastropod faunas from the Katalla district of the Gulf of Alaska are of cosmopolitan, warm-water aspect; Late Eocene faunas of mixed cosmopolitan and North Pacific temperate-water aspect; Early Oligocene faunas of mixed cosmopolitan and North Pacific temperate- and cold-water aspect; and Late Oligocene assemblages of North Pacific cold-water aspect. Furthermore, Middle Eocene faunas from the Tigil and Palana districts of the Kamchatka Peninsula on the other side of the Bering Sea in the far east of the former Soviet Union are of mixed cosmopolitan and North Pacific temperate-water aspect; Late Eocene and Early Oligocene faunas of mixed cosmopolitan and North Pacific temperate- and cold-water aspect.

The exacting ecological requirements and tolerances of many gastropod species, and their rapid response to changing environmental and climatic conditions, render them useful in palaeoclimatology and climatostratigraphy, and in environmental archaeology (Pickford, 1995; Kerney, 1999; Pickford, 1995, 2004; Davies, 2008). Transfer functions based on the environmental distributions and tolerances of terrestrial gastropods are of considerable use in palaeoclimatic interpretation (Moine & Rousseau, 2002). The isotopic compositions of terrestrial gastropod shells have also been used in palaeoclimatic interpretation (Zanchetta *et al.*, 2005). In areas where ground-water is related to rainfall, they can and have been used in the reconstruction of rainfall.

Ammonoids

Ammonoids are interpreted, on the bases of analogy with their living counterparts, functional morphology, and associated fossils and sedimentary facies, as having been at least marginal marine – with at least some ammonites, such as *Placenticeras*, demonstrated on oxygen isotope evidence to have been able to tolerate brachyhaline or slightly brackish waters – and nektonic or nekto-benthic (Tsujita &

Westermann, 1998; Westermann & Tsujita, in Savazzi, 1999; Norman, 2003). Note that it has been argued that the more appropriate analogues for the ammonoids are not the nautiloids, which are essentially intermediate-water or mesopelagic at the present time, but the coleoids, which are essentially, although not exclusively, shallow-water or epipelagic at the present time (Hanlon & Messenger, 1996; Norman, 2003; Boyle & Rodhouse, 2005). Incidentally, at least according to one widely accepted hypothesis, nautiloids evolved in shallow-water environments, but become forced into intermediate-water environments by predation pressure exerted by epipelagic fish and marine reptiles (Packard, 1972). Coleoids then evolved from nautiloids in intermediate-water environments, and from there colonised not only shallow-water but also deep-water environments. Nautiloids appear never to have colonised deep-water, bathypelagic environments, possibly because of limitations imposed by having chambered shells partially filled with low-pressure gas. Interestingly, they are able to use this gas for emergency respiration in dysoxic environments (Ward, 2006).

Ammonites are interpreted to have fed on plankton or slow-swimming prey in the water column, in the case of nektonic forms, or on slow-moving prey – such as foraminifera, ostracods and other crustaceans, bryozoans, corals and brachiopods – on the sea-floor, in the case of nekto-benthic forms. Some forms may have been detrital scavengers, still others active predators on other ammonites. They are interpreted as having been fed on by mosasaurs and other, unidentified predators (Tsujita & Westermann, 1998). Epipelagic *Placenticeras* and *Baculites* exhibit the most, and 'demersal' *Hoploscaphites* and *Jeletskytes* the least, evidence of mosasaur bite marks, perhaps suggesting that mosasaurs preferred to hunt in the upper levels of the water column, where visibility was better. The abundance of ammonites has been used as a measure of carbonate dilution or palaeoproductivity in the Early Cretaceous, Valanginian of the Vocontian basin in southeast France.

Palaeobathymetry. In the Palaeozoic, in the Carboniferous, the goniatite *Phillipsoceras* was apparently shallow epipelagic, living in the surface layer of the water column; and *Anthracacoceras, Ramosites,* *Glaphyrites, Eumorphoceras* and *Prolecanites* deep epipelagic or mesopelagic, living at depth within the water column (Westermann & Tsujita, in Savazzi, 1999).

In the Mesozoic, in the Jurassic–Cretaceous, the ammonites *Haploceras* and *Dactylioceras* were apparently shallow epipelagic, *Stephanoceras, Baculites, Ptychites, Macrocephalites, Desmoceras, Ancyloceras, Psiloceras, Phylloceras* and *Pseudoxybeloceras* deep epipelagic, living to depths of approximately 200 m within the water column; and *Lytoceras* mesopelagic, living to depths of approximately 500 m (Westermann & Tsujita, in Savazzi, 1999). In the Late Cretaceous of the 'Western Interior Seaway' of the United States, *Placenticeras*, with compressed, streamlined 'oxyconic' shells, and adult *Baculites ex gr. compressus*, with elongate 'orthoconic' shells, were apparently shallow epipelagic; juvenile *Baculites ex gr. compressus* and adult *B. ovatus*, with truncate 'orthoconic' shells, deep epipelagic; *Rhaeboceras*, with comparatively inflated, 'elliptosphaerocone' body chambers, mesopelagic; and *Hoploscaphites* and *Jeletskytes*, with compressed, 'scaphitoconic' body chambers, 'demersal', living close to the sea-floor (Tsujita & Westermann, 1998). Juvenile *Baculites ex gr. compressus* and adult *B. ovatus* are conventionally interpreted as having been oriented vertically – downwards – within the water column, which would have been the most hydrodynamically stable position for organisms with wide separations between the centres of mass, at the distal ends of the shells, where the organisms lived, and buoyancy, at the proximal ends, which were filled with air. In contrast, adult *Baculites ex gr. compressus* is interpreted as having been oriented horizontally within the water column, counterbalancing the mass of the organism at the distal end of the shell, and maintaining neutral buoyancy, by replacing the air at the proximal end with 'cameral liquid'. *Baculites ex gr. compressus* evidently underwent a change in orientation within the water column, from vertical to horizontal, within its life cycle. In the Late Cretaceous of Tunisia, the Scaphitidae appear to have occupied shallow marine environments, and the Desmocerataceae and Tetragonitiaceae deep marine environments.

Palaeobiogeography. In the Mesozoic, ammonite distributions can be used to characterise low-latitude Tethyan, and high-latitude southern, or

Austral, and northern, or Boreal, realms. Ammonites are typically characteristic of only one of these realms, for example the phylloceratids of the Boreal realm. Note, though, that the lytoceratids are essentially cosmopolitan. Ammonite distributions can also be used to characterise Boreal–Atlantic, Boreal–Pacific, and Arctic provinces within the Boreal realm, the Arctic province also encompassing the Chukotka–Canada sub-province, characterised by an essentially endemic fauna. Importantly, connection and dispersal between these provinces, and indeed between the Boreal and sub-Boreal/Tethyan realms, was facilitated by marine transgressions. Ammonite distributions point towards the existence of such connections intermittently through the late Middle to Late Jurassic, Callovian, Oxfordian and Kimmeridgian, and in the *Kilmovi, Panderi*, early and late *Nikitini* and *Nodiger* Zones of the latest Jurassic, Volgian. Ammonite distributions also point towards the existence of a connection between the Boreal and sub-Boreal/Tethyan realms in the form of a longitudinal strait in the area west of the Urals in the Early Cretaceous, generally characterised by southward flow of Boreal waters in the Neocomian and by northward flow of Tethyan waters in the Apto-Albian; of a disconnection in the Albo-Cenomanian; and of a re-connection in the area of Turgai in the Turonian. The Turgai Strait connected the Peri-Tethyan basins, the West Siberian Boreal basins, and the so-called 'Western Interior Seaway' of the United States. The distributions of different species of *Libycoceras* point toward the existence of a 'Trans-Saharan Seaway' connecting west and north Africa in the Late Cretaceous, Campanian–Maastrichtian (Zaborski & Morris, 1999). The connection was east of the Hoggar massif in southern Algeria in the Campanian, and west of the Hoggar massif in the Maastrichtian.

Belemnites

Belemnites are interpreted, on the bases of analogy with their living counterparts, functional morphology, and associated fossils and sedimentary facies, as having been marine and nektonic, and as having occupied a range of pelagic environments (Hanlon & Messenger, 1996; Doyle, in Savazzi, 1999; Norman, 2003; Boyle & Rodhouse, 2005). They are further interpreted as having maintained neutral

buoyancy by means of some form of chemical or dynamic lift (Packard, 1972).

Palaeobathymetry. Belemnite rostra commonly exhibit preferred orientation, enabling inference as to the activity and direction of palaeo-currents, not necessarily always in shallow water. Mass accumulations, or 'belemnite battlefields', may be related to mass mortality events taking place after mating, as observed in living coleoids; to other catastrophic events; to concentration by predation, either by regurgitation or by the death of the predator; or to concentration by sedimentological processes, such as winnowing in shallow water, and gravity flow in deep water. Three belemnite assemblages have recently been recognised in the Early Cretaceous of Morocco that have been interpreted as indicative of various bathymetric zones: a *Hibolithes–Duvalia* assemblage indicative of the inner to middle shelf; a *Duvalia–Hibolithes* assemblage indicative of the outer shelf; and a *Duvalia–Pseudobelus* assemblage indicative of the slope to basin (Mutterlose & Wiedenroth, 2008).

Palaeobiogeography. Belemnites appear to have evolved in the Mediterranean region of Tethys, and in the Early and Middle Jurassic were essentially restricted to that region, with only isolated occurrences elsewhere. By the Late Jurassic, though, they had extended their range to cover much of the world. Nonetheless, even at this time, many belemnites had restricted or endemic biogeographic distributions, rendering them of use in the characterisation of palaeobiogeographic realms, and in turn in the constraint of plate tectonic reconstructions. Thus, at this time, the Boreal realm is characterised by the belemnitids and cylindroteuthids; and the Tethyan realm by the docoelitids, duvaliids, *Belemnopsis* and *Hibolites*. Moreover, Arctic and Boreal–Atlantic provinces are characterised within the Boreal realm; and Mediterranean, Ethiopian and Indo-Pacific provinces within the Tethyan realm. The occurrence of similar belemnite assemblages in the Middle–Late Jurassic, Callovian–Kimmeridgian of Madagascar, Antarctica, South America and New Zealand provides evidence of the existence at this time of a 'Trans-Erythrean Seaway' connecting the Ethiopian and Indo-Pacific provinces (Challinor & Hikuroa, 2007). In the Early Cretaceous, the Boreal realm is characterised by cylindroteuthids and newly evolved oxyteuthids; the Tethyan realm by duvaliids, *Belemnopsis, Hibolites* and the newly evolved *Curtohibolites, Mesohibolites, Neohibolites* and *Parahibolites*; and the Austral realm by dimitobelids (Mutterlose & Wiedenroth, 2008). Both characteristically Boreal cylindroteuthids and other taxa and Tethyan duvaliids and other taxa have recently been recorded in the Early Cretaceous of northeast Greenland, leading to the interpretation that the area at this time was situated at a 'crossroads of belemnite migration' (Alsen & Mutterlose, 2009). By the Late Cretaceous, the cylindroteuthids and oxyteuthids were extinct, and the (sub-)Boreal realm was characterised by the newly evolved *Actinocamax, Belemnella, Belemnitella* and *Gonioteuthis*.

Palaeoclimatology. Importantly, belemnites are of considerable use in palaeotemperature and palaeoclimatic as well as in palaeobiogeographic interpretation, palaeobathymetric complications notwithstanding (Price *et al.*, 2000; Price & Mutterlose, 2004; Rexfort & Mutterlose, 2009; Malkoc & Mutterlose, 2010; Nunn & Price, 2010; Nunn *et al.*, 2010; Price, 2010). This is on account of their enhanced resistance to chemical change during diagenesis, and their consequent ability to preserve intact geochemical signals such as oxygen – and strontium – isotope ratios. Indeed, the 'Pee Dee' belemnite standard is that against which such measurements are calibrated.

Tentaculitids

Tentaculitids have historically been interpreted, on the basis of associated fossils and sedimentary facies, as having been marine and nektonic. However, it is also possible to interpret them, on the basis of functional morphological analogy with the microconchids, as having been benthic, and either cemented to a hard substrate or vertically embedded in a soft substrate, with their feeding apparatuses extended into the overlying water column.

Palaeobiogeography. Tentaculitids had a wide geographic distribution in low to moderate latitudes.

Trilobites

Trilobites are interpreted, essentially on the bases of analogy with their living counterparts, namely horseshoe-crabs, and associated fossils and sedimentary facies, as having been at least marginal marine, and either benthic, nekto-benthic, or nektonic.

Trilobites are also interpreted, essentially on the basis of functional morphology, as having been benthic and infaunal, as in the case of the smooth 'illaenimorph' morphotypes, to epifaunal and motile, as in the case of the tuberculate 'phaco-morph' morphotypes, to pelagic (Fortey & Owens, in McNamara, 1990; Fortey & Owens, in Savazzi, 1999). In fact, a further five trilobite morphotypes have been recognised, namely 'atheloptic', 'marginal/cephalic spines', 'miniaturization', 'pitted fringe' and 'olenimorph'. Articulated *Asaphus (Asaphus) raniceps* body fossils have recently been discovered in *Thalassinoides* burrows.

Trilobites are further interpreted, again essentially on the basis of functional morphology, as having practised a variety of feeding strategies, including predation or scavenging, detritus-feeding, and filter-feeding (Babcock, in Kelley *et al.*, 2003). Interpreted predators or scavengers are characterised by claw-like spines at the end of their legs, rigidly attached conterminant or impedent hypostomes with posterior forks, and expanded anterior glabellar folds; and are exemplified by *Colpocoryphe, Dindymene, Illaenopsis* and *Ormathops*. Interpreted detritivores are characterised by detached hypostomes, which may have functioned as scoops; and are exemplified by *Shumardia* and other generalised ptychopariides (and, incidentally, by the *Cruziana* trace-maker). Interpreted filter-feeders are characterised by elevated heads and thoraces, often flanked by extended genal spines; and are exemplified by *Ampyx* and other generalised trinucleiids.

Many trilobites bore spines either as a defence against predators, or as an adaptation to reduce the sinking rate within the water column in the case of pelagic forms, or to maintain stability on, or to facilitate burrowing into, the sea floor, in the case of benthic forms (Babcock, in Kelley *et al.*, 2003).

Palaeobathymetry. In general, shallow marine benthic environments are characterised by low diversity assemblages of sighted, and typically large, trilobites, including *Flexicalymene, Merlinia* and *Neseuretus* (and the locomotion or feeding trace *Cruziana*), in association with Algae, crinoids and gastropods; intermediate-depth benthic environments by high diversity assemblages, including

Selenopeltis, Geragnostus, Chasmops and *Remopleurides*, in association with brachiopods and nautiloids; deep-marine environments by moderate diversity assemblages of blind, and typically small, forms including *Ampyx, Tretaspis, Shumardia* and trinucleiids; and pelagic environments by forms with downward-directed eyes, including *Cyclopyge* (Turvey, 2005). A general model of sighted trilobites in illuminated, shallow-water environments and blind ones in dark, deep-water environments was proposed by Clarkson, in Hallam (1967), and later slightly modified to take into account observations of blind trilobites in interpreted dark, shallow-water environments such as algal glades. The model is complicated by the further observations that blind and sighted species often occur together, in interpreted deep-water as well as shallow-water environments. Interestingly, similar observations have been made on Recent arthropods. Importantly, though, while sighted Recent arthropods from below 200 m have normal-looking eyes specially adapted to the perception of dim light (eyes, incidentally, rejoicing in the wonderful, Wagnerian name of 'Dammerungsaugen'), those from below 500 m have distinctly degenerate eyes, possibly sensitive to bioluminescence. The general model of trilobite distribution is further complicated by evident substrate as well as depth control. Taking this into account, specifically in the Early Ordovician of Laurentia, intertidal environments are characterised by bathyurids, shallow marine limestones by illaenids and cheirurids, deep marine limestones by nileids, and deep marine graptolitic shales by olenids In the Middle Silurian of Gondwana, marginal to shallow marine sandstones are characterised by *Acaste* and *Trimerus*, shallow marine limestones by *Proetus* and *Warburgella*, shallow marine shales by *Dalmanites* and *Raphiophorus*, deep marine limestones by *Radnoria* and *Cornuproteus*, and deep marine graptolitic shales by *Delops* and *Miraspis*.

Palaeobiogeography. In the Cambrian, there were three trilobite realms, a Laurentia–Baltica–Avalonia realm, characterised by olenellids, a Siberia realm, characterised by bigotinids, and a Gondwana realm, characterised by redlichiides. The interpreted pelagic agnostides exhibited essentially cosmopolitan distributions. Throughout most of the Ordovician, there were four trilobite realms,

a Laurentia realm, characterised by bathyurids, a Baltica–Avalonia realm, characterised by megalaspids and asaphids, a low-latitude Gondwana realm, characterised by *Selenopeltis* and dikelokephalinids, and a high-latitude Gondwana realm, characterised by calymenaceans and dalmanitaceans. At this time, Baltica–Avalonia, of which England and Wales were a part, was separated from Laurentia, of which Scotland was a part, by the Iapetus Ocean, a substantial body of deep water that constituted an insuperable barrier to the dispersal of shallow-water benthic trilobites. Baltica–Avalonia was also separated from Gondwana by the Rheic Ocean. Towards the end of the Ordovician, trilobite provincialism generally decreased, and cosmopolitanism increased, as Baltica–Avalonia and Laurentia drifted into one another, thereby uniting Great Britain, albeit as part of a larger northern continent, and as said continent and Gondwana drifted towards one another. Warm-water elements became extinct, and a cosmopolitan cold-water fauna evolved, at the time of the Late Ordovician, Hirnantian glaciation. Cosmopolitanism remained the norm through the succeeding Silurian deglaciation and transgression. In the Devonian, provincialism was re-established. At this time, the Malvino-Kaffric realm, centred on South America, the Falklands or Malvinas, and South Africa, became established for the first time. It is difficult to discern any palaeobiogeographic patterns in the interval from the Late Devonian mass extinction to the end-Permian mass extinction.

Ostracods

Fossil ostracods are interpreted, on the bases of analogy with their living counterparts, functional morphology, and associated fossils and sedimentary facies, as having been occupied a range of – semi-terrestrial and – aquatic, freshwater, brackish and marine, environments (McKenzie *et al.*, in Savazzi, 1999; Iyeka *et al.*, 2001; de Deckker, in Holmes & Chivas, 2002; Frogley, in Holmes & Chivas, 2002; Mischke & Holmes, 2008).

Ostracods appear to have first colonised freshwater environments, by way of coastal plain brackish environments, in the Carboniferous (Williams *et al.*, 2006). Fluctuation in the amount of terrigenous input, in turn related to delta growth, is interpreted to have been the most important factor controlling ostracod distribution in the Early Carboniferous Yoredale series of northern England. Ostracods have been used as indicators of the palaeochemistry, and in particular the salinity and alkalinity, concentration and turbidity, and content and oxygenation, of freshwater lakes, such as those of the Pleistocene–Holocene of the east African rift valley (Frogley, in Holmes & Chivas, 2002). The abundance, diversity and specific composition of assemblages, key species, and ornamentation are all guides. Platycopides such as *Cytherella* have historically been used as indicators of low oxygenation in ancient seas, the model being that they were able to obtain sufficient oxygen to survive even in low-oxygen environments by virtue of the volume of water they circulated around their respiratory surfaces in the course of their normal filter-feeding behaviours (Whatley, 1991; Whatley *et al.*, 1994; Whatley, 1995; Whatley *et al.*, 2003). This model has recently been challenged on the basis of empirical observations on a recently discovered living species of *Cytherella* (Brandao, 2008; Brandao & Horne, 2009). Ostracod shell chemistry, and especially the ratio between strontium and calcium, and the ratio between calcium and magnesium, is useful as an indicator of water chemistry and salinity, and can be used to distinguish non-marine and marine environments (Holmes & Chivas, in Holmes & Chivas, 2002). The interpreted hypersaline species *Simeonella brotzenorum* is an indicator of the 'Carnian Salinity Crisis' of the Levant.

Palaeobathymetry. Fossil ostracods are interpreted, essentially on the bases of analogy with their living counterparts, and associated fossils and sedimentary facies, as having been occupied a range of depth habitats in marine environments. As in the case of their living counterparts, fossil marginal marine forms tend to be torose; shallow marine forms robust, thick-shelled and highly ornamented; and deep marine forms thin-shelled and unornamented, and also blind rather than sighted.

In the Palaeozoic, bairdiids and/or paraparchitids have been interpreted as indicative of shallow marine environments in the Late Devonian to Early Carboniferous of the Nanbiancun section of Guilin in Guangxi province in China, and also in the Permian of the Mid-Continent of the United States.

The ?nektonic myodocopides *Bolbozoe* and *Entomis* have been interpreted as indicative of 'mid-water' environments in the Silurian of France and Bohemia. Spinose podocopides have been interpreted as indicative of deep marine environments in the Late Devonian to Early Carboniferous of the Shizhiling section of Guilin, and also elsewhere in southeast Asia, in Eurasia, especially around Thuringia, and in north Africa. Ostracod assemblages associated with a chemosynthetic community in the Early Carboniferous on the Port au Port Peninsula, Newfoundland are dominated by the paraparchitacean *Chamishella*.

Palaeobiogeography. In the Palaeozoic, in the Cambrian, *Petrianna* is characteristic of north Greenland in Laurentia (in the olenellid trilobite realm), *Cambria* of Siberia (in the bigotinid trilobite realm), and *Cambria, Paracambria, Auriculatella, Chuanbeiella* and *Shangsiella* of China in Gondwana (in the redlichiid trilobite realm) (Williams *et al.*, 2007b). Interestingly, *Isoxys* is found in Laurentia, in Siberia and in Gondwana.

In the Mesozoic, marine ostracod distributions point towards generally distinct Austral, Boreal and Tethyan realms. Note, though, that there was also at least local or intermittent communication between these realms. The presence of the essentially Tethyan ostracod *Orthonotocythere* in the Austral realm in the Early Cretaceous, Barremian indicates a connection between these areas at this time, possibly through the interior rifts of central and west Africa or South America. The presence of distinct suites of ostracods in the ammonite-dated Middle Jurassic, Callovian of Tanzania and Madagascar in east Africa indicates the presence of a barrier to dispersal between these areas at this time, most likely in the form of a deepwater gulf in the area of the modern Mozambique Channel (Grekoff, 1953; Besairie, 1971; Bate, 1975, 1977; Mette, 2004). Non-marine ostracod distributions point toward the development of a single terrestrial palaeobiogeographic realm, namely the Gondwana realm.

In the Cenozoic, Palaeocene–Eocene ostracod distributions point toward the (?continued) existence of a 'Trans-Saharan Seaway' linking Tethys and the South Atlantic (Zaborski & Morris, 1999).

Palaeoclimatology. In the Cenozoic, ostracod distributions point towards the palaeogeographic and palaeoclimatic evolution of the Old World in the Oligocene–Holocene. Magnesium to calcium ratios in deep sea ostracod shells have been used as measures of bottom-water palaeotemperature in the Pliocene (Cronin *et al.*, 2005). Trace element ratios and isotopic compositions have been used as indicators of palaeoclimate in the Miocene, Eggenburgian to Pannonian of the Bavarian Molasse of the North Alpine Foreland of Germany and Austria (Janz & Vennemann, 2005).

The exacting ecological requirements and tolerances of many ostracod species, and their rapid response to changing environmental and climatic conditions, render them especially useful in palaeoclimatology and climatostratigraphy, and in environmental archaeology (Griffiths, 2001; Holmes & Chivas, 2002; Frenzel & Boomer, 2005; Horne, 2007, 2010; Holmes *et al.*, 2010; Horne *et al.*, 2010).

Branchiopods

Fossil branchiopods are interpreted on the basis of analogy with their living counterparts as especially characteristic of ephemeral bodies of fresh water. The genus *Estheria* has been interpreted on the basis of sedimentary facies as having had some tolerance for hypo- or even hyper-saline conditions (living 'brine-shrimps' of the genus *Artemia* have a similar tolerance). *Estheria mangaliensis* occurs in playa lake and flat deposits of the Permo-Triassic Maji ya Chumvi formation of the Mombasa basin of Kenya; *E. greyi* in a poorly dated evaporitic sequence in the Mandawa-7 well in the northern part of the Ruvuma basin in Tanzania (Schluter, 1997; author's unpublished observations).

Insects

Fossil insects are interpreted, essentially on the bases of analogy with their living counterparts, and functional morphology, as having been terrestrial, aquatic and aerial.

The Triassic insect faunas from the Molteno formation of the Karoo group of South Africa are interpreted as representing a range of habitats, including: mature and immature *Dicroidium* riparian forest (bordering channels), characterised by beetles, cockroaches, bugs and dragonflies; *Dicroidium* woodland (on the open flood-plain), by beetles, bugs and cockroaches; *Sphenobaiera* woodland (adjacent to lakes on the flood-plain), by beetles,

cockroaches, bugs, dragonflies, crickets and moths; *Heidiphyllum* thicket (on flood-plains or channel sand-bars, or in areas of high water table), by beetles, cockroaches and bugs; *Equisetum* marsh (on flood-plains), by beetles and bugs; and fern/ *Ginkgophytopsis* meadow (on sand-bars in braided rivers), with no insect faunas. The constituent species are further interpreted as herbivores, carnivores, omnivores and, importantly, pollinators (of gymnosperms, excluding glossopterids, extinct by this time). Insect feeding traces on plants, and inferred insect-plant behavioural interactions, have recently been described and discussed by Scott *et al.* (2004).

Palaeobathymetry. Chironomids and chaoborids have recently been used to infer past lake levels in the Holocene of Finland (Luoto, 2009).

Palaeoclimatology. The exacting ecological requirements and tolerances of many insect species, and their rapid response to changing environmental, and especially climatic, conditions, render them useful in palaeoclimatology and climatostratigraphy, and in environmental archaeology. Coleopterans or beetles are especially useful in these fields, as their strongly sclerotinised wing-cases or elytra have uncommonly good preservation potential (Atkinson *et al.*, 1987; Elias, 1994; Ashworth *et al.*, 1997; Elias, 1997; Coope, 2006; Horne, 2010; Horne *et al.*, 2010). My former lecturer Russell Coope thus developed an 'inordinate fondness' for them; as, incidentally, no doubt for his own reasons, did God, as J.B.S. Haldane once famously remarked. Chironomids are also useful in Quaternary palaeoclimatology (Brooks & Birks, 2001; Brooks, 2006; Rolland *et al.*, 2009; Horne, 2010; Horne *et al.*, 2010). Dipterans have been used in the palaeoclimatic interpretation of the Eocene of Florissant in Colorado (Moe & Smith, 2005). The mutual climate range or MCR methodology resulted in a palaeotemperature estimation of 12–14 °C, consistent with that obtained by palaeobotany.

Crinoids

Echinoderms in general are interpreted, essentially on the bases of analogy with their living counterparts, functional morphology, and associated fossils and sedimentary facies, as having been epifaunal and as having fed by filter-feeding. Morphotypes adapted to 'Proterozoic-style' substrates, that is, those without a well-developed mixed layer, tended to dominate before the time of the Cambrian evolutionary diversification and associated 'substrate revolution'; and morphotypes adapted to 'Phanerozoic-style' substrates, that is, those with a well-developed mixed layer, tended to dominate after this time (Dornbos, 2006). (Interestingly, though, morphotypes adapted to 'Proterozoic-style' substrates also rose to prominence immediately after the Late Devonian mass extinction.) Morphotypes adapted to 'Proterozoic-style' substrates include 'shallow sediment stickers' and 'suctorial sediment attachers'. Morphotypes adapted to 'Phanerozoic-style' substrates include 'hard substrate attachers' and 'holdfast strategists'.

Most fossil crinoids are interpreted, essentially on the basis of analogy with their living counterparts, as having been marine, and either nektonic, nekto-benthic or benthic (Janevsky & Baumiller, 2010). They are further interpreted, on the bases of functional morphology, and associated fossils and sedimentary facies, as having been more typically benthic than their living counterparts (Donovan, in Savazzi, 1999). However, the Jurassic *Seriocrinus*, with its unusual distribution of weight and flexibility, has been interpreted, partly on the basis of functional morphology, and partly on the basis of empirical observation, as having lived suspended upside-down from floating logs; and the Jurassic *Pentacrinites* has since been interpreted as similarly pseudo-planktonic, as has the Triassic *Traumatocrinus* (Hagdorn *et al.*, 2007). Moreover, Silurian scyphocrinitids have been interpreted as wholly planktonic, and as having lived suspended upside-down from floating holdfasts. Furthermore, the Jurassic–Cretaceous *Saccocoma*, which has secondarily lost its stalk, has been interpreted as nektonic, like the living *Antedon* alluded to above (Brodacki, 2006). Interestingly in this context, well-preserved specimens show projections called *schwimmplatten* on the axillary and on some of the brachial plates. The similarly stalkless, although larger, Cretaceous *Uintacrinus* and *Marsupites* have been interpreted as nektonic by some authors, but as benthic by some others. The latest interpretation is that they were benthic, although possibly with planktonic developmental stages, and that they lived with their calices

floating in a sea of soft, chalky substrate, and with their arms outstretched to collect food. Interestingly in this context, *Uintacrinus* is especially characteristic of the Late Cretaceous, Santonian Niobrara Chalk of the Uinta Mountains in Utah and of contiguous Colorado and Kansas in the American mid-west, where it was first discovered by the dinosaur collector Marsh, and both *Uintacrinus* and *Marsupites* of the Santonian part of the Chalk of Great Britain and Northern Ireland.

Palaeobathymetry. Fossil benthic crinoids are interpreted, essentially on the basis of associated fossils and sedimentary structures, as having been more typically shallow marine than their living counterparts.

Palaeoclimatology. Many crinoids appear to have had restricted or endemic biogeographic distributions, rendering them of some use in the characterisation of palaeobiogeographic realms or provinces, and in turn in the constraint of plate tectonic reconstructions, in the Palaeozoic and in the Mesozoic. The occurrence of essentially Tethyan stemless crinoids – saccocomids and rovearinids – in the South Atlantic in the 'Middle' Cretaceous, Albian–Turonian is indicative of a marine connection between the two areas at this time, possibly through the 'Trans-Saharan Seaway' and Gulf of Guinea (Dias-Brito & Ferre, 2001; Ferre & Granier, 2001).

Echinoids

Fossil echinoids are interpreted, on the bases of analogy with their living counterparts, and functional morphology, as having been marine and benthic (Smith, 1984; Donovan, in Savazzi, 1999; Smith, in Briggs, 2005).

Palaeobathymetry. Fossil echinoids are interpreted, on the basis of associated fossils and sedimentary structures, as having been more typically shallow marine than their living counterparts. Note, though, that demonstrable deep marine fossil echinoids are known, for example from fossil cold seeps. Note also that fossil echinoid diversity appears to be highest during times of transgression, at least in the Cretaceous. Seven echinoid biofacies have been recognised in the Early Miocene of Egypt that have in turn been used to identify sedimentary facies and cycles of transgression and regression. The *Phyllacanthus* biofacies represents a shallow-marine carbonate environment; the *Clypeaster martini* biofacies a shallow marine, high-energy environment; the *Parascutella* biofacies a shallow marine, high-energy environment characterised by storm winnowing of sand-dollars; the Mixed biofacies a shallow marine, low-to-moderate energy environment; the Cidaroid–Echinacea biofacies a deeper, low energy environment, characterised by a highly structured habitat and corresponding diversity of regular and irregular sea-urchins; the Spatangoid biofacies a deep, low energy environment characterised by diverse burrowers; and the Transported biofacies a deep marine, high energy environment characterised by transportation of material from updip.

Graptolites

Fossil graptolites are interpretable, essentially on the bases of analogy with their living counterparts, namely *Rhabdopleura* and *Cephalodiscus*, and associated fossils and sedimentary facies, as having been marine and benthic, and either sessile or motile (Boucot & Xu Chen, 2009). However, some are interpreted, on the bases of functional morphology and associated fossils and sedimentary facies, as having been essentially pelagic rather than benthic (Rickards & Rigby, in Savazzi, 1999). Note in this context that graptolite diversity appears to have been highest in shelf–slope break settings characterised by upwelling, or in upper slope settings characterised by low dissolved oxygen concentrations (Finney & Berry, 1997; Zalasiewicz et al., 2007).

Specifically, some dendroids are interpreted as having been benthic and sessile (Rigby & Fortey, in Palmer & Rickards, 1991). These are envisaged as having had the form of small shrubs, with a specially strengthened proximal or sicular end held fast or rooted to the sea bottom, and the distal stipes branching out into the overlying water column. Other dendroids, though, are interpreted as having lived suspended upside-down from floating seaweed, that is, as having been pseudo-planktonic.

Graptoloids have been interpreted as planktonic by some authors (Bulman, 1964; Rickards, 1975), but as nektonic by some others (Kirk, 1969, 1972). Observations on the behaviour of life-size models in columns of water indicated that they would have required some means of auto-mobility, or at least some means of increasing buoyancy, possibly involving the deployment of stored gas or fat, in order to stop themselves simply falling to the bottom (Rigby & Rickards, 1989; Rigby, 1991). The latest interpretation is that they were nektonic, achieving a swimming motion through undulating movements of lateral extensions to the muscular cephalic shields of individual zooids (Melchin & DeMont, 1995). Whether they were planktonic or nektonic, graptoloids were well adapted to a pelagic habit by virtue of their long nemas, hook-like thecae and net-like overall forms, all of which helped to reduce their sinking rate and maintain their position within the water column. Interestingly, some show evidence of having been parasitised.

Palaeobathymetry. Fossil graptolites are interpretable, essentially on the bases of analogy with their living counterparts, and associated fossils and sedimentary facies, as having been essentially deep marine (Boucot & Xu Chen, 2009).

Palaeobiogeography. Many graptolites appear to have had restricted or endemic biogeographic distributions, rendering them of some use in the characterisation of palaeobiogeographic realms, and in turn in the constraint of plate tectonic reconstructions (Fortey, 2009). In the Ordovician, there were two graptolite realms, a high-latitude Gondwanan realm, characterised by *Didymograptus*, and a low-latitude Laurentian realm, characterised by *Tetragraptus*, isograptids, cardiograptids and oncograptids.

Chitinozoans

Fossil chitinozoans are interpreted, essentially on the basis of associated fossils and sedimentary facies, as having been marine and pelagic. They are further interpreted as typically deep marine. Note, though, that some, such as species of *Belenochitina*, appear to be associated with regressive phases, in the Middle Ordovician of the Canning basin of Western Australia (Winchester-Seeto et al., 2000).

3.1.7 Vertebrates

Fossil fish and dinosaurs from the Late Cretaceous of Montana and Alberta are characterised by stable isotope signatures interpreted as indicating a range of coastal, brackish, freshwater and terrestrial environments along the western margin of the Western Interior Seaway (Fricke et al., 2008).

Palaeobiogeography. In the Palaeozoic, the occurrence of essentially identical vertebrate assemblages in the Permian of China, in the Karoo basin of South Africa and in the Russian Urals points toward the complete assembly of the supercontinent of Pangaea by this time.

In the Mesozoic, the occurrence of essentially identical tetrapods and/or tetrapod trackways in the Triassic of the Cuyana basin of Argentina in South America, and of the Karoo basin in South Africa, points toward these areas having been contiguous at this time, within the western part of Gondwana. Q-mode and R-mode cluster analysis of vertebrate microfossil assemblages from the Late Cretaceous of Alberta in Canada revealed two clusters corresponding to interpreted biogeographically distinct communities in the north and south of the study area. *Paratarpon* is more abundant in the north; *Adocus* in the south.

In the Cenozoic, the occurrence of vertebrates with Orinoco basin affinities in Colombia and western Venezuela in the Oligocene to Middle Miocene indicates that the palaeo-Orinoco discharged into the Caribbean rather than into the Atlantic at this time (Kay *et al.*, 1997).

Palaeoclimatology. Importantly, vertebrates are of considerable use not only in in palaeobiogeography but also in palaeoclimatology, climatostratigraphy and environmental archaeology.

In the Cenozoic, analysis of non-avian tetrapods, using a computerised, Geographic Information System (GIS) analogue database on worldwide Recent distributions, enables reconstruction of the palaeoclimate of the Middle Eocene of the Messel Lagerstatte in Germany. Results compare well with those based on other proxies of climate. The abundance of 'cold-blooded', or, more correctly, ectothermic, taxa appears to be directly proportional to, and an accurate measure of, temperature. Overall diversity, though, appears to be a function not only of temperature, but also of other variables, notably the preferred adaptive strategy for procuring energy, which differs from one group of organisms to another.

Fish macrofossils, conodonts, ichthyoliths and otoliths

Fossil fish are interpreted, essentially on the basis of analogy with their living counterparts, as having been aquatic. They are further interpreted, on the basis of associated fossils and sedimentary facies, as having occupied a range of freshwater and marine environments, and on the basis of functional morphology as having occupied a range of pelagic and benthic environments, and as having practised a variety of feeding strategies. Thusly interpreted freshwater fish include, for example, the freshwater herring, stingray, gar, perch, bowfin and paddlefish of the Middle Eocene of the Green River basin of Wyoming, Utah and Colorado; the bowfin, sucker, catfish and perch of the Late Eocene of Florissant in Colorado; and the cichlids of the Oligocene and Miocene of the Great Lakes area of the East African rift system (van Couvering, 1982; Nudds & Selden, 2008). Fish appear to have colonised freshwater environments, by way of estuaries, in the Ordovician (Davies *et al.*, 2007; Park & Gierlowski-Kordesch, 2007). Some (lungfish from the earliest Cretaceous of Bornholm in Denmark) are interpreted, on the basis of empirical observation of burrows in the beds of dried-up coastal lakes, as having aestivated during summer droughts (Surlyk *et al.*, 2008).

Fossil conodont parent animals are interpreted, essentially on the basis of associated fossils and sedimentary facies, as having been deep marine. Most are also interpreted as having been nektonic, some as nekto-benthic or benthic. At least some are interpreted as having been venomous (Szaniawski, 2009).

Fossil ichthyoliths are interpreted, essentially on the basis of analogy with their living parent animals, as having been marine, some probably nektonic, others nekto-benthic or benthic. Abundances, as in the Palaeocene–Eocene, are thought to reflect high palaeoproductivity.

Fossil otoliths are interpreted, again essentially on the basis of analogy with their living parent animals, as having been aquatic (gadid, congrid and ophidiid otoliths as marine; cichlid otoliths as freshwater).

Palaeobathymetry. The palaeobathymetric distributions of conodonts from the Late Ordovician–Early Silurian of Cornwallis Island in the Canadian Arctic and Anticosti Island on the Canadian Atlantic margin have been inferred from community analysis, and used to construct comparative sea-level curves (Zhang *et al.*, 2006). Interestingly, the curves from the two areas differ in detail, in terms of the timing of sea-level change. This has been interpreted as indicating that sea-level change was controlled not only by anticipated synchronous eustatic effects associated with glaciation and deglaciation, but also by diachronous isostatic effects.

Palaeobiogeography. In the Palaeozoic, the occurrence of endemic fish faunas, dominated by galeaspids, in the Silurian and Early Devonian of south China indicates the isolation of the South China micro-plate at that time. In contrast, the occurrence of essentially cosmopolitan fish (and, as noted above, tetrapod) faunas in the Permian indicates the complete assembly of Pangaea by this time.

In the Mesozoic, the occurrence of the shallow marine coelacanth *Mawsonia* in at least approximately age-equivalent 'Middle' Cretaceous, late Aptian–Cenomanian sediments in Brazil in South America and in Congo, Niger, Algeria, Morocco and Egypt in Africa indicates that at this time the areas were still essentially contiguous, and that the rifting apart of the South Atlantic had not proceeded particularly far (de Carvalho & Maisey, in Cavin *et al.*, 2008).

In the Cenozoic, the occurrence of freshwater serrasalmine fish (pacus and piranhas) with Orinoco Basin affinities in the Early Miocene Castillo Formation of Falcon State in northwestern Venezuela indicates that the palaeo-Orinoco discharged into the Caribbean rather than into the Atlantic at this time (Albert *et al.*, 2006). The occurrence of sea catfish with Pacific affinities in the Early Miocene Castillo and Middle–Late Miocene Urumaco Formations of Falcon State in northwestern Venezuela, and in the Late Miocene to Early Pliocene Cubagua of Margarita Island in Nueva Esparta State in northeastern Venezuela, indicates that the Caribbean and Pacific were connected – through the Strait of Panama – at these times.

Palaeoclimatology. Conodonts and otoliths are of use in palaeoclimatic as well as in palaeobiogeographic interpretation. The use of conodonts in palaeotemperature interpretation in the Ordovician in the Iapetus Ocean is discussed by Armstrong and Owen, in Crame and Owen (2002) (see also Wheeley *et al.*, 2009). The use of otoliths in palaeotemperature interpretation in the Palaeogene of the Gulf Coast of the United States is discussed by Ivany *et al.*, in Prothero *et al.* (2003).

Amphibians

Fossil amphibians are interpreted, essentially on the basis of analogy with their living counterparts, as having occupied a range of aquatic and (moist) terrestrial environments in low to moderate latitudes, and as having practised a variety of feeding strategies. Some (lysophorids from the Permian of Kansas) are interpreted, on the basis of empirical observation of burrows associated with dessication surfaces, as having aestivated or as having holed up during more prolonged droughts in terrestrial environments (Hembree *et al.*, 2006). Similar behaviours are exhibited by living amphibians in the southeastern USA (and also, incidentally, by lungfish).

Palaeobiogeography. In the Cenozoic, the occurrence of amphibians with Orinoco basin affinities in the Early or Middle Miocene, Friasian or Santacrucian South American Land Mammal Age, La Venta formation of Colombia indicates that the palaeo-Orinoco discharged into the Caribbean rather than into the Atlantic at this time (Kay *et al.*, 1997).

Palaeoclimatology. Importantly, amphibians are of considerable use in palaeoclimatic as well as in palaeobiogeographic interpretation, as, for example, in the palaeoclimatic interpretation of the Early–Middle Pleistocene archaeological site at Atapuerca in Spain (Blain *et al.*, 2008).

Reptiles and birds

Fossil reptiles and birds are interpreted, essentially on the bases of analogy with their living counterparts, and functional morphology, as having occupied a range of terrestrial, aquatic, freshwater, brackish and marine, and, in the case of birds, aerial, environments, and as having practised a variety of feeding strategies (Cloudsley-Thompson, 2005; Prum, in Briggs, 2005). The feeding strategy and mechanics of the 'bizarre and biomechanically unfeasible' sauropod *Diplodocus* have recently been elucidated using finite-element analysis or FEA (Young *et al.*, 2007).

Palaeobiogeography. In the Palaeozoic, Carboniferous and Permian, reptiles were apparently restricted to essentially equatorial palaeolatitudes in Euramerica or Laurasia. The occurrence of essentially identical reptile assemblages, including *Mesosaurus*, in the coeval Permian, Sakmarian–Kungurian Whitehill formation of the Karoo supergroup of Namibia and South Africa and Irati formation of eastern South

America indicates that at this time these areas were contiguous (Schneider & Marais, 2004).

In the Mesozoic, the occurrence of identical dinosaurs in coeval Late Jurassic to Early Cretaceous, Neocomian? to Aptian sediments in northern Africa in Gondwana and southern Europe in Laurasia indicates that at this time these areas were contiguous, i.e. connected by a land bridge (Upchurch et al., 2002; Canudo et al., 2009). The occurrence of the crocodile *Araripesuchus* in at least approximately age-equivalent Early Cretaceous, Albian sediments in the Araripe basin in Brazil in eastern South America and in the Benue Trough in Niger in west Africa indicates that at this time these areas were also connected by a land bridge. An alternative hypothesis is that they were separated by a seaway at this time, which after all was the time of the rift to drift transition in the South Atlantic, but a seaway narrow enough not to constitute a significant barrier to the dispersal of land animals. The occurrence of identical marine turtles in Cretaceous – and indeed also in Cenozoic – sediments in southern Europe and west Africa indicates that at this time these areas were connected by a seaway, termed the 'Trans-Saharan Seaway' (Moody, in Callaway & Nicholls, 1997).

In the Cenozoic, the occurrence of identical crocodilians in coeval Middle Eocene, Lutetian (MP11) sediments in Purga di Bolca in Italy and Geiseltal and Messel in Germany indicates that at this time these areas were connected. The occurrence of *Crocodylus bambolii* in the Late Miocene, Vallesian/Turolian of Monte Bamboli in Tuscany in Italy, of Sardinia, and of Africa, indicates that at this time these areas were connected. Interestingly, various lines of vertebrate evidence indicate that at this same time Apulia–Abruzzi in Italy was not connected to Tuscany, but belonged to a separate biogeographic province.

Also in the Cenozoic, the occurrence of a gharial with Orinoco Basin affinities in the Early Miocene Castillo Formation of Falcon state in northwestern Venezuela indicates that the palaeo-Orinoco discharged into the Caribbean rather than into the Atlantic at this time (Riff et al., in Hoorn & Wesselingh, 2010).

Palaeoclimatology. Importantly, like fossil amphibians, fossil reptiles are further interpreted as having been typically essentially restricted to low and moderate latitudes, rendering them of use in palaeoclimatic as well as in palaeobiogeographic interpretation (Markwick, 1998; Blain et al., 2008).

Mammals

Fossil mammals are interpreted, essentially on the bases of analogy with their living counterparts, functional morphology (biomechanics), and tooth morphology and wear, as having occupied a range of terrestrial and aquatic, freshwater, brackish and marine environments, and as having practised a variety of feeding strategies (Ungar, 2007; Hublin & Richards, 2009).

For example, fossil mammals that graze or – by analogy – grazed on abrasive grasses, typically in arid environments, are characterised by high-crowned or hypsodont teeth; those that browse or browsed on non-abrasive vegetation, typically in humid environments, by low-crowned or brachydont teeth (Janis, in Thomason, 1995; Fortelius et al., in Reumer & Wessels, 2003; Fortelius et al., 2006). (Incidentally, mean hypsodonty can be used an indicator of aridity or humidity, or, in other words, palaeo-precipitation: Fortelius et al., in Reumer & Wessels, 2003; Fortelius et al., 2006.) Grazing, browsing, mixed and other feeding strategies can be inferred from tooth wear and micro-wear as well as morphology (Green et al., 2005; Merceron et al., 2005; Kaiser & Brinkmann, 2006; Merceron et al., 2006; Rivals et al., 2010; Solounias et al., 2010).

Fossil mammals from the Pleistocene of eastern Beringia are characterised by stable isotope signatures interpreted on the basis of analogy as indicating that they practised a range of feeding strategies (Fox-Dobbs et al., 2008). Thus, horse, bison, yak and mammoth are interpreted as having consumed grasses, sedges and herbaceous plants; and caribou and musk-ox as having consumed tundra plants, including lichen, fungi and mosses. Carnivores, including wolves and various large felids and ursids, are interpreted as having consumed a range of recorded herbivorous species, and possibly some unrecorded species such as forest-dwelling cervids; although one ursid, the short-faced bear, is interpreted as having specialised in predating a single prey species.

The stable isotope signature of small mammal teeth can be deciphered by the direct laser fluorination analysis technique (Grimes et al., 2008).

Palaeobiogeography. In the Cenozoic, Palaeocene–Eocene mammal distributions point toward the palaeogeographic and palaeoclimatic evolution of the North Atlantic (see Box 3.7 below). Oligocene–Holocene mammal distributions point towards the palaeogeographic and palaeoclimatic evolution of the Old World (see Box 3.8 below).

In the Miocene, the occurrence of a megatheriid or mylodontid sloth with Orinoco basin affinities in the Early Miocene of Cerro La Cruz in Lara state in northwestern Venezuela indicates that the palaeo-Orinoco discharged into the Caribbean rather than into the Atlantic at this time. The comparable body size of mammals in the Miocene of North America and southern Central America indicates that the latter area was probably a peninsula rather than an archipelago at this time, immediately prior to the emergence of the Isthmus of Panama in the mid-Pliocene (Kirby & MacFadden, 2005). An archipelago would have been expected to have been characterised by an increase or decrease in body size relative to that of a continental mainland, according to the 'island rule'.

Also in the Miocene, the record of a marine ziphiid whale in the Early/Middle Miocene Turkana grit of the interior of Tanzania points toward a connection to the Indian Ocean at this time, presumably by way of the Anza graben and Lamu embayment (Mead, 1975; Tiercelin & Lezzar, in Odada & Olago, 2002). The whale, represented only by a weathered rostrum or beak, is arguably attributable to either *Proroziphius* and *Belemnoziphius*, both of which ancient genera exhibit similarities to the modern beaked whale *Mesoplodon* and bottle-nosed whale *Hyperoodon* (Mead, 1975; Heyning & Lento, in Hoelzel, 2002). Modern *Mesoplodon* and *Hyperoodon* live in deep water (Martin & Reeves, in Hoelzel, 2002).

In the Middle–Late Miocene and in the Pleistocene, the occurrence of essentially identical suites of mammals in North America and Eurasia indicates that the two areas were connected at these times, by the so-called Bering land bridge (Wang & Tedford, 2008).

Palaeoclimatology. The exacting ecological requirements and tolerances of many mammal species, and their rapid response to changing environmental and climatic conditions, render them especially useful in palaeoenvironmental interpretation,

palaeoclimatology, climatostratigraphy and archaeology (see Verzi & Quintana, 2005). Rodents have recently proved useful in the palaeoclimatic – including palaeotemperature and palaeo-precipitation – interpretation of the late Cenozoic in Europe (Legendre et al., 2005; Fernandez, 2006; Montuire et al., 2006; van Dam, 2006; Fernandez et al., 2007). The resolution of the rodent record over the past 125 000 years, i.e. since the last interglacial, is comparable to that derived from pollen from the Grande Pile peat bog in France, and to that derived from oxygen isotopes from the North Greenland Ice Core Project (NGICP) (Fernandez, 2006).

Trace fossils

A range of sedimentary sub-environments is differentiable on the basis of trace fossils or of trace fossil associations or *ichnofacies*, such as the *Psilonichnus, Skolithos, Cruziana, Zoophycos* and *Nereites* ichnofacies, and the hardground *Trypanites*, woodground *Teredolites*, and firmground *Glossifungites* ichnofacies (Frey & Pemberton, 1985; Fig. 3.6).

In particular, a wide range of marine sub-environments is differentiable, incuding storm versus fair-weather sub-environments in shallow marine environments, and turbiditic versus hemipelagic sub-environments in deep marine submarine fan environments (Knaust, 2009). Some marginal marine sub-environments such as delta plains and estuaries are also locally differentiable. The relationship between ichnofacies and microfacies in marine environments is discussed by Jones, in Jones and Simmons (1999), and Jones et al., in Jones and Simmons (1999). Non-marine sub-environments are becoming better known in terms of their ichnofacies, with five specific tetrapod ichnofacies now described, namely, the *Chelichnus, Grallator, Carichnium, Batrachichnus* and *Characichichnos* ichnofacies, in addition to the generic *Scoyenia* ichnofacies (Hunt & Lucas, in Lucas et al., 2006; Minter et al., 2007). Dinosaur tracks are already sufficiently well known and understood that inferences not only as to gait but also as to migratory behaviour have been drawn from them.

Ichnofacies analysis is of use not only in palaeoenvironmental interpretation, but also in sequence stratigraphic interpretation. Thus, a shoaling-upward sequence from below storm wave-base to above the

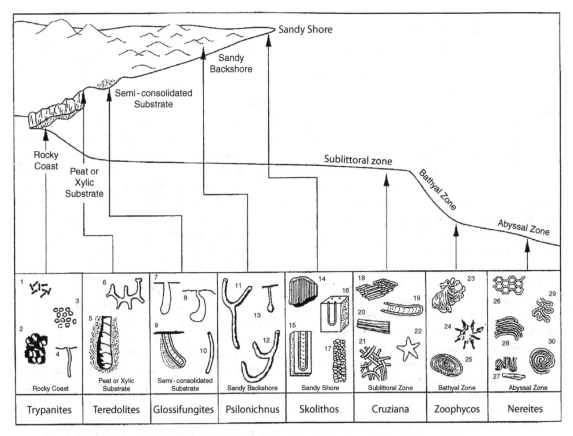

Fig. 3.6 **Ichnofacies**. From Jones (2006) (modified after Frey & Pemberton, 1985).
(1) *Caulostrepsis*; (2) *Entobia*; (3) echinoid borings; (4) *Trypanites*; (5) Teredolites;
(6) *Thalassinoides*; (7–8) *Gastrochaenolites*; (9) *Diplocraterion*; (10) *Skolithos*; (11–12) *Psilonichnus*;
(13) *Macanopsis*; (14) *Skolithos*; (15) *Diplocraterion*; (16) *Arenicolites*; (17) *Ophiomorpha*;
(18) *Phycodes*; (19) *Rhizocorallium*; (20) *Teichichnus*; (21) *Planolites*; (22) *Asteriacites*;
(23) *Zoophycos*; (24) *Lorenzinia*; (25) *Zoophycos*; (26) *Palaeodictyon*; (27) *Taphrhelminthopsis*;
(28) *Helminthoida*; (29) *Cosmoraphe*; (30) *Spiroraphe*.

high-tide mark has been inferred on the basis of the succession from *Zoophycos* ichnofacies through *Cruziana* and *Skolithos* ichnofacies to *Psilonichnus* ichnofacies, in the Cretaceous of the 'Western Interior Seaway' of the United States. (Interestingly, the *Zoophycos* ichnofacies is dominated by grazing and foraging ichnospecies, the *Cruziana* ichnofacies by deposit-feeding ichnospecies, and the *Skolithos* ichnofacies by filter-feeding ichnospecies.) Key omission surfaces such as sequence boundaries (SBs), amalgamated or composite SBs/transgressive surfaces (TSs), also known as transgressive surfaces of erosion (TSEs), and high-energy maximum flooding surfaces (MFSs) have been inferred on the basis of the substrate-controlled firmground *Glossifungites*, hardground *Trypanites* and woodground *Teredolites* ichnofacies.

In my working experience, trace fossils have proved of use in palaeoenvironmental and sequence stratigraphic interpretation in surface outcrop sections in the Cambrian to Silurian of the Welsh basin, the Cretaceous and Palaeogene of the Serrania del Interior in the eastern Venezuelan basin, the Palaeogene of Barbados, of the Polish Carpathians and of the Spanish Pyrenees, and the Neogene to Pleistogene of

Trinidad, and in subsurface core sections in regional exploration and reservoir exploitation in the Palaeocene of the North Sea, the Miocene to Pleistocene of Trinidad, and the Early Pliocene 'Productive Series' of Azerbaijan (they are readily recognisable in subsurface cores, and indeed even on occasion on image logs, as well as in surface outcrops).

Incidentally, as well as being of use in reservoir exploitation, trace fossils can also be of economic value in enhancing reservoir quality (Gordon *et al.*, 2010; Tonkin *et al.*, 2010). In the case of the Cretaceous of the Jeanne d'Arc basin, offshore Newfoundland, Canada, bioturbation by trace fossils can enhance porosity and permeability by up to 600%, the exact figure depending on the type of trace and tracemaking organism (Tonkin *et al.*, 2010).

Palaeobiogeography. The occurrence of identical ichnofossil assemblages in the Silurian of the Middle East and north and west Africa on the one hand, and eastern South America on the other, provides evidence of the existence at this time of a trans-Gondwanan seaway (Seilacher, 2007). Key species include *Arthrophycus alleghaniensis, Cruziana acacensis, C. ancora* and *C. bonariensis.*

3.2 PALAEOBIOLOGICAL, PALAEOECOLOGICAL OR PALAEOENVIRONMENTAL INTERPRETATION

As intimated in Section 3.1 above, the palaeobiological, palaeoecological or palaeoenvironmental interpretation based on fossils is based on analogy with their living counterparts, on functional morphology, and on associated fossils and sedimentary environments (see also below). The palaeoenvironmental interpretation of fossil representatives of extant, atypically Mesozoic to Palaeogene and typically Neogene to Pleistogene, species or higher-level taxa is generally based on analogy with their living counterparts; that of extinct, Mesozoic to Palaeogene, taxa with living functional morphological analogues, or 'eco-homoeomorphs', on functional morphology; and that of extinct, Palaeozoic, taxa with no living counterparts or functional morphological analogues, on associated fossil and sedimentary environments.

At least non-marine and marine environments are normally reasonably readily interpretable on the basis of their fossil content (Jones, 2006; see also 3.2.1 and 3.2.2 below).

Marginal, shallow and deep marine environments or palaeobathymetric zones are normally also reasonably readily interpretable (see Section 3.3 below); as are non-marine and marine palaeobiogeographic realms or provinces (see Section 3.4 below).

Palynofacies

Palynofacies analysis enables at least a degree of palaeoenvironmental interpretation, with non-marine and marine, and proximal and distal marine, environments generally distinguishable, and a range of proximal marine, peri-deltaic sub-environments sometimes so (Jones, 2006; Hardy & Wrenn, 2009; Fig. 3.7; see also 3.2.1, 3.2.2 and 3.3.1 below).

However, it should be stressed that the extent to which the studied palynomorphs and phytoclasts have been transported is not always easy to establish.

Palaeoenvironmental interpretation on the basis of analogy

The palaeoenvironmental interpretation of fossils on the basis of analogy relies on Lyell's 'principle of uniformitarianism' (Jones, 2006; Foote & Miller, 2007; Fox-Dobbs *et al.*, 2008). Fossil representatives of extant, atypically Mesozoic to Palaeogene and typically Neogene to Pleistogene, species or higher-level taxa are assumed to have occupied the same types of environment as their living counterparts. Note that this assumption is not necessarily universally valid. For example, some benthic nodosariide and robertinide foraminifer genera appear, on the basis of associated fossils and sedimentary facies, to have occupied shallower environments in the Jurassic than they do today, arguably becoming displaced by increased competition from evolving rotaliides.

The principle of uniformitarianism

The principle of uniformitarianism posits that processes operating at the present time, including, in this context, those that control environmental and other conditions and thus the environmental distributions of species, have also been operative in the past. Thus, understanding the present, and, in this context, the present environmental distributions of species, enables understanding of the past. Put simply, if not simplistically, *'understanding the present is the key to understanding the past'*.

This is not to say that conditions, or even species, were the same in the past as they are at the present. Indeed, empirical observations clearly indicate that

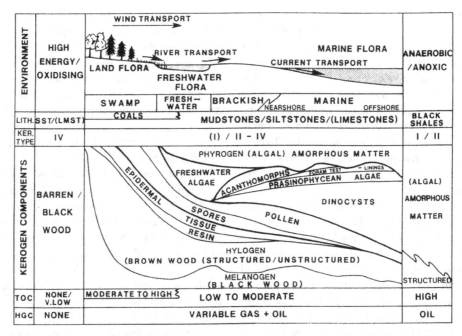

Fig. 3.7 Use of palynofacies in palaeoenvironmental interpretation. From Jones (1996).

they were not. However, the very differences in species are due to the operation of the process of evolution over time, albeit at what appears through the narrow window of historical as against geological time to be at a slow rate. Thus, the observations actually validate the principle rather than invalidate it, as some have argued.

Neither is it to say that rates of change of conditions or of the operations of processes have been uniform through time. Again, empirical observations clearly indicate not. For example, the otherwise apparently steady progress of (macro)evolution has been periodically interrupted by catastrophic mass extinction events over geological time. Importantly, though, uniformitarianism and catastrophism are not mutually exclusive, as some have argued.

Palaeoenvironmental interpretation on the basis of functional morphology

The palaeoenvironmental interpretation of fossils on the basis of functional morphology relies on the principle that morphological form represents an architectural solution to the dynamic problems associated with function, and therefore that the analysis of form informs as to that function (Thomason, 1995; Savazzi, 1999; Jones, 2006; Foote & Miller, 2007; Figs. 3.3 and 3.5; see also Figs. 3.13 and 5.49, below). Note that the functional morphological principle is not necessarily universally applicable. For example, some morphological features fulfil no obvious function, but rather are 'accidents of history' that have not been eliminated by natural selection because they have no actual adverse effect on the organism, while others appear to represent pre-adaptions, that is, adaptations to one function or environment, retained on the change to another. Moreover, certain opportunistic behaviours such as dispersal and colonisation are probably driven more or less excusively by reproductive and dispersal strategies and are essentially independent of morphology.

Life position and feeding strategy

The functional morphological principle works well in the interpretation of the relationship of the fossil to the sedimentary environment, that is to say, for example, whether terrestrial or aquatic; if aquatic, whether benthic (living on the sea bed) or pelagic (living within the water column); if benthic, whether epifaunal (living on the sediment on the sea bed) or infaunal (living in the sediment); if epifaunal, whether motile or sessile; and if pelagic, whether planktonic (free-floating) or nektonic (free-swimming) (Jones, 2006; Figs. 3.3 and 3.5). The principle also works well in the interpretation of leaf function in plants; of locomotion in fish, in aquatic reptiles such as ichthyosaurs and plesiosaurs, and in terrestrial reptiles such as dinosaurs; and of flight in aerial reptiles such as pterosaurs, and in birds (Webb, 1984; Chatterjee & Templin, 2004; Prum, in Briggs, 2005; Unwin, 2005; Jones, 2006).

The functional morphological principle also works well in the interpretation of feeding strategies, for example from morphology in the case of foraminifera, from mouth-part morphology in the case of insects, and from general morphology and tooth morphology and/or wear and micro-wear in the case of vertebrates (Webb, 1984; Janis, in Thomason, 1995; Collin & Janis, in Callaway & Nicholls, 1997; Massare, in Callaway & Nicholls, 1997; Liem et al., 2001; Fortelius et al., in Reumer & Wessels, 2003; Green et al., 2005; Merceron et al., 2005; Fortelius et al., 2006; Jones, 2006; Kaiser & Brinkmann, 2006; Kardong, 2006; Merceron et al., 2006; Rivals et al., 2010; Solounias et al., 2010; Fig. 3.3). Interestingly, there appear to have been morphological constraints on suspension-feeding among the marine reptiles of the Mesozoic (Collin & Janis, in Callaway & Nicholls, 1997).

Importantly, interpretation of the feeding strategies of individual fossil species in turn enables reconstruction of the trophic structure of the entire fossil community – loosely, the 'food pyramid'. Interestingly, there appears to be bathymetric variation in the trophic structure of modern marine communities, with suspension-feeders tending to predominate in shallow waters, and detritus-feeders in deep waters. However, there are also patches of the deep sea that are dominated by suspension-feeders, presumably on account of the lack of availability of organic detritus (other than phytodetritus). Deep-sea feeding strategies are also adapted to the avoidance of wastage.

3.2.1 Non-marine environments

As intimated in Section 3.1 above, modern, and, by analogy, ancient, non-marine, terrestrial environments

are characterised by plants, fungi, gastropods, ostracods, insects, amphibians, reptiles and birds, and mammals. Palaeoaltitudes or palaeoelevations, and elevation or uplift rates, of mountain ranges have been calculated from contained plant macro- and microfossils, as for example in the case of the Alps, Andes and Himalayas (Hooghiemstra, in Vrba et al., 1995; Fauquette et al., 1999; Gregory-Wodzicki, 2000; Hoorn et al., 2000; Hershkovitz et al., 2006; Hooghiemstra et al., 2006; Graham, 2009; Montoya et al., 2010; Mora et al., in Hoorn & Wesselingh, 2010; Xiao-Yan Song et al., 2010); from plants and insects, as in the case of the Caucasus (Mchedlishvili, 1951; Bekker-Migdisova, 1967); or from micro-mammals such as rodents (Montuire et al., 2006). In the case of the Middle Miocene, Karaganian of Vishnevaya Balka in the Caucasus, analogy between contained fossil plants and insects and their living counterparts indicates the development of a range of marginal marine, freshwater and continental ecological habitats, including coastal/lagoonal, lacustrine, fluviatile flood-plain/riparian forest or 'tugai', lowland and highland deciduous and coniferous forest, and wooded grassland and grassland (Mchedlishvili, 1951; Bekker-Migdisova, 1967). Lowland and highland forests are distinguishable by means of elevated incidences in the latter of psyllids, which are inferred to have been associated with pine trees and/or leguminous shrubs. The observed temporal succession of lowland and highland habitats helps constrain the uplift history of the Caucasus (and also, incidentally, its palaeoclimatic evolution). The uplift history of the area has important consequences for its petroleum geology, insofar as it influences the provenance and thus the quality of potential reservoir sandstones.

Modern and, by analogy, ancient, aquatic, freshwater environments are characterised by Cyanobacteria, pennate diatoms, non-calcareous and calcareous chlorophyte Algae, calcareous charophyte Algae, testate Amoebae, sponges, bivalves, gastropods, cypridacean and allied ostracods, branchiopods, insects, fish, amphibians, reptiles and birds, and mammals. Ancient freshwater sub-environments are differentiable on the basis of the distributions and/or shell chemistry of charophyte Algae and ostracods. Ancient water levels in lakes are identifiable on the basis of the distributions of charophytes, as in the case of the Holocene of Lake Tigalmamine in the Moroccan Atlas (Soulie-Marsche *et al.*, 2008), and on the basis of the distributions of chironomids and chaoborids, as in the case of the Holocene of Finland (Luoto, 2009) (see also Section 3.3 below).

Non-marine palaeobiogeographic realms or provinces are normally reasonably readily interpretable on the basis of their fossil content (see Section 3.4 below).

3.2.2 Marine environments

As intimated in Section 3.1 above, modern, and, by analogy, ancient, brackish-water and marine environments are characterised by Cyanobacteria, dinoflagellates, silicoflagellates, centric diatoms, calcareous nannoplankton, calcareous rhodophyte and chlorophyte Algae, acritarchs, *Bolboforma*, testate *Amoebae*, foraminifera, radiolarians, calpionellids, 'ediacarans', 'small shelly fossils', sponges, archaeocyathans, stromatoporoids, corals, brachiopods, bryozoans, bivalves, gastropods, ammonoids, belemnites, tentaculitids, trilobites, cytheracean and allied ostracods, crinoids, echinoids, graptolites, chitinozoans, fish, reptiles and birds, and mammals.

As noted above, marginal, shallow, and deep marine environments or palaeobathymetric zones are normally reasonably readily interpretable on the basis of their fossil content (see Section 3.3 below).

Marine palaeobiogeographic realms or provinces are also normally reasonably readily interpretable on the basis of their fossil content (see Section 3.4 below).

Oxygen-related biofacies in marine environments

Additionally, a number of – essentially depth-independent – oxygen-related environments and biofacies can be differentiated on the basis of the various fossil groups, namely, anaerobic, quasi-anaerobic, exaerobic, dysaerobic and aerobic biofacies (Wignall, 1990; Savrda & Bottjer, in Tyson & Pearson, 1991; Kaiho, 1994; Wignall, 1994). Anaerobic biofacies are characterised by an absence of *in situ* benthonic fossils and of bioturbation, but by the local presence of *ex situ* pelagic fossils and/or faecal material derived from pelagic organisms. Quasi-anaerobic biofacies are characterised by the additional presence of benthonic microfossils such

as foraminifera and of micro-bioturbation. Exaerobic biofacies are characterised by the additional presence of oxygen-independent chemosynthetic benthonic macrofossils and of some macro-bioturbation. Dysaerobic biofacies are characterised by a low diversity of weakly calcified macrofossils and/or macro-bioturbation (i.e. they may contain trace fossils but no body fossils). Aerobic biofacies are characterised by a high diversity of strongly calcified macrofossils and macro-bioturbation.

In the micropalaeontological literature, anaerobic or anoxic, dysaerobic or dysoxic (also known as hypoxic or kenoxic), and aerobic or oxic environments and biofacies are the most widely recognised (and what in the macropalaeontological literature would be classified as quasi-anaerobic biofacies are not widely recognised as distinct from dysaerobic or dysoxic biofacies) (Kaiho, 1994). There are a number of schemes that attempt to define the biofacies in terms of dissolved oxygen concentrations in millilitres per litre or ml/l. Anoxic biofacies are defined in terms of dissolved oxygen concentrations of 0–0.1 ml/l, dysoxic biofacies in terms of concentrations of 0.1–2 ml/l, and oxic biofacies in terms of concentrations of 0.5–2 ml/l or over (overlapping ranges are given because there is no consensus).

Benthic foraminifera are found in anoxic, dysoxic and oxic environments and biofacies (Poag, 1985; Quinterno & Gardner, 1987; Kaiho, 1994; Jorissen *et al.*, 1995; Jones, 2006; Ashckenazi-Polivoda *et al.*, 2010; Nagy *et al.*, 2010). Only certain buliminides are found in anoxic biofacies. Both certain agglutinated benthic foraminifera and certain buliminides are found in dysoxic biofacies. Certain agglutinated foraminifera are characteristic of dysoxic biofacies associated with riverine input, 'freshwater overhang', salinity stratification and bottom-water stagnation in modern and by analogy ancient marginal marine, peri-deltaic and associated environments (Nagy *et al.*, 2010; see also 3.3.1, and Box 3.1, below). Certain buliminides are characteristic of dysoxic biofacies associated with oxygen minimum zones (OMZs) in modern and ancient deep marine, especially upper slope, environments, and have been used as proxies for dysoxia, or elevated organic carbon flux, or both (Poag, 1985; Quinterno & Gardner, 1987; Kaiho, 1994; Jorissen *et al.*, 1995; Jones, 1996; Fontanier *et al.*, 2005; Jones, 2006;

Jorissen *et al.*, in Hillaire-Marcel & de Vernal, 2007; Koho *et al.*, 2007; Koho, 2008a, b; Ashckenazi-Polivoda *et al.*, 2010; see also 3.3.3. below). Essentially all major groups of foraminifera are found in oxic environments, apart from those certain buliminides specifically adapted to anoxic environments.

Dysoxic to anoxic environments are conducive to the preservation of organic carbon and hence of petroleum source-rocks (see Box 5.1 below).

3.3 PALAEOBATHYMETRY

As noted in Section 3.2 above, modern, and, by analogy, ancient, marginal, shallow and deep marine environments are normally reasonably readily interpretable on the basis of their fossil, especially brachiopod, trilobite and ostracod, and most especially foraminiferal, content (see also 3.3.1–3.3.3 below). Foraminiferal palaeobathymetric range charts have been constructed not only for the Pleistogene, Neogene and Palaeogene but also for the Cretaceous (Olsson & Nyong, 1984; Jones, 1996, 2006; Fig. 3.8).

Foraminiferal palaeobathymetric data have been used to model the Cenozoic uplift history of Barbados (Jones, 2009) and Timor (Haig, 2010).

3.3.1 Marginal marine environments

Modern marginal marine or paralic environments are characterised by palynomorphs such as terrestrially derived spores and pollen, freshwater Algae and marine dinoflagellates; by foraminifera such as *Ammoastuta*, *Ammobaculites*, *Ammomarginulina*, *Ammotium*, *Arenoparrella*, *Haplophragmoides*, *Jadammina*, *Miliammina*, *Reophax*, *Textularia*, *Trochammina* and *Trochamminita* (agglutinated foraminifera), *Quinqueloculina* (miliolide), *Bolivina*, *Bulimina* and *Buliminella* (buliminides), and *Ammonia*, *Asterorotalia*, *Discorbis*, *Elphidium*, *Nonion* and *Protelphidium* (rotaliides); and by torose ostracods (Jones, 2006). Various marginal marine sub-environments are differentiable on the basis of the biota, including, generally in relation to depth and tidal range, supratidal, intertidal, including salt-marsh and in low latitudes mangrove swamp, and subtidal; and specifically in relation to sedimentary environment, deltaic and estuarine (Aaron *et al.*, 1999; Jones, 2006; Hardy & Wrenn, 2009; see also Box 3.1 below). (Marginal marine sub-environments are also differentiable on the basis of the shell chemistry of foraminifera.) Sea-

grass communities are characterised by porcelaneous miliolide and hyaline rotaliide foraminifera (Brasier, 1975; Eva, 1980; Jones, 1996; Semeniuk, 2001; Buchan & Lewis, in Demchuk & Gary, 2009). The *Cymodocea* and *Posidonia* sea-grass communities of the Indo-Pacific are characterised by the porcelaneous miliolide genus *Peneroplis*, and the *Thalassia* sea-grass communities of the Indo-Pacific and Caribbean by the porcelaneous miliolide genera *Amphisorus*, *Marginopora* and *Sorites*.

Ancient marginal marine environments are characterised by taxonomically or morphologically similar biotas (Fiorini *et al.*, 2010). Marsh or paludal sub-environments in the Late Cretaceous of the 'Western Interior Seaway' of the United States are characterised by *Trochammina* and *Miliammina*; estuarine sub-environments by *Trochammina* and *Ammobaculites*. (Ostracods are also of use in palaeoenvironmental interpretation here.) Sea-grass communities in the Eocene of Florida are characterised by porcelaneous miliolide foraminifera, macro-invertebrates, including bryozoans, molluscs and echinoderms, and vertebrates, including dugongs (Ivany *et al.*, 1990). Those in the pre-Miocene of the Caribbean are characterised by porcelaneous miliolide foraminifera (Eva, 1980). Those in the Miocene of the Kerala basin on the west coast of India are characterised by the porcelaneous miliolide foraminifer *Pseudotaberina malabarica* (Piller *et al.*, 2010).

Deltas

Deltas are sedimentary bodies formed by the interaction of fluvial and marine processes. Down-dip variation in facies has been observed in modern and ancient delta systems. This is typically manifested by a transition from sand-prone facies associated with fluvially and tidally dominated delta-top ('topset') sub-environments, and by delta-front ('foreset') sub-environments modified by marine shoreface processes, to silt- and mud-prone facies associated with prodelta ('bottomset' or 'toe-set') sub-environments.

3.3.2 Shallow marine environments

Modern shallow marine carbonate environments are diagnosed by calcareous Algae, which photosynthesise within the photic zone in order to produce food, and also by larger benthic foraminifera (LBFs) such as *Alveolinella*, *Amphisorus*, *Archaias*, *Borelis*,

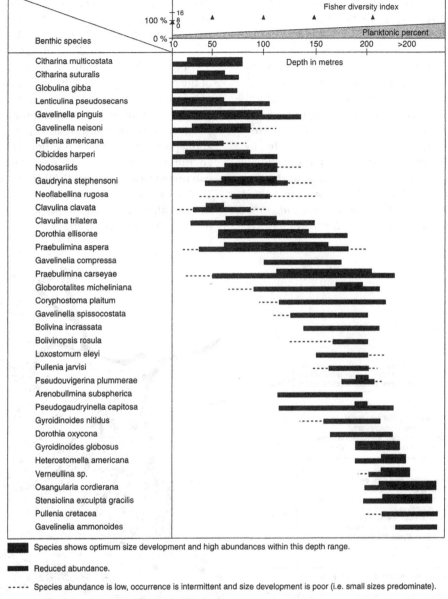

Fig. 3.8 **Palaeobathymetric range chart for Cretaceous foraminifera.** © Cushman Foundation for Foraminiferal Research. Reproduced with permission from Olsson and Nyong (1984).

Cyclorbiculina, Marginopora, Peneroplis and *Sorites* (miliolides) and *Amphistegina, Baculogypsina, Calcarina, Heterocyclina, Heterostegina, Operculina* and *Operculinella* (rotaliides), and z-corals, which harbour algal photosymbionts that photosynthesise in the photic zone (Jones, 2006). Modern shallow marine clastic environments are characterised by foraminifera such as *Clavulina* and *Textularia* (agglutinated

BOX 3.1   PALAEONTOLOGICAL CHARACTERISATION OF MARGINAL MARINE, PERI-DELTAIC SUB-ENVIRONMENTS

Modern and ancient deltas, such as those of the Mahakam, Mississippi, Niger and Orinoco, and ancient deltas, such as those of the Jurassic of the North Sea, are characterised by palynomorphs such as terrestrially derived spores and pollen, freshwater Algae and marine dinoflagellates; by benthic foraminifera; and by ostracods (Williams & Sarjeant, 1967; Haman, in Verdenius *et al.*, 1983; Nagy, 1992; Jones, 1996; Jones, in Ali *et al.*, 1998; van Gorsel, 1988; Armentrout *et al.*, in Jones & Simmons, 1999; Jones *et al.*, in Jones &

Simmons, 1999; van der Zwan & Brigman, in Jones & Simmons, 1999; Jones, 2006; Hardy & Wrenn, 2009; Nagy *et al.*, 2010; Fig. 3.9; see also 5.2.2, 5.2.3, 5.3.2 and 5.3.3, and Figs. 5.37 and 5.47 below). (Note, though, that the distributions of the palynomorphs, and the amount and value of information that can be gleaned from them, are subject to some control by complex and poorly understood transportation pathways and taphonomic processes.) Various sub-environments are differentiable on the basis of palynological assemblages (Williams & Sarjeant, 1967), or of foraminiferal 'morphotypes' (Haman, in Verdenius *et al.*, 1983) or 'morphogroups' (Nagy, 1992; Nagy *et al.*, 2010).

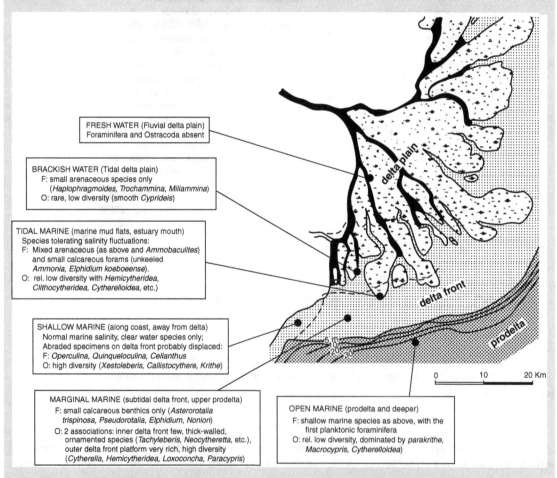

Fig. 3.9 **Micropalaeontological discrimination of modern and ancient peri-deltaic sub-environments.** © The Indonesian Petroleum Association. Reproduced with permission from van Gorsel (1988).

In the case of the modern and ancient Orinoco delta, fluvially dominated delta-top sub-environments are characterised by *Pediastrum*, by palm/gymnosperm/grass/*Bombax*/spore, by fungal spore and by pteridophyte spore (back-mangrove) palynological assemblages (and by the testae amoeban *Centropyxis*); tidally dominated delta-top sub-environments, by *Zonocostites ramonae* (mangrove swamp), by palm (palm swamp), and by palm/*Z. ramonae* (mixed swamp) palynological assemblages, and by *Miliammina* and *Trochammina* foraminiferal assemblages; delta-front sub-environments, by *Buliminella* foraminiferal assemblages; proximal prodelta sub-environments, by *Eggerella* assemblages; and distal prodelta sub-environments by *Glomospira* and *Alveovalvulina/Cyclammina* assemblages (Jones, in Ali *et al.*, 1998; Jones *et al.*, in Jones & Simmons, 1999; see also 5.2.3 and 5.3.3 below; and Figs. 5.37–5.38 below).

In the case of the modern Mississippi delta, channel, levee and interdistributary bay sub-environments are differentiable on the basis of ternary plots of agglutinated foraminiferal 'morphotypes' (Haman, in Verdenius *et al.*, 1983).

In the case of the ancient, Jurassic delta of the Dunlin formation of the North Sea and its onshore equivalents in surface sections of Spitsbergen and Yorkshire, delta-top, delta-front and prodelta sub-environments are differentiable on the basis of ternary plots of agglutinated foraminiferal 'morphogroups' (Nagy, 1992; Nagy *et al.*, 2010; see also 5.2.2 and 5.3.2 below). Interpreted delta-top and delta-front environments are characterised by *Ammodiscus* and *Trochammina*; and interpreted prodelta environments by *Trochammina* and *Verneuilinoides*. The interpreted delta-top and delta-front environments are further interpreted on the basis of analogy with modern environments and biofacies similarly characterised by *Ammodiscus* and *Trochammina* in the Drammensfjord in Norway and Aso-kai lagoon in Japan as not only hyposaline but also hypoxic or dysoxic.

foraminifera), *Quinqueloculina* (miliolide), *Dentalina* and *Lenticulina* (nodosariides), *Bolivina, Bulimina, Cassidulina* and *Uvigerina* (buliminides), *Hoeglundina* (robertinide) and *Ammonia, Cancris, Chilostomella, Cibicides, Discorbis, Elphidium, Eponides, Gyroidina, Hanzawaia, Melonis, Nonion, Nonionella, Pullenia, Rosalina* and *Siphonina* (rotaliides), and by robust, thick-shelled and highly ornamented ostracods. Inner shelf sub-environments are characterised by *Ammonia* and *Hanzawaia* assemblages, and middle–outer shelf sub-environments by *Uvigerina* and *Bolivina floridana* assemblages, in the case of the eastern Venezuelan basin.

Ancient shallow marine environments are diagnosed by taxonomically or morphologically similar biotas. They are also characterised by Bacteria, 'small shelly fossils' (SSFs), sponges, archaeocyathans, stromatoporoids, bryozoans, crinoids and echinoids.

Reefs

Bacterially – or microbially – mediated microbialites, calcareous Algae, LBFs, sponges, SSFs, stromatoporoids, archaeocyathans, corals, brachiopods, bryozoans, bivalves, gastropods, insects and crinoids were all important contributors to the construction of biological reefs or bioherms, not only in shallow marine but also in lacustrine environments, and more rarely in deep marine environments, at one time or another in the Proterozoic, Palaeozoic, Mesozoic and Cenozoic (see below; see also Box 3.2, and 3.8.2, below).

The marine microbialite reefs of the Precambrian of Oman and the lacustrine ones of the Early Cretaceous of Brazil, the marine nummulite banks of the Palaeocene–Eocene of the circum-Mediterranean region, the marine stromatoporoid reefs of the Late Devonian of Alberta in Canada and of the Timan Pechora basin in Russia, the lacustrine bivalve and gastropod 'coquinas' of the Early Cretaceous of Angola and Brazil, and the marine rudist reefs of the Cretaceous of Mexico and of the Middle East are all of considerable commercial significance as petroleum reservoirs (see also Box 3.2, Boxes 5.2–5.3, and 5.3.1).

Bacterially or microbially mediated microbialites were important contributors to the construction of reefs in shallow marine and lacustrine environments in the Proterozoic, Palaeozoic, Mesozoic and Cenozoic (Abell & McClory, in Frostick *et al.*, 1986; Vincens *et al.*, in Frostick *et al.*, 1986; Tiercelin *et al.*, 1992; Bertrand-Sarfati *et al.*, in Bertrand-Sarfati & Monty, 1994; Cohen, 2003; Schneider & Marais, 2004; Adams *et al.*, 2005; Konhauser, 2007; Sanz-Montero *et al.*, 2009; Gandin & Debrenne, 2010; McCall, 2010; author's unpublished observations). Marine microbialite reefs are known from the Proterozoic, for example the Precambrian Nama group of Namibia and Huqf supergroup of Oman; from the Palaeozoic, for example the Cambrian of Siberia, and the latest Devonian, Famennian to Early Carboniferous, Dinantian of Guilin province in China and of north Africa, Europe and North America; and from the Mesozoic, Early Triassic, Griesbachian to Spathian. That from the Precambrian Nama group of Namibia is capable of being mapped out in three dimensions by means of digital surveying technologies. That from the Cambrian of Siberia, which crops out along the Lena, Aldan and Uchur rivers, is capable of being mapped out in three dimensions in the field, and appears to constitute a rare example of a barrier-reef rather than patch-reef complex formed in association with archaeocyathans (see below). Lacustrine microbialite reefs are known from the Mesozoic, for example the Early Cretaceous of the rift basins of Brazil; and from the Cenozoic, for example the undifferentiated Tertiary of the Limagne graben in France, the Miocene of the Madrid and Duero basins in Spain and of Idaho in the USA, and the Pliocene–Pleistocene of Lake Turkana, the Pleistocene–Holocene of Lake Tanganyika and the Holocene of Lake Bogoria, all in the east African rift valley.

Calcareous Algae were sufficiently abundant in some shallow marine environments as to have been important contributors to the construction of reefs, for example in the Palaeozoic, Late Ordovician to Devonian and Late Carboniferous to Permian of the former Soviet Union, the Late Devonian of the Canning basin of Western Australia, and the Permian of the Permian basin of west Texas; in the Mesozoic, Cretaceous of the Middle East; and in the Cenozoic of Papua New Guinea in the Far East. In the Palaeozoic, the Late Devonian mass extinction caused reef ecosystems to collapse, or, in the case of the Canning basin of Western Australia, to re-organise; and

resulted in a 'reef gap' in the rock record from latest Devonian to earliest Carboniferous. This time interval was characterised by the general absence of reef-building organisms and of reefs, but also by the extensive development of mud-dominated reef-like structures or mounds known as 'Waulsortian mounds' or 'Waulsortian-like mounds'. These structures are thought to have resulted partly from algal activity, and are distinguishable from their level-bottom equivalents partly in terms of elevated incidences of Algae, indicating development within the photic zone. Recognisably reef-like structures, characterised in part by the green alga *Koninckopora* and by the red alga *Ungdarella*, reappeared in the Asbian–early Brigantian; true reefs, characterised by the phylloid/ancestral coralline red alga *Archeolithophyllum*, in the late Brigantian. Note that the Asbian–Brigantian was an important time of diversification of calcareous Algae. Not only the red Algae *Ungdarella* and *Archaeophyllum* but also the palaeoberesellid green alga *Kamaenella*, first appeared at this time, to rise to prominence in the reefs of the Late Carboniferous. Early Late Carboniferous, Bashkirian–early Moscovian reefs are characterised by *Dvinella/Donezella*, *Komia*, *Cuneiphycus* and phylloids; late Late Carboniferous, late Moscovian–Gzhelian buildups by erect udotacean phylloid Algae, and, in deeper water, by encrusting red Algae; and Permian reefs by 'baffling' erect phylloid Algae, and 'binding' encrusting red Algae.

Microgranular fusulinide *LBFs* were important contributors to the construction of reefs in the Palaeozoic; and hyaline rotaliide LBFs were important contributors to the construction of reefs in the Cenozoic.

The *SSF Cloudina* has been recorded in low-relief boundstone 'reefs' in Namibia, *Anabarites* in 'mounded aggregations' elsewhere, and orthothecimorph and hyolithomorph hyoliths in association with archaeocyathan reefs in Siberia (Schneider & Marais, 2004).

Sponges were important contributors to the construction of reefs in the Palaeozoic, less so in the Mesozoic, and not at all in the Cenozoic (their decline appearing to mirror the rise of the scleractinian corals with algal photosymbionts). Early Ordovician, Tremadoc reefs are characterised by lithistid sponges; Early Carboniferous, Asbian–early

Brigantian, 'reef-like structures' by *Solenopora*, recently reclassified as a chaetitid sponge rather than a red alga; early Late Carboniferous, Bashkirian–early Moscovian reefs by chaetitid sponges; late Late Carboniferous, late Moscovian–Gzhelian reefs by calcareous sponges; and Permian tropical and, importantly, temperate-water reefs by 'baffling' calcareous sponges and heliosponges.

Archaeocyathans apparently lived in colonies, and at times, and in places, were sufficiently abundant as to be rock- or reef-forming (Debrenne, 2007; Gandin & Debrenne, 2010). They often form reefs in association with microbialite-producing Cyanobacteria, apparently living in reef cavities or crypts, probably because adjacent substrates were covered by microbial biofilms that prevented colonisation. Archaeocyathan–microbialite reefs are typically patch-reefs, between 10 and 30 m in diameter and 3 m in thickness, and exhibit a dominantly outward growth form that has been interpreted as indicating optimisation of exposure to sunlight and facilitation of photosynthesis. One example is known to the author of what appears to be an archaeocyathan–microbialite barrier-reef complex, exposed along the Lena, Aldan and Uchur rivers in Siberia, and capable of being mapped out in three dimensions in the field. This barrier-reef complex built up between the Anabar and Aldan shields in the 'Tommotian–Botoman' regional stages and foundered in the 'Botoman–Toyonian' regional stages, coincident with a major archaeocyathan extinction. Importantly from the petroleum geological point of view, the restricted back-barrier lagoon on the landward side of this reef complex was characterised by the precipitation of evaporites, that now constitute the cap-rocks to underlying reservoirs in the Irkutsk Amphitheatre.

Stromatoporoids were important contributors to the construction of reefs, including core frameworks, in the Ordovician through Devonian of northern Eurasia and North America, the Late Ordovician of south China, and the Late Devonian of the Canning basin of Western Australia, of Alberta in Canada, and of the Timan Pechora basin in the northeastern part of the Russian platform.

Corals were important contributors to the construction of reefs in shallow marine environments to a lesser extent in the Palaeozoic, and to a greater extent in the Mesozoic and Cenozoic, coincident with the rise of the scleractinian corals with algal photosymbionts (Wood, in Aronson, 2007). On Barbados, Pleistocene coral reef systems, interpreted as having formed at or near sea-level during interglacials, have been uplifted to form a series of terraces in the interior of the island (Fairbanks & Matthews, 1978; Schellmann & Radtke, 2004; Jones, 2009). The oldest terraces have been uplifted the most, and the youngest the least. Mean rates of uplift can be demonstrated to have been of the order of 220 m/Ma.

In the Palaeozoic, when they dominated the low-level, filter-feeding benthos, *brachiopods* were sufficiently abundant in shallow marine environments as to be rock-forming, and to contribute to the construction of reefs, as in the Early Permian, Asselian–Sakmarian.

Bryozoans were sufficiently abundant in some shallow marine environments as to have been important contributors to the construction of reefs, or bryoherms, perhaps most particularly in the Palaeozoic. In the Palaeozoic, bryozoans are characteristic of the Early Ordovician, Tremadoc reefs of South China; encrusting bryozoans of Early Carboniferous reefs; and 'baffling' fenestellid and ramose bryozoans and 'binding' fistuliporid bryozoans of Permian reefs. Fenestrate, ramose and encrusting bryozoans are characteristic of the Early Carboniferous, Tournaisian, early *Typicus* Conodont Zone 'Waulsortian-like mounds' of the Alamagordo member of the Lake Valley formation of the Sacramento Mountains of New Mexico, and fenestrate bryozoans of the early to late *Typicus* to *Anchoralis–Latus* Conodont Zone 'Waulsortian mounds' of the Nunn and Terra Blanca members, which differ in their higher relief and greater degree of differentiation from crest to flank (interestingly, bryozoans are much less characteristic of the level-bottom equivalents of the mounds). Bryozoans are also characteristic of the (cool-water) Late Ordovician, Ashgill mud mounds of Libya (Buttler *et al.*, 2007; Cherns & Wheeley, 2007); and the Early Palaeocene, Danian mounds of the Danish basin (Bjerager & Surlyk, 2007). Fenestrate bryozoans are characteristic of the cool-water facies of Late Carboniferous–Early Permian temperate-, as opposed to tropical-, water buildups, such as those of the northwestern United States, Arctic Canada, the Barents Sea, and the

Russian Platform; for example the latest Carboniferous, Gzhelian to earliest Permian, Asselian part of the Kozhim Bank in the Timan–Pechora Basin.

Bivalves and gastropods were sufficiently abundant in the lacustrine environments of the Early Cretaceous of the proto-South Atlantic as to have been rock-forming. Rudist bivalves were important contributors to the construction of reefs in the Mesozoic.

Interestingly, *insects*, specifically caddis-flies, are interpreted as having been partially responsible for the buildup of lacustrine bioherms in the Early Cretaceous Jinju formation of Jahyeri in Korea, the Eocene Green River formation of Little Mesa in Wyoming, and the Miocene of the Massif Central in France (Paik, 2005).

In parts of the Palaeozoic, *crinoids* appear to have congregated together to form 'crinoid gardens' in shallow marine environments, locally in sufficient abundance as ultimately to be rock-forming. Indeed, in parts of the Palaeozoic, crinoids were important contributors to the construction of reefs and reef-like structures in shallow marine environments. For example, they are characteristic of the Early Carboniferous, Brigantian platformal and peri-reefal facies of the Burren, Buttevant and Callan areas of southern and western Ireland. They are also characteristic of the initial stages of development of the Early Carboniferous, Tournaisian 'Muleshoe Mound', a 'Waulsortian mound' in the Sacramento mountains of New Mexico.

BOX 3.2 PALAEONTOLOGICAL CHARACTERISATION OF SHALLOW MARINE, PERI-REEFAL SUB-ENVIRONMENTS

Modern and ancient shallow marine reefs are or were characterised by Bacteria, calcareous Algae, larger benthic foraminifera, sponges, 'small shelly fossils', stromatoporoids, archaeocyathans, z-corals, brachiopods, bryozoans, rudist bivalves, and crinoids. Various sub-environments are differentiable on the basis of algal, foraminiferal and macroinvertebrate assemblages or foraminiferal 'morphogroups' (Reiss & Hottinger, 1984; Hallock & Glenn, 1986; van Gorsel, 1988; Jones, 1996; Jones, 2006; Figs. 3.10–3.11).

In the case of modern and ancient reefs, shallow marine, euphotic, back-reef to reef-crest sub-environments are characterised by green or red calcareous Algae, by miliolide LBFs with green or red algal, dinoflagellate or B-1 diatom photosymbionts, and by elevated coral morphotypes; and deeper marine, sub-euphotic, fore-reef sub-environments by globular to lenticular rotaliide LBFs with B-2 or B-3 diatom photosymbionts, and by depressed coral morphotypes (Reiss & Hottinger, 1984; van Gorsel, 1988).

Wilson's standard sub-environments are differentiable on the basis of cross-plots of foraminiferal 'morphogroups' (Hallock & Glenn, 1986).

Photosynthesis is, and hence photic zone sub-environments are, indicated by the shell chemistry of LBFs, specifically the boron isotopic composition.

Palaeozoic
In the case of the Carboniferous, Asbian Urswick Limestone formation of the southern Lake District in northern England, which has been subject to a semi-quantitative sedimentological analysis, cycle bases, interpreted as representing water depths of <20 m, are characterised by the calcareous Algae *Coelosporella* and *Stacheoides*. Cycle middles are characterised by *Kameana*, *Kamaenella*, *Epistacheoides* and *Ungdarella*. Cycle tops, probably representing water depths of 0–10 m, are characterised by *Koninckopora*, *Anatolipora* and *Polymorphocodium*. Note, though, that there are significant variations in allochem distribution according to palaeogeography as well as bathymetry.

Mesozoic
In the case of certain of the reservoir-prone Mesozoic peri-reefal and platformal carbonate environments of the Middle East, precise palaeoenvironmental and palaeobathymetric interpretations have been made on the basis of the distributions of particular calcareous Algae with particular photosynthetic light requirements and associated palaeobathymetric preferences, and also of LBFs and rudist bivalves (Banner & Simmons, in Simmons, 1994; Jones, 1996; Simmons *et al.*, in Hart *et al.*, 2000; Jones *et al.*, in Bubik & Kaminski,

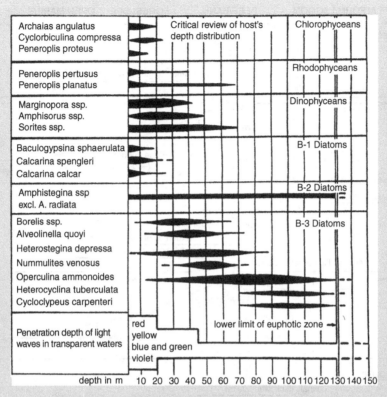

Fig. 3.10 Bathymetric distribution of larger benthic foraminifera in modern platformal and peri-reefal carbonate environments. © Springer. Reproduced with permission after Reiss and Hottinger (1984).

2004; see also 5.2.1). These precise palaeobathymetric interpretations have proved invaluable in reservoir characterisation and correlation (see 5.3.1). They have also proved invaluable in the identification of reservoir 'sweet spots', for example, in the Early Cretaceous Thamama group, Kharaib and Shu'aiba formation reservoirs of certain fields in Abu Dhabi, where reservoir quality can be demonstrated to be best in palaeobathymetries of 10–30 m – characterised by extensive development of porous *Bacinella/ Lithocodium* boundstone facies.

Cenozoic

In the case of the Cenozoic coral reefs of the Indo-Pacific palaeobiogeographic realm, Oligo-Miocene reef systems of the Eil area of northeast Somalia are characterised by a vertical succession of facies from shelf-slope through deep shelf, deep fore-reef

and fore-reef to reef (Bosellini *et al.*, 1987). Reef facies are characterised by diverse corals (initially *Hydnophora regularis*, and later *Goniopora microsideria, Hydnophora insignis, H. solidor, Favia africana, F. preamplior, Plesiastrea grayi, Diploastrea haimei, Leptoporia africana, L. concentrica* etc.). Early Miocene reef systems of the Makran area of southern Iran are characterised by corals that modern analogy indicates are of comparatively deep-water aspect (McCall *et al.*, 1994). These include poritids and faviids, indicating water depths of the order of 5–20 m, and *Leptoseris* and pectiniids, indicating water depths of the order of 100–130 m, that is, near the 'light floor'. (The Oligo-Miocene Qom Formation of the Abadeh area of northern Iran is characterised by what would appear to be a shoaling-upward succession of *Leptoseris–Stylophora* and *Porites*–Faviidae assemblages.) Pleistocene–Holocene reef systems of

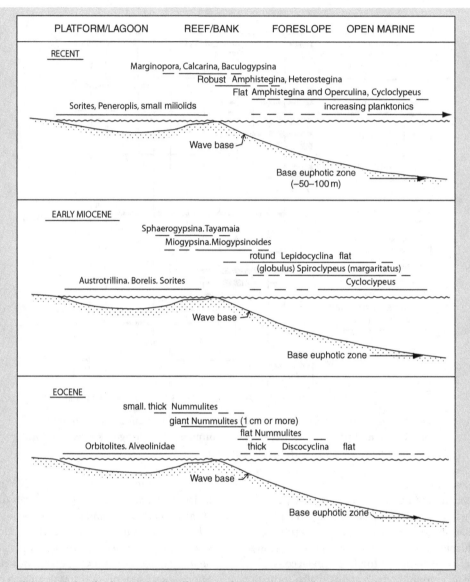

Fig. 3.11 Palaeobathymetric distribution of larger benthic foraminifera in ancient platformal and peri-reefal carbonate environments. © The Indonesian Petroleum Association. Reproduced with permission from van Gorsel (1988).

the Indo-Pacific realm in general are characterised by a horizontal succession of facies from back-reef through reef-flat and reef-crest to fore-reef (Montaggioni, 2005). Back-reef facies are characterised by arborescent, and to a lesser extent foliaceous and encrusting, and domal and tabular branching, coral morphotypes. Reef-flat and reef-crest facies are characterised by domal and

robust branching coral morphotypes, and additionally by coralline Algae in high-energy systems, and by tabular branching and to a lesser extent arborescent corals in low–moderate energy systems. Fore-reef facies are characterised by foliaceous and encrusting coral morphotypes. The arborescent morphotypes include *Acropora aspera, A. divaricata* and *A. muricata;* the foliaceous morphotypes *Pavona cactus*, *P. decussata*, *P. varians*, *Pachyseris rugosa*, *Montipora foliosa* and *Turbinaria*; the encrusting morphotypes *Porites lobata*, *Montipora capitata* and *M. patula*; the domal morphotypes various poritids (*Cyphastrea*, *Goniopora*, *Porites lutea*), faviids (*Diploastrea heliopora*, *Favia speciosa*, *F. stelligera*, *Favites*, *Goniastrea retiformis*, *Platygyra daedala*, *P. sinensis*), acroporids (*Astreopora listeri*) and mussids (*Symphillia*); the tabular branching morphotypes various acroporids (*Acropora hyacinthus*) and pocilloporids; and the robust branching morphotypes various other acroporids

(*Acropora palifera*, *A. robusta*) and pocilloporids (*Pocillopora damicornis*, *Stylophora pistillata*).

In the case of the Pleistocene–Holocene coral reefs of the Atlantic/Caribbean palaeobiogeographic realm, reef-crest sub-environments are characterised by the elk's-horn coral, *Acropora palmata*; fore-reef sub-environments by the stag's-horn coral, *Acropora cervicornis* (Mesolella, 1967; Mesolella *et al.*, 1970; Fairbanks & Matthews, 1978; Schellmann & Radtke, 2004; Spalding, 2004; Pandolfi & Jackson, in Aronson, 2007; Jones, 2009).

3.3.3 Deep marine environments

Modern deep marine environments are characterised by diverse agglutinated foraminifera (Box 3.3), *Spirosigmoilina* (miliolide), *Dentalina*, *Lenticulina* and *Nodosaria* (nodosariides), *Bolivina*, *Francesita*, *Globobulimina*, *Sphaeroidina* and *Stilostomella* (buliminides), *Hoeglundina* (robertinide) and *Alabaminoides*, *Anomalinoides*, *Chiolostomella*, *Cibicidoides*, *Gyroidina*, *Ioanella*, *Oridorsalis*, *Osangularia*, *Osangulariella*, *Planulina* and *Pullenia* (rotaliides), and abundant and diverse planktonic foraminifera (globigerinides), and by typically thin-shelled and unornamented, and also

blind rather than sighted, ostracods (Jones, 2006). Upper to middle slope sub-environments are characterised by buliminides; lower slope environments by rotaliides; and abyssal environments, especially those below the calcite compensation depth, by agglutinated foraminifera. Oxygen minimum zone (OMZ), submarine fan, and hydrothermal vent and cold (hydrocarbon) seep sub-environments may also be differentiable.

Ancient deep marine clastic environments are characterised by taxonomically or morphologically similar faunas.

BOX 3.3   DEEP-WATER AGGLUTINATED FORAMINIFERA (MODIFIED AFTER JONES, 2006)

Introduction

Deep-water agglutinated foraminifera (DWAFs), first described from various classical turbidite or flysch localities in Europe, for example, the Gurnigel–Schlieren flysch in Switzerland, the Peira Cava flysch in France and the Macigno flysch in Italy, are composed almost exclusively of single-chambered or unilocular and uniserial multilocular agglutinating foraminifera, which typically possess non-calcareous cements.

They have been interpreted on the basis of analogy with the agglutinated components of

modern mixed foraminiferal faunas as associated with abyssal and hadal environments of deposition. A notable feature of these types of environment is that they occur below the calcite compensation depth, where secretion or preservation of calcareous tests or cements is inhibited.

Discussion

At the present time, there are considered to be a number of critical factors in the interpretation of DWAF faunas, some of which are inter-related (Jones, 1996; Jones, in Ali *et al.*, 1998).

Geological age. There are few records known to the author of Jurassic or even earliest Cretaceous DWAF faunas, and none at all of any older. This is despite the fact that flysch deposits of these ages are well documented from various localities around the world. The inference is that agglutinated foraminifera did not colonise environments characterised by flysch deposition until the later Cretaceous. Notably, some of the families most characteristic of DWAF faunas in later Cretaceous and younger deposits occur in carbonate shelf deposits of the Welsh Borderlands in the Silurian, but not in coeval 'geosynclinal' deposits of mid Wales.

Geological setting. Ancient DWAF faunas occur in a variety of geological settings. These include both active margins such as the Alpine–Carpathian orogenic belt in Europe, parts of Trinidad in the Caribbean and parts of Borneo and Papua New Guinea in the Far East, and passive margins such as the Labrador and North Seas in the North Atlantic. One factor common to both active and passive margins characterised by DWAF faunas is a restricted basin configuration brought about by structuration, in turn leading to restricted circulation and reduced oxygen concentration at or just below the sediment surface. These factors seem to favour the sustenance and/or preservation of non-calcareous agglutinated foraminifera, and the exclusion and/or dissolution of their and calcareous counterparts, especially at times of lowered sea level, when the input of organic material and hence the release of acids in early diagenetic decomposition reactions is at its highest. Restricted circulation and reduced oxygen concentration at or just below the sediment surface can also be brought about by salinity stratification associated with excess freshwater run-off from a major river system, which, incidentally, would also have the effect of eliminating the normal marine phytoplankton that, as – seasonal – phytodetritus, is known to constitute an important food source for at least some calcareous benthic species.

Water depth. Modern foraminiferal assemblages dominated by agglutinated foraminifera occur in every water depth from the marginal marine to the hadal. However, true DWAF faunas appear to be generally associated with deep rather than shallow environments. Certainly, many authors are of the opinion that the DWAF faunas of the Alpine–Carpathian orogenic belt in Europe were associated with great depth. Some others regard depth as only one of a number of factors controlling the occurrence of the group, but nonetheless infer a minimum depth corresponding to wave-base (200 m).

Sedimentary regime. Somewhat self-evidently, DWAF or 'flysch-type' faunas are associated with flysch or turbidites and hence submarine fan environments. In the case of the Miocene of Trinidad in the Eastern Venezuelan basin, an association with brackish as well as turbid water, and with the proto-Orinoco delta, has been inferred.

Instances of both horizontal variation and vertical variation in microfaunal assemblage composition within fan complexes have been reported in the literature.

Other physico-chemical factors. A number of authors have intimated that calcium carbonate content might be a critical factor controlling the distribution of DWAF faunas, and indeed there does appear to be an inverse correlation between the carbonate content of flysch deposits and the development of DWAF faunas. However, the extent to which this apparent control is primary rather than secondary, that is, essentially a function of dissolution, is unknown.

Some authors have also intimated that the availability of iron (?and silica) and the ability to secrete a ferruginous (?and siliceous) cement might be a critical factor.

Dynamic factors. There appears to be a direct correlation between the development of flysch and of DWAF or 'flysch-type-' faunas and tectonoeustatically-mediated sea-level low-stands and associated low-stand systems tracts (LSTs).

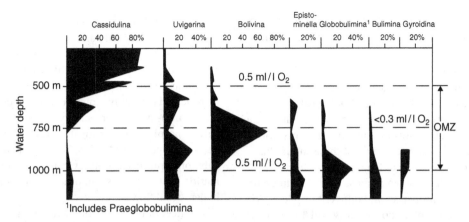

Fig. 3.12 **Distribution of modern Buliminida from offshore California in relation to the oxygen minimum zone.** © Cushman Foundation for Foraminiferal Research. Reproduced with permission from Quinterno and Gardner (1987).

Oxygen minimum zones

Modern oxygen-poor environments, such as, especially, upper slope OMZs and submarine canyon sites rich in refractory organic carbon, are characterised by infaunal buliminides (Poag, 1985; Quinterno & Gardner, 1987; Kaiho, 1994; Jorissen *et al.*, 1995; Jones, 1996; Fontanier *et al.*, 2005; Jones, 2006; Jorissen *et al.*, in Hillaire-Marcel & de Vernal, 2007; Koho *et al.*, 2007; Koho, 2008a, b; Ashckenazi-Polivoda *et al.*, 2010; see also Levin, in Gibson & Atkinson, 2003; Fig. 3.12).

It has recently been discovered that many species in oxygen-poor environments are able to obtain oxygen from nitrates (Risgaard-Petersen *et al.*, 2007; Pina-Ochoa *et al.*, 2010).

Ancient oxygen-poor and organic carbon-rich environments are characterised by taxonomically or morphologically similar benthic foraminiferal faunas, and Cretaceous 'oceanic anoxic events' (OAEs) also by similar benthic foraminiferal faunas, and by planktonic foraminifera with radially elongate chambers (Coccioni *et al.*, 2006; Friedrich & Erbacher, 2006; Jones, 2006; Friedrich, 2010).

Submarine fans

Submarine fans are sedimentary bodies formed in deep water by downslope mass transport under the influence of gravity ('turbidity currents'). Instances of horizontal and vertical variation in facies have been observed in modern and ancient fan systems.

(Horizontal variation is related to vertical variation through Walther's law.) Horizontal variation is typically manifested by a transition from sand-prone facies associated with supra-fan channels, through silt-prone facies associated with channel levees, to mud-prone facies associated with overbanks. Vertical variation is manifested by a transition from sand-prone 'turbiditic' facies at the bases of individual channel-fill units or 'Bouma cycles', through silt-prone 'interturbiditic' facies in the middle parts, to mud-prone 'hemipelagic' facies at the tops. 'Inverse' as opposed to 'normal' grading has been interpreted as indicative of hyperpycnal flow.

Hydrothermal vents, 'nekton falls' and 'cold (hydrocarbon) seeps'

Modern and ancient *hydrothermal vents and 'nekton falls'*, which are restricted to deep marine environments, and *'cold (hydrocarbon) seeps'*, which are not, all support benthic meio- and macro-biotas that ultimately depend for their survival on chemo-autotrophic or chemosynthetic Bacteria rather than photosynthetic plants or Algae, and thus on a geochemical rather than a biochemical energy source (Jones, 2006; Little, 2007). They are essentially only found where geochemical energy sources are available on the sea-floor, for example seeping methane hydrocarbon, metabolised by chemo-autotrophic or chemosynthetic Bacteria, in the case of 'cold (hydrocarbon) seeps'.

BOX 3.4   PALAEONTOLOGICAL CHARACTERISATION OF DEEP MARINE, SUBMARINE FAN SUB-ENVIRONMENTS

Introduction

In recent years there have been several studies on the palaeontological – and ichnological – characterisation of modern and ancient submarine fans, and submarine fan sub-environments (Jones, 2006). One of the objectives of these studies has been to characterise different types of intra-submarine fan reservoir mudstones with different engineering properties and different implications for reservoir exploitation strategy and 'biosteering'.

Material and methods

Studies have been undertaken on the palaeontological – and ichnological – characterisation of modern submarine fans and submarine fan sub-environments in the Gulf of Mexico, the northeast and eastern equatorial Pacific, the northwest and western equatorial Atlantic, the Ionian Sea in the Mediterranean, and the Southern Ocean; and of modern submarine canyons in the northeastern Atlantic (Fontanier *et al.*, 2005; Jones, 2006, and additional references cited therein; Jorissen *et al.*, in Hillaire-Marcel & de Vernal, 2007; Koho *et al.*, 2007; Koho, 2008a and b; Fillon, in Demchuk & Gary, 2009). Studies have also been undertaken on the palaeontological – and ichnological – characterisation of ancient submarine fans and submarine fan sub-environments in Spain, in Austria, in the Polish Carpathians and in the North Sea; and of ancient submarine fans in Spain and Norway (Jones, in Jones & Simmons, 1999; Jones *et al.*, in Powell & Riding, 2005; Jones, 2006, and additional references cited therein; Rogerson *et al.*, 2006; Knaust, 2009). This section summarises the results of some of these studies.

Results and discussion – palaeontological characterisation of submarine fans

Modern and/or ancient submarine fans are characterised by an autochthonous deep marine benthic component; an allochthonous, but contemporaneous, shallow marine benthic and planktonic, and non-marine, component; and, occasionally, an allochthonous, and non-contemporaneous component.

The autochthonous deep marine component is characterised by DWAFs and calcareous benthic foraminifera, and also, at least locally, by ostracods and by macrofaunas. The allochthonous component is characterised by shallow marine benthic and planktonic foraminifera and ostracods, and by non-marine, terrestrially derived palynomorphs such as fungal spores and hyphae, and other spores and pollen. Terrestrially derived palynomorphs occur to the virtual exclusion of marine palynomorphs and other microfossils in hyperpycnal flows.

Instances of both horizontal variation and vertical variation in benthic foraminiferal assemblage composition within fan complexes have been reported in the literature.

Horizontal variation. Horizontal variation is manifested by a transition from essentially allochthonous faunas associated with fan channel sub-environments, to autochthonous faunas associated with overbank sub-environments.

In the case of the modern Mississippi fan in the Gulf of Mexico, channel sites are characterised by the impoverished development, and overbank sites by the richer development, of an autochthonous bathyal benthic foraminiferal assemblage. Allochthonous, but contemporaneous, shelfal benthic components are present in both settings. Allochthonous, and non-contemporaneous, components are also present in both settings, though more obviously so in sand-prone channels, suggesting that reworking is associated with flows essentially confined to the channels. Modern and ancient channel sites of the Monterey and other fans off California in the northeast Pacific, the Baltimore canyon in the northwest Atlantic, and the Amazon fan in the western equatorial Atlantic are also characterised by contemporaneous and non-contemporaneous allochthonous assemblages.

Ancient, Eocene channel axis sites on the Ainsa fan in the Spanish Pyrenees are characterised by generally low-diversity agglutinated and calcareous benthic foraminiferal assemblages; channel off-axis sites by generally high-diversity assemblages; and levee/overbank sites by moderate-diversity assemblages (Jones *et al.*, in Powell & Riding, 2005; Fig. 3.13).

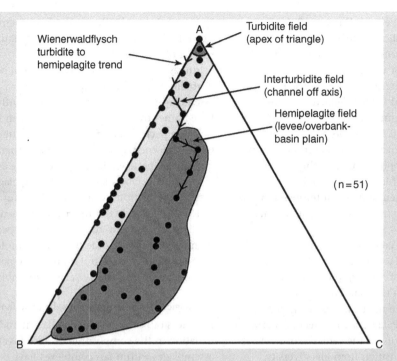

Fig. 3.13 **Triangular plot of agglutinating foraminiferal 'morphogroups' in submarine fan sub-environments.**

Similarly, Miocene canyon axis sites in the El Buho system in the Tabernas basin in southeastern Spain are characterised by low-abundance and low-diversity assemblages; and canyon off-axis and extra-canyon sites by high-abundance and high-diversity assemblages (Rogerson *et al.*, 2006). Here, interestingly, agglutinated benthic foraminifera are significantly more conspicuous at extra-canyon than at canyon sites, as are several species of calcareous benthic foraminifera.

The observation that the highest diversity on the Ainsa fan is at channel off-axis sites is attributed to the presence of both autochthonous deep-water and – contemporaneous – allochthonous shallow-water components. Interestingly, channel off-axis sites are characterised by increased incidences of epifaunal suspension-feeders relative to channel axis and levee/overbank sites ('Morphogroup' A of Jones & Charnock, 1985), at least arguably indicating an ecological preference for such sites rather than a taphonomic artefact – that is to say,

habitation rather than transportation. Canyon off-axis sites in the El Buho system are characterised by increased incidences of epifaunal or shallow infaunal *Cassidulina laevigata*, indicating an ecological preference for such sites, interpreted as characterised by sediment flow activity sufficient to increase the supply of food but insufficient to cause physical disturbance. (Local sites on the El Buho fan, characterised by sediment flow activity sufficient to increase the supply of food to an excess, are dominated by dysoxia-tolerant deep infaunal *Globobulimina* spp.)

Note in this context that sites of turbidite sedimentation on the modern Crati fan in the Ionian Sea are characterised by increased incidences of filter-feeding bivalve morphotypes. Note also that modern submarine fan canyon sites off the coast of California, which act as conduits of ?terrestrial and coastal macrophyte detritus to the deep sea, are characterised by higher macrofaunal abundances and diversities than extra-canyon sites (Vetter & Dayton, 1998). Carbon isotope analyses indicate that at least certain canyon-dwelling species metabolise primarily coastal detritus such as kelp and surf-grass, while non-canyon-dwelling species metabolise primarily pelagic detritus, that is, that derived from the overlying water column. Canyon-dwelling species include the echiuran *Listriolobus pelodes*, which is known to be tolerant not only of eutrophication but also of hypertrophication; and the bivalve *Thyasira flexuosa*, which contains chemosymbionts that oxidise hydrogen sulphide, interpreted in this instance as deriving from decomposing organic matter.

Vertical variation. Vertical variation, as observed, for example, in surface outcrops in the Wienerwaldflysch of Austria, in the Polish Carpathians and on the island of Gadvos off Crete in the eastern Mediterranean, and in subsurface cores cut from the Palaeocene Maureen, Andrew and Forties formations of the central North Sea, is manifested by a transition from low diversity, interpreted essentially allochthonous assemblages associated with turbidite deposits to high diversity, interpreted autochthonous assemblages associated with hemipelagites (Holmes, in Jones & Simmons, 1999; Payne *et al.*, in Jones & Simmons, 1999; Wakefield *et al.*, 2001; Drinia *et al.*, 2007). According to this interpretation, the observed variation through individual turbidite packages, from assemblages characterised by epifaunal suspension-feeders ('Morphogroup' A of Jones & Charnock, 1985) to assemblages characterised by epifaunal and infaunal detritus-feeders ('Morphogroups' B and C), reflects the greater ability of the former group to colonise the sea-floor in between turbidity flows, and to exploit the source of food in the turbid suspension. According to an alternative interpretation, the assemblages associated with hemipelagites, as well as those associated with turbidites, are allochthonous, and the observed vertical variation is attributable to waning flow, and associated hydrodynamic sorting of particles of differing settling velocity. Note in this context, though, that turbidites from the Monterey fan are characterised by hydrodynamic sorting of foraminifera, but hemipelagites are not.

Note also that turbidites from off southern Australia are characterised by hydrodynamic sorting of ostracods.

Interestingly, similar vertical variation between 'turbidite-type' and 'hemipelagite-type' assemblages is also observed over evidently longer time-frames than those represented by individual turbidite packages, as, for example, in the Oligo-Miocene sequences of offshore Angola. Here, 'turbidite-type' assemblages appear to be associated with sequence boundaries, and 'hemipelagite-type' assemblages with maximum flooding surfaces (MFSs), of 'second-' to 'third-order' sequences mediated by tectono- or glacio-eustasy, perhaps pointing toward an underlying medium- to long-term climatic control on both the nature of sedimentation and the associated assemblage type.

Ichnological characterisation of submarine fans

Instances of both horizontal variation and vertical variation in the composition of trace fossil assemblages within fan complexes have been reported in the literature.

Horizontal variation. Horizontal variation is manifested by a transition from low-diversity assemblages dominated by interpreted opportunistic essentially shallow-water taxa such as *Ophiomorpha* in inner fan and canyon sub-environments, to high-diversity assemblages dominated by deep-water taxa in outer fan and fan-fringe sub-environments.

In a recent case study on a submarine fan complex in the Campanian Nise formation of the Norwegian Sea, *Ophiomorpha* ichnofabrics were found to characterise proximal, inner to middle, fan turbiditic sand sub-environments; *Planolites* ichnofabrics distal, middle to outer, fan or inter-channel interbedded turbiditic and interturbiditic sand and silt sub-environments; *Scolicia* ichnofabrics distal fan and proximal overbank interbedded turbiditic and interturbiditic silt and mud sub-environments; *Palaeophycus–Planolites* ichnofabrics distal overbank interturbiditic mud sub-environments; *Phycosiphon* ichnofabrics fan-fringe interturbiditic and hemipelagic mud sub-environments; *Thalassinoides* ichnofabrics basin-plain hemipelagic mud sub-environments; and *Chondrites* ichnofabrics dysoxic basin-plain hemipelagic mud sub-environments (Knaust, 2009).

Vertical variation. Vertical variation is manifested by a transition from low-diversity assemblages dominated by interpreted opportunistic taxa associated with turbidites, to high-diversity assemblages associated with hemipelagites, together with pervasive bioturbation and complex tiering. The rate and pattern of recolonisation appears to vary according not only to physical effects, but also to biological effects, including feeding and reproductive behaviours, such that there is no standard succession. Many of the opportunist species observed exploiting disturbed environments in modern settings are soft-bodied, and have no body-fossil counterparts in ancient settings. Examples include the polychaete worms *Streblospio benedicti*, *Hobsonia florida*, *Polydora ligni* and *Capitella capitata*, which are all deposit-feeders with short reproductive cycles embodying a planktonic larval phase. Polychaetes have been observed to be significantly less impacted than tanaidaceans, isopods and bivalves by experimentally induced physical disturbance in the eastern equatorial Pacific, the principal effect of which was the expansion of the 'semi-liquid' zone at the sediment-water interface (Borowski & Thiel, 1998).

Modern *hydrothermal vents*, which are restricted to deep marine environments, support biotas comprising over 400 species, over 80% of which are endemic to hydrothermal vent sites (van Dover, 2000; Little & Vrijenhook, 2003; Campbell, 2006; Jones, 2006; Leveque *et al.*, 2006; Lietard & Pierre, 2009). The dominant groups are vestimentiferan worms such as *Riftia*, and bivalves such as the giant vesicomyid clam *Calyptogena* and the 'vent-mussel' *Bathymodiolus*, although copepods are also locally important. *Riftia* and *Calyptogena* host chemo-autotrophic bacterial symbionts. *Bathymodiolus* hosts methane-oxidising bacterial symbionts. Modern hydrothermal vents and associated biotas are known from the mid-Atlantic, the eastern and western Pacific and the Indian Ocean. Ancient hydrothermal vents and associated biotas, which also include foraminifera, corals and brachiopods, are known from the Palaeozoic to Cenozoic (Campbell & Bottjer, 1995; Little & Vrijenhook, 2003; Berkowski, 2004; Campbell, 2006; Jones, 2006; Little, 2007; Tyszka *et al.*, 2010).

Modern wood, fish and whale falls or, loosely, 'nekton falls', which are also restricted to deep marine environments, support biotas comprising foraminiferans, bivalve molluscs including the mytilid *Adipicola*, the solemyid *Solemya* and the vesicomyids *Vesicomya* and *Calyptogena*, and gastropods, including *Provanna* (Amano & Little, 2005; Majima *et al.*, 2005; Kiel & Goedert, 2006; Amano *et al.*, 2007; Kiel, 2008; Dominici *et al.*, 2009; Shapiro & Spangler, 2009; McGann *et al.*, 2010). Modern 'nekton falls' and associated biotas have only comparatively recently come to light, such that their distributions are currently probably only incompletely known. Interpreted ancient 'nekton falls' and associated biotas have been described from the Cretaceous and Cenozoic of the Pacific, and from the Cenozoic of the Mediterranean.

Modern 'cold (hydrocarbon) seeps', which are not restricted to deep marine environments, support biotas comprising over 200 species of organisms, the vast majority of which are endemic to 'cold seep' sites, and the remainder of which are only known elsewhere from hydrothermal vent sites (Campbell & Bottjer, 1995; Campbell, 2006; Jones, 2006; Kiel & Little, 2006; Mae *et al.*, 2007; Olu-Le Roy *et al.*, 2007; Sahling *et al.*, 2008; Xiqiu Han *et al.*, 2008; Kaim *et al.*, 2009; Kiel & Dando, 2009; Lietard & Pierre, 2009; Baco *et al.*, 2010; Kiel, 2010; Martin *et al.*, 2010; Thurber *et al.*, 2010). Characteristic groups include vestimentiferan worms, soft-bodied pogonophoran worms and copepods. Cyanobacteria, sponges, corals, brachiopods, bryozoans, bivalves, gastropods and echinoids also occur in association with 'cold seeps'. Modern cold seeps and associated biotas are known from the North Sea, the North and South Atlantic, the Gulf of Mexico, the Barbados accretionary prism in the Caribbean, the North Pacific, the South China Sea and New Zealand. Ancient cold seeps and

BOX 3.5   BENTHIC FORAMINIFERA AND OSTRACODS ASSOCIATED WITH 'COLD (HYDROCARBON) SEEPS' (MODIFIED AFTER JONES, 2006)

Introduction

In recent years there have been several studies on benthic foraminifera and, to a lesser extent, ostracods associated with modern and ancient hydrocarbon seeps. The ultimate objectives of these studies have been to evaluate the potential of benthic foraminifera as indicators of modern and ancient hydrocarbon seepage, and/or to determine the timing and duration of seepage. Respective applications are in providing sense checks on geochemical indications of hydrocarbon seepage, and in determining the timing of hydrocarbon charge in relation to that of trap formation.

Material and methods

Studies have been undertaken on benthic foraminifera and, to a lesser extent, ostracods associated with modern seeps all around the world (Torres *et al.*, 2003; Panieri, 2005; Millo *et al.*, 2006; Panieri, 2006; Sen Gupta *et al.*, 2007; Lobegeier & Sen Gupta, 2008; Panieri & Sen Gupta, 2008; Martin *et al.*, 2010; Panieri & Camerlanghi, 2010; Torres *et al.*, 2010). Studies have also been undertaken on benthic foraminifera associated with ancient seeps in the Miocene, Pliocene and Pleistocene (Martin *et al.*, 2007; Panieri *et al.*, 2009). This box summarises the results of some of these studies. The results are not all directly comparable, as the studies did not involve directly comparable analytical and interpretive procedures. Nonetheless, the following observations and comparisons can be offered.

Results and discussion

Samples from seep and control sites have proved distinct when compared with one another using specific-level similarity indices.

Abundance, diversity, dominance and equitability. In the North Sea and Gulf of Mexico, seep samples generally proved distinct in terms of lower abundance and diversity of foraminifera, higher dominance and lower equitability (that is, evenness of species distribution).

In the Blake Ridge area of the western North Atlantic, and also in the Adriatic, seep samples proved distinct in terms of higher abundance of foraminifera (Panieri, 2006; Panieri & Sen Gupta, 2008). One possible explanation is that the bacterial mat associated with the seeps here provides an additional food source.

In the Porcupine basin in the North Atlantic, seep samples proved distinct in terms of lower abundance, but higher diversity, of foraminifera. One possible explanation for the higher diversity is that the hard substrate provided by the carbonates associated with the seeps provides niches for a wider variety of species. Note in this context that attached species are significantly both more abundant and more diverse in the seep samples. Note also that species attached to vestimentiferan tube-worms well above the sediment–water interface, possibly in order to avoid dysoxic and/or toxic conditions there, have recently been recorded at seep sites in the Gulf of Mexico (Sen Gupta *et al.*, 2007; Lobegeier & Sen Gupta, 2008). Also in the Porcupine basin in the North Atlantic, seep samples proved distinct in terms of increased incidences of the ostracods *Paradoxostoma*, *Paracytherois* and *Propontocypris*. Interestingly, these taxa are also known from hydrothermal vents, and from so-called 'wood-island' habitats.

Assemblage composition. In New Zealand, seep samples proved distinct in terms of proportionately higher abundances of calcareous benthic foraminifera and lower abundances of agglutinated foraminifera. In the North Sea, the North Atlantic and the Gulf of Mexico, seep samples proved distinct in terms of proportionately higher abundances of essentially epifaunal rotaliide and lower abundances of essentially infaunal buliminide calcareous benthic foraminifera.

The epifaunal calcareous benthic rotaliides *Cibicidoides pachyderma*, *Elphidium ex gr. clavatum* and *Hyalinea balthica* appear positively correlated with seepage at more than one site, as also does the epifaunal agglutinated foraminifer *Arenomeandrospira glomerata*, at Recent 'pockmark' sites in the Skagerak and Kattegat in Scandinavia, and at a Miocene site in the Nam Con Son basin in Vietnam.

The infaunal calcareous benthic buliminides *Bulimina marginata* and *Trifarina angulosa* appear negatively correlated with seepage at more than one site. The infaunal rotaliide *Chilostomella oolina* also appears negatively correlated with seepage, at one site in the Gulf of Mexico and another in Japan in the Pacific.

Interestingly, some species of infaunal or 'surficial' buliminide appear positively correlated with seepage, including *Cassidulina laevigata carinata s.l.* in the North Sea, Gulf of Mexico and North Atlantic, and *Rutherfordoides cornuta* in Japan and California in the Pacific. Only these species appear able to tolerate the immediately sub-surface hydrogen sulphide, dysoxia or hypoxia, and hypo- or hypersalinity associated with the seepage. Note in this context that it has been hypothesised that *Cassidulina laevigata carinata s.l.* is a facultative anaerobe or microaerophile.

Carbon isotope analyses of the tests or cytoplasm of foraminifera associated with seepage in the Recent of the Adriatic, of California and Oregon in the Pacific, and of New Zealand, indicate that the infaunal buliminides *Bolivina subargentea* and *Uvigerina peregrina* are actually metabolising methane, or more accurately that the carbon that they contain is derived either directly or indirectly from the sulphate-dependent anaerobic oxidation of methane (Torres *et al.*, 2003; Panieri, 2006; Martin *et al.*, 2010). Analyses of the tests of foraminifera associated with seepage in the Pliocene of the Cascadia accretionary margin in Washington in the Pacific indicate that the epifaunal rotaliides *Cibicidoides mckannai* and *Nonion basispinatum* and the infaunal buliminide *Globobulimina pacifica* appear to have been metabolising methane (Martin *et al.*, 2007).

associated biotas are known from the Palaeozoic of Greenland, Germany, Morocco and Namibia, the Mesozoic of Arctic Canada, California, Argentina, Greenland, the United Kingdom and France, the Mesozoic and Cenozoic of Japan, and the Cenozoic of Washington state in the United States, the Caribbean, Peru, Italy and Vietnam (Goedert & Squires, 1990; Campbell & Bottjer, 1995; Gischler *et al.*, 2003; Goedert *et al.*, 2003; Berkowski, 2004; Majima *et al.*, 2005; Peckmann & Goedert, 2005; Peckmann *et al.*, 2005; Campbell, 2006; Jones, 2006; Cavalazzi, in Coccioni & Marsili, 2007; Little, 2007; Allison *et al.*, 2008; Himmler *et al.*, 2008; Kiel *et al.*, 2008; Jones, 2009; Kaim *et al.*, 2010). There are three types of 'cold seep' biota in the Mesozoic and Cenozoic of Japan (Majima *et al.*, 2005). Type I is characterised by interpreted allochthonous vesicomyid bivalves and/or tube-worms associated with diapirism, faulting and slumping in deep-water environments (>1000 m). Type II is dominated by interpreted autochthonous vesicomyids associated with shallower-water environments (<1000 m). Type III is dominated by interpreted autochthonous *Lucinoma* and/or *Conchocele* associated with shallow-water environments (<300 m).

3.4 PALAEOBIOGEOGRAPHY

As noted above, dinoflagellates, diatoms, calcareous nannoplankton, calcareous Algae, acritarchs, foraminifera, radiolarians, calpionellids, plants, ediacarans, SSFs, sponges, archaeocyathans, stromatoporoids, corals, brachiopods, bryozoans, bivalves, gastropods, ammonoids, belemnites, tentaculitids, trilobites, ostracods, crinoids, echinoids, graptolites, fish, amphibians, reptiles and birds, mammals and trace fossils all appear to have had restricted or endemic biogeographic distributions at one time or another, rendering them of use in the characterisation of palaeobiogeographic realms or provinces, and in turn in the constraint of plate tectonic and terrane reconstructions (Hallam, 1973; Meyerhoff *et al.*, 1996; Jones, 2006; Figs. 3.14–3.22; see also 3.4.1–3.4.4 below).

Importantly, fossils provide at least some measure of longitude as well as latitude (not provided by palaeomagnetism). The palaeobiogeographic dispersal of terrestrial organisms ordinarily occurs principally during low-stands of sea level, and that of marine and oceanic organisms during high-stands (Gheerbrant & Rage, 2006). Oceanic dispersal of terrestrial organisms has recently been reviewed by de Queiroz (2005). The disjunct

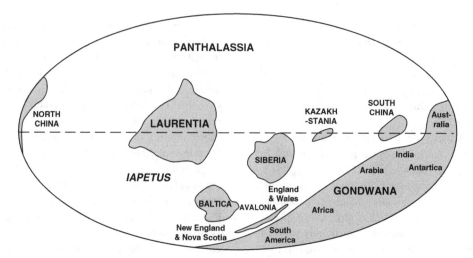

Fig. 3.14 **Cambrian palaeogeography and palaeobiogeography.** From Jones (2006).

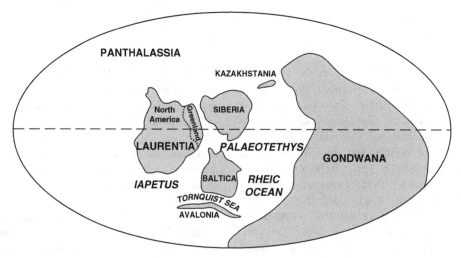

Fig. 3.15 **Ordovician palaeogeography and palaeobiogeography.** From Jones (2006).

populations of a wide range of terrestrial plant and animal taxa on either side of the present Atlantic, Indian and Pacific oceans have been re-interpreted as explicable in terms of oceanic dispersal rather than plate tectonic vicariance. The re-interpretation is supported by molecular dating of lineage divergences.

Palinspastically restored palaeogeographic maps for the Phanerozoic, constrained in part by palaeobiogeographic distribution data, can be found at Chris Scotese's website (www.paleomap.org).

3.4.1 Proterozoic

'Ediacarans' and SSFs appear to have had restricted or endemic biogeographic distributions in the Proterozoic, rendering them of use in the characterisation of Proterozoic palaeobiogeographic realms or provinces, and in turn in the constraint of Proterozoic plate tectonic and terrane reconstructions (Jones, 2006).

Palaeobiogeographic realms or provinces

The bilaterally symmetrical 'edicarans' appear to have had a provincial or endemic distribution, in

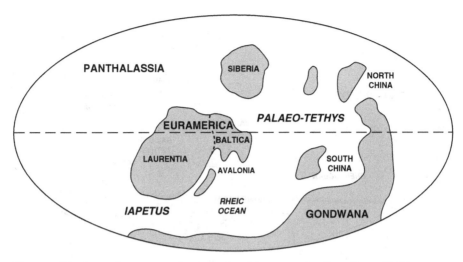

Fig. 3.16 **Silurian palaeogeography and palaeobiogeography**. From Jones (2006).

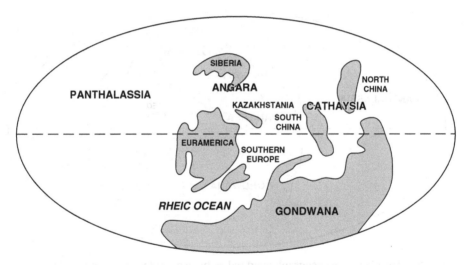

Fig. 3.17 **Devonian palaeogeography and palaeobiogeography**. From Jones (2006).

equatorial palaeolatitudes. Parsimony analysis of endemism (PAE) points toward separate 'ediacaran' provinces centred on Avalonia, the White Sea and Namibia. The SSF *Cloudina* appears to have been essentially endemic to proto-Gondwana, and *Anabarites* to proto-Laurasia.

3.4.2 Palaeozoic

Calcareous Algae, acritarchs, foraminifera, radiolarians, plants, SSFs, sponges, archaeocyathans, stromatoporoids, corals, brachiopods, bryozoans, bivalves, gastropods, ammonoids, tentaculitids, trilobites, ostracods, crinoids, echinoids, graptolites, fish, amphibians, reptiles and birds, mammals and trace fossils all appear to have had restricted or endemic biogeographic distributions in the Palaeozoic, rendering them of use in the characterisation of Palaeozoic palaeobiogeographic realms or provinces, and in turn in the constraint of Palaeozoic plate tectonic and terrane reconstructions (Hallam,

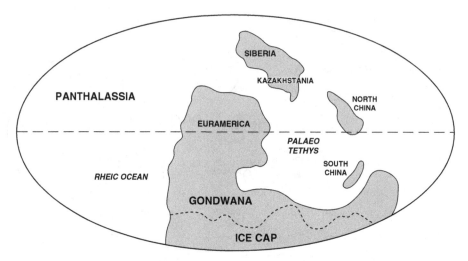

Fig. 3.18 **Carboniferous palaeogeography and palaeobiogeography.** From Jones (2006).

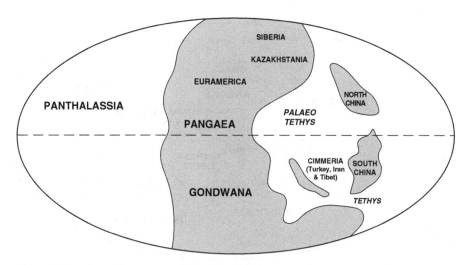

Fig. 3.19 **Permian palaeogeography and palaeobiogeography.** From Jones (2006).

1973; McKerrow & Scotese, 1990; Meyerhoff *et al.*, 1996; Cocks & Torsvik, 2005; Jones, 2006; Raymond *et al.*, 2006; Figs. 3.14–3.19; see also below).

Palaeobiogeographic realms or provinces

Marine palaeobiogeographic realms or provinces. Fossil, especially brachiopod, bivalve, trilobite, ostracod and graptolite, distributions point toward the existence of three marine palaeobiogeographic realms in the Cambrian, namely the Laurentia–Baltica–Avalonia, Siberia, and Gondwana realms; and four realms in the Ordovician, namely the Laurentia, Baltica–Avalonia, low-latitude Gondwana and high-latitude Gondwana realms. In the Ordovician, Baltica–Avalonia, of which England and Wales were a part, was separated from Laurentia, of which Scotland was a part, by the Iapetus Ocean, a substantial body of deep water that constituted an insuperable barrier to the dispersal of shallow-water benthic organisms. Baltica–Avalonia was also separated from Gondwana by the Rheic Ocean. Towards the end of the Ordovician, provincialism generally decreased, and cosmopolitanism increased, as Baltica–Avalonia, Laurentia and Gondwana drifted towards one another. The cold-adapted *Hirnantia* brachiopod fauna appears to have had an essentially cosmopolitan distribution during the Late Ordovician, late Ashgillian, Hirnantian glaciation, and warm-adapted taxa appear to have had cosmopolitan distributions during the succeeding Early Silurian, Llandoverian deglaciation. Provincialism became re-established in the Devonian, and the Malvino-Kaffric province, centred on South America, the Falklands or Malvinas, and South Africa, became established for the first time, within Gondwana. Further marine provinces became established by the Permian. The cold-adapted opportunistic r-strategist *Pachycyrtella omanensis* was the dominant taxon in the early stages of the Permian deglaciation in Oman. The *Pachycyrtella omanensis* community was succeeded by a more diverse and stable warm-adapted *Reedoconcha permixta–Punctocyrtella spinosa* community, also known from India, Afghanistan, the Himalayas and Thailand (the so-called 'Westralian province', possibly part of a still larger 'Indoralian province', also including eastern Australia), in the later stages of the deglaciation.

Trace fossil distributions indicate that there was a marine connection, or 'Trans-Gondwana Seaway', between the Middle East and north and west Africa on the one hand, and eastern South America on the other, in the Silurian (Seilacher, 2007). Sponge and coral distributions indicate that there was an open marine connection, or 'Iberian–Midcontinent Seaway', between western Palaeotethys and eastern Panthalassa, in the Moscovian, that was closed by the Kasimovian (Garcia-Bellido & Rodriguez, 2005).

Non-marine palaeobiogeographic realms or provinces. Plant and vertebrate distributions point toward the existence of four generally separate non-marine, terrestrial palaeobiogeographic realms in the Devonian–Permian, namely the Angara, Euramerica, Cathaysia and Gondwana realms, but only one amalgamated one by the end of the Permian, namely the Pangaea realm. In the Carboniferous-Permian, the Gondwana realm was characterised by the plant *Glossopteris* (Anderson *et al.*, 1999). Note, though, that part of Euramerica was also characterised by *Glossopteris*, specifically *Glossopteris anatolica* (Berthelin *et al.*, 2006).

Terrane reconstruction

In terms of terrane reconstruction, the occurrence of the calcareous Alga *Palaeoaplysina* in the Klamath Mountains of California indicates a palaeobiogeographic link with Idaho, British Columbia, Yukon, and Ellesmere and Axel Heiberg Islands elsewhere in North America in the Late Carboniferous–Early Permian, and further indicates that the Klamath Mountains do not form part of a far-travelled terrane, as has been hypothesised. *Palaeoaplysina* is also found in Svalbard, Bjornoya, Timan–Pechora and the Urals.

3.4.3 Mesozoic

Dinoflagellates, diatoms, calcareous nannoplankton, calcareous Algae, acritarchs, foraminifera, radiolarians, calpionellids, plants, sponges, stromatoporoids, corals, brachiopods, bryozoans, bivalves, gastropods, ammonoids, belemnites, ostracods, crinoids, echinoids, fish, amphibians, reptiles and birds, and mammals all appear to have had restricted or endemic biogeographic distributions in the Mesozoic, rendering them of use in the characterisation of Mesozoic palaeobiogeographic realms or provinces, and in turn in the constraint of Mesozoic plate tectonic and terrane reconstructions (Hallam, 1973; Meyerhoff *et al.*, 1996; Jones, 2006; Figs. 3.20–3.22; see also below).

Palaeobiogeographic realms or provinces

Marine palaeobiogeographic realms or provinces. Fossil, especially dinoflagellate, foraminiferal, radiolarian, bivalve, ammonoid, belemnite and ostracod distributions point toward the existence of three marine palaeobiogeographic realms in the Triassic–Cretaceous, namely the Tethyan, Boreal and Austral realms (Masure & Vrielynck, 2009). As a general rule, the low-latitude Tethyan realm was distinguishable by the presence, and the high-latitude Boreal and Austral realms by the absence, of algal–coral–rudist reefs. Note, though, that there were provinces within each of these three realms,

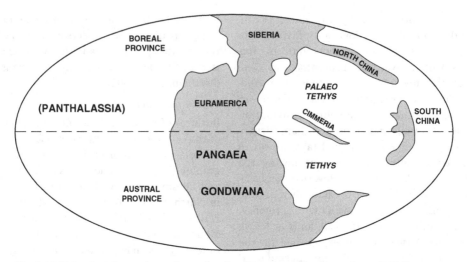

Fig. 3.20 **Triassic palaeogeography and palaeobiogeography.** From Jones (2006).

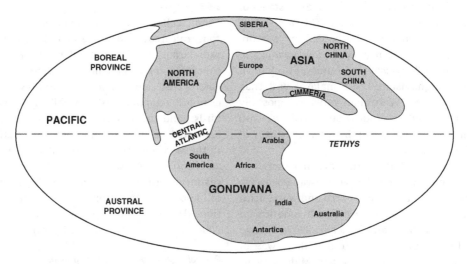

Fig. 3.21 **Jurassic palaeogeography and palaeobiogeography.** From Jones (2006).

for example, the Arctic, Boreal–Atlantic and Boreal–Pacific provinces in the Boreal realm; and also that the boundaries between the various provinces and realms changed through time in response to tectonic, climatic and tectono- or glacio-eustatic sea-level changes, and the consequent creation or destruction of dispersal routes.

Ammonite distributions indicate that there was an intermittent marine connection, or 'Turgai Strait', between the Tethyan and Boreal realms, in

the 'Middle' Cretaceous. Ammonite and marine turtle distributions indicate that there was also a marine connection, or 'Trans-Saharan Seaway', between the Tethyan and Austral realms in the Late Cretaceous (Moody, in Callaway & Nicholls, 1997; Zaborski & Morris, 1999).

Non-marine palaeobiogeographic realms or provinces. Non-marine ostracod, and terrestrial plant and vertebrate distributions point toward the existence of a single non-marine, terrestrial palaeobiogeographic

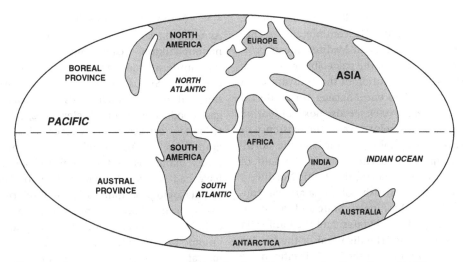

Fig. 3.22 **Cretaceous palaeogeography and palaeobiogeography.** From Jones (2006).

realm throughout much of the Mesozoic, namely the Gondwana realm (Upchurch *et al.*, 2002; Canudo *et al.*, 2009). However, in the Cretaceous, provincialism became established, as the eastern Indian Ocean began to rift open between Greater India and Australia, and the South Atlantic began to rift open between Africa and South America, forming barriers to the dispersal of terrestrial organisms (McLoughlin & Pott, 2009). (During the Cretaceous, the configuration of the continents and oceans changed from one that can be thought of as characteristically Mesozoic to one that can be thought of as characteristically Cenozoic.) Plant and pollen distributions point toward the existence of a distinct West Africa/South America or WASA province in the Gondwana realm in the Early Cretaceous; and distinct *Aquilapollenites* and *Normapolles* provinces in the Euramerican realm, and a distinct *Nothofagidites* province in the Gondwana realm, in the Late Cretaceous (Traverse, 2008)'. The WASA province in the Gondwana realm in the Early Cretaceous is characterised by *Classopollis* pollen and its parent plant, an extinct cheirolepidacean (Alvin, 1983; Seward, in McCabe & Parrish, 1992; Xiao-Yu Jang, 2008).

Terrane reconstruction

In terms of terrane reconstruction, the occurrence of identical sauropod dinosaurs in the Late Jurassic–Early Cretaceous of Apulia in Italy and of

north Africa indicates that the two areas were contiguous at this time, necessitating a re-evaluation of the previously accepted geodynamic model for the opening of the eastern Mediterranean (Bosellini, 2002).

3.4.4 Cenozoic

Dinoflagellates, diatoms, calcareous nannoplankton, calcareous Algae, acritarchs, foraminifera, radiolarians, plants, sponges, stromatoporoids, corals, brachiopods, bryozoans, bivalves, gastropods, ostracods, crinoids, echinoids, fish, amphibians, reptiles and birds, and mammals all appear to have had restricted or endemic biogeographic distributions in the Cenozoic, rendering them of use in the characterisation of Cenozoic palaeobiogeographic realms or provinces, and in turn in the constraint of Cenozoic plate tectonic and terrane reconstructions (Hallam, 1973; Meyerhoff *et al.*, 1996; Jones, 2006; see also below).

Palaeobiogeographic realms or provinces

Marine palaeobiogeographic realms or provinces. Larger benthic foraminifer (LBF) and z-coral distributions point toward the existence of three marine palaeogeographic provinces within the Tethyan realm, namely the Atlantic/Caribbean, Mediterranean, and Indo-Pacific provinces, in the Cenozoic (Jones, in Agusti *et al.*, 1999; Rosen, in Agusti *et al.*,

1999; Chaix & Cahuzac, 2005; Renema, in Renema, 2007; Renema et al., 2008). The commonality of LBFs between the Indo-Pacific and Mediterranean provinces is a key indicator of the palaeo(bio)geographic and palaeoclimatic evolution of the Tethyan realm over the critical Oligocene–Miocene time interval (Jones, in Agusti et al., 1999; see also Box 3.8). Marine bivalve distributions point toward the existence of three palaeobiogeographic provinces in the northeast Pacific in the Neogene, namely the Gulf of Alaska province to the north; the Pacific Northwest province, centred on British Columbia and Vancouver in Canada, and Washington State and Oregon in the conterminous United States; and the California province to the south. The dispersal of the bivalve *Astarte* from the Arctic into the North Pacific in the Pliocene demonstrates that the Bering Strait had opened by this time. Marine gastropod distributions demonstrate that the Bering Strait was closed by the Pleistocene, when vertebrate evidence indicates that the 'Bering land bridge' was open (see below).

Diatom and radiolarian distributions indicate that there was an intermittent marine connection, or 'Obik Sea' or 'Turgai Strait', between the Tethyan and Boreal realms, in the Palaeogene. Ostracod distributions indicate that there was a continuing connection, or 'Trans-Saharan Seaway', between the Tethyan and Austral realms, in the Palaeogene (Zaborski & Morris, 1999).

Non-marine palaeobiogeographic realms or provinces. Plant and vertebrate distributions point toward the palaeogeographic and palaeoclimatic evolution of the North Atlantic in the Palaeocene–Eocene (see also Box 3.7). Plant and vertebrate distributions are key indicators of the palaeo(bio)geographic and palaeoclimatic evolution of Eurasia over the critical Oligocene–Miocene time interval (Jones, in Agusti et al., 1999; see also Box 3.8).

Vertebrate distributions indicate that there was an intermittent land connection, or 'Thulean land bridge', between North America and Eurasia in the Eocene (see also Box 3.7); and another, the 'Bering land bridge', in the Miocene and Pleistocene, although not in the Pliocene or Holocene, when mollusc evidence indicates that the Bering Strait was in being (Wang & Tedford, 2008; see also above).

3.5 PALAEOCLIMATOLOGY

As noted above, dinoflagellates, diatoms, calcareous nannoplankton, calcareous Algae, acritarchs, foraminifera, radiolarians, calpionellids, plants, ediacarans, SSFs, sponges, archaeocyathans, stromatoporoids, corals, brachiopods, bryozoans, bivalves, gastropods, ammonoids, belemnites, tentaculitids, trilobites, ostracods, insects, crinoids, echinoids, graptolites, fish, amphibians, reptiles and birds, mammals and trace fossils all appear to have had restricted palaeobathymetric/altitudinal or palaeobiogeographic/latitudinal distributions at one time or another, rendering them of use not only in palaeobathymetric and palaeobiogeographic interpretation but also, alongside non-biological, lithological and isotopic proxies, in palaeotemperature and palaeoclimatic interpretation, in *palaeoclimatology*, and hence in studies of global change (Jones, 2006; Cronin, 2010; see also Section 9.4 below). The mutual climate range (MCR), transfer functions based on fossil distributions, and other quantitative techniques are also of use.

Plant macro- and microfossils are of particular use in palaeoclimatic interpretation. Climate has been inferred from plant macrofossil distribution, abundance and diversity, and also from functional morphology and leaf physiognomy (for example through the use of leaf margin analysis or LMA, and of the Climate Leaf Analysis Multivariate Program or CLAMP; tropical rainforest from a high incidence of plants with 'drip tips' at the ends of their leaves, etc. Climate has also been inferred from plant microfossil (spore and pollen) distribution, abundance and diversity. Atmospheric carbon dioxide concentration has been inferred from stomatal density. Not only have plant distributions been used to model palaeoclimates; independently reconstructed palaeoclimates have been used to model plant distributions and vegetation cover, for example for the Carboniferous, Jurassic, Cretaceous, Eocene, Pliocene and Quaternary (Beerling & Woodward, 2001; Haywood & Valdes, 2006). The computer programmes used in this modelling include the Hadley Centre (for Climate Prediction and Research: part of the UK Meteorological Office) Atmospheric General Circulation Model Version 3 or HadAM3, the Dynamic Global Vegetation Model or DGVM, and the Top-down Representation of Interactive Flora and Foliage Including Dynamics or TRIFFID (named, after a certain amount of contrivance, after John Wyndham's novel 'The Day of the Triffids', in which plant-like aliens take over the world).

Importantly, foraminifera, brachiopods, belemnites, fish, reptiles and mammals are useful in the direct measurement rather than indirect interpretation of palaeotemperature, based on the oxygen isotopic record of changing temperature or ice volume that they preserve in their shells, bones or teeth (note in this context that certain other isotopic records that they preserve are increasingly also being used in palaeothermometry). The theory is that the heavier isotope of oxygen, ^{18}O, is proportionately commoner in the atmosphere and oceans, and in the teeth, bones or shells of terrestrial and marine organisms, during glaciations, on account of differential evaporation and sequestration into ice-buildups of the lighter isotope, ^{16}O (Jones, 2006). (Incidentally, recent research has shown that long-term climatic maxima or so-called 'greenhouse' episodes are associated with atmospheric carbon dioxide concentrations of >1000 ppm, and climatic minima or 'ice-house' episodes, with the exception of the end-Ordovician one, with concentrations of <500 ppm: Royer, 2006.) High-resolution palaeotemperature curves have been constructed on the basis of the marine oxygen isotope record that also serve, when calibrated against biostratigraphy, magnetostratigraphy or absolute chronostratigrahy, as the basis of a workable 'marine isotope stage' (MIS) or 'oxygen isotope stage' (OIS) climatostratigraphy, at least for the Tertiary and Quaternary. Belemnites are particularly important in palaeotemperature measurement, leastwise in the Jurassic–Cretaceous (Price et al., 2000; Price & Mutterlose, 2004; Jones, 2006; Rexfort & Mutterlose, 2009; Malkoc & Mutterlose, 2010; Nunn & Price, 2010; Nunn et al., 2010; Price, 2010). This is on account of their enhanced resistance to chemical change during diagenesis, and their consequent ability to preserve intact geochemical signals such as oxygen – and strontium – isotope ratios. Indeed, the 'Pee Dee' belemnite standard is that against which such measurements are calibrated.

Certain molecular biological indicators are also useful in palaeotemperature interpretation, for example TEX86, an index of isoprenoid glycerol dibiphytanyl glycerol tetra-ethers (GDGTs) in lipids derived from different types of terrestrial and marine crenarchaeote 'Archaebacteria' with different temperature tolerances (Schouten et al., 2002; Powers et al., 2004; Weijers et al., 2006; Rosell-Mele & McClymont, in Hillaire-Marcel & de Vernal, 2007; Jung-Hyn Kim et al., 2008).

3.5.1 Proterozoic

The Neoproterozoic, Cryogenian (former Riphean to early Vendian, Varangerian) has been interpreted, essentially on the basis of lithogical and isotopic proxies, as a period of extensive – 'Snowball Earth' – glaciation (see also 3.8.1 below).

3.5.2 Palaeozoic

Direct data and indirect inferences on foraminifera, plants, corals, brachiopods, trilobites and fish are important in the palaeoclimatic interpretation of the Palaeozoic (Armstrong & Owen, in Crame & Owen, 2002; Jones, 2006; Tabor & Poulsen, 2008; Taylor et al., 2009; Wheeley et al., 2009; Cronin, 2010). The plant macrofossils have been analysed using LMA and CLAMP.

Plant macrofossils and microfossils, and brachiopods, have recently been used in multidisciplinary studies of climate change in the Permian of Oman following the glaciation of Gondwana and the opening of the Neotethyan Ocean. Brachiopods from the early stages of the deglaciation are cold-adapted, while those from the later stages of the deglaciation are warm-adapted. Plants from the later stages of the deglaciation are of essentially tropical – rainforest – aspect.

3.5.3 Mesozoic

Direct data and indirect inferences on dinoflagellates, silicoflagellates, diatoms, calcareous nannofossils, calcareous Algae, foraminifera, radiolarians, plants, corals, brachiopods, bivalves, gastropods, belemnites, ostracods, insects, echinoids, fish, amphibians and reptiles are important in the palaeoclimatic interpretation of the Mesozoic (Jones, 2006; Tremolada et al., 2006; Sellwood & Valdes, in Williams et al., 2007a; Taylor et al., 2009; Cronin, 2010). The plant macrofossils have been analysed using LMA and CLAMP. Mesozoic palaeoclimate has been modelled, using the Hadley Centre Atmospheric General Circulation Model, by Sellwood and Valdes, in Williams *et al.*, (2007a). Recent research on isotope signals has shown that the Cretaceous, conventionally interpreted as representing an unusually long-term climatic maximum or 'super-greenhouse', appears in fact to have been characterised by intermittent glaciation (Bornemann *et al.*, 2008).

The thus-far incomplete belemnite palaeotemperature record reveals a cooling to warming cycle in the late Toarcian to Aalenian, with temperatures first apparently cooling down from 22 to 8 °C, and then warming back up to 12 °C, at studied sites. That from the Callovian to Kimmeridgian reveals an overall warming from 12 °C to 20 °C. That from the Volgian to the Valanginian reveals a significant cooling, apparently culminating in sub-zero temperatures in polar latitudes. That from the Hauterivian reveals a warming to cooling cycle, with temperatures first warming up from 11 °C to 15 °C, and then cooling back down to 11 °C. That from the Barremian reveals a warming to cooling to warming cycle, with temperatures first warming up to 20 °C, then cooling back down to 14 °C in the *Elegans* Zone, coincident with, and arguably as a consequence of, volcanism on the Ontong–Java Plateau, and finally warming back up again to 16 °C. Interestingly, although reflecting the same trends, the oxygen isotope ratio recorded by *Hibolites* is consistently more positive than that of *Acroteuthis*, possibly on account of vital or bathymetric effects (differing depth habitats). Importantly, in *Acroteuthis* and in *Aulacoteuthis*, although not in *Hibolites*, the oxygen isotope ratio is positively correlated with the content of the trace elements sodium, strontium and magnesium, implying that the latter can also be used as a proxy of palaeotemperature.

3.5.4 Cenozoic

Direct data and indirect inferences on dinoflagellates, silicoflagellates, diatoms, calcareous nannofossils, calcareous Algae, foraminifera, radiolarians, plants, corals, brachiopods, bivalves, gastropods, ostracods, insects, echinoids, fish, amphibians, reptiles and mammals are important

in the palaeoclimatic interpretation of the Cenozoic (Atkinson *et al.*, 1987; Elias, 1994, 1997; Ivanov *et al.*, 2002; Ivany *et al.*, in Prothero *et al.*, 2003; Battarbee *et al.*, 2004; de Man & van Simaeys, 2004; Pfeiffer *et al.*, 2004; Cortese *et al.*, 2005; Moe & Smith, 2005; Yu *et al.*, 2005; Jones, 2006; Haywood *et al.*, in Williams *et al.*, 2007a; Horne, 2007; Ivanov *et al.*, 2007; Jing-Xian Xu *et al.*, 2008; Ke Xia *et al.*, 2009 ; Taylor *et al.*, 2009; Cronin, 2010; Holmes *et al.*, 2010; Horne, 2010; Horne *et al.*, 2010; Tao Su *et al.*, 2010; Fernandez-Jalvo *et al.*, in press; Horsfield, 2010). The plant macrofossils have been analysed using not only LMA and CLAMP, but also the co-existence approach or CoA. The ostracods and insects have been analysed using the MCR and related methods. The radiolarian data has been analysed using artificial neural networks.

Long-term climatic maxima are indicated directly by data or indirectly by inference in the Palaeocene–Middle Eocene, Early to early Middle Miocene, and Early Pliocene; minima in the Late Eocene–Oligocene, late Middle to Late Miocene, and Late Pliocene–Pleistocene (Jones, 2006). Medium- to short-term climatic maxima and minima are recorded in the interglacials and interstadials, and glacials and stadials, respectively, of the Pleistocene (and Holocene). (Ultra-short-term climatic changes are recorded in the Dansgaard–Oeschger and Heinrich events of the Pleistocene.) According to power spectrum analyses, the variance in the medium- to short-term data is attributable to 100 ka (50%), 40 ka (25%) and 20 ka (10%) Milankovitch cycles, of which the 40 ka is associated with variation in the obliquity of the earth's axis, and the 20 ka with variation in the orientation or precession in the earth's axis.

These climatic cycles are interpreted as having had important consequences for mammalian evolution (Vrba *et al.*, 1995; Calvin, 2002; Jones, 2006; see also Box 3.8 below). Note in this context that there is also a (macro)evolutionary trend toward an increase in the test size of globigerinide foraminifera through the Cenozoic, which is most marked during the climatic cooling following the Early–Middle Miocene temperature maximum, and which has been interpreted as having been driven by climatic cooling (Schmidt *et al.*, 2004).

3.6 PALAEO-OCEANOGRAPHY

As noted above, dinoflagellates, diatoms, calcareous nannoplankton, foraminifera and radiolarians are of considerable use in palaeo-oceanography, for example as tracers of water-masses with particular temperature and salinity characteristics, and as indicators of palaeoproductivity and organic carbon flux to the sea floor, and of oxygenation levels in the bottom sediment (see also Jones, 2006; Reynolds, 2006; Falkowski & Knoll, 2007; Hillaire-Marcel & de Vernal, 2007).

The ratio of autotrophic gonyaualacoid to heterotrophic peridinioid dinoflagellates, the abundance and diversity of calcareous nannofossils, and certain benthic foraminifera have all been used as indicators of and proxies for palaeoproductivity; and diatoms, planktonic foraminifera and radiolarians, as indicators of upwelling. Certain infaunal agglutinated and buliminide calcareous benthic foraminifera have been used as indicators of oxygenation levels in the bottom sediment (see 3.2.2 and 3.3.3 above). Dysoxic to anoxic sediments are conducive to the preservation of organic carbon and hence of petroleum source-rocks (see Box 5.1).

3.7 QUANTITATIVE AND OTHER TECHNIQUES IN PALAEOBIOLOGY

A number of quantitative techniques have been applied in the fields of pure and applied palaeobiology, palaeoecology and palaeoenvironmental interpretation (Robinson & Kohl, 1978; Lesslar, 1987; Hammer & Harper, 2006; Jones, 2006). In my working experience in the petroleum industry, palaeobathymetric interpretation techniques, palaeobiogeographic and palaeoclimatological interpretation techniques, and various forms of cluster analysis and associated techniques, have proved of particular use (Jones, 2006). These techniques are discussed, in turn, below.

3.7.1 Palaeobathymetric interpretation techniques

The ratio of planktonic to benthic foraminifera (P:B) can be used as a measure of bathymetry (Hayward, 1990; Jones, 1996, 2006, Fig. 3.23).

However, the trend toward a higher ratio in deeper water is non-universal and commonly locally reversed, such that values are not

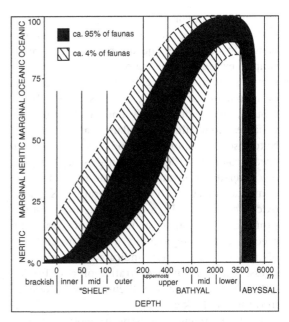

Fig. 3.23 **Variation in the ratio of planktonic to benthic foraminifera with depth.** © Cushman Foundation for Foraminiferal Research. Reproduced with permission after Hayward (1990).

necessarily unique to any particular bathymetry or bathymetric zone. For example, the ratio is anomalously low in deep waters off the mouths of major rivers, owing to low planktonic productivity in the freshwater plume; on the upper slope, owing to low predation and hence high benthic productivity – of appropriately adapted forms – in the oxygen minimum zone; and also on the abyssal plain, owing to preferential dissolution of planktonics below the calcite compensation depth. The trend toward a higher ratio is also non-linear, such that the degree of bathymetric differentiation or resolution that it allows is variable.

Transfer functions based on the environmental distributions and tolerances of diatoms as well as foraminifera can also be used in sea-level reconstructions (Horton & Edwards, 2006; Jones, 2006).

Cross-plots of benthic foraminiferal 'morphogroups' allow the differentiation of bathymetric zones and/or sedimentary sub-environments (Jones, 1996, 2006; see also Fig. 3.13 above, and Fig. 5.49). Abundance, diversity, dominance and equitability are also useful guides to bathymetry, environment and/or systems tract (ST). Abundance, that is, the number of specimens per unit volume or weight of sample, and diversity, that is, most simply expressed, the number of species per sample, are typically highest in deep water and/or in maximum flooding surfaces (MFSs). Note, though, that depth is only one of a number of variables controlling abundance and diversity. Note also that abundance and diversity are not always co-variant. High abundance associated with low diversity often indicates some form of environmental stress, such as temperature, salinity, oxicity or toxicity. High dominance, most simply expressed as relative abundance of dominant species, and low equitability, that is, evenness of distribution, also often indicate environmental stress. The bio-sequence stratigraphic utility of 'SHE' diversity analysis is discussed by Wakefield, in Olson and Leckie (2003).

In the petroleum industry, palaeobathymetric interpretation is now capable of being automated by means of the use of a proprietary data entry, manipulation and display software package that contains links to bathymetric distribution datasets and 'look-up' tables, as in the case of the Mesozoic and Cenozoic of the Indus basin of Pakistan (Wakefield & Monteil, 2002; see also 5.2.5 below). The interpretation is typically displayed in the form of a palaeobathymetric curve, capable of being imported into the work-station environment for integration with well log and seismic data. Excursions in the palaeobathymetric curve can be used alongside the well log and seismic data to identify facies dislocations and/or stratigraphic discontinuities. Running displays of correlation coefficients between samples (cosine theta and Otsuka plots) can also be used to identify facies dislocations and/or stratigraphic discontinuities (Olson et al., in Olson & Leckie, 2003; Moss et al., 2004).

3.7.2 Palaeobiogeographic and palaeoclimatological interpretation techniques

Transfer functions are also of use in palaeobiogeographic and palaeoclimatological interpretation. For example, transfer functions based on the environmental distributions and tolerances of terrestrial and marine plant macrofossils and microfossils (palynomorphs) have been used in establishing terrestrial and marine climate, hardiness, moisture needs, and salinity tolerances (Jones, 2006). Aspects of plant leaf physiognomy and the percentage of entire-margined species have been used as proxies for mean annual temperature and mean annual range of temperature. Cladistic techniques have been used in the palaeobiogeographic interpretation of the Early Ordovician of Gondwana. Geographic Information System (GIS) databases of fossil brachiopod and bivalve distributions have been used in the palaeobiogeographic interpretation of the Late Ordovician and Middle–Late Devonian (Jones, 2006; Stigall, 2006, 2010). GIS databases of Recent non-avian tetrapod distributions have been used in the palaeoclimatic interpretation of the Middle Eocene Messel Lagerstatte in Germany (Jones, 2006). The palaeobiogeographic implications of Early Triassic ammonite and Cenozoic mammal distributions have been quantitatively assessed using Interactive Data Language (IDL) technology (Brayard et al., 2004).

3.7.3 Cluster analysis and 'fuzzy C means' cluster analysis

Various forms of cluster analysis and associated techniques have proved of use in palaeoenvironmental interpretation, for example, hierarchical cluster analysis in the differentiation of restricted and fully marine environments in the Miocene, Eggenburgian of Austria (Jones, 2006).

In academia, cluster analysis of palynomorph assemblages has been used in the palaeoenvironmental and palaeoclimatic interpretation of the 'late' Aptian–Albian of the Sergipe basin of Brazil (Carvalho, 2004). Here, Palynomorph Assemblage or PA 1, identified on the basis of cluster analysis, is characterised by marine palynomorphs; PA 2, PA 3 and PA 4 by terrestrial palynomorphs. PA 2 is characterised by a preponderance of ephedroid pollen, derived from gnetophyte gymnosperm parent plants such as the extant Ephedra, Gnetum and Welwitschia, and interpreted on the basis of modern and ancient analogy as indicating arid environments. Such plants occur in such environments in the modern Namib desert of Namibia, and also in the ancient Araripe basin of Brazil (Dilcher et al., 2005; Mohr et al., in Martill et al., 2007). PA 3 is characterised by a preponderance of Classopollis pollen, derived from an extinct cheirolepidacean parent plant interpreted on the basis of its distribution and association as indicating seasonally arid environments (Batten, 1975; Alvin, 1982, 1983; Seward, in McCabe & Parrish, 1992; Xiao-Yu Jang, 2008). The parent plant of the *Classopollis* pollen is further interpreted as having had some tolerance to elevated salinity as well as aridity, and as having constituted one of the principal components of a 'halophytic' salt-marsh plant community that occupied the same ecological niche in the Mesozoic as mangroves in the Cenozoic. PA 4 is characterised by a preponderance of *Araucariacites* pollen, derived from araucariacean parent plants, and of *Cicatricosisporites* and *Cyathidites* spores, derived from pteridophytes, and interpreted as indicating generally humid environments.

In the petroleum industry, in the case of the Jurassic reservoir in Hawkins field in the North Sea, 'fuzzy C means' (FCM) clustering has been applied in the quantification of previous qualitative observations on 'sporomorph eco-group' or SEG distributions, palaeoenvironmental interpretation, and, ultimately, reservoir characterisation for the purposes of exploitation (Gary *et al.*, in Demchuk & Gary, 2009; see also Section 5.3 below). Specifically, it has been applied to the identification and quantification of four distinct clusters or assemblages, each with a distinct significance in terms of climate, namely: the cool, wet *Perinopollenites elatoides* assemblage; the transitional *Lycopodiumsporites* assemblage; the warm, dry *Cerebropollenites mesozoicus* assemblage; and the warm, wet *Callialasporites* assemblage. It has been further applied to the construction of curves of assemblage distributions to serve as proxies for climatically mediated sea level, facilitating palaeoenvironmental interpretation and the recognition of maximum flooding surfaces and sequence boundaries.

3.7.4 'Fuzzy logic'

In the petroleum industry, in the case of the Palaeocene reservoir in Fleming field in the North Sea, 'fuzzy logic' has been applied in the quantification of previous qualitative observations agglutinated foraminiferal 'morphogroup' distributions, palaeoenvironmental interpretation, and, ultimately, reservoir characterisation for the

purposes of exploitation (Wakefield *et al.*, 2001; see also Section 5.3 below). Specifically, it has been applied to the characterisation by means of contained agglutinated foraminiferal 'morphogroups' of intra-reservoir mudstones constituting potential barriers and baffles to fluid flow. Importantly, it has been used to develop not only alternative reservoir layering schemes, but also alternative reservoir simulation models that are testable on the basis of production history matching. In consequence, it has had a significant impact on production strategy and well planning.

3.8 KEY BIOLOGICAL EVENTS IN EARTH HISTORY

This section deals with the generalities and specifics of the evolutionary and extinction events and trends that have controlled past and present, and will control future, biodiversity on Earth (including biodiversity crises such as mass extinction events).

Readers interested in further details of a general nature are referred to Gaston and Spicer (2004), Stanley (2005), Jones (2006), Falkowski and Knoll (2007), Foote and Miller (2007), Markov and Korotayev (2007), Benton and Harper (2009) and Lieberman and Kaesler (2010).

Evolution and extinction

The history of life on Earth, insofar as it can be captured in words, can be said to have been one of general evolutionary diversification intermittently interrupted by mass extinction and recovery (Jones, 2006; Benton & Emerson, 2007). Certain of the difficulties in description arise from difficulties in quantifying diversity, in part in turn arising from uncertainties as to the representativeness of the fossil record, and from biases and inconsistencies in sampling and interpreting same. There are particular problems associated with taxonomic bias and inconsistency, that is, the definition of taxa across different fossil groups. The two main models that have been proposed to date for the evolution of diversity are the equilibrium model, which assumes – logistic – diversification only up to and not beyond a critical point, and the expansion model, which assumes continuous – either additive or exponential – diversification.

Evolution

This is not the appropriate forum for a detailed discussion of the complex and contentious subject of evolution, that is, the process whereby species and higher-level taxa evolve – or are transformed – into and are replaced by others. Suffice it to say here that empirical observations of the fossil record should leave little doubt, even in the mind of the most ardent creationist, that evolution has manifestly evidently been taking place throughout geological time. The difficulties that sceptics have in understanding and accepting evolution perhaps arise in part from the fact that the process cannot actually be observed, other than under artificial laboratory conditions, as it occurs over generations (Jones, 2006).

The means or mode of evolution is 'natural selection' (Wallace, 1858; Darwin, 1859). Put simply, if not over-simplistically, natural selection acts through some form of pressure, such as environmental pressure, on a naturally varying population. Over generations, this pressure selects for those individuals that possess characters or genetic mutations that are both beneficial and heritable, and against those that do not; and ultimately results in a new set of characters, and a new species. The currently most widely accepted version of evolutionary theory, 'Neodarwinism', has replaced Darwin's original concept of 'blending' inheritance with Mendel's concept of 'particulate' or genetic inheritance. (Thankfully, the so-called 'Social Darwinism' of the late nineteenth century and the related 'eugenics' programme of the early twentieth, with their connotations of class and racial supremacy, are now recognised as among the worst ideas in the dark and lamentable catalogue of human history.) It is now believed that the process of evolution is facilitated by the reproductive isolation of parts of the population (at least in the case of so-called allopatric speciation). Such isolation is most commonly brought about by a geographical barrier, such as a rising mountain range, or a propagating rift, or a rising sea level; or by a climatic change, and associated changes to distribution fronts.

The timing or tempo of evolution remains disputed even within the scientific community. To me, though, the two main models that have been proposed, the progressive 'phyletic gradualism'

attributable to Darwin, although not named as such by him, and the effectively – within the limits of temporal resolution – instantaneous 'punctuated equilibrium' of Eldredge and Gould, in Schopf (1972) are not mutually exclusive, as some have argued. Either can be pre-eminent at a given place at a given time; the other at the same place but at a different time, or at the same time but at a different place.

Readers interested in details of the scientific as against the religious view of evolutionary biology, or in the products as against the processes thereof, are referred to, for example, Skelton (1993), Patterson (1999), Rothschild and Lister (2003) and Ruse (2005), or to the numerous more or less popularised accounts, notably those provided by Richard Dawkins, Stephen Jay Gould and Simon Conway Morris. In contrast to Gould, who emphasised the importance of contingent events in evolution, Conway Morris emphasises the importance of convergence; and cites it as evidence that there is a comparatively limited number of evolutionary pathways open to organisms. He interprets us as '*inevitable* humans …' (my italics), albeit '… in a lonely universe' (Conway Morris, 2003).

Those interested in the specifics of individual (macro)evolutionary diversification events are referred to the appropriate sections below.

Evolutionary events

Key evolutionary events in the history of life include: the origin of life (prokaryotes); the evolution of complex life (eukaryotes); the evolution of multicellularity; the evolution of mineralised skeletons, and the Cambrian evolutionary diversification (the 'Cambrian explosion'); the evolution of reefs; the Ordovician evolutionary diversification; the evolution of life on land; the evolution of vertebrates; the evolution of trees and forests; the evolution of flight; the Mesozoic evolutionary diversification; the evolution of flowering plants; the evolution of grasslands and grassland animals; and the evolution of humans (Jones, 2006). Each of these events is dealt with, in turn, in its stratigraphic context, below.

In the marine realm, certain evolutionary events appear to be associated with niche diversification during times of transgression, and hence with

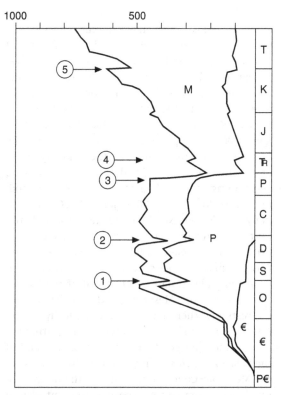

Fig. 3.24 **Evolutionary biotas and mass extinction events**. Modified after Sepkoski (1981). Units on horizontal axis are millions of years; units on vertical axis refer to numbers of families. €, Cambrian evolutionary biota; P, Palaeozoic evolutionary biota; M, Mesozoic evolutionary biota. 1, end-Ordovician mass extinction event; 2, Late Devonian mass extinction event; 3, end-Permian mass extinction event; 4, end-Triassic mass extinction event; 5, end-Cretaceous mass extinction event.

maximum flooding surfaces; certain extinction events with habitat destruction during times of regression, and hence with sequence boundaries.

Evolutionary biotas

Evolutionary biotas are sets of higher taxa that have similar histories of diversification, dominance and decline over geological time (Sepkoski, 1981; Jones, 2006; Fig. 3.24).

The principal three, as recognised on the basis of factor analysis, are the Cambrian, the Palaeozoic, and the Mesozoic–Cenozoic – or modern – evolutionary biotas. The succession of these evolutionary biotas records long-term change in the nature and diversity of life on earth.

'Ecological evolutionary units' are somewhat different in nature, as they represent reorganisation on the time-scale of a few million years rather than several tens or hundreds of millions of years. They appear to represent short-term adjustments to mass extinction events.

The Cambrian evolutionary biota includes SSFs, inarticulate brachiopods, trilobites, eocrinoids and other higher taxa. It appeared in the Late Precambrian, late Cryogenian–Ediacaran (Vendian) and ranges through to the Recent. However, it was most important in the Cambrian, representing the principal constituent of the evolutionary diversification or explosion of that time, and has been in decline thereafter, the decline accentuated by the end-Ordovician and Late Devonian mass extinctions.

The Palaeozoic evolutionary biota comprises articulate brachiopods, anthozoans, cephalopods, crinoids, graptolites and other higher taxa. It appeared in the Cambrian and ranges through to the Recent, although it was most important in the Ordovician–Devonian, representing the principal constituent of the evolutionary diversification of that time, and has been in decline thereafter, the decline accentuated by the rise of the Mesozoic–Cenozoic biota in the Carboniferous and Permian, and by the end-Permian mass extinction. The Palaeozoic fauna of the terrestrial realm includes the primitive labyrinthodonts, anaspids and synapsids that dominated the Devonian–Permian (Benton, 1985). The Palaeozoic flora of the terrestrial realm includes the early vascular plants that dominated the Silurian–Devonian and the pteridophytes that dominated the Carboniferous–Permian (Niklas *et al.*, 1983).

The Mesozoic–Cenozoic or modern evolutionary biota comprises bivalves, gastropods, echinoids, vertebrates and other higher taxa. It appeared in the Palaeozoic and ranges through to the present day, and is most important in the Post-Palaeozoic, representing the principal constituent of the evolutionary diversification of that time. The Mesozoic–Cenozoic fauna of the terrestrial realm includes the dinosaurs and pterosaurs that dominated the

Mesozoic and died out at the time of the end-Cretaceous mass extinction, and the reptiles, birds and mammals that have dominated the Cenozoic (Benton, 1985). The Mesozoic–Cenozoic flora of the terrestrial realm includes the gymnosperms or seed plants that dominated the Mesozoic and the angiosperms or flowering plants that have dominated the Cenozoic (Niklas *et al.*, 1983).

Extinction

Background extinction, like evolution, has evidently been taking place throughout geological time, although mean rates are similarly difficult to quantify (Jones, 2006). Interestingly, recent research indicates that extinction rates were highest, and biodiversity lowest, during 'greenhouse' phases (Mayhew *et al.*, 2007).

In addition to background extinction events through time, there have been a number of mass extinction events, during which extinction rates were significantly elevated above the background level (and, obviously, above coeval evolution rates). The effects of these events were severe, and they were experienced over short time frames and over wide geographical areas. Indeed, they were apparently essentially global and non-selective. Readers interested in further details of extinction and mass extinction are referred to, for example, Raup and Sepkosi (1982), Lawton and May (1995), Hart (1996), Hallam and Wignall (1997), Courtillot (1999), Erickson (2001), Hallam (2004), Taylor (2004), Twitchett (2006), Ward (2007) and Elewa (2008). Those interested in the specifics of individual mass extinction events are referred to the appropriate sections below.

Mass extinction events

Major mass extinction events include the Late Precambrian, Early Cambrian, Late Cambrian, end-Ordovician, Late Devonian, end-Permian, end-Triassic, Late Cretaceous (Cenomanian–Turonian), end-Cretaceous, end-Palaeocene, end-Eocene, Pleistocene and Holocene events, which are discussed in detail, in context, below (Jones, 2006; Twitchett, 2006; Ward, 2007; Elewa, 2008). (Comparatively minor mass extinction events include the Early Jurassic, Middle Jurassic, end-Jurassic, Early Cretaceous, Oligocene, Middle Miocene and Pliocene events, which are not discussed.)

The end-Ordovician, Late Devonian, end-Permian, end-Triassic and end-Cretaceous mass extinction events constitute the 'big five' of Raup and Sepkoski (1982) and other authors. The 'big five' appear to differ from other mass extinction events, and indeed from background extinctions, only in degree, and not in kind (MacLeod, in Rothschild & Lister, 2003; MacLeod, in Beaudoin & Head, 2004; Jones, 2006). All types of extinction can therefore be interpreted as the effects of the same types of underlying cause, 'mediated through the waxing and waning of … ecological hierarchies' (MacLeod, in Rothschild & Lister, 2003; Jones, 2006). Interestingly in this context, there are statistically significant relationships between the intensity of mass extinction and sea-level fall in the Palaeozoic, and between mass extinction and flood-basalt eruption from the end-Permian event at end of the Palaeozoic through the Mesozoic to the Cenozoic. Also interestingly, there has been a statistically significant decrease in the intensity of mass extinction events since the end-Permian event at the end of the Palaeozoic. This decrease mirrors an increase in the Mesozoic–Cenozoic not only in the delivery of nutrients to shallow marine habitats, but also in the number of recycler niches and in the length of food chains, which would appear to have had some sort of buffering effect.

Interestingly, the taxonomic and ecological impacts of the 'big five' mass extinction events were apparently at least to an extent decoupled. The end-Ordovician mass extinction was extremely significant in terms of taxonomic impact, but not in terms of ecological impact. Conversely, the end-Cretaceous mass extinction was comparatively insignificant, leastwise the least significant of the big five, in terms of taxonomic impact, but the second most significant after the end-Permian mass extinction in terms of ecological impact. It seems that the component species and structure of an ecosystem are at least as important as diversity in maintaining its integrity, and that the loss of those species with the highest ecological value can lead to ecological crises.

Effects of mass extinction events

Selectivity. There is evidence that the effects of some mass extinction events were selective, by trait

(Jones, 2006). Large species appear to have been disproportionately susceptible to mass extinction, as do species or communities occupying restricted geographical ranges or specialised ecological niches, those with limited capabilities to invade new niches, and those exhibiting low or unstable population densities (Stigall Rode & Lieberman, in Over et al., 2005; Jones, 2006; Kiessling, 2006; Stigall, 2006). Low-diversity communities also appear to have been disproportionately susceptible to mass extinction, leading to the suggestion that high diversity might somehow buffer the effects of the events (Jones, 2006). Moreover, marine invertebrate communities characterised by low metabolic rates also appear to have been disproportionately susceptible (Knoll et al., 1996; Bambach et al., 2002; Erwin, 2006). This has been interpreted as due to the likely comparative inability of their respiratory and circulatory systems to buffer the effects of the dysoxia or hypercapnia – that is, low oxygen or high carbon dioxide concentration – often invoked as among the causes of mass extinction (Ward, 2006, 2007).

Recovery. Kauffman and Harries have proposed a general model for recovery following mass extinction events, in which the aftermath is characterised by an initial survival interval and a later recovery interval proper (Jones, 2006). Interestingly, there are certain similarities between the recovery following mass extinction events and the succession following non-lethal ecological disturbance. The survival interval is characterised initially by blooms of 'disaster' species and of 'opportunistic' species, and later by increases of pre-adapted survivors and ecological generalists, and, importantly, by the evolution and/or radiation of crisis-adapted 'progenitor taxa'. The recovery interval proper is characterised by further evolutionary diversification, initially within surviving and newly established crisis progenitor lineages, and later within new lineages. It is also characterised by the reappearance of so-called 'Lazarus taxa' from refugia.

Recovery time appears approximately proportional to the percentage loss sustained. Interestingly in this context, although it took 10 Ma for diversity to stabilise after the end-Cretaceous mass extinction event, the initial recovery was actually to a higher level, indicating some kind of rebound effect (of the sort that financial analysts are wont to term a 'dead cat bounce'). Note in this context that recovery time appears to vary across trophic groups, with primary producers (re)appearing first, followed by secondary consumers, that is, herbivores and carnivores. Detritivores appear able to survive even drastic declines in primary production. Note also that recovery time appears to vary across entire ecosystems. For example, recovery following the end-Permian mass extinction was effectively geologically instantaneous in terrestrial but not in marine ecosystems.

Periodicity of mass extinction events

Raup and Sepkoski (1984) have proposed that there is a 26 Ma periodicity associated with mass extinction events. The proposed periodicity has been elegantly explained by extrinsic, extraterrestrial causes, according to the 'Nemesis' and 'Planet X' hypotheses. The former hypothesis involves the existence of a dim binary companion star to the sun, Nemesis, and the latter, a tenth planet in the solar system, 'Planet X', both with eccentric orbits, which would cause passage through the Oort cloud and a comet shower every 26–28 Ma. Note, though, that the existence of Nemesis has never been proven, and that simulations suggest that if it did exist it would probably be unstable and easily thrown off course. Note also that the existence of 'Planet X' has never been proven either, and that even if it did exist it might not have sufficient mass to to cause a comet shower. The proposed periodicity associated with mass extinction events has also been explained by intrinsic causes, such as dynamic instability in the mantle – and associated increased mantle plume activity – on a periodicity of approximately 30 Ma (coincident with that of magnetic field reversal).

Many authors have questioned whether there is actually any periodicity associated with mass extinction events. For example, Whatley, in Whatley and Maybury (1990), states: 'The present author's data for the extinction of Mesozoic Ostracoda ... clearly indicates ... intervals between ... Jurassic extinction peaks ... very close to Raup & Sepkoski's 26 MY interval ...' However, he also adds, 'In the Cretaceous, the interval is less regular', and '... no ... clear support is forthcoming for the Cainozoic data'. He concludes that 'No doubt the data could be "massaged" in order to improve ... fit to existing dogma. The author, however, prefers to leave this task to a professional data manipulator.' MacLeod, in Rothschild and Lister (2003), has also eloquently argued that there is no periodic component to mass extinction events. His Monte Carlo simulations of Phanerozoic extinction data – and comparison of the simulation results with the Fourier transformation of an extinction intensity time series – appear to demonstrate that the proposed 26 Ma periodicity is not statistically significant.

Causes of mass extinction events

Intrinsic, earth-bound causes that have been invoked as responsible for mass extinction events include sea-level rise and fall, volcanism and climate (Jones, 2006; Kidder & Worsley, 2010). As noted above, there are statistically significant relationships between mass extinction events and on the one hand, in the Palaeozoic, sea-level fall, and on the other hand, in the terminal Palaeozoic, Mesozoic and Cenozoic, volcanism. Volcanism can in turn affect climate, and sea-level, through the eruption of particulate matter into the atmosphere, resulting in short-term blackout and (global) cooling, and through the release of the greenhouse gas carbon dioxide, resulting in longer-term (global) warming. (Interestingly, biotic forcing could also be capable of affecting climate, as suggested by James Lovelock's 'Gaia' hypothesis. For example, high marine phytoplankton or land plant productivity could result in the abstraction of sufficient amounts of carbon dioxide in the atmosphere to cause warming, while low productivity could result in the retention of sufficient carbon dioxide in the atmosphere to cause cooling: Strother *et al.*, in Vecoli *et al.*, 2010.)

More recently, a significant decrease in atmospheric and oceanic oxygen concentration, a significant increase in atmospheric and oceanic carbon dioxide concentration, a significant increase in atmospheric and oceanic hydrogen sulphide concentration, and a significant increase in radiation caused by breakdown of the protective ozone layer in the atmosphere have been invoked (Ward, 2006; Beerling, 2007; Ward, 2007; see also Beerling & Woodward, 2001). Significantly, at least the Late Cambrian, end-Ordovician, Late Devonian, end-Permian, end-Triassic, Early Jurassic, end-Jurassic,

Late Cretaceous (Cenomanian–Turonian) and end-Palaeocene mass extinction events can be explained as having been caused by the effects of atmospheric carbon dioxide concentrations above, or well above, 1000 ppm, and the end-Cretaceous event at least partly by such effects (Ward, 2007; see also Berner, in Gensel & Edwards, 2001, Berner & Kothavala, 2001 and Berner, 2006). This amount of carbon dioxide in the atmosphere would be sufficient to cause global warming (and also the destabilisation of methane hydrates, the resultant release of methane, which, like carbon dioxide, is a greenhouse gas, into the atmosphere, and further warming). The warming would cause anoxia, lethal to all life forms other than anaerobic Bacteria, in the oceans, by slowing down and eventually stopping circulation. And the activity of anaerobic sulphate-reducing Bacteria would in turn cause an increase in the concentration of hydrogen sulphide, which is also lethal, in the oceans and in the atmosphere (and could also, together with methane, cause a breakdown of the ozone layer in the atmosphere). Evidence for an increase in hydrogen sulphide concentration is provided by an increased incidence of biomarkers derived from sulphate-reducing Bacteria (Ward, 2007); that for an increase in radiation by an increased incidence of mutations in plants (Beerling, 2007).

Extrinsic, extraterrestrial causes include bolide impacts and associated effects such as darkness and cooling ('impact winter'), greenhouse warming and acid rain. Evidence for extraterrestrial forcing comes from cratering, iridium anomalies, shocked quartz, microtektites (melt-glass) etc.

Interestingly, it has recently been hypothesised that extraterrestrial bolide impacts can cause earth-bound volcanism (Elkins-Tanton & Hager, 2005). It has also been hypothesised that they can cause volcanism on the opposite side of the earth from the impact site, through a process called 'antipodal forcing'.

3.8.1 Proterozoic

The origin of life (prokaryotes)

The most widely accepted hypothesis is that life on earth originated spontaneously through the natural biochemical synthesis of complex organic compounds under ambient conditions (Jones, 2006;

BOX 3.6   FORAMINIFERAL DIVERSITY TRENDS THROUGH TIME (MODIFIED AFTER JONES, 2006)

Introduction

Foraminifera are a class of sarcodine protists characterised by granulo-reticulose protoplasm. Most known modern forms are also characterised by agglutinated or secreted shells or tests (the soft-bodied allogromiids excepted). Ancient – preservable – forms have an extensive and comparatively well-documented fossil record, extending from latest Precambrian/earliest Cambrian – in the case of *Platysolenites* – to Recent. Foraminifera constitute part of the Mesozoic–Cenozoic or modern evolutionary biota of Sepkoski (1981). Little cladistic work has been done on the group.

The fossil record of the foraminifera is sufficiently good, and sufficiently well documented, to allow detailed observations to be made on the evolution and control of the diversity of the group through time, some of which may be of more general applicability.

Discussion – diversity trends through time

Familial-level diversity data from a database based on Haynes's (1981) scheme has been plotted against time, enabling observations on trends through time (Jones, 2006; Fig. 3.25).

Diversity data from Loeblich and Tappan's (1987) scheme has also been plotted for comparative purposes. 'Taxonomic bias' is evident only in the case of the apparent end-Carboniferous event in the Loeblich and Tappan data (when a large number of mono- or oligogeneric 'families' died out), and other than in this case can be effectively eliminated. Other potential sources of bias remain, including 'preservational bias' (that to do with the representativeness of the fossil record), 'sampling bias' (palaeontologist interest) and 'interpretive bias' (lack of normalisation of studied time intervals).

Palaeozoic. Empirical observations indicate that foraminifera underwent a slow expansion of diversity in the Early Palaeozoic, followed by a rapid expansion in the Carboniferous, especially of the specialist shallow marine larger benthic fusulinides (Groves & Lee, 2008). Sixty-four per cent of all families were wiped out at the time of the

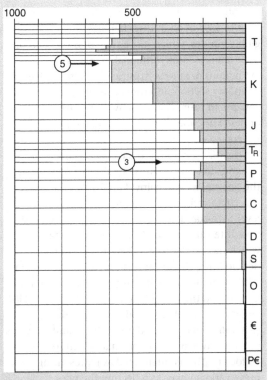

Fig. 3.25 **Foraminiferal diversity trends through time**. From Jones (2006). Units on horizontal axis are numbers of families. 3, End-Permian mass extinction event; 5, end-Cretaceous mass extinction event.

end-Permian mass extinction (one of the 'big five' of Raup & Sepkoski, 1982), all of them fusulinides. Nintey-one per cent of all genera of calcareous benthic foraminifera were wiped out at this time (Groves & Altiner, 2005; Marquez, 2005) Recovery of pre-extinction diversity took until the Jurassic. However, the 'recovery interval' may be said to have initiated in the early Middle Triassic, Anisian, approximately 15 Ma after the end of the Permian, when there was a significant increase in the rate of recovery. The Early Triassic represents the 'survival interval'.

Mesozoic. There was another period of slow expansion in the Jurassic, followed by rapid expansion in the Cretaceous, especially of the specialist shallow marine larger benthic rotaliides and the planktonic globigerinides. Nineteen per cent of all families were then wiped out at the time of the end-Cretaceous mass extinction (another of the 'big five'), all of them rotaliides, which lost a total of 43% of all of their families, and globigerinides, which lost a total of 86% of all of their families. Many infaunal taxa underwent extinctions or at least 'Lazarus' extinctions over the boundary section, interpreted as indicating a collapse in primary productivity. Recovery of pre-extinction diversity took until the Eocene.

Cenozoic. According to data based on Haynes's scheme, diversity attained an all-time maximum in the Eocene, which, coincidentally or otherwise, was characterised by a climatic optimum and sea-level high-stand, and has been declining over the time interval from the Oligocene to the present, which has been characterised by generally deteriorating climate and falling sea level. In detail, 13% of families have become extinct between the Eocene and the present time, all of them specialist larger benthic rotaliids and globigerinides. (According to data based on Loeblich and Tappan's scheme, diversity attained a maximum in the Holocene. This is interpreted as representing preservational bias, as reflecting the poor preservational potential of some of the features Loeblich and Tappan used to distinguish some of their families.)

Incidentally, there is a (macro)evolutionary trend toward an increase in the test size of globigerinides through the Cenozoic (Schmidt *et al.*, 2004). This evolutionary trend is most marked during the time of climatic cooling following the Early–Middle Miocene temperature maximum, and has been interpreted as having been driven by climatic cooling and associated intensification of surface water stratification.

Observations on foraminiferal diversity trends

The rapid expansion of the specialist shallow marine larger benthic fusulinides in the Carboniferous may have been on account of the occupation of niches vacated by stromatoporoids and tabulate corals during the Late Devonian mass extinction. That of the specialist shallow marine larger benthic rotaliides and planktonic globigerinides in the Cretaceous may have been on account of the occupation of niches newly created by the sea-level rise of the time. (Note also in this context that the diversity of the globigerinides has been independently interpreted as exhibiting cyclicity paralleling coeval cyclic changes in the environment, such as sea-level change, perturbed by intermittent extinction events.)

It appears that specialist groups that have just undergone rapid expansion suffer disproportionately in mass extinction events, as has recently been described in the case of the miliolide LBFs in the Cenomanian–Turonian boundary mass extinction event (Parente *et al.*, 2008). This can be interpreted as indicating either that rapid expansion is unsustainable and an unsuccessful life strategy in the longer term, or that it is the specialisation itself that is ultimately unsuccessful. Interestingly in this context, many if not all modern specialist LBFs appear to be essentially extreme K-strategists as opposed to r-strategists, characterised by a facultative or obligate symbiotic relationship with photosynthetic Algae, a degree of dependence rendering them extremely vulnerable to mass extinction during any change of conditions or relationships. Analogy with their living counterparts indicates that many, if not all, fossil LBFs also harboured symbionts (as does carbon isotope evidence).

Observations of general applicability

The observation that recovery time following a mass extinction is proportional to the percentage loss sustained may apply more generally. Note, though, that the recovery from the end-Triassic mass extinction appears to have been unusually slow, possibly in account of the low diversity going in to the event, in turn inherited from the end-Permian mass extinction.

The observation that specialist groups suffer disproportionately may also apply more generally. It may even apply to entire communities of specialist organisms, such as reef communities. This may be on account of limited spatial distribution or population density (Kiessling, 2006).

Luisi, 2006; Konhauser, 2007; Wacey, 2009). Importantly, a series of elegant laboratory experiments conducted by Stanley Miller in the 1950s demonstrated that at least amino acids could indeed be spontaneously generated under simulated Precambrian, Hadean oceanic and atmospheric conditions, although the precise mechanism by which these proto-proteins aggregated into actual life forms remains unclear. Interesting alternative hypotheses are that life originated in hydrothermal vent systems or similarly extreme environments (Little & Vrijenhook, 2003), or through the introduction of organic compounds in carbonaceous chondrites of extraterrestrial origin (Cockell & Bland, 2005). The development of conditions for life on earth and elsewhere in the universe is reviewed by Jastrow and Rampino (2008).

Molecular biological data indicates that 'life on earth arose from a single source, ~4400–4200 million years ago (Ma), and quickly achieved a prokaryotic level of complexity' (Hedges, in Hedges & Kumar, 2009). Note, though, that there are issues surrounding the calibration of the 'molecular clock' (Benton et al., in Hedges & Kumar, 2009). Isotopic signals arguably indicating the existence of life have been recorded from as long ago as approximately 3800 Ma (Jones, 2006; Konhauser, 2007; Taylor et al., 2009). However, unequivocal fossils confirming the existence of life only appeared in the rock record approximately 3500 Ma, in the form of cyanobacterial stratomatolites in the Pilbara craton of Western Australia and in the Barberton greenstone belt of South Africa and Swaziland (Brasier et al., 2005; Jones, 2006; Allwood et al., 2007; Konhauser, 2007; Schopf et al., 2007; van Kranendonk et al., 2008; Phillipot et al., 2009; Schopf, 2009; Taylor et al., 2009). The microbialites of the Pilbara craton have been interpreted as having been associated with hydrothermal springs (van Kranendonk et al., 2008; Phillipot et al., 2009). The associated so-called 'ambient inclusion trails' detected through the use of nanometre-scale secondary ion mass spectrometry (NanoSIMS) have been interpreted as having been produced by the decomposition of organic material of biogenic origin (Wacey et al., 2008).

At least some primitive prokaryotes apparently subsisted by photosynthesising, which process produced free oxygen, and ultimately made the surface of the planet habitable by other groups of organisms for the first time (Jones, 2006; Ward, 2006; Blankenship et al., in Falkowski & Knoll, 2007; Knoll et al., in Falkowski & Knoll, 2007; Taylor et al., 2009). Atmospheric oxygen concentration through the Phanerozoic has been charted by Berner and Canfield (1989), Graham et al. (1995), Berner (2001) and Algeo and Ingall (2007) (atmospheric carbon dioxide concentration by Graham et al., 1995, Berner, in Gensel & Edwards, 2001, Berner & Kothavala, 2001 and Berner, 2006).

The evolution of complex life (eukaryotes)

The most widely accepted hypothesis is that eukaryotes evolved through the development of a symbiotic relationship between two or more different types of prokaryotes, or by the incidental – and non-lethal – ingestion of one or more, by another (Jones, 2006; Konhauser, 2007; Martin, in Falkowski & Knoll, 2007; Jastrow & Rampino, 2008; Bhattacharya et al., in Hedges & Kumar, 2009; Hedges, in Hedges & Kumar, 2009; Taylor et al., 2009). Eukaryotes differ from prokaryotes in their possession of membrane-bounded nuclei and organelles, and in their capability to reproduce sexually, and correspondingly greater capacity to mutate and to evolve. Eukaryotes first appeared in the rock record, in the form of Algae of uncertain affinity in rocks dating to 2100 Ma, and of sphaeromorph acritarchs in rocks dating to 1900 Ma. Molecular evidence in the form of arguably eukaryote-derived side-chain alkylated steranes in oil source-rocks and in oils indicates that they might have actually evolved as long ago as 2800–2700 Ma. However, the oldest unequivocally eukaryote-derived steranes, from the Matienda formation of Elliot Lake in Canada and the FA formation of the Franceville basin in Gabon, date to only 2200–2100 Ma (George et al., 2007).

The evolution of multicellularity

Multicellular organisms or metazoans are interpreted as having evolved through the aggregation of multiple eukaryotic unicells, and their organisation into different organs and tissue types with different functions (Jones, 2006; Jastrow & Rampino, 2008; Brasier, 2009). (Interestingly, multicellularity also appears to have arisen independently in a number of evolutionary lineages.)

The oldest known interpreted multicellular organisms have recently been described from the Palaeoproterozoic of Gabon, from rocks as old as 2.1 Ga, and thus immediately post-dating the so-called 'great oxidation event'of 2.5–2.3 Ga (El-Albani et al., 2010). With one possible exception, multicellular organisms had previously only been recorded from rocks as old as Mesoproterozoic, 1.6–1.0 Ga (El-Albani et al., 2010). A diverse array appeared in the so-called 'Precambrian surge' of the late Neoproterozoic, late Cryogenian–Ediacaran (Vendian), from approximately 650 Ma, including not only 'ediacarans' and SSFs (see below), but also, in the Doushantuo Formation of China, a number of enigmatic forms, exceptionally well preserved in three dimensions (Yuan Xunlai, 2002; Jones, 2006; Jun-Yuan Chen et al., 2009; Halverson et al., in Gaucher et al., 2010).

The oldest 'ediacarans' thus far known are from the Ediacaran, immediately above Cryogenian, 'Marinoan' or 'Varangerian' tillites dated to 650–620 Ma (Jones, 2006; McCall, 2006; Vickers-Rich & Komarower, 2007; Maruyama & Santosh, 2008). Incidentally, it has been hypothesised that certain metazoans could have sought refuge at deep marine hydrothermal vent sites during the 'Marinoan' or 'Varangerian' glaciation, and only emerged to recolonise shallow marine environments in the Cambrian, contributing to an apparent rather than real evolutionary diversification. The abundance and diversity of the 'ediacaran' metazoans already by the time of the Ediacaran is arguably, albeit indirectly, indicative of a still older origin, in the Cryogenian (Jones, 2006). Another indication that this was the case comes from the observation that microbialites began to decline in the Cryogenian, from around 1000 Ma, which some authors have suggested was due to an early rise of 'ediacarans', and associated excessive grazing activity. Note, though, that some other authors have hypothesised that the rise of the 'ediacarans' and decline of the microbialites at this time was brought about by environmental change associated with a series of glaciations, ultimately resulting in a so-called 'Snowball Earth' in the Cryogenian

(Etienne et al., 2006; Maruyama & Santosh, 2008; Hoffman & Li, 2009; Chumakov, in Gaucher et al., 2010; Kaufman et al., in Gaucher et al., 2010). Incidentally in this context, recent calculations have shown that weathering of the Laurentian flood basalts extruded in the Tonian could have resulted in sufficient consumption of greenhouse carbon dioxide to cause global cooling and initiate a 'snowball' glaciation.

The oldest calcareous and phosphatic SSFs thus far known are from the late Ediacaran, approximately 560 Ma (Jones, 2006; Gaucher & Germs, in Gaucher et al., 2010). Calcareous SSFs are also known as 'weakly calcifying metazoans' or WCMs (Halverson et al., in Gaucher et al., 2010).

The Late Precambrian mass extinction

There was a series of mass extinctions at or near the end of the Precambrian, Ediacaran, although the precise number and timing, and indeed the severity, of the events remain somewhat poorly constrained (Corsetti et al., 2006; Jones, 2006; Gaucher & Sprechmann, in Gaucher et al., 2010; Halverson et al., in Gaucher et al., 2010). The principal groups to have been affected were the Cyanobacteria, the acritarchs, and the 'ediacarans'. A number of ichnogenera also became extinct.

In the case of the Cyanobacteria, there may not have been a mass extinction at all, but rather a long-term decline over the Cryogenian–Ediacaran. Various causal mechanisms have been proposed for this decline in Cyanobacteria, including excessive grazing by early 'ediacarans', and environmental change associated with a series of glaciations.

In the case of the acritarchs, there were discrete mid-Cryogenian, 770–740 Ma and late Cryogenian, 700–635 Ma mass extinctions pre-dating the observed event at the end of the Ediacaran, c. 560–542 Ma.

In the case of the 'ediacarans', too, there appears to have been at least one Cryogenian mass extinction pre-dating the Ediacaran, evidence for which is provided by the absence of taxa such as Charnia and Charniodiscus in the Ediacaran Kotlin Horizon of the Russian Platform. The group may or may not have been affected by another mass extinction at the end of the Ediacaran. Some authors have argued that its apparent disappearance at this time is essentially an artefact of poor preservation, and that in fact it ranged through to the Cambrian, an argument seemingly supported by the recent discovery of 'ediacaran'-like forms in Cambrian Lagerstatten.

3.8.2 Palaeozoic

Note that there was an additional, minor, evolutionary diversification event in the Devonian, possibly initiated by the creation of new niches during the transgression of the time. Brachiopods, ammonoids, trilobites and crinoids all evolved or diversified (Jones, 2006).

Note also that there was an additional, minor extinction event in the Silurian, Wenlock, and another in the Silurian, Ludlow, both apparently more or less localised to Baltica (Jones, 2006; Eriksson et al., 2009). The extinction event in the Wenlock, also known as the Lau event, took place in the Upper *Polygnathoides siluricus* to Icriodontid conodont zones, and principally affected vertebrates. The pre-extinction event vertebrate fauna is dominated by jawed acanthodian fish; the post-extinction fauna by jawless thelodonts. The event in the Ludlow took place in the *Lundgreni, Testis* and *Flemingii–Dubius* zones, and principally affected graptolites. Phytoplankton abundance and diversity patterns, total organic carbon (TOC) fluctuations, and strong positive carbon and oxygen isotope anomalies indicate that the productivity regime was unstable during the extinction event. There appears to be a relationship between the extinction event and eutrophication and anoxia associated with the prolonged global sea-level high-stand of the time, and/or with enhanced Rheic Ocean upwelling.

The Cambrian evolutionary diversification

There appears to have been a major evolutionary diversification in the earliest Cambrian, 'Nemakit–Daldynian', the so-called 'Cambrian evolutionary diversification' – or 'Cambrian explosion' (Jones, 2006; Foote & Miller, 2007; Jastrow & Rampino, 2008; Brasier, 2009; Vannier, 2009; Halverson et al., in Gaucher et al., 2010).

Much of the observed diversification in the Cambrian is of organisms with strongly mineralised skeletons with enhanced preservation potential,

and as such could represent an apparent – preservational – rather than a real phenomenon, although, importantly, diversification is also observed among trace fossils, including vertically burrowing types, which appear for the first time at this time (Bengtson, in Briggs, 2005; Seilacher et al., 2005; Jones, 2006; Seilacher, 2007). (Incidentally, the process of mineralisation appears to have been facilitated by a change in the chemistry of the ocean, forcing organisms to ingest and excrete increased quantities of minerals, and/or an increase in the amount of oxygen in the atmosphere, enabling them to precipitate minerals more easily: Jones, 2006; Stanley, 2006; Ward, 2006.)

Both the strong mineralisation and the deep burrowing observed in the Cambrian could represent evolutionary responses to selection pressure exerted by increased predation at this time, perhaps in turn associated with the evolution of the eye (Jones, 2006; Dzik, in Vickers-Rich & Komarower, 2007; Seilacher, 2007). Importantly, representatives of all the known trophic types, including primary producers (phytoplanktonic acritarchs), secondary consumers (zooplanktonic filter-feeding arthropods), and tertiary and higher-level consumers (benthic filter-feeding hyolithids, benthic priapulids, nekto-benthic trilobites and nektonic anomalocarids) are observed in the Cambrian.

The (co-)evolution of predator–prey systems, whereby prey and predator orgamisms evolve ever-better defensive and offensive strategies respectively, became a characteristic feature not only of the Cambrian, but also of the Phanerozoic as a whole (Kelley et al., 2003; Harper, 2006; Jones, 2006).

The evolution of reefs

Reefs are more or less rigorously defined as resistant organic frameworks forming raised relief on the sea floor (Jones, 2006). (Interestingly, the alternative usage of the word is as a danger to shipping.) At the present time, reefs are essentially restricted to shallow waters in low latitudes, as indeed they appear to have been throughout their geological history. They are important reservoirs of biodiversity.

Reefs range from the Early Cambrian to the present time. Different types of organisms have constructed, or have contributed to the construction of, reefs over time, as plate configurations and climates have changed, and as evolution and extinction have proceeded (see 3.3.2 and Box 3.2 above). There have even been times, following mass extinction events, when there have been no reef-building organisms, and consequently no reefs.

In the Palaeozoic, Early Cambrian reefs were constructed chiefly by microbialites and archaeocyathans (Gandin & Debrenne, 2010). The Middle–Late Cambrian, following on from the Early Cambrian mass extinction, was a time during which there was essentially no reef construction. Ordovician–Devonian reefs were constructed chiefly by sponges, stromatoporoids and rugose and tabulate corals. The latest Devonian to Early Carboniferous, following the Late Devonian mass extinction, was another time during which there were essentially no reefs, but rather mud-dominated reef-like mounds known as 'Waulsortian mounds' which did not form rigid frameworks (Riding, 2005; Jones, 2006). Late Carboniferous–Permian reefs were constructed chiefly by calcareous Algae.

In the Mesozoic, Triassic reefs were constructed chiefly by calcareous Algae and scleractinian corals; Jurassic reefs by corals; and Cretaceous reefs by scleractinian corals and rudist bivalves. The Early Triassic, following the end-Permian mass extinction, was a time during which there were essentially no reefs, but rather microbial mounds (Riding, 2005; Jones, 2006). The Early Jurassic, following the end-Triassic mass extinction, was another such time.

In the Cenozoic, reefs were constructed chiefly by scleractinian corals – as indeed they are at the present time. The Palaeocene, following the end-Cretaceous mass extinction, was a time during which there was essentially no reef construction (but see Baceta et al., 2005).

The Early Cambrian mass extinction

There was a mass extinction at the end of the Early Cambrian, sometimes known as the 'Botoman–Toyonian crisis' (Jones, 2006). The principal group to have been affected was the archaeocyathans, the result being that archaeocyathan reefs, which had previously been widespread, for example in east Siberia, foundered.

The Late Cambrian mass extinction

There was another mass extinction, in fact in detail consisting of a series of as many as five distinct extinction events, in the Late Cambrian (Jones, 2006). The principal groups to have been affected were the inarticulate brachiopods, and, in North America and around the world, the trilobites.

The Ordovician evolutionary diversification

There was a major evolutionary diversification in the Ordovician, possibly initiated by the competition for the niches evacuated by the earlier extinction event(s) (Miller, 1997; Harper, 2006; Jones, 2006; Munnecke & Servais, 2007). Rugose and tabulate corals, articulate brachiopods, stenolaemate bryozoans, molluscs, trilobites, echinoderms and graptolites all evolved or diversified at this time.

The evolution of vertebrates

Molecular biological data indicates that vertebrates evolved in the late Precambrian, ~580–542 Ma (Janvier, 2003; Hedges, in Hedges & Kumar, 2009). Note, though, that there are issues surrounding the calibration of the 'molecular clock' (Benton et al., in Hedges & Kumar, 2009). Unequivocal vertebrate fossils first appeared in the rock record in either the Cambrian or the earlier Ordovician, in the form of marine fish interpreted as having evolved from a primitive cephalochordate, such as the lancelet-like Pikaia (Janvier, 2003; Jones, 2006; Janvier, in Anderson & Sues, 2007; Jastrow & Rampino, 2008; Janvier, 2009). However, they did not diversify until the later Ordovician, by which time they had apparently colonised non-marine environments (Sansom, 2005; Davies et al., 2007; Park & Gierlowski-Kordesch, 2007). Amphibians are interpreted as having evolved, from air-breathing fish, in the Devonian; reptiles and birds, from amphibians, in the Carboniferous; and mammals, from mammal-like reptiles, in the Triassic (Jastrow & Rampino, 2008).

The evolution of life on land

The first recognisable land plant microfossils appeared in the rock record in the Ordovician; the first land plant macrofossils in the Silurian (Klitzsch et al., 1973; Boureau et al., 1978; Douglas &

Lejal-Nicol, 1981; Gray *et al.*, 1982; Strother *et al.*, 1996; Edwards & Wellman, in Gensel & Edwards, 2001; Wellman *et al.*, 2003; Steemans & Wellman, in Webby *et al.*, 2004; Labandeira, 2005; Wellman, 2005; Jones, 2006; Beerling, 2007; le Herisse *et al.*, 2007; Jastrow & Rampino, 2008; Taylor *et al.*, 2009; Davies & Gibling, 2010; Vecoli *et al.*, 2010). ('Enigmatic, spore-like organic-walled microfossils' have recently been recorded in the Cambrian: Vecoli *et al.*, 2007.) These early plants – probably together with fungi – generated their own soils, providing fertile new ground for further colonisation, and also, importantly, stabilising the substrate, slowing down the rate of erosion, and ultimately, between Cambrian 'Vegetation Stage' or 'VS' 2 and Devonian 'VS' 6, causing a change in the dominant style of sedimentation in alluvial systems from braided to meandering (Jones, 2006; Davies & Gibling, 2010). Seed-plants are interpreted as having evolved, from club-mosses, ferns and allied forms, in the Devonian; flowering plants, from seed-plants, in the Cretaceous.

Arthropods, molluscs and a range of soft-bodied invertebrate animals appear to have colonised freshwater environments, by way of estuaries, in the Cambrian (Park & Gierlowski-Kordesch, 2007; Vecoli *et al.*, 2010). The first land invertebrates to appear, at least as long ago as the Devonian, were arthropods, whose flexible exoskeletons were pre-adapted well to life in the miniature forests of the time (Shear & Selden, in Gensel & Edwards, 2001; Giorgiani *et al.*, in Briggs, 2005; Jones, 2006; Nudds & Selden, 2008; Taylor *et al.*, 2009). Indeed, there is some trace fossil evidence to suggest that they might have appeared in the Ordovician or Silurian. Land gastropods appeared in the Carboniferous.

Fish appear to have colonised freshwater environments in the Ordovician, by way of estuaries (Davies *et al.*, 2007; Park & Gierlowski-Kordesch, 2007; Vecoli *et al.*, 2010). The first land vertebrates to appear – to exploit the growing plant and invertebrate food sources – were amphibian tetrapods, in the Devonian. Complete terrestrialisation may be said to have occurred in the Carboniferous, when reptile tetrapods laying eggs with shells appeared, and the former tie to the aquatic environment was finally severed.

The end-Ordovician mass extinction

There was a significant mass extinction at the end of the Ordovician, resulting in the extinction of somewhere between 70 and 85% of all species, and between 22 and 33% of all families, although the degree of severity has been questioned (Sheehan, 2001; Jones, 2006; Nicholls, 2009). In fact, in detail, there appear to have been two distinct events, separated in time by some 0.5–1 Ma, the first in the *Normalograptus extraordinarius* – Graptolite – Zone, at the beginning of the Hirnantian, and the second in the *Normalograptus persculptus* Zone, within the Hirnantian (Sutcliffe *et al.*, 2000, 2001; Jones, 2006). The first event accelerated an already evident decline in some planktonic groups, such as acritarchs, chitinozoans and graptolites; and caused a dramatic decline in many shallow-marine, benthic groups, leaving only a low-diversity fauna, dominated everywhere by the newly evolved and interpreted opportunistic brachiopod *Hirnantia* (Sutcliffe *et al.*, 2001; Rong Jia-Yu *et al.*, 2002; Jones, 2006). The second event eliminated many elements of the *Hirnantia* fauna, and many remaining elements of the pre-extinction fauna (Sutcliffe *et al.*, 2001; Rong Jia-Yu *et al.*, 2002; Bourahrouh *et al.*, 2004; Jones, 2006).

The principal groups to have been affected were the acritarchs, corals, brachiopods, stenolaemate bryozoans, nautiloids, ostracods, trilobites, echinoderms, graptolites and chitinozoans (and also trace fossils) (Jones, 2006). Entire reef communities were also affected. Interestingly, however, the structure of the ecosystem as a whole was but little affected.

The cause that has been most often invoked for the end-Ordovician mass extinction is global cooling – of up to 8 °C – associated with glaciation; rapidly followed by global warming, accompanied by transgression and oceanic anoxia, associated with deglaciation (Jones, 2006). A 'super-plume' event has also been implicated, as has a significant reduction in atmospheric oxygen concentration, and cosmic gamma-ray radiation (Jones, 2006; Ward, 2006; Melott & Thomas, 2009).

There is abundant chemical evidence for glaciation and deglaciation in the form of carbon and oxygen isotope excursions, worldwide, and physical evidence in the form of erosional glacial valleys and syn- and post-glacial deposits, throughout Gondwana (Jones, 2006). Global cooling associated with glaciation, and accompanying glacio-eustatic sea-level fall and shallow marine habitat destruction, could certainly account for certain of the observed effects on the benthic fauna at the time of the first extinction event, including the rise of the opportunistic – cold-adapted – *Hirnantia* fauna (Sutcliffe *et al.*, 2000, 2001; Jones, 2006). Accompanying changes in oceanic circulation, including advection of toxic, hypertrophic bottom waters to the surface, could account for the observed effects on the plankton. Succeeding global warming associated with deglaciation, and accompanying transgression and – S-State – oceanic anoxia, could account for the remaining observed effects on the pre-extinction benthic faunas, and the *Hirnantia* fauna, at the time of the second extinction event.

The Late Devonian mass extinction

There was a significant mass extinction in the Late Devonian, resulting in the extinction of somewhere between 70 and 80% of all species, and 20% of all families (Jones, 2006). In fact, in detail, there appear to have been five distinct extinction events, separated in time by as much as 0.8 Ma from first to last (Racki, in Over *et al.*, 2005; Jones, 2006). The first was the so-called lower Kellwasser event, at 364.7 Ma, within the Late *Rhenana* – Conodont – Zone of the Frasnian; the second, third and fourth, the upper Kellwasser events at 364.2 Ma, 364.1 Ma and 364.0 Ma, within the *Linguiformis* Zone of the Frasnian; and the fifth, the Homoctenid event, at 363.9 Ma, within the Early *Triangularis* Zone of the Famennian.

The principal groups to have been affected were the cyanophytes, phytoplankton, calcareous Algae, plants, stromatoporoids, rugose and tabulate corals, brachiopods, molluscs, ostracods, trilobites, echinoids, crinoids, fish and amphibians (Jones, 2006; Riegel, 2008). Tentaculitid molluscs entirely disappeared. In the marine realm, entire reef ecosystems either collapsed, as substantial groups of reef-building organisms disappeared, at least temporarily; or, in the case of the Canning basin of Western Australia, re-organised. In the Timan–Pechora basin in the northeastern part of the Russian platform, solitary corals were the first group to

disappear, in the Frasnian, followed by stromatoporoids and colonial corals, later in the Frasnian, and, finally, red calcareous Algae, in the Famennian. Stromatoporoids and rugose and tabulate corals never recovered their pre-extinction diversity at any time in their subsequent evolutionary history. The phytoplankton did not recover their pre-extinction diversity until the Late Triassic, 130 million years after the event. Ultimately, acritarchs became displaced from their pre-eminent position among the phytoplankton by dinoflagellates; brachiopods from their pre-eminent position among the filter-feeding benthos by bivalve molluscs. Recovery of reef ecosystems took until the Kinderhookian sub-stage of the Tournaisian in Australia, Asia and eastern Europe, and until the Visean in western Europe and North America, with the earlier parts of the Carboniferous represented by a gap – the 'reef gap' – in the marine fossil record (Webb, 1998). Recovery of terrestrial ecosystems took until the Courceyan sub-stage of the Tournaisian, with the earlier part of the Tournaisian represented by a similar gap – 'Romer's gap' – in the terrestrial fossil record (Smithson & Wood, 2009).

The causes that have been invoked for the Late Devonian mass extinction are intrinsic, and associated with global cooling and associated sea-level fall, or global warming and associated sea-level rise, oceanic anoxia, atmospheric dysoxia and hypercapnia; or extrinsic, and associated with impacts of extraterrestrial bodies; or some combination thereof (Daizhao Chen & Tucker, 2004; Bond & Wignall, in Over *et al.*, 2005; Racki, in Over *et al.*, 2005; Schieber & Over, in Over *et al.*, 2005; Jones, 2006; Ward, 2006; Bond & Wignall, 2008; Gutak *et al.*, 2008). The holistic 'Devonian plant hypothesis' links the extinction to a combination of long-term global cooling caused by drawdown of atmospheric carbon dioxide, and short-term anoxia caused by intensified pedogenesis (soil formation) and associated increased nutrient flux and eutrophication, both processes being driven by the evolutionary diversification of plants, particularly gymnosperms (Algeo *et al.*, 1995; Algeo & Scheckler, 1998; Algeo *et al.*, in Gensel & Edwards, 2001; Strother *et al.*, in Vecoli *et al.*, 2010).

Global cooling could account for all the observed effects on marine and terrestrial faunas and floras not (pre-)adapted to cold conditions, including the apparently increasing restriction of their range towards the tropics. Importantly, glaciation in the Late Devonian, although within the Famennian rather than Frasnian, has been hypothesised by Algeo et al. (1995), Algeo and Scheckler (1998) and Algeo et al., in Gensel and Edwards (2001) as part of the 'Devonian plant hypothesis' alluded to above, and also by Caplan and Bustin (1999), Streel et al. (2000), Brand et al. (2004), McGhee, in Over et al. (2005), Racki, in Over et al. (2005) and Kaiser et al. (2006). Glaciation within the Early-Middle Devonian has been hypothesised by Elrick et al. (2009).

Global warming, to temperatures indicated by isotope evidence to have been as high as 34 °C, could have killed off a wide range of shallow marine carbonate-producing organisms, as, incidentally, high-temperature El Niño–Southern Oscillation (ENSO) events do at the present time. Associated sea-level rise and oceanic anoxia could account for the observed effects of the lower Kellwasser event and the first of the upper Kellwasser events, which were apparently principally on shallow marine benthic faunas not (pre-)adapted to low-oxygen conditions. Oceanic anoxia and overturn of the type invoked in the case of the end-Ordovician mass extinction could account for the observed effects of the second and third of the upper Kellwasser events and of the Homoctenid event, which were apparently principally on marine planktonic faunas and floras. Obviously, oceanic anoxia could not account for any observed effects on terrestrial faunas and floras, although atmospheric dysoxia and hypercapnia (high carbon dioxide concentration) could. Note in this context that carbon dioxide concentration would have been high at the time of the mass extinction, on account of all of the rift-related volcanism in eastern Laurussia and northern Gondwana (Mahmudy Gharaie et al., 2004).

The impacts of extraterrestrial bodies, especially in combination with – ?consequent – global cooling, could account for all the observed effects of all the events. Perhaps significantly, there is reliable evidence of impact in Belgium at the time of the Homoctenid event, in the form of microtektite, or melt-glass. There is also less reliable evidence of impacts in Austria and China at the times of the Kellwasser events, in the form of microspherules,

interpreted as fragmentary microtektites, or of iridium anomalies (possibly the result of biological concentration). Intriguingly, though, there is also evidence of impacts immediately pre- and post-dating the mass extinction event(s), which apparently had no obvious effect on the biota, in the *punctata* Conodont Zone of the Frasnian, and the *crepida* Conodont Zone of the Famennian, respectively.

The evolution of forests

Forests first appeared in the Middle Devonian, as evidenced by the Gilboa Lagerstatte of up-state New York, although they only became widespread after the diversification of gymnosperms in the Late Devonian to Carboniferous (Algeo et al., in Gensel & Edwards, 2001; Jacobs, 2004; Cleal & Thomas, 2005; Jones, 2006; Beerling, 2007; Nudds & Selden, 2008; Cleal & Thomas, 2009; Taylor et al., 2009; Mintz et al., 2010; Vecoli et al., 2010). Vertical tiering of forest habitats evidently first appeared in the Carboniferous, enabling the evolutionary diversification not only of plants, but also of invertebrates and vertebrates, in essentially entirely new niches. Recognisably modern rainforests first appeared after the diversification of the angiosperms in the Cretaceous.

The evolution of flight

The first flying animals to appear – to exploit the growing plant food source – were winged insects, interpreted as having evolved from wingless insects, in the Carboniferous (Jones, 2006). Flight has since arisen independently in a number of lineages of vertebrates. Reptiles capable of powered flight, as opposed to gliding, appeared, in the form of pterosaurs, in the Triassic; birds in the Jurassic; and flying mammals, in the form of bats, in the Palaeocene. The full range of flying or gliding vertebrates today includes flying fish, frogs, lizards and snakes, birds, and marsupial and placental mammals.

The end-Permian mass extinction

There was a significant mass extinction, the most significant of all the mass extinction events, at the end of the Permian, resulting in the extinction of somewhere between 70 and 95% of all species, and 50–60% of all families (Ward, 2004; Erwin, 2006;

Jones, 2006; Foote & Miller, 2007; Hongfu Yin *et al.*, 2007; Ward, 2007). In fact, in detail, there appear to have been at least two distinct extinction events, separated in time by as much as 8–10 Ma – the actual figure varying from one time-scale to another (Racki & Wignall, in Over *et al.*, 2005; Jones, 2006; Hongfu Yin *et al.*, 2007; Bond & Wignall, 2009; Isokazi, 2009a, b; Isokazi & Aljinovic, 2009; Wignall *et al.*, 2009a; Wignall *et al.*, 2009b; Bond *et al.*, 2010). The first event was at the end of the Middle Permian, Guadalupian, Capitanian, at approximately 260 Ma, and particularly affected marine faunas in low latitudes, most particularly the 'tropical trio' of verbeekinid fusulinide foraminifera, waagenophyllid corals and giant alatoconchid bivalves; and the second, more severe, event, or series of events, took place towards the end of the Late Permian, Lopingian, Changhsingian or Changxingian, at approximately 251 Ma, and affected almost the entire range of marine and terrestrial faunas and floras worldwide.

The principal groups affected were the phytoplankton, plants, foraminifera, radiolarians, sponges, rugose and tabulate corals, articulate brachiopods, stenoloaemate bryozoans, goniatite ammonoids, trilobites, insects, echinoderms and tetrapods (and also trace fossils) (Erwin, 2006; Jones, 2006; Bond & Wignall, 2009; Isokazi & Aljinovic, 2009; Wignall *et al.*, 2009a; Wignall *et al.*, 2009b). In the marine benthic realm, the rugose and tabulate corals ultimately disappeared entirely. The brachiopods and crinoids were also badly affected, and never recovered their pre-extinction diversity. The bryozoans underwent significant turnover, as did the bivalves and echinoids, with formerly insignificant groups rising to replace those lost. In the pelagic realm, the zooplanktonic radiolarians experienced almost complete extinction, possibly on account of the effects on their phytoplanktonic food sources (carbon isotope evidence indicating a collapse of primary productivity in the oceans at the end of the Permian). On land, a wide range of plant groups were adversely affected, and the former *Glossopteris* floras of Gondwana, and the equivalent *Cordaites* forests of the northern hemisphere, were killed off, although fungi were evidently able to flourish, possibly on decaying plant matter. Animals were also adversely affected, and a number of early insect groups were killed off, as were a number of early tetrapod groups, including the herbivorous pareiasaurs, the carnivorous gorgonopsids, and the omnivorous millerettids. The most important net effects of the end-Permian mass extinction in the marine benthic realm were: firstly, the effective elimination of the former sessile, low-metabolic-rate epifauna of reef-forming sponges, corals, brachiopods, bryozoans and crinoids, and its replacement by a motile, high-metabolic-rate epifauna and infauna of bivalves and gastropods; and, secondly, the resultant effective elimination of reefs, which were only able to become re-established in the Middle Triassic (Knoll *et al.*, 1996; Bambach *et al.*, 2002; Fraiser *et al.*, 2005; Pruss & Bottjer, 2005; Erwin, 2006; Jones, 2006; Pruss *et al.*, 2006).

In terms of the recovery from the end-Permian mass extinction, in the marine benthic realm, there was a succession over the recovery interval of biotas dominated by *Lingula*, *Claraia*, microgastropods, crinoids, and Cyanobacteria (Fraiser *et al.*, 2005; Erwin, 2006; Jones, 2006; Zhong-Qiang Chen *et al.*, 2010). Among the brachiopods, the most successful survivors were generalists with wide bathymetric and biogeographic distributions (Zhong-Qiang Chen *et al.*, 2005a, b). In the terrestrial realm, there was a succession of floras characterised by the lycopsid *Pleuromeia* in the Early Triassic, conifers in the early Middle Triassic, early Anisian, and cycadophytes and pteridosperms in the early Middle Triassic, late Anisian (Grauvogel-Stamm & Ash, 2005). In the Karoo basin of South Africa, there was a succession of faunas from a 'survivor fauna' characterised by *Ictidosuchoides*, *Moschorhinus* and *Tetracynodon*, to a 'recovery fauna' characterised by medium-sized dicynodont reptiles, including the opportunistic *Lystrosaurus*, by small proterosuchian archosauromorph, procolophonid and insectivorous cynodont reptiles, and by small amphibians (Smith & Botha, 2005). The faunal recovery here appears to have been relatively rapid, and indeed may have taken as little as 40–50 000 years (Botha & Smith, 2006).

Historically, the causes that have been most often invoked for the end-Permian mass extinction are volcanism, climatic change, sea-level change, anoxia, the dissociation of methane from deep-sea hydrates, or, according to the '*Murder on the Orient Express*' hypothesis, some combination thereof (Racki & Wignall, in Over *et al.*, 2005; Erwin, 2006; Jones, 2006; Shu Zhong Shen *et al.*, 2006; Heydari *et al.*, 2008; Isokazi, 2009a, b; Wignall *et al.*, 2009a). (Impacts of extraterrestrial bodies have also been invoked, and indeed fullerenes of possible, although not proven, extraterrestrial origin have been reported from the event horizon: Jones, 2006.) There is evidence of volcanism at the time of the extinction event at the end of the Capitanian, in the form of the explosive volcanics of the Emeishan Traps of China, which have been dated to 259 Ma, and at the time of the extinction event at the end of the Changhsingian, in the form of the eruptive volcanics of the Siberian Traps of Siberia, which have been dated to 250 Ma (Mei-Fu Zhou *et al.*, 2002; Jones, 2006; Isokazi, 2009a, b; Reichow *et al.*, 2009; Wignall *et al.*, 2009a). The amount of volcanism represented by the Siberian Traps would almost certainly have resulted in the release of significant quantities of ash into the atmosphere, which could have blocked out the sun, prevented photosynthesis, and caused a collapse in primary productivity on land and at sea, thus accounting for at least some of the observed effects of the mass extinction (Jones, 2006). There is carbon isotopic evidence of significant climate change at the end of the Capitanian, when there appears to have been a short-term ('Kamura') cooling event coincident with the extinction of the warm-water verbeekinid foraminifera, waagenophyllid corals and alatoconchid bivalves (Isokazi & Aljinovic, 2009). There is also carbon isotopic evidence of a significant climatic change at the time of extinction event at the end of the Changhsingian, when there appears to have been a warming of perhaps as much as 5–6 °C (Jacobsen *et al.*, 2009). And there is palaeopedological or palaeosol evidence from the Karoo basin of South Africa, from eastern Australia and from Antarctica of a change at this time from a humid climatic regime characterised on land by the extensive development of flood-plains, to a much more arid regime characterised by calcretes, even at high palaeolatitudes, with low palaeolatitudes probably at least bordering on uninhabitable (Jones, 2006). There is palaeontological evidence of a major sea-level fall in some parts of China at the time of the extinction event at the end of the Capitanian, and a minor fall, or a series of minor falls, followed by a major rise, at least arguably associated with global warming, at the time of the extinction event at the end of the Changhsingian (Jones, 2006). Note, though, there is no palaeontological evidence for such sea-level change in other parts of China, or in Japan, where the event horizon is marked by a calcimicrobial framestone of enigmatic origin, possibly related to an 'anomalous oceanic event'. There is also evidence for 'superanoxia' affecting not only the deep oceans but also the shallow seas at the time of the extinction event at the end of the Changhsingian, which, as noted above, was also a time of transgression (Jones, 2006). The expansion of the anoxic zone through the water column is recorded by a widespread cerium anomaly (Kakuwa & Matsomoto, 2006). Finally, there is evidence, too, of disassociation of methane hydrates in the deep sea at the time of the extinction event (Kidder & Worsley, 2004; Jones, 2006). The associated release into the atmosphere of the 'greenhouse gas' methane would have caused global warming, through the much-discussed 'greenhouse effect' (compounded by the release into the atmosphere of carbon dioxide associated with Siberian Trap volcanism). According to Kidder and Worsley (2004), in the oceans, the global warming caused transgression through melting of polar ice, a weakened pole-to-equator thermal gradient, weakened wind-driven upwelling, sluggish circulation, and ultimately widespread anoxia and mass extinction. Also according to Kidder and Worsley (2004), on land, it caused the extinction of the extensive coal-forests of the Permian, resulting in the release into the atmosphere of carbon dioxide no longer capable of being consumed by the forest plants, and reinforcing the 'greenhouse effect'. (The extinction in the oceans may have had a similar effect, resulting in the release into the atmosphere of carbon dioxide no longer capable of being consumed by the phytoplankton.)

More recently, a significant decrease in atmospheric and oceanic oxygen concentration, a significant increase in atmospheric and oceanic carbon dioxide concentration, a significant increase in atmospheric and oceanic hydrogen sulphide concentration, and a significant increase in radiation caused by breakdown of the protective ozone layer in the atmosphere have been implicated (Ward, 2006; Beerling, 2007; Ward, 2007). Evidence for an increase in hydrogen sulphide concentration is provided by an observed increased incidence of biomarkers derived from sulphate-reducing Bacteria (Kump et al., 2005; Ward, 2007); that for radiation by an observed increased incidence of mutations in plants and spores (Visscher et al., 2004; Looy et al., 2005; Beerling, 2007).

Whatever the cause, and the details do still remain to be worked out, the end-Permian mass extinction was on a scale never encountered before or since, probably because of the catastrophic effects on the primary producers forming the basis of the trophic structure in terrestrial and marine communities.

3.8.3 Mesozoic

Note that there were additional, minor, mass extinction events in the Early, Middle and Late Jurassic and in the Early Cretaceous (Jones, 2006). The Early and Late Jurassic events involved extinctions of benthic brachiopods, bivalves and gastropods, and nektonic ammonites apparently as a result of anoxia, although essentially only in Europe. Many other marine animals were unaffected, and terrestrial animals appear to have been entirely unaffected. The Middle Jurassic event is poorly constrained, but appears to have involved extinctions of cephalopods. The Early Cretaceous event is also poorly defined, but appears to have involved extinctions as a result of an oceanic anoxic event. The causes that have been most often invoked for the Early Jurassic mass extinction are the dissociation of methane from deep-sea hydrates, global warming, anoxia and volcanic eruption, or some combination thereof; and the impact of an extraterrestrial body.

The Mesozoic evolutionary diversification

There was a major evolutionary diversification in the Mesozoic, possibly initiated by the competition for the niches evacuated by the earlier extinction event. In the seas, cyclostomate and gymnolaemate bryozoans, scleractinian corals, bivalves, ammonites, belemnites and echinoids all evolved or diversified at this time, the time of so-called 'Mesozoic marine revolution' (Jones, 2006). Echinoids replaced crinoids as the dominant echinoderms. On land, dinosaurs, crocodiles, pterosaurs, turtles, lizards, frogs and salamanders, and mammals replaced labyrinthodonts, thecodonts, procolophonids, prolacertiforms, rhyncosaurs and mammal-like reptiles as the dominant tetrapods (Fraser, 2006; Jones, 2006). New groups apparently replaced old ones either by simply occupying the niches evacuated by them, or by out-competing them. For example, the dinosaurs appear to have out-competed the thecodonts and rhyncosaurs through the acquisition of superior locomotory and other skills, including the ability to self-regulate body temperature and metabolic rate (Fraser, 2006; Jones, 2006; Ward, 2006).

Other causes that have been invoked for the Mesozoic evolutionary diversification include increased dispersal and endemism associated with the break-up of Gondwana, increased predation pressure, and the co-evolution of predator–prey systems. The evolutionary diversification of the planktonic foraminifera in the Jurassic and Cretaceous has been linked to oceanic anoxic events.

The end-Triassic mass extinction

There was apparently a significant mass extinction event at the end of the Triassic, resulting in the extinction of approximately 80% of all species, and 20% of all families, although the degree of severity has been questioned (Jones, 2006; Ward, 2007; McElwain et al., 2009). In detail, there were apparently at least two distinct extinction events. There was one event in the Carnian at the beginning of the Late Triassic, and one in the Rhaetian at the end, separated in time by as much as 28 Ma (the actual figure varying from one time-scale to another).

The principal groups affected were the plants, sponges, corals, brachiopods, stenolaemate bryozoans, bivalves, gastropods, insects, echinoderms, amphibians and reptiles (and also, at least locally, trace fossils). Interestingly, the groups affected by the first, Carnian event were both marine and terrestrial, whereas those affected by the second, Rhaetian event were mostly marine. In the marine realm, the ceratite ammonoids and conodonts disappeared entirely over the course of the Late Triassic, as did the strophomenid brachiopods, conulariids, nothosaurs and placodonts, and the bivalves lost an estimated 92% of all of their species, and 42% of all of their genera. Reefs were effectively eliminated by the end of the Late Triassic, and only recovered towards the end of the Early Jurassic. In the terrestrial realm, the formerly dominant tetrapods, that is, the labyrinthodonts, the mammal-like reptiles, the thecodonts, the procolophonids, the prolacertiforms and the rhyncosaurs, disappeared entirely over the course of the Late Triassic, and new groups, including the dinosaurs, crocodiles, pterosaurs, turtles, lizards, frogs and salamanders, and mammals, came to rise to prominence. Note, though, that it has been argued that the observed turnover was less the result of a single mass extinction event than a series of minor extinction events, or even that it was the result of environmental change rather than extinction. (New groups could have replaced old ones by simply occupying the niches vacated by them, for example the crocodiles replacing the phytosaurs, or the lizards the procolophonids and prolacertiforms; or by outcompeting them, for example, the dinosaurs outcompeting the mammal-like reptiles, rhyncosaurs and thecodonts.) Interestingly, plant macrofossils indicate a mass extinction event, while plant microfossils do not, although this may be because many evidently distinct biological species of parent plant appear to have produced indistinguishable morphological species of spore (Mander et al., 2010).

Historically, the causes that have been most often invoked for the end-Triassic mass extinction(s) are climatic change, sea-level change, volcanic eruption, and the dissociation of methane from deep-sea hydrates. (Impacts of extraterrestrial bodies have also been implicated: Jones, 2006; Simms, 2007. Indeed, there is evidence for impact in the form of the Manicouagan impact crater in Canada, dated to 206–213 Ma, although there is no evidence for associated shocked quartz or an iridium anomaly actually at the mass extinction event horizon. There is also evidence for either impact or earthquake in the Cotham member of the Penarth group of the UK, dated to Rhaetian.) There is abundant evidence of climatic change at the time of the extinction event(s), associated with the onset of rifting of Pangaea and the subsequent drifting of Gondwana and Laurasia away from the tropical belt, although the relationship, if any, between the same and the extinction event(s)

remains unclear. It may be stated, though, that there was apparently a change to a more arid climate over the period of the extinctions, as evidenced on land by the replacement of the old, Gondwanan, *Dicroidium* Flora by a new, essentially cosmopolitan, conifer-benettitalean flora, and that this would have stressed at least land animal populations, and could account for the observed extinctions of the same. Palaeobotanical evidence also indicates greenhouse warming of as much as 3–4 °C (McElwain et al., 1999). There is also evidence of associated sea-level change, although again the relationship to the extinction event(s) remains unclear. It may be stated, though, that sea-level over the period of the extinction event(s) was apparently generally extremely low. This would have resulted in habitat loss in the marine realm. It would also have resulted in the removal of barriers between populations in the terrestrial realm, and hence reduction in the rate of speciation, if not actual extinction. There is evidence, too, of flood-basalt volcanic eruption in the Central Atlantic Magmatic Province or CAMP around the end of the Triassic, and associated rising atmospheric carbon dioxide levels and other environmental effects associated with this volcanism have been implicated in the mass extinction (Marzoli et al., 2004; Huynh & Poulsen, 2005; Verati et al., 2007). Note, though, that the peak of the volcanic activity, at 199 Ma, appears to post-date the extinction event, at 202 Ma (Nomade et al., 2007; Verati et al., 2007; Whiteside et al., 2007). Lastly, there is evidence – in the form of a significant carbon isotope excursion – of dissociation of methane from deep-sea hydrates, or of some other perturbation to the carbon cycle, at, at least approximately, the same time as the end-Triassic mass extinction event (Hesselbo et al., 2002). Early Jurassic stromatolitic bioherms in the Neuquen basin in Argentina have been interpreted as related to methane seepage.

More recently, a significant increase in atmospheric and oceanic hydrogen sulphide concentration has been implicated (Ward, 2007), evidence for which is provided by an observed increased incidence of isorenieratine, a biomarker derived from sulphate-reducing Bacteria (Kump et al., 2005).

Incidentally, there is also evidence of global climatic change, sea-level change, volcanic eruption and dissociation of methane from deep-sea hydrates at the same time as the slightly later Early

Jurassic mass extinction (Hesselbo *et al.*, 2000; Wignall *et al.*, 2005).

The evolution of flowering plants

Flowering plants first appeared in the latest Jurassic, and recognisably modern rainforests in the Cretaceous (Soltis *et al.*, 2005; Jones, 2006; Beerling, 2007).

The Late Cretaceous (Cenomanian–Turonian) mass extinction

There was a mass extinction in the Late Cretaceous, at the Cenomanian–Turonian boundary, at the end of the *Rotalipora cushmani* Zone (Jones, 2006; Ward, 2007). The principal groups affected were the dinoflagellates, calcareous nannoplankton, foraminifera, sponges, caprinid rudists, cephalopods, echinoids, bony fish and ichthyosaurs. In terms of recovery from the Cenomanian–Turonian mass extinction, in Tibet, there is evidence of the existence of a 'survival' interval in the *Whiteinella archaeocretacea* Zone at the base of the Turonian and a 'rapid recovery' interval thereafter, the latter characterised by the development of a diverse community of planktonic foraminifera including both shallow and deep water-column dwellers. The extinction and survival intervals are characterised by dysaerobic facies. The causes that have been most often invoked for the Cenomanian–Turonian mass extinction are global climatic change, sea-level change, anoxia, volcanic eruption or some combination thereof. Impacts of extraterrestrial bodies have also been implicated. There is good evidence of global warming, sea-level rise and an oceanic anoxic event. There is also good evidence of oceanic plateau-forming volcanic eruption in the Ontong–Java Plateau, Caribbean/Colombian Plateau, Madagascar Rise and Broken Ridge, totalling some million square kilometres. Associated emission of large quantities of carbon dioxide would have led to greenhouse warming. It would also have led to enhanced productivity in surface waters, which would in turn have exacerbated the anoxia in bottom waters.

The end-Cretaceous mass extinction

There was a significant mass extinction event at the Cretaceous–Tertiary (K–T) boundary, resulting in the extinction of somewhere between 40 and 76% of all species, and 14% of all families (Jones, 2006; Ward, 2007; Nichols & Johnson, 2008; Schulte *et al.*, 2010).

The principal groups affected were, in the seas, the foraminifera, diatoms, cyclostomate and some gymnolaemate bryozoans, inoceramid and rudist bivalves, ammonites, belemnites, and marine reptiles/mosasaurs and plesiosaurs; and on land, the plants, dinosaurs and pterosaurs. (Note in this context, though, that the inoceramids, rudists and dinosaurs all appear to have been declining before the event, although this may simply be due to an artefact of sampling.) Interestingly, the effects of the event appear to have been selective across taxonomic groups, with some scarcely affected at all, such as the brachiopods and echinoids; and some, such as ferns, even flourishing, as they do today in areas of disturbance. The effects also appear to have been selective across size ranges within groups, with large taxa tending to die or be killed off, and small taxa to survive.

The initial recovery from the end-Cretaceous mass extinction event, in the form of renewed diversification, took place between 1 ka and 10 ka in the case of one of the most intensively studied groups, the planktonic foraminifera, and 40–400 ka in the case of another, the calcareous nannoplankton. The final recovery of pre-extinction diversity took anything up to 15 Ma for the marine invertebrate biota as a whole, and up to 20 Ma for certain groups, such as the bivalves and gastropods. The recovery of reef ecosystems appears to have taken place relatively rapidly, in as little as 2 Ma, possibly on account of the rapid emergence of new groups of organisms in the niche formerly occupied by the rudists (Baceta *et al.*, 2005). The recovery of terrestrial ecosystems also appears to have taken place relatively rapidly, possibly in this instance on account of the rapid emergence, and, incidentally, size increase, of the mammals in the niche formerly occupied by the dinosaurs (Jones, 2006). Thus, the ultimate long-term effect of the mass extinction event was the redirection of the interaction between different groups of organisms.

The cause that has been most often invoked for the end-Cretaceous mass extinction is the catastrophic impact of an extraterrestrial bolide, possibly combined with flood volcanism (Jones, 2006), although numerous other potential causes have also been invoked, including a significant reduction in atmospheric oxygen concentration (Ward,

2006). As is now well established, there is abundant evidence for an impact, in the form of a crater at Chicxulub on the Yucatan Peninsula in Mexico. Apart from its immediate effects (wild fires on land, tsunamis at sea), an impact could also ultimately have resulted in the release of sufficient particulate matter into the atmosphere to block out the sun for months or years, causing global cooling, or a so-called 'impact winter'. The ultimate result could have been a prevention of photosynthesis and a collapse of primary production on land and in the sea. Carbon isotope evidence indeed indicates a major perturbation in the carbon cycle in the ocean, such as would be expected from a collapse of primary phytoplankton – and secondary zooplankton – production. The resulting, essentially lifeless ocean has been dubbed the 'Strangelove Ocean', in reference to remarks on the effects of nuclear weapons in Stanley Kubrick's film 'Dr Strangelove'. The longer-term after-effects of the impact could have included global warming associated with the 'greenhouse effect', perhaps accounting for the evidence from leaf physiognomy of warming following the 'impact winter' in the Western Interior of the United States. It is possible, though, that climate change proceeded entirely independently of any extraterrestrial event.

3.8.4 Cenozoic

Note that there were additional evolutionary or diversification events through the course of the Cenozoic (Jones, 2006). There were also minor mass extinction events in the Oligocene, Miocene and Pliocene. The Oligocene event involved extinctions of mammals, although apparently only in North America. The Miocene event involved extinctions of plankton. The Pliocene event also involved extinctions of plankton, and of benthic bivalves and gastropods, although apparently essentially only in the tropics. Note, though, that there is also a distinct turnover of mammals between the Chapadmalalan and Barrancalobian Land Mammal Ages of the Pampean region of Argentina in South America, dated to approximately 3.3 Ma. Approximately 37% of all genera and 53% of all species became extinct at this time. Palaeoecological analysis indicates that the Chapadmalalan fauna is balanced from the point of view of trophic structure, whereas the Barrancalobian, although less well known, appears to contain disproportionately few carnivores, especially in the medium to large size category. Perhaps significantly, there is evidence in the succession of an impact event.

The end-Palaeocene mass extinction

There was a mass extinction at the end of the Palaeocene (Jones, 2006; Foote & Miller, 2007; Ward, 2007). The cause that has been most often invoked is global warming, possibly in turn triggered by methane hydrate dissociation.

The end-Eocene mass extinction

There was another mass extinction at the end of the Eocene (Jones, 2006). The principal groups affected were the plankton and open-water bony fish in the sea, and mammals on land, at least in Europe. The cause that has been most often invoked is global cooling. Impacts of extraterrestrial bodies have also been implicated. An impact crater has been found at Chesapeake Bay.

The evolution of grasslands and grassland animals

Grasses first appeared in the Cretaceous or Palaeocene, while C4 grasses and grasslands, which are tolerant of arid climates, first appeared in the Middle Miocene and first abounded in the Late Miocene, coincident with the development of arid climatic conditions worldwide (Jacobs et al., 1999; Jacobs, 2004; Jones, 2006; Beerling, 2007; Stromberg et al., 2007).

Grassland animals first appeared in the Cretaceous, and first abounded in the Miocene. Termites, lagomorphs and ungulates, modern representatives of which groups contain symbiotic gut biotas enabling them to digest the cellulose in the cell walls of grasses, first appeared in the Cretaceous, Palaeocene and Eocene respectively; other herbivores whose masticatory morphology indicates an adaptation to grazing, in the Eocene or Oligocene.

The evolution of humans

'Anatomically modern' human beings evolved in Africa approximately 150 000 years ago, and dispersed 'Out of Africa' approximately 85 000 years ago, to colonise the remainder of the world (Jones, 2006; Jastrow & Rampino, 2008). They, or rather we, are characterised by enhanced encephalisation, arguably in consequence of selection pressure associated with climatic and environmental forcing.

However, neither our ability to manufacture tools, nor our ability to communicate, is unique in the animal kingdom, as was once thought. What – we think – sets us aside from all other animals that have ever lived, including our immediate ancestors, is our consciousness: that is, our awareness of ourselves; our overview of ourselves in space and time; and our individual and collective ability to process information on our world past and present to plan for the future, and, for good or ill, to influence our own destiny.

The Pleistocene mass extinction

There was yet another mass extinction in the Pleistocene, albeit accounting for the extinction of only 1% of all species (Martin, 2005; Johnson, 2006; Jones, 2006; Foote & Miller, 2007).

The principal groups affected were large land mammals, that is, those with mean adult body weights in excess of – and in many cases considerably in excess of – 44 kg, such as mammoths, mastodons and woolly rhinoceroses. The effects were also selective across geographic regions, with 73% of large mammal genera becoming extinct in North America, and 79% in South America, but only 29% in Europe, and 14% in Africa. The effects were such as to change the structures of communities in the affected areas forever.

The causes that have been most often invoked are global warming and environmental change associated with the end of the Pleistocene ice age, human hunting activity, and some combination thereof (Burney & Flannery, 2005).

The change from a glacial to an interglacial climate could certainly have stressed animal populations, not least on account of the accompanying effects on the structure and distribution of plant communities, and could account for the observed end-Pleistocene mass extinction. For example, according to the so-called 'co-evolutionary disequilibrium' model, the herbivores that had evolved to feed on the mosaic of vegetation that characterised the glacial periods would not have been able to survive the change to the climatically zoned vegetation that characterised the interglacial periods, in consequence of which they could have died out, as could the carnivores and scavengers that depended on them. The problem with this model is that it does not adequately account for the apparent absence of mass extinctions associated with changes from glacial to interglacial climates earlier in the Pleistocene. The principal biotic response to glaciation and deglaciation appears to have been a simple shift in geographic distribution.

Human hunting activity could also have stressed animal populations, especially those taking a long time to attain reproductive maturity, or producing only small numbers of offspring, and could also account for the observed end-Pleistocene mass extinction (Klein, in Vrba & Schaller, 2000). Circumstantial evidence in support of the so-called 'overkill hypothesis' comes from the coincidence in time of the observed mass extinction at the end of the Pleistocene with the first appearance of anatomically modern humans, known as the Clovis people, in North America. One problem with this hypothesis is that it does not in itself account for the apparent absence of a mass extinction associated with the first appearance of anatomically modern humans in Eurasia 35 000 years BP. Another is distinguishing any anthropogenic effects from coeval climatic effects.

The Holocene mass extinction

One mass extinction that is often ignored is that apparently ongoing at the present day (Jones, 2006; Turvey, 2009). However, for birds, it is thought that 116 species (1% of the total number) have become extinct since reasonably reliable records began to be kept in the seventeenth century, and that a further 1029 species (11%) are threatened or endangered, and could disappear within the next century. In the case of mammals, 59 species (1%) have become extinct since records began, and a further 505 species (11%) – or, according to more pessimistic estimates, a further 1130 species (24%) – are threatened or endangered. If these rates are scaled appropriately, they equal those found in ancient mass extinctions. The most likely cause of the ongoing mass extinction is natural or artificial – anthropogenic – climatic and associated environmental change, or human hunting activity or habitat destruction, or some combination thereof. Note in this context that anthropogenic global warming – or at least an offsetting of global cooling – may have been ongoing for thousands of years, ever since forests were first cleared for farm-land, and significant quantities of the greenhouse gas carbon dioxide were released into the atmosphere. Note also, though, that the local mid-Holocene extinction of the black-flanked rock-wallaby (*Petrogale lateralis*) at the Tunnel Cave archaeological site in southwestern Australia is now attributed not to human agency, as was once the case, but rather to natural climatic and vegetational change, specifically a post-glacial increase in rainfall and an associated encroachment of closed habitat at the expense of open grassland and grazing areas.

BOX 3.7 NEW EVIDENCE FOR LAND MAMMAL DISPERSAL ACROSS THE NORTHERN NORTH ATLANTIC IN THE EARLY EOCENE (MODIFIED AFTER JONES, 2006)

Introduction

In the northern hemisphere, there is a well known and close generic and specific level commonality between North American and European land animals, including amphibians, reptiles and birds, and mammals, and, arguably, freshwater fish, together with land plants, in the Early Eocene. Land mammal groups appearing in both North America and Europe for the first time at this time include even-toed ungulates or perissodactyls, odd-toed ungulates or artiodactyls, and primates.

The commonality is closest in the earliest Eocene, 'early' Wasatchian North American Land Mammal Age or NALMA (Wa0 to Wa4) and equivalent 'early' Neustrian European Land Mammal Age or ELMA (PEI to 'early' PEIII), which, according to the most recent calibrations of the NALMAs and ELMAs against the geomagnetic time-scale, represents a comparatively short time time interval in 'late' Chron C24r, between 55 and 53.5 Ma (Fig. 3.26).

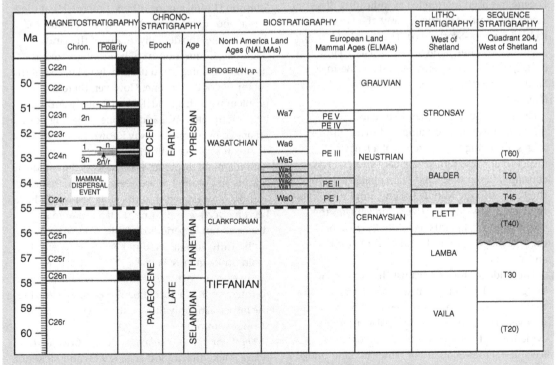

Fig. 3.26 **Eocene stratigraphy of Europe and North America.** From Jones (2006).

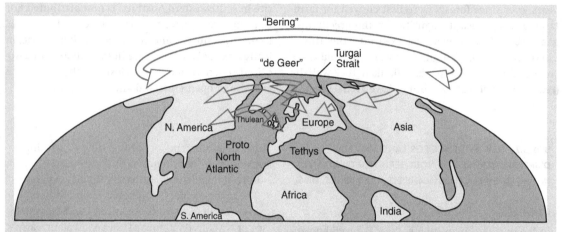

Fig. 3.27 **Eocene palaeogeography of the North Atlantic, showing proposed land bridges and mammal dispersal routes.** From Jones (2006).

The commonality between North American and European land animals and plants in the earliest Eocene has been interpreted as indicating dispersal across the North Atlantic, by way of a Greenland–Iceland–Faeroes–Scotland or 'Thulean' land bridge, a Norwegian–Greenland Sea or 'de Geer' land bridge, or a North America–Eurasia or 'Bering' land bridge at this time, between 55 and 53.5 Ma (Fig. 3.27).

Commonality among land animals is markedly lower in preceding and succeeding stages, implying the existence of a barrier to dispersal at these times; whereas, interestingly, commonality among land plants persists from the Early into the Middle Eocene, implying that the barrier was not effective in preventing the dispersal of land plants, possibly because of the role played by winds and currents (or birds and insects).

The evidence for and against the various land bridges and land animal/mammal dispersal routes that have been proposed is presented below, and new geological and supportive palaeontological evidence for the existence of the Greenland–Iceland–Faeroes–Scotland or 'Thulean' land bridge is discussed.

Land bridges and land mammal dispersal routes

The North America–Eurasia or 'Bering' land bridge. Simpson (1946), among others, proposed that there was a land bridge connecting North America to Asia – where, incidentally, many of the land mammals in question have been interpreted as having originated – and ultimately to Europe, at the beginning of the Eocene. However, there is a problem with this model, in that many authors have interpreted Asia and Europe as having been separated by a broad seaway known as the Obik Sea or Turgai Strait, constituting a significant barrier to land mammal dispersal, until at least the middle if not the end of the Eocene. Note, though, that other authors have argued that the Obik Sea would have been ineffective as a barrier to land mammal dispersal during periods of sea-level low-stand even in the Early Eocene (Godinot & de Lapparent de Broin, in Reumer & Wessels, 2003; Hooker & Dashvezeg, in Wing *et al.*, 2003). Perhaps it is best to interpret it as a partial barrier or filter, allowing the dispersal of only those terrestrial taxa capable of strong swimming.

The Norwegian–Greenland Sea or 'de Geer' land bridge. McKenna, in Bott *et al.* (1983) and McKenna (1983), among others, proposed that there was a land bridge connecting Greenland to the European land-mass by way of Svalbard. However, there is a problem with this model, too, in that the actual land connection would only have extended as far as the Voring Plateau during the dispersal interval. The Hammerfest–Nordkapp basin between the Voring Plateau and Norway would have been a seaway several tens of kilometres wide and several hundreds of metres deep during the dispersal interval, constituting a significant barrier to land mammal dispersal.

The Greenland–Iceland–Faeroes–Scotland or 'Thulean' land bridge. McKenna, in Bott *et al.* (1983) and McKenna (1983), among others, proposed that this was the most likely land bridge across the North Atlantic in the Early Eocene. However, there are problems with this model, too, at least as it stands. Firstly, the brevity of the earliest Eocene land mammal dispersal interval (55–53.5 Ma, so <2 Ma) is inconsistent with the longevity of subaerial exposure indicated by the Early Eocene (Chron C24r, 55 Ma) ages of the formerly contiguous subaerial basalts in Greenland and the Faeroes, and the Late Eocene (33.5 Ma) age of the subaerial laterites on the Iceland–Faeroe Ridge (55–33.5 Ma, so >20 Ma). The most likely explanation for this discrepancy is that there were at least intermittent barriers along this putative dispersal route. Secondly, and more seriously, the actual land connection would only have extended from Greenland to the Faeroes during the dispersal interval. The Faeroe–Shetland basin between the Faeroes and Scotland would have been a seaway several tens of kilometres wide and several hundreds of metres deep during the dispersal interval, constituting a significant barrier to land mammal dispersal.

Discussion – new evidence for the existence of the 'Thulean' land bridge

New 3D seismic data combined with well data from the Faeroe–Shetland basin provide evidence for the closure of the former seaway in the basin, and thus the establishment of a land connection and potential land mammal dispersal route between North America and Europe, at precisely the time of the observed dispersal, 55–53.5 Ma (Naylor

et al., in Fleet & Boldy, 1999; Roberts *et al.*, in Fleet & Boldy, 1999).

A 3D seismic image of a time-slice through Early Eocene Sequence T45/50 in the Faeroe–Shetland basin clearly shows dendritic drainage on an interpreted subaerial delta-top and also channelisation on an interpreted submarine delta-front. The delta progrades to the northeast, onlaps the subaerial Upper Series basalts of the Faeroes, and closes the former seaway of the Faeroe–Shetland basin. The role, if any, of eustasy in this remains poorly understood, although conceivably any one or more of a number of proposed eustatic sea-level falls around the Palaeocene/Eocene boundary could have contributed.

Well log, sedimentological and biostratigraphic data pertaining to 204/24–1A in the Foinaven field in the Faeroe–Shetland basin confirms a general shallowing upward from deep marine environments in the Late Palaeocene to marginal to non-marine environments in the Early Eocene. The Late Palaeocene is characterised by submarine fan sandstones; the Early Eocene by peri-deltaic sandstones interbedded with coals (and tuffs). Biostratigraphic data further indicates that Early Eocene sequence T45/50, comprising the upper part of the Flett formation and the Balder formation, is characterised in 204/24–1A by exclusively non-marine palynomorph assemblages containing abundant *Taxodiaceaepollenites* (cypress and/or swamp-cypress pollen), and is confirmed as having been deposited in a subaerial delta-top environment. (Recently acquired proprietary biostratigraphic data indicates that Late Palaeocene sequence T40, comprising the lower part of the Flett formation and the upper part of the Lamba formation, is absent.)

This is significant because stratigraphic correlation enables the locally demonstrably non-marine sequence T45/50 to be dated to precisely the time of the mammal dispersal interval, earliest Eocene, 55–53.5 Ma. There is a seismostratigraphic correlation between sequence T45/50 of the area West of Shetland and the sequence containing the Sele and Balder formations of the North Sea; and a biostratigraphic correlation between the Sele and Balder formations of the North Sea and the Woolwich & Reading and Harwich formations of the London Basin in southeast England, which have been independently magnetostratigraphically dated to 'late' Chron C24r, and which, moreover, contain 'early' Neustrian (PEI to 'early' PEIII) land mammals (Hooker & Millbank, 2001). The biostratigraphic correlation between the Sele and Balder formations and the Woolwich & Reading and Harwich formations is on the basis of the abundant occurrence of the – paratropical – dinoflagellate *Apectodinium* in the lower part of the Sele and in the Woolwich & Reading, and on the occurrence of the diatom *Fenestrella antiqua* [*Coscinodiscus* sp. 1] in the upper part of the Sele and Balder and in the Oldhaven member of the Harwich.

The evident land connection between the Faeroes and Scotland described above, in other words the Greenland–Iceland–Faeroe–Scotland or Thulean land bridge and potential land mammal dispersal route between North America and Europe, was short-lived, lasting for only the <2 Ma of the dispersal interval between 55 and 53.5 Ma. The connection was severed by marine transgression associated with thermal subsidence at the time of sea-floor spreading in Chron C24n time, 53.5 Ma (Jolley & Bell, in Jolley & Bell, 2002). This transgression is marked in wells in the Faeroe–Shetland basin by the return of marine palynomorph assemblages in the Stronsay formation. (Incidentally, it appears that while the land bridge was in place, it presented at least a partial barrier to the interchange of marine organisms between the Norwegian Sea to the north and the proto-Atlantic to the south.)

A possible mechanism for the facilitation of the land mammal dispersal across the 'Thulean' land bridge

The evident land mammal dispersal, at least arguably associated with the land connection described above, could have been facilitated by an essentially coincident and similarly short-lived (?0.2 Ma) climatic optimum, or Palaeocene–Eocene Thermal Maximum, before and after which land mammals might have found it less easy to access the high-latitude area where the land connection is interpreted to have been (Slujis *et al.*, in Williams *et al.*, 2007a). This climatic optimum has been interpreted as having been generated by a runaway 'greenhouse effect' caused by dissociation of methane hydrates in the deep sea, possibly in turn brought about by regional uplift (MacLennan & Jones, 2006; Lunt *et al.*, 2010). It has also been interpreted as having been generated by a comet impact. Direct supportive palaeontological evidence of a climatic optimum derives from the observations that the structure, composition and diversity of land mammal faunas from North America and Europe from the time of dispersal closely resemble those from the present-day tropical rainforest, with coeval land plants also of essentially tropical aspect (Utescher & Mosbrugger, 2007).

Indirect evidence derives from observations of strong negative carbon isotope or positive oxygen isotope excursions in tooth enamel from mammals, shell material from freshwater bivalve molluscs and carbonate from soils from the 'early' Wasatchian (Wa0) of the Bighorn and adjacent basins in Wyoming in North America; from age-equivalent sections in the San Juan basin in New Mexico to the south and from Ellesmere Island in Arctic Canada to the north; and from the 'early' Neustrian (PEI) of the Paris basin in Europe. Such excursions confirm a significant warming event at this time.

(Incidentally, the carbon isotope excursion associated with this warming event has recently been recommended as the global standard for the Palaeocene–Eocene boundary, essentially because it is globally recognisable, that is, not only within the terrestrial but also within the marine record.)

One might speculate that on account of the essentially tropical climate obtaining at the time, there may well have been abundant vegetation available in the mires of the delta-top of the Balder formation of Scotland to nourish the land mammals after their journey across the recently volcanised landscape of the Greenland–Faeroes Ridge!

BOX 3.8   ASPECTS OF THE PALAEOGEOGRAPHY AND PALAEOCLIMATE OF THE OLIGOCENE–HOLOCENE OF THE OLD WORLD, AND CONSEQUENCES FOR LAND MAMMAL EVOLUTION AND DISPERSAL (MODIFIED AFTER JONES, 2006)

Introduction

Old World palaeogeography and palaeoclimate over the Oligocene–Holocene interval were affected by plate movements and mountain-building (tectonism) and associated effects on atmospheric and oceanic circulation (for example the convergence of India and Eurasia and the formation of the Himalayas; and the divergence of South America and Australasia from Antarctica and the formation of a circum-Antarctic seaway, and hence a source of cold, dense water with which to power the psychrosphere), and by cyclic fluctuations in sea-level (glacio-eustasy). Palaeoclimate fluctuated cyclically, generally, though never irreversibly, deteriorating until the Pleistocene, which was characterised by widespread glaciation.

Before the closure of the Tethyan Seaway in the Middle Miocene, marine faunas, the best documented being invertebrates such as larger benthic and planktonic foraminifera, corals and molluscs, were able to disperse between the Indo-Pacific and Mediterranean; and afterwards, terrestrial faunas, the best documented being vertebrates such as mammals, were able to disperse between Afro-Arabia and Eurasia.

All available evidence pertinent to the palaeogeographic and palaeoclimatic evolution of the Old World over the Oligocene–Holocene is collated below, and the consequences for land mammal evolution and dispersal are described, in the form of a series of time-slice maps (Jones, in Agusti *et al.*, 1999; see also Calvin, 2002 and Lehman & Fleagle, 2006).

The time-slices selected for palaeogeographical mapping are: Early Oligocene (Rupelian), Late Oligocene to earliest Miocene (Chattian–Aquitanian), late Early to early Middle Miocene (Burdigalian–Langhian), Middle Miocene (early to middle Serravallian), late Middle to early Late Miocene (late Serravallian–Tortonian), latest Miocene (Messinian), Early Pliocene (Zanclian), Late Pliocene (Piacenzian) and Pleistocene–Holocene (an Eocene time-slice is also included for the purposes of comparison: Fig. 3.28).

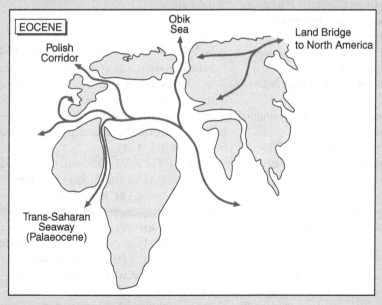

Fig. 3.28 **Eocene palaeogeography of the Old World**. Grey arrows indicate dispersal routes. From Jones (2006).

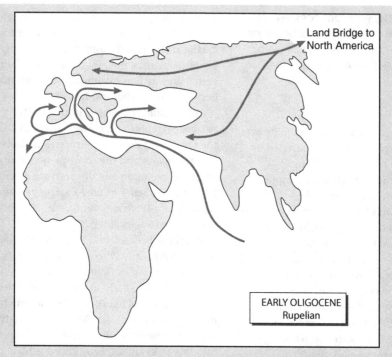

Fig. 3.29 **Early Oligocene, Rupelian palaeogeography of the Old World**. From Jones (2006).

Importantly, the mean duration of these time-slices, dictated by biostratigraphic resolution, is much lower than the measured frequency of climate change over the same time interval, such that both climate and climatically induced sea-level could have changed within each time-slice, and each time-slice is best regarded as a composite.

The time-slice palaeogeographies are constrained in part by larger benthic foraminiferal palaeobiogeographic distribution data. Palaeoclimate is constrained by LBF diversity data, and by coral diversity data (Chaix & Cahuzac, 2005; Karabiyikoglu *et al.*, 2005; Bosellini & Perrin, 2008); and, on land, by land plant – palaeobotanical and palynological – data (Collinson & Hooker, in Reumer & Wessels, 2003; Kovar-Eder, in Reumer & Wessels, 2003; Kovar-Eder *et al.*, 2006), and by land animal data (Fortelius *et al.*, in Reumer & Wessels, 2003; Fortelius *et al.*, 2006; Blain *et al.*, 2008; Costeur & Legendre, 2008). A 'Neogene Old World' or 'NOW' land animal distribution database is available on Mikael

Fortelius's website (www.helsinki.fi/science/now/database.htm).

Discussion

Early Oligocene (Rupelian), approximately 34–28 Ma

Palaeogeography. LBF commonality data indicates that at this time the Tethyan Seaway extended from the Indo-Pacific in the east, through the area of the modern Mid-East Gulf and the Mediterranean basin, and into the Aquitaine Basin and Atlantic in the west (Fig. 3.29).

The Gulf of Aden was open, the Red Sea unopened, Africa and Arabia contiguous.

Palaeoclimate. LBF diversity data indicates a range of sea-surface temperatures (SSTs) from a low of 15 °C in Australasia, thought to have been located almost entirely in extra-tropical palaeolatitudes, to a high of 25 °C in the northern Mediterranean. Note, though, that LBF diversity is sensitive not only to temperature but also other factors such as facies, evolution, dispersal/migration, duration of studied interval and/or size of studied area, sampling artefact and taxonomic artefact.

Palynological and palaeobotanical evidence points toward a warm temperate to tropical humid climate. There was evidently extensive development of boreo-tropical elements in Europe, including mixed mesophytic forest, marsh, and aquatic vegetation, in turn including mangroves, in Germany; although there were also separate northern, central and southern European provinces. There was also evidently extensive development of tropical rainforest in equatorial Africa, the Indian subcontinent, parts of southeast Asia and northermost Australasia.

Land mammal evolution and dispersal. Camels evolved in the Oligocene, in North America, and dispersed to their present-day heartlands in the deserts of north Africa, the Middle East and central Asia, and became adapted to the extreme aridity there, in the Miocene (van der Made *et al.*, 2002). Indeed, the dispersals of various mammals from the Americas into Eurasia, by way of land bridges across the Bering Sea and Obik Sea/Turgai Strait, appear to have taken place at this time. This event is sometimes referred to as the 'grande coupure'.

Late Oligocene to earliest Miocene

(Chattian–Aquitanian/Agenian European Land Mammal Age, Mein Zones MN1–MN2, approximately 28–20 Ma.)

Palaeogeography. LBF commonality data indicates that at this time the Tethyan Seaway from the Indo-Pacific to the Mediterranean was probably still more often open, at times of relative high-stand of sea level, than closed, at times of relative low-stand (Fig. 3.30).

However, there may also have been an at least intermittent land bridge between Italy, Sicily and Africa. By the end of this time interval, the Red Sea had rifted open.

Planktonic foraminifera confirm a connection between Tethys and Paratethys. Foraminiferal and ostracod data indicate that there are three palaeobiogeographic units in the Late Oligocene to Early Miocene of central Europe. These are: (1) a northern unit, (2) the upper Rhine sub-province or URSP, comprising the Mainz basin, northern upper Rhine graben and Hanau basin/Wetterau, and (3) the western Paratethys. Progressive isolation

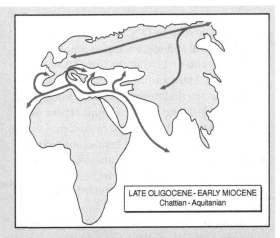

Fig. 3.30 **Late Oligocene to Early Miocene, Chattian–Aquitanian palaeogeography of the Old World.** From Jones (2006).

of the URSP through time is indicated by progressive reduction in similarity between it and the other sub-provinces from the basal Miocene onwards. (Similar observations can be made on the basis of molluscs and fish.)

Palaeoclimate. LBF diversity data indicate a range of SSTs from a low of 15 °C in Australasia and east Africa to a high of 25 °C in southeast Asia in the Chattian, and a low of 15 °C in central Europe, thought to have been located at approximately 40° N, that is, close to the limit of tolerance of modern LBFs, to a high of 25 °C in southeast Asia in the Aquitanian. Note, though, that diversity is sensitive not only to temperature but also to other factors.

Palaeontological and palaeopedological (palaeosol) evidence points toward a warm temperate to tropical, (sub-)humid climate, although with significant provincial differences. Palaeobotanical and palynological evidence indicates extensive development of boreo-tropical or Palaeotropic elements in Eurasia (Utescher *et al.*, 2002; Bruch & Zhilin, 2007; Jimenez-Moreno *et al.*, 2009; Jin-Feng Li *et al.*, 2009). Mean coldest month temperatures were 6 °C in the Chattian and 13 °C in the Aquitanian in the Lower Rhine Basin, and up to 6 °C in the undifferentiated Early Miocene of the Weichang district of north China; mean annual temperatures 15 °C in the Aquitanian of Kazakhstan, and up to 15 °C in the Early Miocene of north China; and mean warmest month temperatures up to 25 °C in the Early Miocene of north China. Mean annual precipitation was up to 1400 mm in the undifferentiated Early Miocene of the Weichang district of north China. There was multi-tiered forest characterised by canopy, sub-canopy, liana screen and herbaceous ground cover, similar to that of the southeastern United States wetlands, in the Weisselster basin in Germany in the Chattian; and acid-tolerant upland riparian forest, swamp-forest and aquatic vegetation in northern Bohemia in the Aquitanian. There was subtropical to tropical so-called mosaic vegetation, with dry, open landscapes adjacent to humid forests, in the circum-Mediterranean. There was tropical rainforest in equatorial Africa, the Indian subcontinent, parts of southeast Asia and northernmost Australasia. There was ever-wet and cool, but frost-free, southern beech or *Nothofagus* forest in New Zealand. Palaeobotanical and palaeopedological evidence points toward a mosaic vegetation of rainforest, dry peripheral forest (possibly precursor Afromontane forest), riparian woodland and 'miambo', and dry deciduous woodland and seasonally waterlogged 'dambo' or 'kiewo' in Africa. Gastropod evidence also points toward a mosaic vegetation in east Africa (Pickford, 1995).

Oxygen, strontium and neodymium isotopic evidence from shark teeth indicates a cooling of as much as 4 °C between 22 and 17 Ma in parts of Paratethys, a local effect of the Alpine orogeny.

Land mammal evolution and dispersal. Apes or hominoids evolved from Old World monkeys, probably in Africa, sometime around the end of the Oligocene, and diversified through the Miocene. Primitive forms, characterised by comparatively small brain-cases, include the African genus *Proconsul*, found in association with the 'dambo' or 'kiewo'. *Proconsul* is a quadruped, interpreted as having moved about on all fours, either in trees or on the ground (seeking sanctuary in trees when threatened). It is further interpreted as having been a frugivore, that is, as having subsisted largely on fruit.

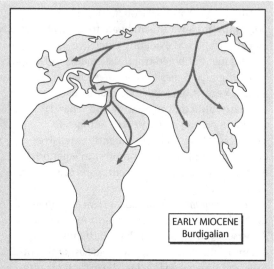

EARLY MIOCENE
Burdigalian

Fig. 3.31 **Early Miocene, Burdigalian palaeogeography of the Old World**. From Jones (2006).

The dispersals of various mammals from the Americas into Eurasia and Africa, again by way of the Bering land bridge, appear to have taken place during the Agenian European Land Mammal Age (Mein Zones MN1–MN2).

Late Early to early Middle Miocene

(Burdigalian–Langhian/Orleanian European Land Mammal Age, Mein Zones MN3–MN5, approximately 20–14 Ma.)

Palaeogeography. LBF commonality data indicates that the Tethyan Seaway was closed in the Burdigalian, and open in the Langhian (Figs. 3.31–3.32).

Planktonic foraminifera confirm a connection between Tethys and Paratethys, as do bivalve and gastropod molluscs. The distributions of cool and warm water species indicate a warming with respect to the Aquitanian.

Palaeoclimate. LBF diversity data indicate a range of SSTs from a low of 15 °C in the southern Mediterranean to a high of 25 °C in southeast Asia. Note again, though, that diversity is sensitive not only to temperature but also to other factors.

Palaeobotanical, palynological and palaeopedological evidence points toward a warm temperate to tropical climate. There was extensive

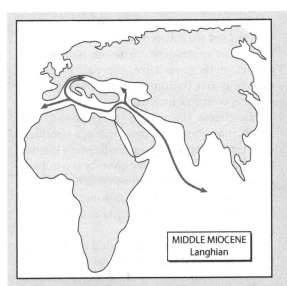

MIDDLE MIOCENE
Langhian

Fig. 3.32 **Middle Miocene, Langhian palaeogeography of the Old World**. From Jones (2006).

development of warm temperate, sub-humid vegetation, locally including mangroves, indicating cold month mean temperatures of 7–13 °C and mean annual temperatures of >20 °C, in central Europe; of subtropical to tropical, sub-arid and seasonally dry vegetation in the circum-Mediterranean and parts of Africa; of tropical rain-forest in equatorial Africa, the Indian subcontinent, southeast Asia, and northern Australasia; and of broad-leaved forest in the Tian Shan mountains of northwest China (Utescher *et al.*, 2002; Jimenez-Moreno *et al.*, 2005; Jimenez-Moreno, 2006; Fauquette *et al.*, in Williams *et al.*, 2007a; Jimenez-Moreno & Suc, 2007; Jimenez-Moreno *et al.*, in Williams *et al.*, 2007a; Jimin Sun & Zhenqing Zhang, 2008). There was also seasonally dry vegetation, characterised by eucalypts and palms, subject to bush fires, in New Zealand.

Independent isotopic evidence indicates that there was a climatic optimum at this time, associated with increased atmospheric carbon dioxide concentration. The evolution of various LBFs at this time is interpreted as related to niche diversification during the sea-level high-stand

associated with this climatic optimum. LBFs were present as far south as the Port Phillip basin in southeastern Australia, which was situated in a palaeolatitude of 55° S, at the time of the climatic optimum (Gourley & Gallagher, 2005). As well as diverse LBFs, other subtropical to tropical climatic indicators such as coral reefs, indicating averaging temperatures of >16 °C, were present as far north as central Europe, which was situated in a palaeolatitude of 45° N. Palaeotemperatures in the Styrian basin in Austria have been revealed by analysis of the isotopic composition of brachiopod and bivalve shells.

Land mammal evolution and dispersal. The primitive hominoid *Griphopithecus* made its first appearance, in Slovakia and Turkey, either in the 'late' Orleanian European Land Mammal Age (Mein Zone MN5), approximately 17.3 Ma, or in the 'early' Astaracian ELMA (MN6) (Begun *et al.*, in Reumer & Wessels, 2003).

Further dispersals of mammals from the Americas into Eurasia and Africa appear to have taken place during the Orleanian (MN3–MN5), again by way of the Bering land bridge.

Dispersals of various mammals, including primitive hominoids, between Africa and Eurasia appear to have taken place at this time, by way of the so-called *Gomphotherium* land bridge, as did dispersals of mammals within Eurasia (Begun *et al.*, in Reumer & Wessels, 2003). In Eurasia, pliopithecid hominoids of this age are recorded in China to the east and throughout central and western Europe to the west.

Incidentally, dispersals of freshwater fish, and of the crocodilian *Tomistoma*, between Africa and Eurasia, also appear to have taken place at this time.

Middle Miocene

(Early–middle Serravallian/Astaracian European Land Mammal Age, Mein Zones MN6–MN8, approximately 14–11 Ma.)

Palaeogeography. LBF commonality data indicates that at this time the Tethyan Seaway was probably more often closed than open. Planktonic foraminiferal and other data indicates that it was closed in the early Serravallian, and open in the middle Serravallian (Figs. 3.33–3.34).

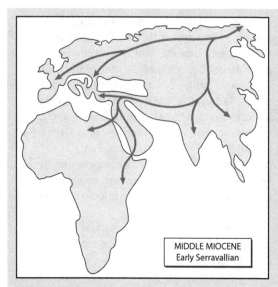

Fig. 3.33 **Middle Miocene, early Serravallian palaeogeography of the Old World**. From Jones (2006).

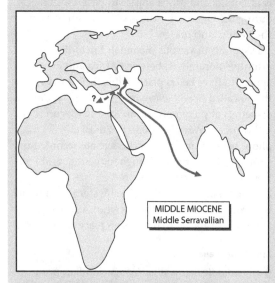

Fig. 3.34 **Middle Miocene, middle Serravallian palaeogeography of the Old World**. From Jones (2006).

Planktonic foraminifera also indicate an at least intermittent connection between Tethys and Paratethys.

Note, though, that circulation in the Mid-East Gulf and Red Sea and in parts of Central Paratethys was at least intermittently extremely restricted, resulting in the formation of hypersaline waters that could have constituted a barrier to the dispersal of marine invertebrates (Peryt, 2006).

Palaeoclimate. LBF diversity data indicate SSTs of <20 °C throughout the region, that is, a significant cooling with respect to the preceding time interval. Note yet again, though, that diversity is sensitive not only to temperature but also to other factors.

The relative abundances of cool- and warm-water species of planktonic foraminifera also indicate a cooling.

Palaeobotanical and palynological evidence points also toward a cooling and/or drying. There was development not only of Palaeotropic but also of Arctotertiary vegetation in central Europe and of 'altitudinal' and non-arboreal elements in the circum-Mediterranean (Utescher *et al.*, 2002; Jimenez-Moreno *et al.*, 2005; Jimenez-Moreno, 2006; Fauquette *et al.*, in Williams *et al.*, 2007a; Jimenez-Moreno & Suc, 2007; Jimenez-Moreno *et al.*, in Williams *et al.*, 2007a). There was also comparatively dry vegetation, characterised by casuarinaceans, in New Zealand, from 14–15 Ma to possibly as late as 10 Ma.

Palaeobotanical and palaeopedological evidence points toward a mosaic vegetation of Afromontane forest, lowland riparian woodland, grassy woodland, wooded grassland and seasonally waterlogged grassland or 'dambo' in Kenya in east Africa, and indeed mixed forest and C3 grassland elsewhere in Africa (Feakins & deMenocal, in Werdelin & Sanders, 2010; Jacobs *et al.*, in Werdelin & Sanders, 2010). The East African Plateau was clearly in evidence at this time (Wichura *et al.*, 2010).

Land mammal evolution and dispersal. The evolution of numerous mammals, including hominoids, in the Astaracian European Land Mammal Age (Mein Zones MN6–MN8) is regarded by some authors as related to selection pressure associated with the change alluded to above to a cooler and/or more arid climate, which caused its arboreal ancestor to move out of its contracting forest habitat and onto an expanding savannah.

Pierolapithecus catalanicus, which exhibits some advanced ape-like morphological traits, and has been interpreted as close to the last common ancestor of great apes and humans, made its first appearance in the 'middle' to 'late' Astaracian ELMA (MN7–MN8) of Pierola in Catalunya (13–12.5 Ma).

Dispersals of various mammals, including hominoids, within Eurasia appear to have taken place at this time. Here, *Sivapithecus* is recorded from the Chinji, Nagri and Dhok Pathan stages of the Siwalik series in India and Pakistan; *Ramapithecus* also from India and Pakistan, from China to the east, and from Turkey and Greece and along the banks of the Danube to the west; and *Dryopithecus* also from India and Pakistan, from China to the east, and from Turkey, the Caucasus, Slovakia and Poland, along the banks of the Danube, Eber and Rhone, and France and Spain to the west. Isotopic variability within the teeth of equids associated with *Sivapithecus* in Pakistan indicates a seasonal climate, with the dry season, during which fruit would have been a scarce food resource, lasting for five to six months, longer than the four months that can be tolerated by modern apes (Nelson, 2005). The implication is that *Sivapithecus* was less reliant on fruit as a resource than are modern apes. Alternatively, the ancient forest may have differed in terms of the spatio-temporal availability of fruit compared with modern analogues.

Late Middle to early Late Miocene

(Late Serravallian–Tortonian, Vallesian to 'middle' Turolian European Land Mammal Ages, Mein Zones MN9–MN12, approximately 11–7 Ma.)

Palaeogeography. By now, the Tethyan Seaway was probably completely and permanently closed (Fig. 3.35).

Palaeoclimate. The permanent closure of Tethys, together with at least approximately age-equivalent uplift in the Himalayas, apparently led to changes in oceanic and atmospheric circulation resulting in the development of a distinctly cooler, drier, more seasonal – monsoonal – regional climate at or around this time (Hoorn *et al.*, 2000; Molnar, 2005; Zhang Zhongshi *et al.*, 2007; Pound *et al.*, 2009;

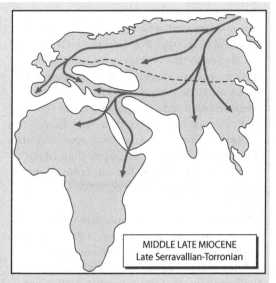

MIDDLE LATE MIOCENE
Late Serravallian-Torronian

Fig. 3.35 **Middle–Late Miocene, late Serravallian–Tortonian palaeogeography of the Old World.** Dashed line indicates approximate northern limit of Pikerman province. From Jones (2006).

Sanyal *et al.*, 2010). Note, though, that the elevation history of the Tibetan Plateau indicates that the monsoon may have originated as long ago as the Early Miocene (Harris, 2006). Palaeontological and palaeopedological evidence from China indicates that the East Asian monsoon may have originated either as long ago as the Early Miocene, or as recently as the Pleistocene, depending on local conditions (Xiangjun Sun & Pinxian Wang, 2005; Yang Wang & Tao Deng, 2005; Hanchao Jiang & Zhongli Ding, 2008; see also Lunt *et al.*, in Clift *et al.*, 2010).

Palaeontological and palaeopedological evidence points toward not only a cool, dry climate but also a change to a cooler, drier, climate, and an accompanying change from a dominantly C3 woodland and grassland to a dominantly C4 grassland type of vegetation, over this time interval (Jacobs *et al.*, 1999; Utescher *et al.*, 2002; Beerling, 2007; Feakins & deMenocal, in Werdelin & Sanders, 2010; Jacobs *et al.*, in Werdelin & Sanders, 2010; Sanyal *et al.*, 2010). Note, though, that the floras of the so-called Pikermian province (Greece, Turkey, Bulgaria, Romania, Moldavia,

Ukraine, Iran and parts of China) have recently been re-interpreted as having been of mixed woodland (C3) rather than savannah (C4) aspect – and the faunas that eventually migrated from it into Africa as pre-adapted rather than adapted to the savannah (Stromberg *et al.*, 2007). The re-interpretation is further supported by the masticatory morphology of the mammals, patterns of tooth micro-wear, and tooth carbonate $\delta^{13}C$ data (a function of dietary $\delta^{13}C$ intake), all of which indicate not only grazing but also browsing and mixed feeding (Merceron *et al.*, 2006; Solounias *et al.*, 2010).

Palaeobotanical and palynological evidence indicates extensive development of Arctotertiary vegetation in Central Europe, and of 'altitudinal' and non-arboreal elements in the circum-Mediterranean, over this time interval (Jimenez-Moreno *et al.*, 2005; Jimenez-Moreno, 2006; Fauquette *et al.*, in Williams *et al.*, 2007a; Jimenez-Moreno & Suc, 2007; Jimenez-Moreno *et al.*, in Williams *et al.*, 2007a). The coexistence approach (CoA) has been used to infer a significantly lower mean annual temperature (MAT) of 13.3–17 °C and a mean annual precipitation (MAP) of 652–759 mm in the Late Miocene, upper Sarmatian (Chersonian) than in the Middle Miocene, Badenian to middle Sarmatian (Bessarabian) of Bulgaria (Ivanov *et al.*, 2002, 2007). There was tropical forest, monsoon forest, riparian woodland, stream margin scrub, seasonally dry swamp, seasonally waterlogged grassland, and grassland, in Pakistan, and a change from dominantly moist monsoon forest to dry monsoon forest and ultimately to grassland vegetation on the Potwar plateau. There was a change from subtropical to temperate broad-leaved forest to woodland and ultimately grassland vegetation in Nepal. Increases in discharge related to intensification of the monsoonal system resulted in increased incidences of the aquatic taxon *Potamogeton*. Localised development of lakes on the overbanks resulted in increased incidences of the alga *Spirogyra*, and of pteridophyte or fern spores. There was dry vegetation, characterised by chenopods and asteraceans, in New Zealand.

Micro-mammal evidence from various European sites indicates a temperature drop of about 7–9 °C at around 13.5–12 Ma, and a significant aridification at around 9–8 Ma (Legendre *et al.*, 2005; van Dam, 2006). That from the Western Desert of Egypt indicates a change from a humid to an arid climate between 11 and 7 Ma (Pickford *et al.*, 2006). That from the Potwar plateau in Pakistan supports the interpretation alluded to above of a change from forest to grassland vegetation by 8 Ma (Flynn, in Reumer & Wessels, 2003). That from the Chindigarh area of India indicates an increase in seasonality, aridity and unpredictability associated with an intensification of the monsoon, with survival-oriented murid rodents becoming replaced by reproduction-oriented cricetids, and non-grazers by grazers.

Interestingly, climate change, or more accurately the effect of climate change on vegetation, appears to have been the driving force behind the evolution of specialist feeding rodents such as *Progonomys–Stephanomys*, but not generalists such as *Apodemus*, over the last 12 Ma (Renaud *et al.*, 2005).

Land mammal evolution and dispersal. A variety of ruminants, adapted to extract the maximum amount of nourishment from poor sources such as grasses, evolved during the Vallesian to 'early' to 'middle' Turolian European Land Mammal Ages (Mein Zones MN9–MN12), which, as noted above, were characterised by an increasingly widespread development of grasslands. Horse teeth evolved from short-crowned, and adapted to browsing, to long-crowned, and adapted to grazing. Terrestrial mammal tooth morphology, specifically mean hypsodonty, points toward the development of an east–west palaeo-precipitation gradient in western Eurasia and Africa, with more arid conditions to the continental interior to the east and more humid ones to the maritime border to the west (Fortelius *et al.*, 2006).

The dispersals of numerous terrestrial mammals and other vertebrates between Africa and Eurasia appear to have taken place at this time. Six terrestrial vertebrate palaeobiogeographic (sub-) provinces existed in southern Europe at this time, namely: the northwestern; the north Dacian; the Balkano-Iranian; the Greek Macedonian; the Eastern Aegean; and the Anatolian (Geraads *et al.*, 2005). The forested Eastern Aegean may have acted as a barrier to the dispersal of terrestrial mammals of open-country aspect. A further three (sub-) provinces existed in Italy, namely: the Abruzzi-Apulian, on the Adriatic side; the Calabria-Sicilian, on the Mediterranean side; and the Tusco-Sardinian, on the Mediterranean side (Rook *et al.*, 2006). Terrestrial mammals were evidently able to disperse between Africa and the Calabria-Sicilian province, which is characterised by 'cosmopolitan' species, but not between Africa and the Abruzzi-Apulian and Calabria-Sicilian provinces, which are characterised by endemic species. Interestingly, semi-aquatic crocodilian reptiles were able to disperse between Africa and the Tusco-Sardinian province.

Significantly, if sampling and preservational effects can be excluded, most hominoids disappeared from Eurasia toward the end of the Vallesian ELMA, although some persisted into the Turolian in southerly refuges such as the Tusco-Sardinian and Pikermian provinces. This has been interpreted as a direct or indirect response to significant climatic and vegetational change, perhaps variously delayed and moderated by geographic and biotic filters.

Vallesian European Land Mammal Age

(Mein Zones MN9–MN10, approximately 11–9 Ma.)

Anapithecus made its first appearance, in Austria and Hungary, in the 'early' Vallesian European Land Mammal Age (Mein Zone MN9). *Ankarapithecus* made its first appearance, in Turkey, also in the 'early' Vallesian ELMA (9.9–9.6 Ma) (Kappelman, Duncan *et al.*, in Fortelius *et al.*, 2003; Kappelman, Richmond *et al.*, in Fortelius *et al.*, 2003). *Graecopithecus* or *Ouranopithecus* made its first appearance, in Greece, in the 'late' Vallesian (9.6–8.7 Ma) (Bernor, 2007). The recently discovered species *Chororapithecus abyssinicus*, *Nakalipithecus nakayamai* and *Samburupithecus kiptalami* all made their first appearances, in Ethiopia or in Kenya in east Africa, in the Vallesian equivalent (10.5–10 Ma, 9.9–9.8 Ma and 9.6 Ma respectively) (Bernor, 2007).

The effectively instantaneous dispersal of the three-toed horse '*Hipparion*' from the Americas into Eurasia and Africa, by way of the Bering land bridge, appears to have taken place at this time

(Vislobokova *et al.*, in Reumer & Wessels, 2003). In Eurasia, '*Hipparion*' is recorded in China, Mongolia and Kazakhstan, in the Nagri and Dhok Pathan stages of the Siwalik series in India and Pakistan, in the United Arab Emirates, Iran, Crete, Greece, Turkey and the Caucasus, in the Ukraine, and in Bulgaria, Hungary, Austria, Germany, France and Spain; in Africa, it is recorded in Tunisia, Algeria, Morocco and Kenya. The '*Hipparion*' event has been independently linked to the coincident glacio-eustatic sea-level low-stand at approximately 11.1 Ma (10.7 Ma according to the recent revision of Kappelman, Duncan *et al.*, in Fortelius *et al.*, 2003). The feeding strategies and dietary regimes of two populations of the extinct hipparionine horse *Hippotherium primigenium* from the 'early' Vallesian (MN9) of southern Germany, one from palaeo-Rhine deposits of Eppelsheim, and the other from lacustrine deposits of Howenegg, have been established through analysis of tooth 'mesowear' and 'extended mesowear'. The feeding strategies of the Eppelsheim population have also been established through analysis of tooth wear using an innovative technique, involving a diamond stylus profiling instrument (Kaiser & Brinkmann, 2006). The Eppelsheim population has been interpreted on the basis of analogy with the extant bushbuck *Tragelaphus strepsiceros*, or the common waterbuck, which exhibit similar patterns of wear, as having fed essentially by grazing and foraging in reed beds and fringing woodlands. The Howenegg population has been interpreted on the basis of analogy with the extant Sumatran rhinoceros as having fed by browsing in subtropical mesophytic forests.

'Early' to 'Middle' Turolian European Land Mammal Age,

(Mein Zones MN11–MN12, approximately 9–7 Ma.)

The cercopithecid (Old World monkey) *Mesopithecus* made its first appearance in Turkey, Greece and Hungary (Pikermian province) in the 'early' to 'middle' Turolian European Land Mammal Age (Mein Zones MN11–MN12). Intriguingly, a recently discovered but as yet unnamed hominoid species from the Turolian ELMA (MN11–MN12) of Turkey has been described as exhibiting affinities not only with the Eurasian *Ouranopithecus* but also with the African *Ardipithecus* and *Australopithecus* (Sevim *et al.*, 2001; Begun *et al.*, in Reumer & Wessels, 2003). It is currently unclear whether the clade represented by this species – potentially the ape–human clade – actually evolved in Eurasia or in Africa (Bernor, 2007).

The dispersals of numerous mammals between the Americas and Eurasia and within Eurasia appear to have taken place at this time (Vislobokova *et al.*, in Reumer & Wessels, 2003). The preponderance of grassland over woodland elements in Eurasia provides evidence of continuing cooling and/or drying. Development of grassland in Afghanistan is indicated by micro-wear patterns on the teeth of fossil artiodactyls, which are similar to those exhibited by the living Przewalski's horse, *Equus przewalskii*. Local development of woodland is indicated by the carbon isotopic signal in the tooth enamel of the bovid *Tragoportax*.

Latest Miocene

(Messinian/'late' Turolian European Land Mammal Age, Mein Zone MN13, approximately 7–5 Ma.)

Palaeogeography. At this time, due to climatic change (cooling/drying), base levels in the Mediterranean and Paratethyan basins fell to such an extent that they became isolated not only from one another but also from the rest of the world's oceans, and, ultimately, during the so-called 'Messinian Salinity Crisis', desiccated (Cagatay *et al.*, 2006; Flecker & Ellam, 2006; Popov *et al.*, 2006; Fig. 3.36).

A multidisciplinary study of the Messinian of the Velona basin in central Italy has recently been undertaken, with a view to establishing its palaeoenvironmental and palaeogeographic context and evolution. This involved detailed geochemical analyses on biogenic carbonates, sedimentological analyses, and palaeontological analyses (charophytes, plant pollen, molluscs, ostracods and mammals). The succession is interpreted as representing a brackish lacustrine or paludal environment rich in aquatic vegetation, surrounded by lowland swamp-forests dominated by swamp-cypresses (Taxodiaceae), and uplands dominated by warm-temperate deciduous forests. The lake or swamp was not connected to the

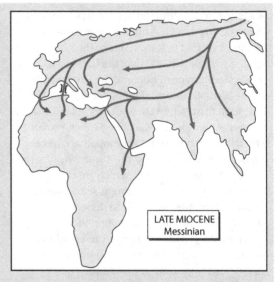

Fig. 3.36 Late Miocene, Messinian palaeogeography of the Old World. From Jones (2006).

Mediterranean Sea, the previous Tusco-Sardinian palaeo-bioprovince already disrupted. The molluscs, ostracods and mammals are of central European palaeobiogeographic affinity. The ostracods show no affinity with those of Paratethys.

Palaeoclimate. Palaeobotanical and palynological evidence again points toward a generally cool?, dry climate. There was continuing extensive development of Arctotertiary vegetation in central Europe, and of non-arboreal elements, lacking mangroves, in the circum-Mediterranean (Utescher *et al.*, 2002; Bertini, 2006; Kovar-Eder *et al.*, 2006; Favre *et al.*, 2007). There was also development, for the first time, of semi-desert elements in the circum-Mediterranean, and of an arid-adapted – Saharan – flora in Africa (Fauquette *et al.*, in Williams *et al.*, 2007a; Jimenez-Moreno *et al.*, in Williams *et al.*, 2007a). Steppe palynofloras developed on the Loess Plateau in China between 6.2–5.8 Ma (Wang *et al.*, 2006). Palaeopedological evidence indicates increasing aridification on the loess plateau from 6.8 Ma (Kaakinen *et al.*, 2006).

Dental micro-wear patterns in bovids from Greece indicate feeding on scrubland and grassland vegetation (Merceron *et al.*, 2005). Mean hypsodonty points toward a change from an east–west to a north–south palaeo-precipitation gradient in western Eurasia and north Africa from approximately 6.8 Ma, with more humid conditions to the north and more arid ones to the south (Fortelius *et al.*, 2006). Micro-mammal evidence indicates a predominance of open herbaceous habitats in the Grenada basin in southern Spain (Garcia-Alix *et al.*, 2008).

Land mammal evolution and dispersal. The oldest candidate representatives of the human lineage, *Sahelanthropus tchadensis, Orrorin tugenensis* and *Ardipithecus kadabba*, appear to have evolved at this time (Sarmiento *et al.*, 2007; Brunet, 2009; Haile-Selassie & WoldeGabriel, 2009; Brunet, 2010; MacLatchy *et al.*, in Werdelin & Sanders, 2010). *Sahelanthropus tchadensis*, nicknamed Toumai, is known from the latest Miocene, approximately 7–6 Ma, of Toros-Menalla in the Djurab Desert of northern Chad in central Africa; *Orrorin tugenensis*, nicknamed 'Millennium Man', from the latest Miocene, 6.2–5.7 Ma, of Kenya in east Africa; and *Ardipithecus kadabba*, from the latest Miocene, 5.8–5.5 Ma, of Ethiopia also in east Africa. *Sahelanthropus tchadensis* appears to exhibit a combination of primitive and advanced characters, and thus to indicate that the divergence between the hominids and their ape ancestors took place earlier than indicated by molecular data. The fauna and flora found in association with it indicate desert, savannah, and gallery forest environments in the vicinity, together with lakes.

The dispersals of numerous arid grassland mammals, including camels, between the Americas and Eurasia appear to have taken place during the 'late' Turolian European Land Mammal Age (Mein Zone MN13), which was evidently a time of continuing climatic deterioration (Vislobokova *et al.*, in Reumer & Wessels, 2003). The dispersals of mammals between Africa and Eurasia, by way of a land bridge at the foot of the Iberian Peninsula, an island chain across the Mediterranean, or, ultimately, the dried-up sea-bed of the Mediterranean, also appear to have taken place at this time (Favre *et al.*, 2007; Minwer-Barakat *et al.*, 2009). Successive muroid, *Paraethomys* and gerbil events are locally distinguishable (Agusti *et al.*, 2006). The cercopithecids *Dolichopithecus* and

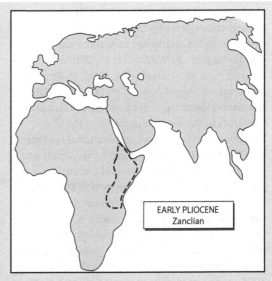

Fig. 3.37 **Early Pliocene, Zanclian palaeogeography of the Old World**. From Jones (2006). Dashed line indicates known range of *Ardipithecus* and early *Australopithecus* species.

Macaca also made their first appearances in Europe at this time.

Early Pliocene

(Zanclian/Ruscinian European Land Mammal Age, Mein Zones MN14–MN15, approximately 5–4 Ma.)

Palaeogeography. At this time, the Mediterranean and Paratethyan basins re-opened and re-filled from the Atlantic (Popov *et al.*, 2006). Non-marine, 'molasse-type' sedimentation then came to characterise the Caspian borderlands and much of the Middle East. Eurasia thus assumed something approaching its present day geography (Fig. 3.37).

Palaeoclimate. Palaeobotanical, palynological, and invertebrate and vertebrate palaeontological evidence points toward a temporary amelioration of climate in the earliest Pliocene, from approximately 5.5–4.5 Ma (Utescher *et al.*, 2002; Feakins & deMenocal, in Werdelin & Sanders, 2010; Jacobs *et al.*, in Werdelin & Sanders, 2010). At this time, warm, at least seasonally humid conditions, evidenced by warm-temperate to subtropical elements, prevailed in the northwest

Mediterranean, and warm, arid conditions, evidenced by open vegetational and grassland elements, in the southwest Mediterranean (Fauquette *et al.*, in Williams *et al.*, 2007a; Jimenez-Moreno *et al.*, in Williams *et al.*, 2007a). Tropical rainforest developed over much of Africa, with some subtropical elements and some grassland or savannah in the north and south in the Langebaanian. Gastropod and mammal evidence points toward the development of savannah and riparian forest in the area around Laetoli in Tanzania (Pickford, 1995; Kingston & Harrison, 2007). Temperate forest palynofloras developed on the Loess Plateau in China between 5.8–4.2 Ma (Wang *et al.*, 2006). Mean hypsodonty indicates a relative climatic amelioration or humidification in Eurasia and Africa from 4.9–4.2 Ma (Fortelius *et al.*, 2006). It also continues to point toward a north–south palaeo-precipitation gradient in Eurasia and Africa, with more humid conditions to the north and more arid ones to the south. In contrast, micro-mammal data points toward an east–west temperature gradient in the 'late' Ruscinian (Mein Zone MN15), with temperatures in France and Spain of 18.3–20.6 °C, and those in Slovakia of only 14.1 °C (Montuire *et al.*, 2006).

Palaeontological evidence also points toward a regional climatic deterioration, although also at least a local climatic amelioration in Africa, from approximately 4.5 Ma (Feakings & deMenocal, in Werdelin & Sanders, 2010). In the Shanxi Plateau of central China, an increase in *Picea* (spruce) and *Abies* (fir) indicates cooling commencing at 4.4 Ma, possibly in response to an intensification of the east Asian monsoon. Steppe palynofloras developed on the Loess Plateau between 4.2–2.4 Ma (Wang *et al.*, 2006). Mean hypsodonty indicates a relative climatic deterioration or aridification in Eurasia and Africa from 4.2 to 3.2 Ma (Fortelius *et al.*, 2006).

Land mammal evolution and dispersal. The oldest widely accepted representative of the human lineage is *Ardipithecus ramidus* from the latest Miocene to Early Pliocene Aramis member of the Sagantole formation of the middle Awash valley of the Afar rift in Ethiopia in east Africa, dated to 5.8–4.4 Ma, and from at least approximately coeval deposits in the Baringo region of Kenya (Sarmiento *et al.*, 2007; Barham & Mitchell, 2008; Gibbons, 2009; Louchart *et al.*, 2009; Lovejoy, 2009; Lovejoy *et al.*, 2009a; Lovejoy *et al.*, 2009b; Lovejoy, *et al.*, 2009c; Lovejoy *et al.*, 2009d; Suwa *et al.*, 2009a; Suwa *et al.*, 2009b; White *et al.*, 2009a; White *et al.*, 2009b; WoldeGabriel *et al.*, 2009; MacLatchy *et al.*, in Werdelin & Sanders, 2010). (*Kenyanthropus platyops*, also from the Early Pliocene of east Africa, is another candidate.) *Ardipithecus ramidus* is represented by numerous fragmentary remains and by a partial skeleton (ARA-VP-6/500) with the skull, pelvis and limbs preserved. The skeleton indicates that the species grew to a height of approximately 1.2 m and had a brain capacity of 300–350cm^3, similar to that of modern bonobos and female chimpanzees. Its skull and teeth indicate that, like its ape ancestors, it ate mainly C3 or woodland vegetation, including fruit, but also that, unlike its ape ancestors, it also ate C4 or grassland vegetation, indicating that the habitat in which it lived was characterised by grassland or savannah as well as woodland. Its pelvis, limbs and appendages indicate that, again like its ape ancestors, it was able to climb, or at least clamber around in, trees, but also that, unlike its ape ancestors, it was also able to walk upright on its hind-limbs, albeit probably not very well, and with its fore-limbs free (*Kenyanthropus platyops, Ardipithecus kadabba* and *Orrorin tugensis* may have been habitually bipedal even earlier: Pickford *et al.*, 2002.) *Ardipithecus ramidus* is interpreted as an essentially terrestrial ape that evolved in response to selection pressure associated with the observed change from a woodland- to a grassland-dominated environment (which same change restricted arboreal apes to rainforest refuges). The environmental change appears to have been forced partly by the change from a warm, humid, to a cool, arid climate, and partly by the coeval opening of the east African rift valley, both leading to a partitioning of the former rainforest. In response, *Ardipithecus* climbed down from the trees and walked out onto the savannah, scrutinising the expanding horizon for danger and for opportunity.

The dispersals of numerous small mammals between Africa and Eurasia appear to have taken place during the Ruscinian European Land Mammal Age (Mein Zones MN14–MN15).

Late Pliocene

(Piacenzian/'early' to 'middle' Villafranchian European Land Mammal Age, Mein Zones MN16–MN18), approximately 4–2.6 Ma.)

Palaeogeography and palaeoclimate. At least initially, the Late Pliocene, like the early part of the Early Pliocene, was essentially a period of climatic amelioration, with temperatures for the most part higher than today, as evidenced by the widespread occurrence of 'cold-blooded' amphibians and reptiles (Feakins & deMenocal, in Werdelin & Sanders, 2010; Jacobs *et al.*, in Werdelin & Sanders, 2010). Palaeobotanical and palynological evidence indicates that warm, humid forests expanded and cool, arid grasslands contracted at this time. However, the Late Pliocene was also a period of high-frequency climatic oscillation, and by its end, of significant climatic deterioration, with widespread evidence of high-latitude glaciation from approximately 3 Ma. The glaciation has been attributed by some authors to changes in oceanic and atmospheric circulation associated with the closure of the Strait of Panama between 4.5 and 3 Ma and the closure of the Indonesian seaway or gateway between 4 and 3 Ma (see also Cane & Molnar, 2001, Smith & Pickering, 2003, and Hall, 2006). Desiccation associated with the glaciation is indicated by the increased incidence of wind-transported terrigenous material, or loess.

Charophyte algal, mollusc, ostracod, fish, amphibian, reptile and micro-mammal evidence from the Chindigarh area of India points toward the development of freshwater pond, pond bank and wooded grassland habitats at 4 Ma. The presence of the gerbil rodent *Tatera* points toward the local development of semi-arid or intermittently arid sandy plains.

Palaeobotanical, palynological, invertebrate and vertebrate palaeontological and palaeopedological evidence indicates that arid grasslands expanded and humid forests contracted in response to cooling or drying after the time of the onset of glaciation, approximately 3 Ma (Utescher *et al.*,

2002; Bobe *et al.*, 2007; Fauquette *et al.*, in Williams *et al.*, 2007a; Jimenez-Moreno *et al.*, in Williams *et al.*, 2007a; Feakins & deMenocal, in Werdelin & Sanders, 2010; Jacobs *et al.*, in Werdelin & Sanders, 2010). Mean hypsodonty indicates further climatic deterioration or aridification in Eurasia and Africa at this time, and also continues to point toward a north–south palaeo-precipitation gradient in Eurasia and Africa, with more humid conditions to the north and more arid ones to the south (Fortelius *et al.*, 2006). In Africa, the effects of cooling and drying were exagerrated by the development of a rain-shadow in the lee of the rising Ethiopian–Kenyan dome. There was evidently extensive development of arid savannah in the north and south of Africa in the Makapanian, from 3 Ma. Microvertebrate dental micro-wear and carbonate isotopic evidence has confirmed the presence of C4 grasses into the Makapansgat limeworks area of the high veld of South Africa in the Makapanian (Hopley *et al.*, 2006). (Note, though, that there would not appear to have been a specialist grazer among the herbivorous rodent population of the time, and that the palaeoenvironment would appear to have been more wooded than the woodland–savannah mosaic environment that obtains today.) Plant, charophyte algal, mollusc, ostracod, fish, amphibian, reptile and mammal evidence from the Chindigarh area of India points toward a further increase in seasonality and aridity associated with an intensification of the monsoon from approximately 2.5 Ma. Arid plants such as *Artemisia*, Chenopodiaceae and *Ephedra* became conspicuously more common on the Shanxi Plateau of central China at the same time.

Land mammal evolution and dispersal. *Australopithecus* appears to have evolved in east Africa in the Pliocene, approximately 4 Ma (Sarmiento *et al.*, 2007; Barham & Mitchell, 2008; MacLatchy *et al.*, in Werdelin & Sanders, 2010). The genus then diversified into a number of gracile species in east and south Africa in the later Pliocene, including, in addition to *A. anamensis*, *A. afarensis* and *A. africanus* (see below), *A. bahrelghazali* and *A. garhi* (Sarmiento *et al.*, 2007; Barham & Mitchell, 2008; Berger *et al.*, 2010;

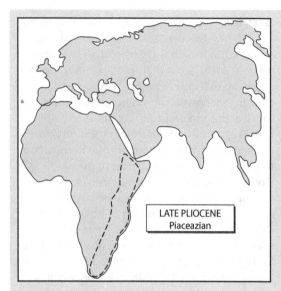

Fig. 3.38 **Late Pliocene, Piacenzian palaeogeography of the Old World**. From Jones (2006). Dashed line indicates known range of late *Australopithecus* and early *Homo* species (*H. habilis*).

Brunet, 2010; Dirks *et al.*, 2010; MacLatchy *et al.*, in Werdelin & Sanders, 2010; Fig. 3.38).

Australopithecus anamensis is represented by fragmentary skeletal remains, including a knee joint from the shores of Lake Turkana in northern Kenya dated to approximately 4 Ma, the so-called 'carrying angle' of which indicates that it was able to walk upright on its hind-limbs, like *Ardipithecus ramidus*, but probably rather better (MacLatchy *et al.*, in Werdelin & Sanders, 2010).

The oldest skeletal remains of *Australopithecus afarensis* are from an unnamed formation overlying the Sagantole in Ethiopia in east Africa, dated to 3.39 Ma, although this species is also thought to have been responsible for the famous trackways at Laetoli in Tanzania, dated to 3.7–3.8 Ma (MacLatchy *et al.*, in Werdelin & Sanders, 2010). *Australopithecus afarensis* is perhaps best known through the substantially complete skeletons of 'Lucy', from Ethiopia, dated to 3.2 Ma, and the recently discovered 'Lucy's baby', also from Ethiopia, dated to 3.3 Ma (Alemseged *et al.*, 2006; Wong, 2006; Wynn *et al.*, 2006; MacLatchy *et al.*, in Werdelin & Sanders, 2010). ('Lucy' was named after the Beatles song 'Lucy in the sky with diamonds',

which was played at the party celebrating her discovery.) 'Lucy' has a pelvis and hind-limb structure of human-like aspect, but a brain of ape-like aspect. Her pelvis is short and horizontal, rather than long and vertical, as in arboreal apes, and her toes are unsuited to grasping; but her brain has a capacity of 415 cm^3, which, for her height of 1–1.2 m, recalls the ratio in modern chimpanzees. The skull morphology of Lucy's species has recently been discussed in detail by Kimbel *et al.* (2004).

The gracile australopithecine *Australopithecus africanus* is perhaps best known through the skull of the 'Taung child', discovered in the Buxton limeworks near Taung in the Cape Province of South Africa, and dated to approximately 3 Ma (MacLatchy *et al.*, in Werdelin & Sanders, 2010). The skull of the 'Taung child' is exquisitely preserved, and still bears brain impressions in its interior. However, it does show some damage, consistent with an attack by a large raptor such as an African crowned eagle. Adult individuals of *Australopithecus africanus*, represented, for example, by the synonymous *Plesianthropus transvaalensis* or 'Mrs Ples', from the Sterkfontein limeworks near Krugersdorp, grew to a height of 1.3 m – apparently with disproportionately short legs and long arms – and had a brain capacity of 480 cm^3.

The dispersals of numerous large browsing mammals from the Americas and Africa into Eurasia appear to have taken place during the 'early' to 'middle' Villafranchian European Land Mammal Age (Mein Zones MN16–MN18) (Vislobokova *et al.*, in Reumer & Wessels, 2003).

Pleistocene–Holocene

('Late' Villafranchian European Land Mammal Age, approximately 2.6 Ma to present.)

Palaeogeography and palaeoclimate. The Pleistocene was a time of significant climatic deterioration marked by extensive – although intermittent – glaciation in high latitudes and at high altitudes in low latitudes. Glacial stages alternated with interglacials, during which the climate was as warm as or warmer than that at the present day. During glacials, vegetation belts in the northern hemisphere were displaced southward by up to as much as 1500–2000 km in the former

Soviet Union. Biogeographic provinces in the oceans were displaced southward by similar amounts in the northeast Atlantic, presumably on account of the associated southward displacement of the North Atlantic drift current or 'Gulf Stream', which, incidentally, also had a marked effect on the maritime provinces of northwest Europe.

As intimated above, palaeobotanical, palynological, invertebrate and vertebrate palaeontological and palaeopedological evidence indicates that arid grasslands expanded and humid forests contracted during glacials (Figs. 3.39–3.40).

(Note, though, that recent evidence from molecular isotope stratigraphy of long-chain n-alkanes suggests that there was no particular aridity during the last glaciation in Sunda Land in southeast Asia.) In Africa as a whole, there was extensive development of tropical rainforest only in – western – equatorial regions, with significant development of savannah elements in the north and south, and dry desert in the extreme north and south.

Importantly, evidence from diatoms in lake sediments in Ethiopia, Kenya and Tanzania in east Africa indicates temporary humidification events associated with 400 000 yr eccentricity maxima at approximately 2.6 Ma, 1.8 Ma and 1 Ma superimposed on the overall long-term aridification (Trauth et al., 2005; Maslin & Trauth, in Grine et al., 2009; Trauth et al., 2009). These events have been interpreted as having had important impacts on the evolution and dispersal of mammals, including early humans. Pollen data indicates a mosaic of open and closed habitat vegetation in the area around Gona in Afar in Ethiopia at approximately 2.6 Ma (Lopez-Saez & Dominguez-Rodrigo, 2009). Significantly, stone tools attributed to early humans have been found in this area, although actual human remains have not.

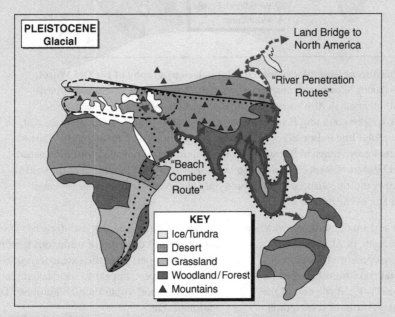

Fig. 3.39 **Pleistocene, glacial/stadial palaeogeography of the Old World**. From Jones (2006), showing reconstructed vegetation at the time of a representative glacial/stadial. Also shown are the distributions of the early *Homo* species *H. erectus*, *H. heidelbergensis* and *H. neanderthalensis*. Dotted line indicates known range of *Homo erectus*. Solid line indicates approximate northern limit of *H. heidelbergensis*. Dashed line indicates known range of *H. neanderthalensis*. Solid arrows indicate possible dispersal routes used by early *Homo* species; and also by *H. sapiens*, in its colonisation of the world. Dashed arrows indicate dispersal routes used only by *H. sapiens*, in its colonisation of Australasia and of the Americas.

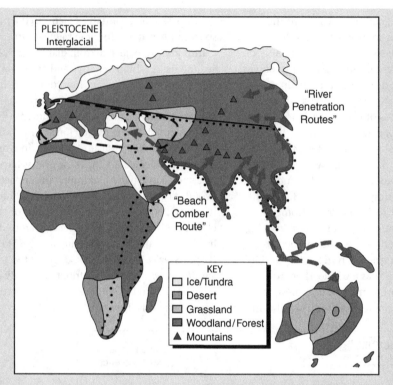

Fig. 3.40 **Pleistocene, interglacial/interstadial palaeogeography of the Old World.**
From Jones (2006). Showing reconstructed vegetation at the time of a representative
interglacial/interstadial. Also shown are the distributions of the early *Homo* species
H. erectus, *H. heidelbergensis* and *H. neanderthalensis*. Dotted line indicates known range of
Homo erectus. Solid line indicates approximate northern limit of *H. heidelbergensis*. Dashed
line indicates known range of *H. neanderthalensis*. Solid arrows indicate possible dispersal
routes used by early *Homo* species; and also by *H. sapiens*, in its colonisation of the world.
Dashed arrows indicate dispersal routes used only by *H. sapiens*, in its colonisation of
Australasia and of the Americas.

Amphibian, reptile and small mammal evidence
from the archaeological site of Atapuerca in Spain
points toward the development of warm, wet,
wooded environmental conditions between
Oxygen Isotope Stages 21–19, at the end of the
Early Pleistocene, and an onset of cooler, drier,
more open conditions at around 780 000 yr
BP, at the beginning of the Middle Pleistocene
(Cuenca-Bescos *et al.*, in Head & Gibbard, 2005;
Blain *et al.*, 2008). Coincidentally or otherwise,
the beginning of the Middle Pleistocene was
when the 100 000 yr Milankovitch cyclicity came
to dominate over the 40 000 yr cyclicity (McNabb,
in Head & Gibbard, 2005).

Land mammal evolution and dispersal. The
evolution and dispersal of numerous mammals
of modern aspect, and the extinctions of those
of archaic aspect, appear to have taken place
during the 'late' Villafranchian European Land
Mammal Age.

The genus *Australopithecus* continued to diversify
into robust and gracile species in east and south
Africa in the earliest Pleistocene, including the
robust species *Australopithecus (Paranthropus)
aethiopicus*, from unnamed deposits in Baringo in
Kenya in east Africa, dated to approximately
2.5 Ma, *A.(P) boisei* and *A.(P.) robustus* (MacLatchy
et al., in Werdelin & Sanders, 2010); and the

recently discovered and as yet still comparatively poorly known gracile australopithecine *Australopithecus sediba* from Malapa in South Africa, dated to 1.95–1.78 Ma (Berger *et al.*, 2010; Dirks *et al.*, 2010).

Our own genus, *Homo*, then evolved, from a gracile australopithecine ancestor, in east Africa, arguably again in response to climatic and environmental forcing (Bromage & Schrenk, 1999; Jolly, 1999; Calvin, 2002; Gunding, 2005; Holmes, 2007; Sarmiento *et al.*, 2007; Barham & Mitchell, 2008; Grine *et al.*, 2009; Hetherington & Reid, 2010; MacLatchy *et al.*, in Werdelin & Sanders, 2010). (The 'aquatic ape' hypothesis remains an interesting alternative.) This selection pressure would have favoured the socialisation and intelligence necessary for efficient food gathering, and hence encephalisation and evolution. Interestingly, though, the bipedalism exhibited by *Homo* is regarded as a pre-adaptation rather than an adaptation to mosaic grasslands, and hence independent of climatic change (Kingdon, 2003). (The role, if any, of predation in human evolution remains unclear: Hart & Sussman, 2005.)

Homo in turn diversified into a number of species in the Pleistocene (Sarmiento *et al.*, 2007; Barham & Mitchell, 2008; MacLatchy *et al.*, in Werdelin & Sanders, 2010). The interpretation of their evolutionary relationships is still controversial, and the argument remains unresolved as to how a number of nominal species – beyond those considered here – fit in (MacLatchy *et al.*, in Werdelin & Sanders, 2010). Nonetheless, the consensus seems to be that *Homo erectus/ergaster* evolved from *H. habilis*, and evolved into *H. heidelbergensis*, or 'archaic *H. sapiens*'. Certainly, this interpreted lineage seems to have been characterised by the most rapid proportional brain growth, although there is still something of a 'muddle in the middle' in between *Homo erectus/ergaster* and *H. sapiens*. DNA evidence indicates that *Homo neanderthalensis* is not part of the lineage leading to *H. sapiens*.

Homo habilis, which has as yet only been recorded from the Early Pleistocene, 2.4–1.5 Ma, of Africa, is represented mainly by a number of more-or-less complete skulls, but also by

approximately 300 skull and skeletal fragments interpreted as representing a single individual from the Olduvai Gorge in Tanzania – Olduvai hominid or OH 62 (Sarmiento *et al.*, 2007; Barham & Mitchell, 2008; MacLatchy *et al.*, in Werdelin & Sanders, 2010). It apparently grew to a height of 1.3 m, and had a brain capacity of 630–800 cm^3 (600 cm^3 representing the somewhat arbitrary criterion – the so-called 'cerebral Rubicon' – used to distinguish *Homo* from *Australopithecus*). There is abundant circumstantial evidence, in the form of association with worked stones, that *Homo habilis* was the first creature ever to use tools (incidentally, its name translates as 'handy man'). Its diet appears to have differed from that of its ancestors in including nuts and seeds of arid scrubland habitats (Ungar, 2007; Hublin & Richards, 2009). The presently comparatively poorly known *Homo rudolfensis* appears similar to *H. habilis*, and is arguably synonymous with it. There is as yet no positive evidence that *Homo habilis* ever dispersed out of Africa, although equally there is only negative evidence that it did not, and such absence of evidence does not constitute evidence of absence (Kohn, 2006).

Homo erectus/ergaster, which has been recorded from Eurasia ('Asian *Homo erectus*') as well as Africa ('African *Homo erectus*' or *Homo ergaster* of some authors), is interpreted as having evolved from *H. habilis* in the Early Pleistocene, approximately 1.9 Ma, and as having become extinct some time after 0.25 Ma or 250 000 years ago (Sarmiento *et al.*, 2007; Barham & Mitchell, 2008; Gilbert & Asfaw, 2008; Dennell, 2009; MacLatchy *et al.*, in Werdelin & Sanders, 2010). It is perhaps best known through the virtually complete skeleton of the 'Nariokotome boy' or 'Turkana boy', discovered at Nariokotome on the west side of Lake Turkana in Kenya in Africa, and dated to approximately 1.6–1.9 Ma; but it is also known through remains in Dmanisi in Georgia, in Riwat in Pakistan, in Longgupo and Nihewan in China, and in Sangiran and Mojokerto in Java in Indonesia, all in Eurasia, and all also dated to 1.6–1.9 Ma. (Some authors refer the remains from Georgia to *Homo georgicus*, and those from China to *H. pekinensis*.) *Homo erectus/ergaster* grew to a height of 1.6 m, and had a brain capacity of 850–1100 cm³. The large, robust skull of the species is interpreted as having fulfilled one or more of a number of possible functions, including: providing sites of attachment for stronger chewing muscles; encasing a larger brain; cooling the brain; and protecting the brain, spinal cord and eyes. Interestingly, there is some palaeopathological evidence to support the interpretation that it fulfilled a primarily protective function: skulls have been found with depressed fractures that had healed rather than proved fatal. Protection may have been against attack from other individuals, which is hypothesised as having been the preferred method of conflict resolution! *Homo erectus/ergaster* evidently used both weapons and – Oldowan – tools, and had mastered the use of fire, for cooking. It apparently occupied semi-permanent settlements, although it may also have migrated seasonally and/or over the longer-term (possibly to track game). It may even have evolved a basic, tribe-like societal structure, and a means of communication through either speech or sign language. *Homo erectus/ergaster* is interpreted by most authors as having evolved in and dispersed out of Africa, perhaps in response to further climatic and environmental forcing, although it is equally possible that it evolved in and dispersed out of Asia (Cameron & Groves, 2004; Kohn, 2006; Dennell, 2009). By approximately 1.9 Ma, owing to climatic change, the formerly impenetrable forests of Africa no longer presented a barrier to its dispersal, having given way to treeless grassland and desert: significantly, as intimated above, its diet appears to have differed from that of its ancestors in including meat from scavenging and hunting animals of grassland habitats (Ungar, 2007; Hublin & Richards, 2009). In Eurasia, the former shelf seas of the south no longer presented a barrier to dispersal either, having retreated and given way to coastal plains. However, the intermittently glaciated highlands and periglacial lowlands of the north – the so-called 'cold wall' – did. Moreover, the 'Wallace Line', delineating the boundary between the essentially continental land-mass of Eurasia to the west and the essentially oceanic archipelago of southeast Asia and Australasia to the east, also still presented a barrier to dispersal to the east of Java.

Incidentally, the event that emplaced the lower Lahar in the Sangiran formation of central Java at 1.9 Ma transformed previously marine environments into estuarine and paludal ones, immediately prior to the immigration of *Homo erectus/ergaster*.

Homo floresiensis, from the island of Flores in Indonesia in southeast Asia, is interpreted as having evolved from *H. erectus/ergaster* in the Late Pleistocene, some time before 38 000 years BP, and as having become extinct 18 000 years BP (Fischman, 2005; Morwood *et al.*, 2005; Morwood & Oosterzee, 2007; Sarmiento *et al.*, 2007; Dennell, 2009). It grew to a height of only 1 m, and, although it apparently manufactured stone tools (with which to hunt the pygmy elephant *Stegodon*), had a brain capacity of only 380 cm³ – less than the 600 cm³ conventionally used to distinguish *Homo* from *Australopithecus*.

Homo heidelbergensis, or 'archaic' *H. sapiens*, from Africa and Eurasia, is interpreted as having evolved independently from *H. erectus/ergaster* in the early Middle Pleistocene, approximately 600 000 years ago, and as having become extinct approximately 150 000 years ago (Sarmiento *et al.*, 2007; Barham & Mitchell, 2008; Dennell, 2009; MacLatchy *et al.*, in Werdelin & Sanders, 2010). It is perhaps best known through the skulls from Broken Hill in Zambia, Bodo d'Ar in Ethiopia, Arago in France, Steinheim in Germany, and Petralona in Greece, the lower jaw from Mauer, near Heidelberg, in Germany, and the recently discovered tooth and leg bone from a raised beach at Boxgrove on the Sussex coastal plain on the south coast of England. (Some authors refer the remains from Zambia and Ethiopia to *Homo rhodesinensis*.) Although it is impossible to be certain from the somewhat fragmentary fossil record, *Homo heidelbergensis* apparently grew to a height of 1.8–1.9 m, and had a brain capacity of 1300 cm³. Circumstantial evidence in the form of associated finds from the Boxgrove site indicates that it used weapons, for hunting, as well as (Acheulian) tools. *Homo heidelbergensis* remains, and/or Acheulian tools, are found throughout Africa and Eurasia.

Homo neanderthalensis is interpreted as having evolved from *H. heidelbergensis* in Eurasia, in the late Middle Pleistocene, approximately 250 000–300 000 years ago, and as having become extinct approximately 27 500 years ago (Harvati & Harrison, 2007; Sarmiento *et al.*, 2007; Dennell, 2009). Interestingly, *Homo neanderthalensis* evidently coexisted with *H. sapiens* between 35 000–27 500 years ago in southern France and the Iberian Peninsula, which at the time would have formed refuges from the glacial ice further north (van Andel & Davies, 2003). *Homo neanderthalensis* was evidently comparatively short and stocky, on account of adaptation to the generally cold climate obtaining at the time – an adaptation also exhibited by modern Inuit peoples – but had a brain capacity averaging 1400 cm^3. It appears to have evolved an advanced societal structure, characterised by communal hunting, the preparation and wearing of clothes of animal skins, care of the elderly and infirm, and, arguably, such respect for the dead, including burial, as to suggest some form of religious belief system (Mellars, 1995). Incidentally, burial of the dead led to a significant improvement of fossilisation potential, by denying scavengers their former freedoms. *Homo neanderthalensis* remains, and/or Mousterian tools, are found throughout the area bounded by southern England, Spain, the Levant and Uzbekistan.

Our own species, 'anatomically modern' *Homo sapiens* is interpreted as having evolved independently from *H. heidelbergensis*, or 'archaic' *H. sapiens*, in Africa, in the Late Pleistocene, approximately 200 000 years ago (Sarmiento *et al.*, 2007; Willoughby, 2007; Barham & Mitchell, 2008; MacLatchy *et al.*, in Werdelin & Sanders, 2010). We have a brain capacity averaging 1360 cm^3 – marginally smaller than that of *Homo neanderthalensis*! We have supposedly evolved still more advanced societies and cultures, the latter manifested in the adornment of the body in life and after death, and in the making of images, from at least 35 000 years ago, and possibly as long as 70 000 years ago. According to the most widely accepted hypothesis, supported by molecular biological (mitochondrial DNA) evidence, *Homo sapiens* dispersed 'out of Africa' and into southwest Asia, by inference, by way of the Bab al Mandab (or 'gate of grief') and the Levantine corridor approximately 85 000 years ago (Goren-Inbar & Speth, 2004; Strachan & Read, 2004; Petraglia, in Head & Gibbard, 2005; Dov Por & Dimentman, 2006; Hoffecker, in Fagan, 2009; Roberts, 2009). By 75 000 years ago, it had reached the then-emergent 'Sundaland' in southeast Asia, by way of the coastal so-called 'beachcomber route'; and by 65 000 years ago, eastern Indonesia, New Guinea and Australia, by 'island-hopping' (Demeter *et al.*, 2003). (Some authors interpret it as having reached Australia at least 130 000 years ago, during Oxygen Isotope Stage OIS6: Webb, 2006.) By 50 000 years ago, it had reached Europe, by way of the overland so-called 'fertile crescent' or 'Caucasus' routes; by 40 000 years ago, the Indian subcontinent and central Asia, also, obviously, by overland, or by 'river penetration', routes; and by 30 000 years ago, it had reached Siberia, in northeast Asia, and by 25 000?–15 000 years ago, Alaska, in northernmost North America, by inference, by way of the then-emergent 'Beringia' (Meltzer, 2009). Alternatively, it may have entered the Americas by way of a Pacific crossing, or, according to the 'Solutrean hypothesis', by way of an Atlantic crossing from southwestern France or Spain (Bonnischen *et al.*, 2006). (Note in this context that recent modelling indicates that an Atlantic crossing might have taken as little as 72 days, and a Pacific crossing as little as 83 days: Montenegro *et al.*, 2006.) By 15 000 years ago, it had reached southernmost South America. The first permanent – as opposed to seasonal, nomadic – settlements date to the time of the earliest implementation of agriculture approximately 11 000 years ago. The subsequent widespread implementation of agriculture from approximately 9000 years ago appears to have been driven by inland migrations of populations of farmers in response to loss of coastal farmland at the time of the early Holocene sea-level rise (Turney & Brown, 2007).

4 • Sequence stratigraphy

Sequence stratigraphy attempts to subdivide the rock record into genetically related – unconformity-bounded – rock units or sequences (Emery & Myers, 1996; Jones, 1996; Holland, in Briggs & Crowther, 2001; Sharland *et al.*, 2001; Coe, 2003; Catuneanu, 2006; Jones, 2006; Simmons *et al.*, 2007; Catuneanu *et al.*, 2009; Miall, 2010). The methodology lends itself well to the interpretation of seismic data, on which unconformities are readily identifiable on the basis of – real or apparent – reflector termin-ations (erosional truncation below, and transgres-sive onlap and associated landward facies shift, above, on basin margins). The unconformities are generated by base-level fall, in turn driven by glaci-ation or glacio-eustasy, and/or by structuration or tectonism. Global coastal onlap and ?glacio-eustatic sea-level charts have been constructed for the Carboniferous–Recent. Sea-level curves have also been constructed for the Cambrian, Ordovician, Silurian and Devonian.

4.1 DEFINITIONS

Sequences

A Vailian sequence is defined as a 'stratigraphic unit composed of a relatively conformable succes-sion of genetically related strata bounded at its top and base by unconformities or their correlative con-formities' (Vail *et al.*, 1977). As intimated above, Vailian sequence boundaries (SBs) (bounding unconformities) are readily identifiable on seismic stratigraphic criteria (truncation below, and onlap and associated landward facies shift, above, on basin margins; and development of deep-sea fans in basin centres). They are also recognisable on palaeontological, sedimentological and petrophysi-cal criteria (truncation below, and landward facies shift above, on basin margins; development of deep-sea fans in basin centres; and associated lithological indications on wireline logs).

A Gallowayan genetic stratigraphic sequence is defined as 'a package of sediments recording a sig-nificant episode of basin-margin outbuilding and basin-filling' (Galloway, 1989). Gallowayan SBs (maximum flooding surfaces (MFSs)) are also readily identifiable on seismic stratigraphic criteria (onlap and associated landward facies shift below; and downlap and associated basinward facies shift above), and on palaeontological, sedimentological and petrophysical criteria (turnaround in direction of facies shift from landward to basinward; devel-opment of maximum flooding or palaeobathyme-try; and associated lithological indications on wireline logs).

The spatio-temporal distribution of sequences is dictated by *accommodation space*, which is defined as 'the space ... available for potential sediment accu-mulation'. Accommodation space is in turn dictated by the complex interplay of a number of dynamically variable processes, including *(glacio-)eustasy*, *tectonism*, *sediment compaction* and *sediment supply*. *Eustasy*, that is, rising or falling absolute sea-level, can be either creative or destructive of accommodation space. *Tec-tonism*, that is (independent) subsidence and falling, or uplift and rising, absolute land-level, can also be either creative or destructive of accommodation space. *Sediment compaction* is always creative. *Sediment supply* is always destructive. At any given point in a basin, if the rate of sediment supply exceeds that of the creation of accommodation space by the pro-cesses described above, there will be a basinward shift in bathymetrically ordered facies belts. This process is termed *progradation*. Conversely, if the rate of sediment supply exceeds that of the creation of accommodation space, there will be a landward shift in facies belts. This process is termed *retrogradation*. If the rates of sediment supply and the creation of accommodation space are in equilibrium, facies belts will remain in place, and build vertically. This process is termed *aggradation*.

Systems tracts

Sequences can be internally subdivided into so-called systems tracts (STs). A systems tract is defined as 'a linkage of contemporaneous depositional systems' or bathymetrically ordered facies belts (for example continental–paralic–shelfal–bathyal–abyssal). The principal types are the *low-stand systems tract (LST)*, the *shelf-margin systems tract or shelf-margin wedge (SMST or SMW)*, the *transgressive systems tract (TST)* and the *high-stand systems tract (HST)*. The LST can be further subdivided into a lower, *low-stand fan (LSF)* and an upper, *low-stand wedge or low-stand pro-grading complex (LSW or LPC)*. (Note, though, that marine geologists interpret LSFs and LSWs as attached rather than detached systems: Nelson & Damuth, 2003.) The LST and TST are separated by the *transgressive surface (TS)*, or essentially equivalent *maximum regressive surface (MRS)*. The TST and HST are separated by the *maximum flooding – or transgressive – surface (MFS)*. *Condensed sections (CSs)* commonly occur at the TS and MFS.

The LST is defined as the lowermost systems tract in a depositional sequence if it lies directly on a Type 1 Sequence Boundary, and the related SMST or SMW as the lowermost systems tract in a depositional sequence associated with a Type 2 Sequence Boundary; the TST as the middle systems tract of both Type 1 and Type 2 sequences; and the HST as the upper systems tract in either a Type 1 or a Type 2 sequence. Note that a Type 1 SB is generated by a sea-level fall to a position basinward, and a Type 2 SB by a sea-level fall to a position landward, of the shelf-edge or 'offlap break' of the preceding sequence. The TS is defined as the first significant marine flooding surface within a sequence. The MFS is defined as the surface corresponding to the time of maximum flooding. CSs are defined as thin marine strati-graphic units consisting of pelagic to hemipelagic sediments characterised by very low sedimentation rates most areally extensive at the time of max-imum regional transgression. They commonly con-tain abundant and diverse planktonic and benthic microfossil assemblages, which enable easy identifi-cation, and also, importantly, calibration against global standard biostratigraphic zonation schemes.

Gamma log trends are useful in the characterisa-tion of STs. LSTs and SMSTs are typically character-ised by low gamma values; TSTs by increasing values; and HSTs by decreasing values. CSs, that is, TSs and MFSs, are typically characterised by high gamma values. Note, though, that CSs are atypically characterised by low gamma values, where

represented by limestones rather than shales (Jones *et al.*, in Jenkins, 1993).

4.2 GENERAL AND CLASTIC SEQUENCE STRATIGRAPHY

Haq *et al.* (1987) have generated a conceptual model of the development of sequences in relation to cycles of sea-level change (Fig. 4.1).

This model, or modifications of it, can be used to predict the spatio-temporal distribution of lithofa-cies, even in tectonically active areas as the compressional Californian Borderland or the exten-sional Gulf of Mexico growth-fault province, or in marginal marine to entirely non-marine settings (Holz *et al.*, 2002; Jones, 2006).

Vail, Haq and others have argued that at least the so-called 'third-order' cyclicity is, essentially, (glacio-)eustatically rather than tectonically medi-ated, and that the (glacio-)eustatic signal, while it may be strengthened or weakened by the tectonic signal, is always stronger than it. This argument has been criticised by a number of authors, on the grounds that it fails to explain the discrepancy between the frequencies of the modelled 'third-order' cycles (of the order of a million years) and the observed incontrovertibly glacio-eustatically mediated cycles (of the order of tens of thousands to hundreds of thousands of years) of the 'ice-house' world of the Pleistocene; and that it also fails to explain the observed cyclicity in the 'greenhouse' world of the Mesozoic (when obviously there would have been no glaciation or glacio-eustasy). This criti-cism has in turn been countered by arguments that 'fourth- to sixth-order' cycles are in fact resolvable, although only in areas of high sedimentation rate such as the Plio-Pleistocene of the Gulf of Mexico, the Niger delta or the south Caspian; and that 'third-order' cycles are mediated not only by (glacio-)eustasy but also by tectonism, in the form of – episodic – local and regional stress release.

4.3 CARBONATE SEQUENCE STRATIGRAPHY

Sarg, in Wilgus *et al.* (1988), and Handford and Loucks, in Loucks and Sarg (1993), have generated conceptual models of the development of carbonate sequences in relation to cycles of sea-level change.

Handford and Loucks's models are comprehen-sive, and take into account criticisms that earlier models of carbonate sequences relied on analogy

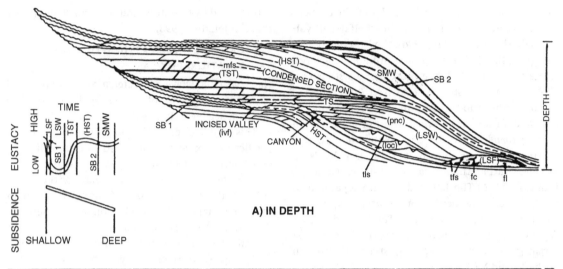

A) IN DEPTH

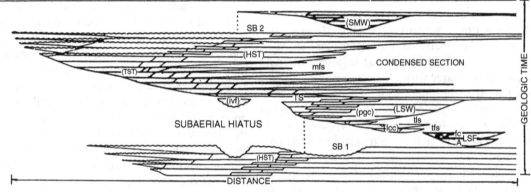

B) IN GEOLOGIC TIME

LEGEND

SURFACES

(SB) SEQUENCE BOUNDARIES
   (SB 1) = TYPE 1
   (SB 2) = TYPE 2

(DLS) DOWNLAP SURFACES
   (mfs) = maximum flooding surface
   (tfs) = top fan surface
   (tls) = top leveed channel surface

(TS) TRANSGRESSIVE SURFACE
   (First flooding surface above maximum regrssion)

Z FACIES BOUNDARIES
   (Proximal at left: Distal at right)

SYSTEMS TRACTS

HST = HIGHSTAND SYSTEMS TRACT
TST = TRANSGRESSIVE SYSTEMS TRACT
LSW = LOWSTAND WEDGE SYSTEMS TRACT
   ivf = incised valley fill
   pgc = prograding complex
   lcc = leveed channel complex
LSF = LOWSTAND FAN SYSTEMS TRACT
   fc = fan channels
   fl = fan lobes
SMW = SHELF MARGIN WEDGE SYSTEMS TRACT

Fig. 4.1 **Clastic sequence stratigraphic model.** © American Association for the Advancement of Science. Reproduced with permission from Haq *et al.* (1987).

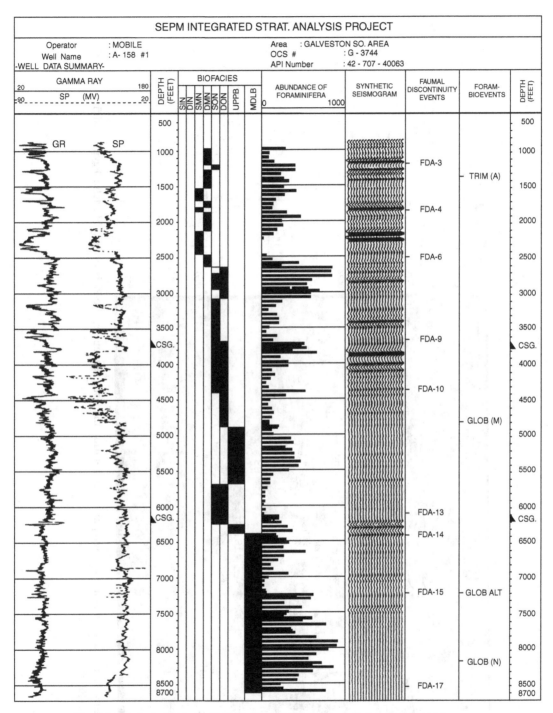

Fig. 4.2 **Integrated micropalaeontological bio- and sequence-stratigraphic interpretation in clastic sequences, A-158 N1, Pleistogene, Gulf of Mexico.** © Springer. Reproduced with permission from Armentrout, in Weimer and Link (1991).

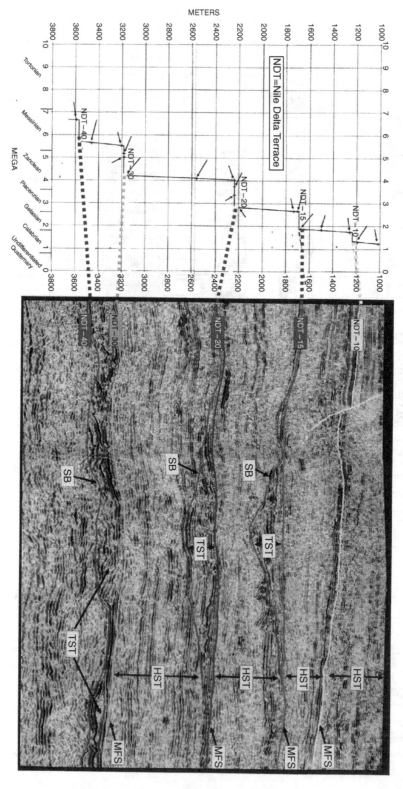

Fig. 4.3 Integrated bio- and sequence-stratigraphic interpretation in clastic sequences, Plio-Pleistogene, Nile delta, Egypt. Modified after Wescott et al. (1998).

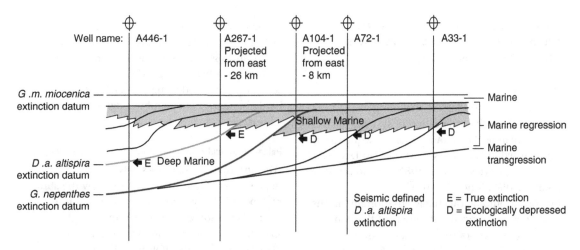

Fig. 4.4 **Integrated micropalaeontological bio- and sequence-stratigraphic interpretation in clastic sequences, Plio-Pleistogene, Gulf of Mexico.** Modified after Armentrout, in Weimer and Link (1991). Note 'ecologically depressed' tops (Ds) located at facies boundaries below the 'time-lines' marking the true extinctions (Es).

with clastic sequences, which, in view of the more complex controls on carbonate sequences, is probably only partly appropriate. In particular, they take into account observations that modern and ancient carbonate systems, unlike clastic systems, deposit comparatively insignificant volumes of sediment during LSTs, when they are subject to chemical as against physical erosion, and much more significant volumes during HSTs.

4.4 MIXED SEQUENCE STRATIGRAPHY

McLaughlin *et al.* (2004) have generated models of the development of mixed clastic–carbonate sequences in the Ordovician of Kentucky and Ohio. They have also argued that the primary control on sequence architecture is (glacio-)eustatic, with only a secondary tectonic component associated with the Taconian orogeny.

4.5 SEISMIC FACIES ANALYSIS

As intimated above, seismic reflection terminations form the basis for sequence stratigraphic analysis. For example, onlap defines Vailian SBs; downlap, Gallowayan SBs or MFSs. Reflection patterns form the basis for seismic facies analysis. To simplify, reflection geometry (i.e. whether clinoform etc.) provides an indication of the depositional process. Reflection amplitude provides an indication of the product of the depositional process, that is, lithology, and also

porosity and pore fluid content. Time-slices through 3D seismic volumes allow the identification of individual sedimentary – and reservoir – bodies, such as deltas and submarine fans. Other seismic attributes such as spectral decomposition and coherency can also be used in palaeoenvironmental interpretation.

4.6 INTEGRATION OF PALAEONTOLOGICAL DATA

It is important to integrate palaeontological, well and seismic data into sequence stratigraphic interpretations (Armentrout, in Weimer & Link, 1991; Jones, 1996, 2006; Fig. 4.2).

Integration is nowadays generally undertaken in the work-station environment (Wescott *et al.*, 1998; Fig. 4.3).

It enables the ready recognition of apparent or ecologically depressed biostratigraphic extinctions or 'tops', and thus prevents mis-correlations (Armentrout, in Weimer & Link, 1991; Fig. 4.4).

Integration also enables the identification of master surfaces such as SBs and MFSs, and the characterisation of STs, and hence the definition of meaningful time-slices for gross depositional environment, and source-, reservoir- and cap-rock common risk segment, mapping as part of the play fairway analysis or evaluation process (see Box 4.1 below; see also Section 5.2).

BOX 4.1   PALAEONTOLOGICAL INPUTS INTO THE CHARACTERISATION OF SYSTEMS TRACTS (MODIFIED AFTER JONES, 2006)

Introduction

The micropalaeontological, including palynological, characterisation of systems tracts (STs) in marine clastic and carbonate sequences is summarised in Figs. 4.5–4.7 (McNeil, in Dixon *et al.*, 1985; Davies & Bujak, 1987; van Gorsel, 1988; Poumot, 1989; Jones, 1996). The micropalaeontological characterisation, by means of freshwater ostracods, of the STs of the non-marine sequences of west Africa's pre-salt Cretaceous is described by Bate, in Cameron *et al.* (1999). Here, lake level is measurable by means of ostracod diversity. Incidentally, it is highest in the Barremian, coincident with the widespread development of lacustrine source-rock.

The palaeontological characterisation of CSs, which commonly occur at TSs and MFSs, is typically by means of measures of abundance and diversity, not only of microfossils (Vecoli *et al.*, 1999; Hart, 2000; Wakefield & Monteil, 2002; Olson & Thompson, in Koutsoukos, 2005) but also of macrofossils (Sharland *et al.*, 2001; Fursich & Pandey, 2003; Holz & Simoes, in Koutsoukos, 2005; Parras & Casadio, 2005; Cantalamessa *et al.*, 2005; Botquelen *et al.*, 2006). In the Late Jurassic to Early Cretaceous of Kutch in western India, TSs are characterised by sorting and preferred – convex-up – orientation of bivalve shells, by a high degree of shell disarticulation, and by a moderate degree of time-averaging; MFSs by preservation essentially in life position, by a lower degree of shell disarticulation, and by a higher degree of time-averaging, having accumulated during times of lower energy and sedimentation rate (Fursich & Pandey, 2003). In the Late Ordovician of Sardinia and Early Devonian of Britanny in France, multivariate analysis has revealed that sequences are characterised by ecological successions of macrofossil associations and 'mega-guilds'; TSs and MFSs by significant components of the pelagic 'mega-guild' as well as the suspension-feeding benthic 'mega-guild' (Botquelen *et al.*, 2006).

The ichnological characterisation of key omission surfaces such as SBs, amalgamated or composite SBs/TSs ('transgressive surfaces of erosion') and high-energy MFSs is by means of substrate-controlled firmground *Glossifungites*, hardground *Trypanites* and woodground *Teredolites* ichnofacies.

Palaeontological characterisation of systems tracts

(Vailian) sequence boundaries. SBs in clastic sequences are typically characterised by the abrupt juxtaposition of younger, more proximal fossil assemblages, sometimes containing reworked specimens, on older, more distal ones ('faunal discontinuity events') (Olson & Thompson, in Koutsoukos, 2005; Raju, in Raju *et al.*, 2005). Note, though, that the contrast in assemblages above and below the SB varies along depositional dip, typically being more pronounced in proximal, updip settings, where a biostratigraphically resolvable unconformity may also be in evidence, and less so in distal, downdip settings, where the relationship may be closer to that of a correlative conformity. Note also that in the most proximal, updip settings, the assemblage above the SB is actually more distal than that below, as it is not directly associated with the SB but rather with the succeeding TS.

Low-stand systems tracts. LSTs in clastic sequences are characterised by erosion in updip settings, and by the depositional of submarine fans and associated sediments in downdip settings. Fossil assemblages in the fans are characterised by an autochthonous deep marine benthic component, an allochthonous, but contemporaneous, shallow marine benthic and planktonic, and non-marine, component, and, occasionally, an allochthonous, and non-contemporaneous component (Olson & Thompson, in Koutsoukos, 2005). The deep-water component is of shallower aspect than that of the underlying sequence, although the difference is not necessarily always resolvable (because bathymetric resolution is comparatively poor – of the order of hundreds of metres – in deep water).

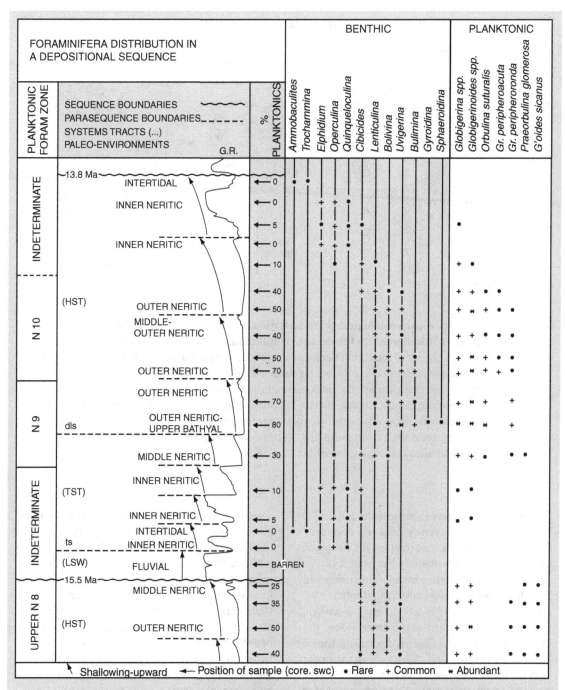

Fig. 4.5 **Hypothetical micropalaeontological characterisation of systems tracts in clastic sequences, Neogene, southeast Asia.** © The Indonesian Petroleum Association. Reproduced with permission from van Gorsel (1988).

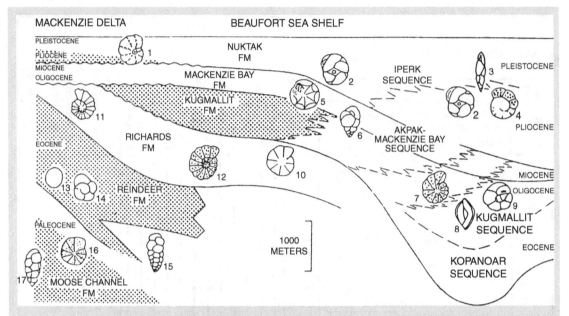

Fig. 4.6 **Micropalaeontological characterisation of systems tracts in clastic sequences, Palaeogene, Mackenzie delta and Beaufort shelf, Arctic Canada.** © The Canadian Society of Petroleum Geologists. Reproduced with the permission of the CSPG, whose permission is required for further use, from McNeil, in Dixon *et al.* (1985). 1, *Elphidium clavatum*; 2, *Cassidulina teretis*; 3, '*Virgulina' loeblichi*; 4, *Cibicides grossa*; 5, *Asterigerina gurichi*; 6, *Turrilina alsatica*; 7, *Alveolophragmium (Reticulophragmium) sp.*; 8, *Spirosigmoilinella sp.*; 9, *Recurvoides sp.*; 10, *Jadammina statuminis*; 11, *Cyclammina sp.*; 12, *Haplophragmoides sp.*; 13, *Saccammina sp.*; 14, *Trochammina sp.*; 15, *Verneuilina sp.*; 16, *Alveolophragmium (Reticulophragmium) sp.*; 17, *Verneuilinoides sp.*

LSTs in the Neogene of southeast Asia, the Palaeogene of the Mackenzie delta and Beaufort shelf in Arctic Canada, and the post-rift Cretaceous of the Pletmos and Bredasdorp basins of South Africa are all characterised micropalaeontologically by a dominance of deep-water agglutinated foraminifera (Jones, 1996, 2006; Figs. 4.5–4.6).

LSTs in the Pleistogene of the Amazon Fan, the Plio-Pleistogene of the Gulf of Mexico and the Niger delta, the Neogene–Pleistogene of the tropics, including southeast Asia, the latest Jurassic to earliest Cretaceous of Papua New Guinea and of the Vocontian basin of southeast France, and the Jurassic of the North Sea are all characterised palynologically by a dominance of terrestrially derived palynomorphs or 'sporomorph eco-groups' (SEGs) (Jones, 1996, 2006; Abbink, 1998; Abbink *et al.*, 2004a, b; Traverse, 2008).

Highland and lowland pollen, fungal spores derived from soils, fern spores derived from riverine habitats, and freshwater algae are all represented. Grass cuticle and pollen is often common, and often charred, suggesting widespread development of at least seasonally dry climatic conditions subject to bush fires. Palynomorph abundance is typically high, and diversity low, indicating environmental stress.

LSTs in the earliest Cretaceous Barrow Group sequences of the northwest shelf of Australia are characterised by zero to moderate microfaunal

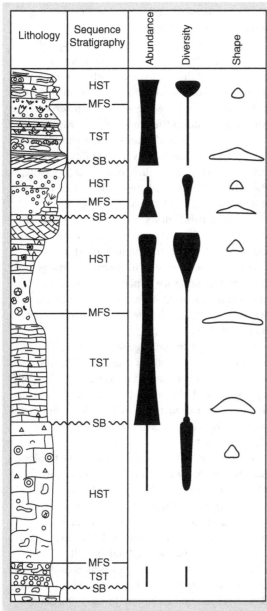

Fig. 4.7 **Micropalaeontological characterisation of systems tracts in carbonate sequences, 'Middle' Cretaceous, Middle East.** © Grzybowski Foundation. Reproduced with permission from Jones *et al.*, in Bubik and Kaminski (2004). SB, sequence boundary; TST, transgressive systems tract; MFS, maximum flooding surface; HST, high-stand systems tract.

abundance and diversity, and by low to high palynofloral diversity, leastwise on the palaeo-prodelta. Agglutinated and calcareous benthic foraminifera occur in approximately equal proportions. The percentage of terrestrially derived palynomorphs is generally >90% at proximal sites, and 60–90% at distal sites.

LSTs in the Early Jurassic mixed clastic–carbonate sequences of Quercy in southwest France are characterised micropalaeontologically by low abundances and diversities of benthic foraminifera, and by upwardly increasing proportions of uncoiling morphotypes of *Lenticulina*.

Transgressive surfaces. TSs in clastic sequences are characterised by condensed sections (CSs), and typically by abundant and diverse fossil assemblages. They are further characterised by the juxtaposition of more distal assemblages over more proximal ones, irrespective of location along depositional dip.

TSs atop low-stand fans in the post-rift Cretaceous sequences of the Pletmos and Bredasdorp basins of South Africa are characterised micropalaeontologically by an acme of the foraminifer *Trochammina*, a possible phytodetritivore.

TSs – and TSTs – in the Early Cretaceous of the Bredasdorp basin of South Africa are characterised by palynomorph abundance maxima and diversity minima. This is attributable to the blooming of opportunistic species in newly created niches.

Transgressive systems tracts. TSTs in clastic sequences are characterised by overall retrogradation, and by fossil and trace fossil assemblages indicating an upward deepening (Olson & Thompson, in Koutsoukos, 2005). Those in coaly sequences are characterised by palynomorph assemblages indicating an upward transition from swamp, through fen, marsh and salt-marsh to marine (Holz *et al.*, 2002).

TSTs in the Neogene of southeast Asia are characterised micropalaeontologically by 'clear-water' calcareous benthic foraminifera, including larger benthic foraminifera (LBFs) thought to be adapted to oligotrophic conditions (Jones, 1996, 2006; Fig. 4.5). Those in the post-rift

Cretaceous sequences of the Pletmos and Bredasdorp basins of South Africa are characterised by upward increases in abundance of deep-water foraminifera. Those in the Late Cretaceous of the Basco-Cantabrian basin in Spain are characterised by the foraminifer *Tritaxia*.

TSTs in the Plio-Pleistogene of the Gulf of Mexico, the Neogene–Pleistogene of the tropics, the latest Jurassic to earliest Cretaceous of the Vocontian basin of southeast France, the Jurassic of the North Sea, and the Early Carboniferous of Britain are all characterised palynologically by terrestrially derived to, increasingly, marine palynomorphs or SEGs (Jones, 1996, 2006; Morley, in Caughey *et al.*, 1996; Abbink, 1998; Abbink *et al.*, 2004a, b; Traverse, 2008). Highland and dry lowland palynomorphs are typically rare; and wet lowland forms such as coastal mangrove and palm pollen, and freshwater and marine forms, including dinoflagellates and acitarchs, typically common. The angularity as well as the abundance of terrestrially derived plant fragments typically decreases up-section.

TSTs in the earliest Cretaceous Barrow group sequences of the northwest shelf of Australia are characterised by zero microfaunal abundance and diversity, and low to moderate palynofloral diversity.

TSTs in carbonate sequences in the 'Middle' Cretaceous of the Middle East and central Asia are characterised micropalaeontologically by upward increases in incidence of orbitolinid LBFs and planktonic foraminifera (Fig. 4.7).

In the Middle East, they are also typically characterised by discoidal or flattened orbitolinids (Simmons *et al.*, in Hart *et al.*, 2000). Note, though, that analogy with modern LBFs indicates that the flattening of the orbitolinids probably represents an adaptation to the maximisation of exposure to light to enable interpreted hosted algal symbionts to photosynthesise, and as such is determined not so much by water depth as by water clarity.

TSTs in the Early Jurassic mixed clastic–carbonate sequences of Quercy in southwest France are characterised by moderate abundances and diversities of benthic foraminifera, and by an upward increase in the proportion of uncoiling morphotypes of *Lenticulina*.

Maximum flooding surfaces. MFSs are typically characterised by condensed sections, and by abundant and diverse fossil assemblages, including a wide range of macrofossils as well as microfossils, as in the Palaeozoic, Mesozoic and Cenozoic of the Middle East and in the Marine Bands (MBs) of the Carboniferous Coal Measures of mainland Great Britain and the southern North Sea (Sharland *et al.*, 2001; Cole *et al.*, in Collinson *et al.*, 2005). Note, though, that high fossil abundance and diversity is not in itself diagnostic of MFSs (it can also characterise debris flows, for example). MFSs are diagnosed by fossil assemblages indicating maximum flooding: for example maximum bathymetry; and, at least in downdip settings, the maximum abundance of plankton, enabling calibration against global standard biozonation schemes established in the open ocean (Raju, in Raju *et al.*, 2005).

MFSs in the Plio-Pleistogene of the Gulf of Mexico, in the Neogene of southeast Asia and elsewhere are characterised micropalaeontologically by benthic and planktonic foraminiferal abundance and diversity maxima, and by benthic foraminiferal palaeobathymetric maxima (Jones, 1996, 2006; Figs. 4.3 and 4.7).

MFSs in the Neogene–Pleistogene of the tropics, the Miocene of central Tunisia, the Oligocene–Miocene of Denmark, the Early Cretaceous of the Bredasdorp basin of South Africa, the latest Jurassic to earliest Cretaceous of the Vocontian basin of southeast France, the Jurassic of the North Sea and the Early Carboniferous of Britain are all characterised palynologically by peak abundances of coastal and marine palynomorphs or SEGs, including dinoflagellates and acritarchs (Jones, 1996, 2006; Abbink, 1998; Abbink *et al.*, 2004a, b; Traverse, 2008).

MFSs in the earliest Cretaceous Barrow group sequences of the northwest shelf of Australia are characterised by zero to moderate microfaunal abundance and diversity, and by low to moderate palynofloral diversity. Agglutinated and calcareous benthic foraminifera occur in approximately equal proportions. MFSs in the post-rift Cretaceous sequences of the Pletmos and

Bredasdorp basins of South Africa are characterised by microfaunal and palynofloral and diversity peaks, and by an acme of the deep-water foraminifer *Verneuilina*.

MFSs in carbonate sequences in the Early Cretaceous of the Middle East and central Asia are characterised micropalaeontologically by peaks in abundance of planktonic foraminifera (Fig. 4.7).

MFSs in the Early Jurassic mixed clastic–carbonate sequences of Quercy in southwest France are characterised by peak abundances and diversities of benthic foraminifera, and of uncoiling morphotypes of *Lenticulina*.

High-stand systems tracts. HSTs in clastic sequences are characterised by overall progradation, and by fossil and trace fossil assemblages indicating an upward shoaling (Moss *et al.*, 2004; Olson & Thompson, in Koutsoukos, 2005). Those in coaly sequences are characterised by palynomorph assemblages indicating an upward transition from marine, through salt-marsh, marsh and fen to swamp (Holz *et al.*, 2002).

HSTs in the Neogene of southeast Asia are characterised micropalaeontologically by 'turbid-water' calcareous benthic foraminifera, including buliminides, thought to be adapted to eutrophic conditions (Fig. 4.5). Those in the post-rift Cretaceous sequences of the Pletmos and Bredasdorp basins of South Africa are characterised by upward decreases in deep-water foraminifera such as *Verneuilina* and radiolarians such as *Dictyomitra*, and increases in shallow-water foraminifera such as *Conorotalites*. Those in the Late Cretaceous of the Basco-Cantabrian basin in Spain are characterised by the foraminifera *Praebulimina*, *Arenobulimina* and *Frondicularia*.

HSTs in the Pleistogene of the Amazon Fan, the Plio-Pleistogene of the Gulf of Mexico, the Neogene–Pleistogene of the tropics, the Miocene of central Tunisia, the latest Jurassic to earliest Cretaceous of of Papua New Guinea and the Vocontian basin of southeast France, and the Jurassic of the North Sea are all characterised palynologically by marine to, increasingly, lowland and highland terrestrially derived palynomorphs or SEGs (Jones, 1996, 2006; Morley,

in Caughey *et al.*, 1996; Abbink, 1998; Abbink *et al.*, 2004a, b; Traverse, 2008). The angularity as well as the abundance of terrestrially derived plant fragments typically increases up-section (both through the HST as a whole and through individual constituent coarsening-upward or progradational parasequences). Early HSTs in the Plio-Pleistogene of the Gulf of Mexico are characterised by pollen of warm/wet aspect; late HSTs by pollen of cool/wet aspect. Early HSTs in the Neogene–Pleistogene of the tropics are characterised by terrestrially derived Rubiaceae and Euphorbiaceae (open forest and swamp plant) pollen; late HSTs by Gramineae (grass) or *Casuarina* (a littoral plant) pollen.

HSTs in the earliest Cretaceous Barrow group sequences of the northwest shelf of Australia are characterised by zero to moderate microfaunal abundance and diversity, and by low to high palynofloral diversity. Agglutinated and calcareous benthic foraminifera occur in approximately equal proportions. The percentage of terrestrially derived palynomorphs is generally 60–90% at both the proximal site and the distal site.

Early HSTs in carbonate sequences in the Early Cretaceous of the Middle East and central Asia are characterised by algal (*Lithocodium*) reef facies; late HSTs by aggradational to progradational rudist–coral reef facies, orbitolinid–valvulinid–echinoderm wackestone and miliolid–dasycladacean (*Hensonella*) wackestone–packstone back-reef facies. In the Middle East, they are also typically characterised by conical orbitolinids (Simmons *et al.*, in Hart *et al.*, 2000) (Fig. 4.7).

HSTs in reef sequences in the Oligo-Miocene of the Horn of Africa are characterised by a succession of facies from shelf-slope through deep shelf, deep fore-reef and fore-reef to reef (Bosellini *et al.*, 1987). Shelf-slope facies are characterised by mudstones and by breccias containing transported corals (*Leptoporia africana*, *Porites*), red Algae (rhodoliths) and LBFs (*Lepidocyclina*). Deep shelf, deep fore-reef and fore-reef facies are characterised by LBFs (*Lepidocyclina*), and locally by bivalve molluscs (*Chlamys labadyei*). Reef facies are characterised by diverse corals (initially *Hydnophora regularis*, and later *Goniopora microsideria*, *Hydnophora insignis*, *H. solidor*, *Favia africana*, *F. preamplior*, *Plesiastrea grayi*, *Diploastrea haimei*, *Leptoporia africana*, *L. concentrica* etc.).

HSTs in the Early Jurassic mixed clastic–carbonate sequences of Quercy in southwest France are characterised by moderate abundances and diversities of benthic foraminifera, and by an upward decrease in the proportion of uncoiling morphotypes of *Lenticulina*.

4.7 CHRONOSTRATIGRAPHIC DIAGRAMS

Chronostratigraphic diagrams describe and predict the distributions of sequences of rocks, and especially of source-, reservoir- and cap-rocks, in time, conventionally displayed on the vertical axis, and in space, conventionally displayed on the horizontal axis (see Fig. 4.1B above; see also Figs. 5.41 and 5.43). Distributions in time are constrained in part by biostratigraphy; and distributions in space in part by palaeobiology. SBs, MFSs and STs are constrained in part by both biostratigraphy and palaeobiology.

5 • Petroleum geology

This chapter deals with petroleum geology. Readers interested in further details are referred to Selley (1998), Gluyas and Swarbrick (2004), Tabak (2009) and Bjorlykke (2010). Those interested in further details of the environmental impact of petroleum geology are referred to Chapter 9.

In petroleum geology, the principal application of palaeontology is in the description and prediction of source-, reservoir- and cap-rock distributions, in time and in space, in a sequence stratigraphic framework, through the biostratigraphic age-dating and the palaeobiological, palaeoecological or palaeoenvironmental interpretation of samples acquired during field geological mapping and drilling; and thereby in the reduction of risk and of uncertainty, all along the value chain from regional exploration (see Section 5.2, below) to reservoir exploitation (see Section 5.3, below). It also has important applications in well-site operations (see Section 5.4, below).

5.1 PETROLEUM SOURCE-ROCKS AND SYSTEMS, RESERVOIR-ROCKS, AND CAP-ROCKS AND TRAPS

There are essentially three components to a successful play, namely, a working petroleum source-rock and system, to provide the petroleum charge; a reservoir-rock, to store, and ultimately to flow, same; and a cap-rock and trap, to prevent it from escaping (Jones, 2006).

5.1.1 • Petroleum source-rocks and systems

Petroleum source-rocks

Petroleum source-rocks are those from which petroleum is derived (Jones, 2006). They typically consist of abundant organic material or kerogen, which can be of either terrestrial land plant or humic, or marine algal or sapropelic, origin. This is transformed, or matured, into oil and gas by heating on burial. The product will depend on the nature of the source-rock, and on the transformation ratio, or degree of maturation, in turn determined by burial history and by basin type. Typically, oil is generated from

so-called 'type I' or 'type II' source-rocks, whose organic material is of lacustrine or marine algal origin; and gas from 'type III' source-rocks, whose organic material is of terrestrial land plant origin. Note, though, that gas can be generated from 'type I' or 'type II' source-rocks subjected to a high degree of maturation; and that oil can be generated from 'type III' source-rocks such as – perhydrous – coals (Khorasani, in Scott, 1987; Murchison, in Scott, 1987; Fleet & Scott, in Scott & Fleet, 1994; Poinar et al., 2004). Maturation – and migration – can be modelled using various publicly available and proprietary 1D, 2D and 3D basin-modelling software packages.

Petroleum systems

A petroleum system has been variously defined as 'a dynamic petroleum generating and concentrating system functioning in geological space and time', or 'a pod of active source rock and the resulting oil and gas accumulations' (Jones, 2006). The petroleum system concept is 'a means of formalising the relationship between the geologic elements in time and space that are required for the development of a commercial petroleum accumulation'.

Petroleum systems analysis is concerned with the presence of these elements, that is, source-, reservoir- and cap-rock and trap, and with their effectiveness, and may also be said to be concerned with how much petroleum a basin has generated, where it has migrated, and where it is trapped, and in what quantity and phase. The relationships between the elements are conventionally presented in the form of a series of structural cross-sections showing not only the fill of the basin but also the fluid flow therein, and, importantly, the progression of the petroleum migration front through time, and hence the limits of the system. An understanding of the limits of petroleum systems within basins is critical to an evaluation of their potential prospectivity, with exploration risk increasing in proportion to the distance from an established petroleum system. For example, in the North Sea basin, where migration is almost exclusively vertical, the limit of prospectivity is essentially controlled by the limit of the mature Kimmeridge clay formation source-rock. An understanding of the certainty with which limits can be defined is also important. For example, in the Campos basin, offshore Brazil, the limit of the petroleum system and of prospectivity was recently significantly extended when advances in drilling technology enabled exploration in deep water for the first time.

Petroleum systems are conventionally named after the source- rather than the reservoir-rock, since one source may charge only one or more than one reservoir. There may be only one petroleum system in a basin, or there may be more than one, in which case they may be stratigraphically or geographically discrete, or they may be overlapping. Identification of individual petroleum systems in basins where more than one is operative requires typing of reservoired petroleum to the source-rock. This is nowadays relatively easily achievable using a range of geochemical techniques including, for example, biomarker and diamondoid analysis.

There are certain empirical observations with regard to the distribution of oil and gas reserves within a basin that constitute general rules (although there are also specific exceptions, especially when

---

BOX 5.1   PALAEONTOLOGICAL INPUTS INTO PETROLEUM SYSTEMS ANALYSIS (MODIFIED AFTER JONES, 2006)

This section includes sub-sections on biostratigraphy and chronostratigraphy; palaeobiology and palaeoenvironmental interpretation; and thermal alteration indices. The sub-section on palaeobiology and palaeoenvironmental interpretation covers palaeobathymetry and palaeotemperature. Palaeobiological controls on – biogenic – source-rock development are discussed on pp. 277–8.

Biostratigraphy and chronostratigraphy

One key input in petroleum systems analysis is absolute age data, which is used to establish the ages of the source-rock and overburden, and, critically, to model the timing of migration as against that of trap formation. Chronostratigraphic data is in turn an output of routine palaeontological or biostratigraphic analysis, or of graphic correlation. Resolution is typically best – better than 1 Ma – in marine environments. However, unconventional biostratigraphic and/or non-biostratigraphic techniques can also provide precise and accurate ages in non-marine environments, for example vertebrate palaeontology and magnetostratigraphy in the 'molasse' sequences of Transcaucasia and Iran.

Interestingly, some sort of stratigraphy can also be established on the basis of 'biomarkers'. For example, C11–C19 paraffins derived from the cyanophyte *Gloeocapsomorpha prisca* indicate an essentially

Ordovician age. Nordiacholestane, a probable diatom derivative, indicates a Late Palaeozoic or younger age (and, when in abundance, Cretaceous or younger) (Kooistra *et al.*, in Falkowski & Knoll, 2007). Highly branched isoprenoid alkanes, probable rhizosolenid diatom derivatives, indicate a Cretaceous or younger age (Damste *et al.*, 2004; Fildani *et al.*, 2005; Kooistra *et al.*, in Falkowski & Knoll, 2007). Oleanane, a higher land plant derivative, indicates a Cretaceous or younger age (and, when in abundance, Cenozoic) (Moldowan *et al.*, 1994).

Palaeobiology and palaeoenvironmental interpretation

Another key input in petroleum systems analysis is palaeoenvironmental interpretation based on palaeobiology, which is used to infer palaeobathymetry and palaeotemperature for modelling purposes. Lithology away from areas of control is inferred from a descriptive and predictive sequence stratigraphic model into which palaeoenvironmental interpretation has been integrated.

Palaeobathymetry. Palaeobathymetric interpretation – essentially based on proxy living benthic foraminiferal distribution data – enables the characterisation of a range of depth environments. Resolution is typically good enough to enable recognition of the following depth environments: non-marine (obviously, on the basis of freshwater Algae, diatoms and ostracods, terrestrially derived spores and pollen, or land plants and animals, rather than foraminifera); marginal marine; shallow

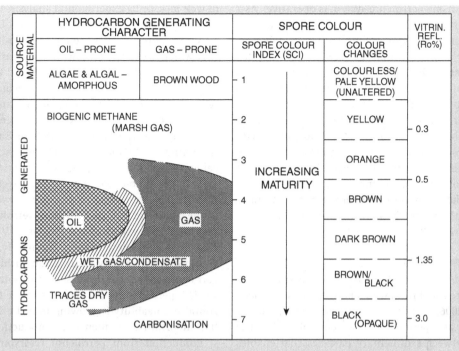

SOURCE MATERIAL	HYDROCARBON GENERATING CHARACTER		SPORE COLOUR		VITRIN. REFL. (Ro%)
	OIL – PRONE	GAS – PRONE	SPORE COLOUR INDEX (SCI)	COLOUR CHANGES	
	ALGAE & ALGAL – AMORPHOUS	BROWN WOOD	1	COLOURLESS/ PALE YELLOW (UNALTERED)	
GENERATED / HYDROCARBONS	BIOGENIC METHANE (MARSH GAS)		2	YELLOW	0.3
			3	ORANGE	0.5
	OIL	GAS	4	BROWN	
			5	DARK BROWN	
	WET GAS/CONDENSATE		6	BROWN/ BLACK	1.35
	TRACES DRY GAS				
	CARBONISATION		7	BLACK (OPAQUE)	3.0

INCREASING MATURITY

Fig. 5.1 **Use of palynofacies in thermal maturity assessment.** From Jones (1996).

marine, inner shelf or neritic (0–50 m); middle shelf (50–100 m); outer shelf (100–200 m); deep marine, upper slope or bathyal (200–1000 m); middle slope (1000–1500 m); lower slope (1500–2000 m); and abyssal plain (>2000 m). (Note that the resolution is of the order of 10s of metres in shallow marine environments, but only of the order of 100s or even 1000s of metres in deep marine environments.) Importantly, high-resolution palaeobathymetric curves can be constructed and imported into work-station environments through software packages (currently proprietary but hopefully ultimately publicly available) containing links to bathymetric distribution datasets or 'look-up' tables.

Additionally, essentially depth-independent dysoxic and anoxic environments can be characterised micropalaeontologically or palynologically (see 3.2.2 and 3.3.3 above). For example, oxygen-poor upper-slope oxygen minimum zones (OMZs), and submarine canyon sites rich in refractory organic-carbon, can be characterised by infaunal buliminide (bolivinid–buliminid–uvigerinid) benthic foraminiferal assemblages.

Palaeotemperature. Palaeobiogeographic data can provide indications as to the palaeolatitude, palaeoclimate and palaeotemperature history of a basin. For example, the palaeotemperature history of the Middle East and north Africa has been established partly on the basis of biological proxies, such as essentially tropical larger benthic foraminifera.

Thermal alteration indices

An independent measure of thermal maturity is provided by the 'spore colour index' (Jones, 1996; Fig. 5.1).

It is also provided by the acritarch and chitinozoan colour indices (Marshall, in Briggs & Crowther, 1990), the foraminiferal colour index (McNeil et al., 1996), the ostracod colour index (Ainsworth et al., 1990), and the conodont colour index (Epstein et al., 1977; Voldman et al., 2010). The spore, foraminiferal and ostracod colour indices have been independently calibrated against vitrinite reflectance. The acritarch, chitinozoan and conodont colour indices have not, as these groups of organisms substantially pre-date the land plants that constitute the source of the vitrinite.

more than one petroleum system is operative). As a general rule, oil gravity decreases with depth. In other words, heavy oil is found at shallow depths, light oil and condensate at intermediate depths, and gas at depth. This is a function of the combined effects of source-rock maturity and of biodegradation, or rather the lack thereof at depth, where temperature conditions are inimical to the Bacteria responsible. Also as a general rule, oil gravity decreases towards the centre of the basin. In other words, heavy oil is found at the basin margin, light oil and condensate in an intermediate position, and gas in the basin centre (where it can sometimes apparently be stratigraphically rather than structurally trapped).

5.1.2   Reservoir-rocks

Reservoir-rocks are those in which petroleum is reservoired, and from which petroleum is produced (Jones, 2006). They require two properties, namely porosity and permeability. *Porosity* (phi, φ) is the space between sedimentary particles that enables the reservoir to store pore fluid, including petroleum. It is measured by point counting of petrographic thin-sections or by petrophysical analysis

of well logs, or by experimental analysis. Values are quoted as percentages or porosity units. Obviously, only open pores constitute effective porosity, that is, that which is capable of flowing hydrocarbons. Porosity can be either primary, and related to the original depositional facies or fabric, or secondary, and related to diagenesis. Primary porosity can be either enhanced or occluded by secondary diagenetic processes. Typically, dissolution, dolomitisation and fracturing are enhancive, and compaction and cementation occlusive, processes. Importantly, an early hydrocarbon charge can cause effective cessation of occlusive diagenetic processes. *Permeability* (k) is the ability of the rock to flow petroleum. It is measured by experimental analysis. Values are measured in darcies, or, more commonly, millidarcies (md). Permeability distribution within a reservoir is commonly characterised by anisotropy, with vertical permeability (k_v) typically lower than horizontal permeability (k_h), owing to bed effects. This has important consequences for production.

Production from petroleum reservoirs can be enhanced by the application of so-called secondary and tertiary recovery technologies (see Section 5.3 below).

BOX 5.2   NUMMULITE BANKS AND RESERVOIRS

Nummulites or 'coin-stones' constitute an extant sub-group of hyaline rotaliide larger benthic foraminiferans or LBFs, ranging in age from Late Palaeocene to Recent. The best known of the approximately 300 nominal species, *Nummulites gizehensis*, occurred in sufficient abundance in the Middle Eocene of Egypt as to have formed the building stone of which the Great Pyramids of Gizeh were later built.

Palaeobathymetry. Modern nummulites host diatom symbionts, and thus characterise sub-euphotic, commonly 'reefal' and 'fore-reefal' carbonate environments.

Palaeobiogeography. Modern nummulites characterise low latitudes. Ancient ones exhibited essentially Pan-Tethyan distributions. The most northerly occurrences were on Rockall Bank in the Late Palaeocene ('*Nummulites*' *rockallensis*), and in the Hampshire basin in the Eocene (*Nummulites*

planulatus, N. laevigatus, N. variolarius, N. prestwichianus and *N. rectus*).

Nummulite banks

As intimated above, nummulites were locally sufficiently abundant, essentially throughout the Early–Middle Eocene of the circum-Mediterranean, as to have formed accumulations called 'banks' (Racey, in Al-Husseini, 1995; Racey, 2001; Papazzoni, 2010). Nummulite banks are constituted of variable proportions of unbroken nummulites and other microfossils, broken and fragmented nummulites ('nummulithoclastic debris'), lime mud and cement. They tend to be dominated either by comparatively large individuals of a single species, generally 'B' forms generated by sexual reproduction, or by small individuals, generally 'A' forms generated by asexual reproduction. Carbonate sedimentologists have historically tended to interpret the size sorting as evidence of hydrodynamic sorting, the

'B'-form-dominated accumulations as winnowed and essentially autochthonous, and the 'A'-form-dominated accumulations as re-deposited and allochthonous (Aigner, in Einsele & Seilacher, 1982; Aigner, 1983, 1985). Note in this context that the hydrodynamic behaviour of nummulites, as demonstrated by flume-tank experiments, is more closely comparable to that of silt than of comparably sized sand particles, on account of their high internal porosities (a function of their functional morphology), and low pick-up velocities (Futterer, in Einsele & Seilacher, 1982; Briguglio & Hohenegger, 2009). (Note also, though, that it is also possible to interpret 'A'-form-dominated accumulations as representing unmodified death assemblages of populations reproducing exclusively asexually under conditions of extreme environmental stress, and as reflecting reproductive strategy rather than taphonomy: Wells, 1986; Loucks *et al.*, 1998, in MacGregor *et al.*, 1998.)

Nummulite reservoirs

Nummulite banks are locally of commercial significance as petroleum reservoirs: for example in the Bourri field, offshore Libya; in the Ashtart and Hasdrubal fields, offshore Tunisia; and in the Kerkennah and adjacent fields, onshore Tunisia (Racey, 2001). Unfortunately, the reservoirs are not easy to exploit, partly because the permeability distribution within, and therefore the producibility of, nummulite banks is difficult to describe and predict. Permeability distribution within nummulite bank reservoirs appears to be controlled by a complex combination of primary, depositional and secondary, diagenetic factors (Loucks *et al.*, in MacGregor *et al.*, 1998; Carrillat *et al.*, 1999; Anketell & Mriheel, 2000; Racey, 2001; author's unpublished observations). Primary permeability is low (<10 md) in most nummulite bank reservoirs, in which the primary porosity is typically *within* individual nummulites (intra-particular), and therefore unconnected and ineffective; although it can also be high (>1000 md), for example in high-energy, well-sorted 'B'-form-dominated grainstones, in which the primary porosity is atypically *between* individual nummulites (inter-particular), and therefore connected and effective. Secondary enhancement of primary

permeability has been described as having been brought about by dolomitisation, dissolution or leaching, and fracturing; secondary occlusion by compaction, pressure solution, stylolitisation and cementation. Importantly, the occlusive effects of cementation can be negated by an early emplacement of hydrocarbon, and actually reversed by dissolution by acidic basinal fluids associated with the emplacement of hydrocarbon.

Bourri field, Libya. Bourri field is situated in the Pelagian basin, offshore northwestern Libya (Bernasconi *et al.*, in Salem & Belaid, 1991; Anketell & Mriheel, 2000; Racey, 2001). The Early Eocene Jdeir formation part of the reservoir is approximately 125–175 m thick, and is interpreted as a nummulite bank, deposited in a rimmed shelf setting. Porosity within the nummulite bank ranges up to 21%, and averages 16%. It is interpreted as having been secondarily enhanced primarily by dissolution or leaching. Oil reserves for the Bourri field have been quoted as within the range 1–3 billion barrels.

Ashtart field, Tunisia. Ashtart field is situated in the Gulf of Gabes, offshore eastern Tunisia (Anz & Ellouz, 1985; Racey, 2001). The Early Eocene El Garia formation reservoir is approximately 90 m thick, and is interpreted as a re-deposited and allochthonous nummulite bank. Porosity within the cored part averages 17.2%. Permeability ranges from 0.2 to 1000 md, and averages 50 md. Both are interpreted as having been secondarily enhanced by dissolution and by fracturing. Oil production rates have been quoted as approximately 3000 barrels per day per well. Reserves have been quoted as within the range 350–400 million barrels.

Hasdrubal field, Tunisia. Hasdrubal field is also situated in the Gulf of Gabes, offshore eastern Tunisia (Macaulay *et al.*, 2001; Racey, 2001; Racey *et al.*, 2001). The Early Eocene El Garia formation reservoir is approximately 90 m thick, and is interpreted as a re-deposited and allochthonous nummulite bank. Porosity ranges from 0.5 to 23%, and averages 10.5%. Permeability ranges from 0.02 to 60 md, and averages 0.5 md. Both are interpreted as having been secondarily enhanced by dolomitisation and by fracturing, and also secondarily occluded by cementation. Gas production rates have been quoted on the BG website as within the range 5–20 million standard cubic feet per day per well. Reserves have been quoted also on the BG website as 250 billion cubic feet of gas and 20 million barrels of oil/condensate.

Kerkennah and adjacent fields, Tunisia. Kerkennah and adjacent fields are situated onshore eastern Tunisia (Racey, 2001). The Early Eocene El Garia formation reservoir is interpreted on the basis of analogy with equivalent three-dimensionally exposed outcrops elsewhere in eastern Tunisia as a nummulite bank, deposited in a mid-ramp setting, below fair-weather wave-base but above storm wave-base (Moody, in Hart, 1987; Loucks *et al.*,

1998, in MacGregor *et al.*, 1998; Racey, 2001; Jorry *et al.*, 2003a, b; Vennin *et al.*, 2003; Beavington-Penney *et al.*, 2005). The outcrops demonstrate that the grain-prone mid-ramp bank facies passes updip into a mud-prone inner ramp back-bank facies, and downdip into a mud-prone outer ramp fore-bank and basinal facies, within less than 3 km (less than the conventional well spacing). Oil production rates for the Kerkennah field have been quoted as approximately 300 barrels per day per well. Reserves for the Kerkennah and adjacent fields have been quoted as <100 million barrels of oil.

BOX 5.3   RUDIST 'REEFS' AND RESERVOIRS (MODIFIED AFTER JONES, 2006)

Rudists constitute an extinct, Late Jurassic–Cretaceous, sub-group of heterodont bivalves, characterised by the attachment of one valve and the uncoiling of the other, and by accretion along the entire mantle margin thereof, and hence the construction of aberrant, essentially tubular, shells, often of large size, with lid-like upper valves (Jones, 2006; Fig. 5.2).

Requienid rudists were attached by the left valve; caprotinids, caprinids, radiolitids and hippuritids by the right valve. The novel technique of X-ray computed tomography (CT) has enabled the internal structure and ontogenetic development of caprinid rudists to be elucidated without the former need for sectioning and thus destruction of specimens (Molineux *et al.*, 2007).

Palaeobathymetry. Rudists are interpreted, essentially on the bases of associated fossils and sedimentary facies, functional morphology, and

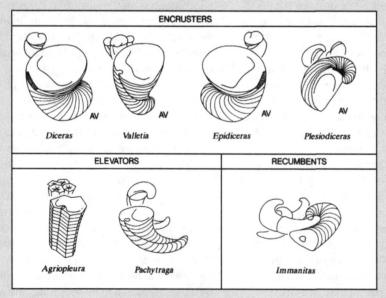

Fig. 5.2 **Hard-part morphology of rudist bivalves**. From Jones, 2006 (reproduced with the permission of Wiley from Doyle, 1996).

Fig. 5.3 **Rudist reef, Late Cretaceous, Saiwan, Oman Mountains**. From Jones (2006).

analogy with the living 'fan-mussel' *Pinna* of the Great Pearl Bank Barrier in the Middle East Gulf, as having been shallow marine, benthic and epifaunal, either embedded in or reclining or recumbent on the sediment surface, clinging to or encrusting it, or elevated above it. It has been hypothesised that they harboured algal symbionts (Kauffman & Johnson, 1988), although the hypothesis has not been universally accepted (Rojas *et al.*, 1995). Worked examples of rudist-based palaeobathymetric interpretation and palaeocommunity reconstruction in the peri-tidal carbonate environments of the Early Cretaceous, 'Urgonian' of Provence in southeast France have been provided in the literature. Quantitative estimates of rudist 'reef' body dimensions and distributions, using transition probabilities (Markov chain analysis), have also been provided (Fenerci-Masse *et al.*, 2005).

Palaeobiogeography. Rudists exhibited essentially Pan-Tethyan distributions, and indeed their occurrence serves to distinguish the Tethyan from the Boreal realm. Some sub-groups exhibited much more restricted distributions, related to continental plate reconfigurations, such that their occurrence serves to distinguish, for example, Afro-Arabian–Indian and American–Caribbean–eastern Pacific provinces within the Tethyan realm (the eastern Pacific and Caribbean having been

contiguous prior to the closure of the Strait of Panama in the Cenozoic). The occurrence of some species in both the Afro-Arabian–Indian and the American–Caribbean–eastern Pacific provinces has been attributed to locally widespread dispersal of larvae by surface currents (Philip, 2003). Incidentally, the most northerly occurrence of a rudist known to the author is that of *Durania borealis* in the Late Cretaceous, Cenomanian part of the Hibernian greensand of Portmuck, near Islandmagee in Co. Antrim in Northern Ireland, some specimens of which are housed in the Ulster Museum in Belfast. This species is colloquially known as the 'cornetto animal' by the sweet-toothed Northern Irish, of whom my wife is one, in reference to its resemblance to the ice-cream cone of that name!

Rudist 'reefs'

Rudists were locally sufficiently abundant in shallow marine environments as to have constructed so-called 'reefs', with elevator morphotypes providing the framework, and encrusters and recliners the binding, as in the circum-Mediterranean and Middle East, and also the southern United States, Central and northern South America, and the Gulf of Mexico and Caribbean at the western end of Tethys (Al-Ameri & Lawa, 1986; Masse & Fenerci-Masse, 2008; Fig. 5.3).

Note, though, that it is questionable whether these constructions actually represent 'reefs' in the conventionally accepted sense, as they are essentially sediment-supported rather than bound, and constratal or low-relief rather than superstratal or high-relief, and probably did not form wave-resistant frameworks: it is perhaps more appropriate to visualise them as 'meadows'. Nonetheless, it is unquestionable that the volume of carbonate produced by the rudist 'factory', if I might mix industrial and agricultural metaphors, was significant.

Interestingly, it has been observed that corals seldom occur in rudist 'reefs'. On the basis of this observation, it has been hypothesised that corals were competitively excluded from the 'reefs' by rudists. However, it has also been counter-hypothesised that the two simply had different ecological preferences or tolerances, especially as regards depth or substrate (Masse & Fenerci-Masse, 2008).

Rudist reservoirs

Rudist 'reefs' are of considerable importance as hosts to mineral deposits, for example in the Early Cretaceous, 'Urgonian' of the La Troya lead–zinc mine in the Basco-Cantabrian basin in the Basque country of Spain (see 6.1.1 below).

Rudist 'reefs' are also of considerable commercial importance as petroleum reservoirs in the Cretaceous of, for example, Mexico and the Middle East. In Mexico, 'Middle' Cretaceous El Abra formation rudist reefs form important reservoirs in the Faja del Oro, or Golden Lane, trend; and Tamabra formation fore-reef facies, reservoirs in the Poza Rica trend (Enos, in Roehl & Choquette, 1985; Moran-Zenteno, 1994; Garcia-Barrera, in Vega *et al.*, 2006). In the Middle East, Neocomian, Ratawi formation rudist reefs form important reservoirs in the Wafra field in the Kuwait–Saudi Arabia partitioned neutral zone; Aptian, Shu'aiba formation reefs, reservoirs in the Bu Hasa field in

Fig. 5.4 **Photomicrograph of rudist debris, Mauddud formation reservoir, north Kuwait.** From Jones (2006).

Abu Dhabi in the United Arab Emirates (UAE) and in the Shaybah field in Saudi Arabia; Albian, Mauddud formation reefs, reservoirs in the Raudhatain and Sabiriyah fields in Kuwait and in the Fahud field in Oman; Cenomanian, Mishrif formation reefs, reservoirs in the Fateh field in Dubai in the UAE, the Fahud field in Oman, and numerous fields in Iran; and Campanian–Maastrichtian, Bekhme and Aqra Limestone formation reefs, reservoirs in Iraq (Al-Ameri & Lawa, 1986; Sadooni, 2005; Aqrawi *et al.*, 2010b; Cross *et al.*, 2010). Interestingly, a not insignificant proportion of the porosity, or storage capacity, in these and other reservoirs is in rudist or in rudist debris facies: radiolitid debris is particularly porous, although it only occurs in reservoirs of Aptian and later age, as the radiolitids had not evolved earlier (Fig. 5.4).

Importantly, rudists are of use in the characterisation of many of the aforementioned reservoirs (Hughes, in Bubik & Kaminski, 2004; Jones *et al.*, in Bubik & Kaminski, 2004; Hughes, in Powell & Riding, 2005; see also 5.3.1 below). They are recognisable in cores, in CT scans of cores, and on image logs.

5.1.3 Cap-rocks and traps

Cap-rocks

Cap-rocks are those beneath which petroleum is trapped (Jones, 2006). They are also known as seals. They require a low permeability and a high sealing capacity, quantified using capillary entry pressure measurements. They can be of any lithology, although mudstones are the commonest, and evaporites such as anhydrite and halite the most effective.

BOX 5.4   PALAEONTOLOGICAL CHARACTERISATION OF CAP-ROCKS (MODIFIED AFTER JONES, 2006)

Introduction

Various types of mudstones from the Oligo-Miocene sections of approximately 30 wells from offshore Angola have been analysed micropalaeontologically, in order to establish whether those that constitute barriers or baffles to fluid flow can be characterised.

Results

Condensed sections (CSs) in the Oligo-Miocene of offshore Angola, interpreted as representing hemipelagites, have been distinguished from expanded sections, interpreted as representing turbidites, on the basis of rock accumulation rate data derived from thickness/age relationships and/or graphic correlation (Fig. 5.5).

The former have been distinguished essentially on the basis of rock accumulation rates of less than

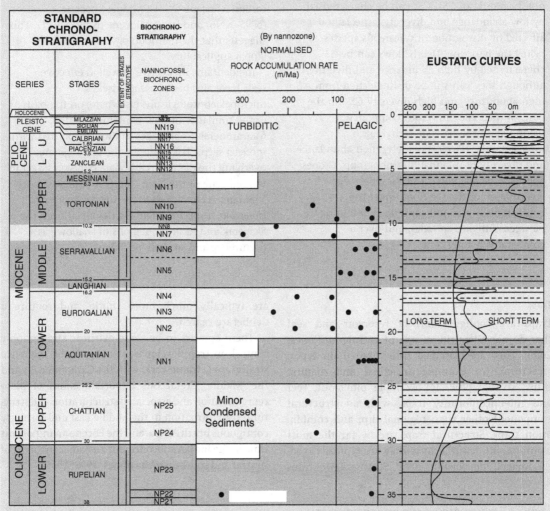

Fig. 5.5 **Mud-rock seal capacity, Cenozoic, Angola.** From Jones (2006). Shading indicates stratigraphic position of condensed sections interpreted as regional seals in Early Oligocene and Early, Middle and Late Miocene. Condensed sections identified principally on basis of normalized rock accumulation rates; seal capacity borne out by capillary entry pressure and other data. Note coincidence between condensed sections/seals and ?glacio-eustatically mediated second-order sea-level high-stands.

100 m/Ma, comparable to those exhibited by (uncompacted) Recent pelagites from from comparable depths in the North Atlantic. CSs have also been distinguished on the basis of maximum benthic foraminiferal abundance and diversity, especially of infaunal morphotypes, and of maximum bathymetry. High abundance and diversity is characteristic of CSs or maximum flooding surfaces (MFSs). Note, however, that high abundance and diversity is not diagnostic of CSs. CSs can be characterised by low abundance and diversity, especially if affected by dissolution of calcareous species below the lysocline. Debris flows can be characterised by high abundance and diversity, although they can still be distinguished from high abundance and high diversity CSs on the basis of their allochthonous components, identified on the basis of bathymetric interpretation. (Proportionately) high abundance and diversity of infaunal 'morphogroup' C does appear to be diagnostic of CSs, reflecting colonisation of the sea floor under the comparatively tranquil conditions obtaining in between turbidite episodes, and over a comparatively long time.

Discussion

Importantly, the micropalaeontologically characterised CSs in the Oligo-Miocene of offshore west Africa correlate well with regional seals identified on seismic facies, wireline log and capillary entry pressure data. Interestingly, they are also coincident in time with tectono-eustatically or glacio-eustatically mediated second-order sea-level high-stands in the Early Oligocene (global standard calcareous nannoplankton zones NP21–NP23), earliest Miocene (NN1), Middle Miocene (NN5–NN6) and latest Miocene (NN9–NN11). This suggests that the methodology has essentially global applicability.

Incidentally, ineffective as well as effective seals have also been characterised micropalaeontologically, elsewhere, on the basis of the similarity of their contained benthic foraminiferal assemblages to those associated with modern seeps. For example, the seal sequence overlying the platform carbonate reservoir target in the Miocene of the Nam Con Son basin, offshore Vietnam, is characterised by *Arenomeandrospira glomerata*, a species originally described from the Skagerak and Kattegat, regions of shallow gas occurrence and sea-bed 'pockmarks'.

Traps

Traps are configurations of reservoir and seal that do not allow the escape of petroleum (Jones, 2006). They are classified into three main types: structural, for example anticlinal and diapiric traps; stratigraphic, for example pinch-out, reef and sub-unconformity traps, with no structural component other than regional dip; and combination traps. Structural traps are by far the most common, not least because they are typically readily imaged on seismic data. Stratigraphic traps

are typically much more subtle, and require a deliberate search.

There appears to be some extrinsic stratigraphic control on trap development outwith the sequence stratigraphic framework, with the Carboniferous and the 'Middle' Cretaceous important times of plate reorganisation and associated structuration and structural trap formation in the Middle East and formerly contiguous north Africa, and the Eocene an important time in the Indian subcontinent and adjacent areas of central and southeast Asia (Jones, 1996, 2006).

BOX 5.5   STRATIGRAPHIC AND PALAEOBIOLOGICAL CONTROLS ON SOURCE-, RESERVOIR- AND CAP-ROCK DISTRIBUTIONS (MODIFIED AFTER JONES, 2006)

Stratigraphic controls

Source-rocks. There are both intrinsic and extrinsic stratigraphic controls on source-rock development

(Jones, 1996). There is intrinsic control within the sequence stratigraphic framework, with preferential development during transgressive systems tracts or TSTs (Creaney & Passey, 1993).

There is also extrinsic control outwith this framework, with preferential development during the 30% of Phanerozoic time represented by the

Silurian, Late Devonian–Early Carboniferous, Late Carboniferous–Early Permian, Middle–Late Jurassic, 'Middle' Cretaceous and Oligo-Miocene accounting for a disproportionate 90% or more of the world's discovered oil (Klemme & Ulmishek, 1991; Huc *et al.*, in Harris, 2005).

The preferential development of source-rocks at these times indicates a considerable excess of production of organic material over consumption.

Reservoir-rocks. There are also both intrinsic and extrinsic stratigraphic controls on reservoir-rock development (Jones, 1996). There is intrinsic control within the sequence stratigraphic framework, with preferential development of deep marine clastic reservoirs during low-stand systems tracts or LSTs, and shallow marine clastic and carbonate reservoirs during high-stand systems tracts or HSTs. (Secondary enhancement of the quality of shallow marine carbonate reservoirs takes place preferentially during LSTs in the case of dissolution, and during TSTs in the case of – brine-reflux – dolomitisation.) There is also some extrinsic stratigraphic control on reservoir-rock development outwith this framework. The primary reservoir properties of the chalk reservoirs of the North Sea and the nummulite bank reservoirs of the circum-Mediterranean are controlled in part by the evolution of the fossils that constitute them.

Cap-rocks. There is some intrinsic stratigraphic control on cap-rock development within the sequence stratigraphic framework, with preferential development during TSTs and MFSs (Jones, 1996).

Palaeobiological controls

Source-rocks. There are (palaeo)biological controls on biogenic and thermogenic source-rock development, as well as the stratigraphic controls alluded to above (Bohacs *et al.*, in Harris, 2005; Katz, in Harris, 2005; Tyson, in Harris, 2005; see also 5.2.5 below).

Biogenic source-rock development is dependent on an excess of production over consumption of the organic matter (OM) that constitutes the feed-stock for the methanogenic 'Archaebacteria' that produce biogenic methane.

Production of OM is controlled essentially by primary biological productivity in the terrestrial and marine realms. Note also in this context that there is a significant input of OM of terrigenous

origin into the marine realm (Rullkotter, in Schulz & Zabel, 2000; Burdige, 2006; Emerson & Hedges, 2008) associated not only with wind- and river-borne sediments (McKee *et al.*, 2004; Seiter *et al.*, 2004); but also with turbidites (Saller *et al.*, 2006); and with hyperpycnal turbidites generated by earthquakes and landslides on steep-sided oceanic islands, and by storms (Dadson, 2008).

In the terrestrial realm, the production of OM, controlled by the productivity of land plants, is typically highest in places or at times of optimal climate (Beerling & Woodward, 2001).

In the marine realm, the production of OM, controlled by the productivity of marine Algae, is highest in places or at times of optimal climate, or of enhanced nutrient supply associated with upwelling. Present-day upwelling is indicated by phytoplankton blooms on satellite images. Past upwelling can also be indicated by palaeontological evidence, and can be modelled using palaeoclimatological computer software. There are various models for upwelling. *Zonal upwelling* is driven by currents, for example along the 'equatorial divergence' in low latitudes or 'polar divergence' in high latitudes. *Meridional upwelling* is driven by winds, for example in the trade-wind and monsoon belts in low latitudes. The northeasterly trades drive NW-directed surface currents in the northern hemisphere, and the southeasterly trades SW-directed surface currents in the southern hemisphere, as a result of a combination of Coriolis and Ekman (frictional drag) effects. The westerly directed component of the surface currents in turn drives easterly directed upwelling of deep currents along the western margins of land-masses in both continents.

Preservation of OM is controlled essentially by the availability of oxygen (Rullkotter, in Schulz & Zabel, 2000; Burdige, 2006). Consumption is lowest, and preservation highest, in places of dysoxia or anoxia, for example the Black Sea or the Indian Ocean; or at times of dysoxia or anoxia, for example the oceanic anoxic events (OAEs) of the Cretaceous (OAE1 in the Barremian–Aptian, OAE2 in the Cenomanian–Turonian, and OAE3 in the Senonian), such as have recently been simulated by a 3D biogeochemical general circulation model (Misumi & Yamanaka, 2008). There is enhanced preservation of OM in the Black Sea at the present time because of anoxia below the depth of the sill in the Bosporus. There is enhanced preservation in the Indian Ocean because of dysoxia in the peculiarly well-developed oxygen minimum zone (OMZ) there. OMZs are widely developed below the photic zone on the upper parts of the continental slope, because of consumption of oxygen by zooplankton, and the absence of production of oxygen by phytoplankton. That in the Indian Ocean is peculiarly well developed because this is an area of excess consumption of oxygen caused by excess production of OM associated with trade-wind- and monsoon-driven upwelling.

5.2 APPLICATIONS AND CASE STUDIES IN PETROLEUM EXPLORATION

In petroleum exploration, applied palaeontology assists generally, in the description and prediction of source-, reservoir- and cap-rock distributions; and specifically, for example in play fairway analysis or evaluation (see Fig. 1.1b; see also below).

Petroleum exploration

Petroleum exploration involves a number of geological, geophysical and integrated techniques (Gluyas & Swarbrick, 2004; Jones, 2006; see also below). Well-site operations in petroleum exploration involve a number of drilling, petrophysical and testing technologies (Jones, 2006; see also below, and Section 5.4).

Geological, geophysical and integrated techniques

Geological techniques. Geological techniques include field geological mapping and sampling, and office-based petroleum systems analysis (see above).

Field geological mapping and sampling is undertaken principally to provide data and interpretations on the stratigraphic and structural distributions of source-, reservoir- and cap-rocks, and the structural configuration on the surface and in the subsurface; and to identify potentially hydrocarbon-bearing structures, and hydrocarbon seeps or the alteration products thereof. It is particularly important in areas in which the acquisition of seismic geophysical data is difficult, for example in mountainous areas, in areas of structural complexity and/or steep dips, or in areas where the surface outcrops are of limestones or evaporites.

Geophysical techniques. Geophysical techniques include seismic surveying, gravity and magnetics surveying, and other remote sensing techniques.

Seismic surveying is undertaken principally to provide further data and interpretations on structural configuration; and to identify potentially hydrocarbon-bearing structures. Acquisition of seismic data on land and at sea is achieved by detonating small explosive charges at the surface and recording the time taken for the shock waves generated by them to reflect back from the various subsurface layers. Processing of seismic data leads to a product that is essentially a structural cross-section through the subsurface, albeit in time rather than depth (until converted to depth using velocity data). Additionally, seismic rock property data and other attributes can be used to infer lithology, porosity and pore fluid content, and to identify 'direct hydrocarbon indications' or DHIs. Three-dimensional seismic data can be used to infer the geometry of individual sedimentary bodies within reservoirs.

Gravity and magnetics surveying is also undertaken principally to provide further data and interpretations on structural configuration on the surface; and to identify potentially hydrocarbon-bearing structures. Acquistion of gravity and magnetic data is achieved by land, marine or airborne surveying. Processing leads to a product that is essentially a map of the local gravity or magnetic field, which highlights density and magnetic contrasts, such as those between the basement and basin fill, and allows interpretations to be made as to the structure of the basement. It also allows interpretations to be made as to the composition of the basement, which is an important consideration in petroleum systems analysis.

Other remote sensing techniques include high-resolution aerial photography, satellite, radar and multi-spectral sensing. Processing of remote sensing data leads to images or maps highlighting particular aspects of culture and geology, such as, for example, potentially hydrocarbon-bearing structures, and hydrocarbon seeps or the alteration products thereof.

Integrated techniques. Integrated techniques include basin analysis and play fairway analysis.

Basin analysis involves the integration of geological and geophysical data and interpretations, and the development of exploration models and strategies. Interestingly, empirical observations in intensively explored basins suggest some relationship between basin type and contained petroleum reserves. The richest in terms of proven reserves are the 'type IV' basins or 'continental borderland down-warps' of the Middle East. Note, though, that some of the largest reserves proven recently have been in 'type VIII' basins or 'Tertiary deltas' such as those of the deep-water Gulf of Mexico, and Brazil and west Africa in the South Atlantic, which have only recently become drillable through advances in technology.

Play fairway analysis or evaluation is a methodology for mapping play component presence and effectiveness, play component presence and effectiveness risk, and composite play risk, in basin exploration (Grant *et al.*, in Dore & Sinding-Larsen, 1996). It is becoming standard practice for the mapping and associated databasing to be undertaken using Geographic Information System (GIS) technology. Play component presence maps are essentially gross depositional environment (GDE) – or palaeogeographic and facies – maps highlighting areas of proven or interpreted potential petroleum source-, reservoir- or cap-rock development (for example, maps of TSTs highlighting transgressive marine source- or reservoir-rock development). Play component effectiveness maps are those highlighting areas of proven or interpreted potential effective petroleum source-, reservoir- or cap-rock development, determined by burial depth etc. Common risk segment (CRS) – or play component presence and effectiveness risk – maps colour code highlighted areas by perceived risk, conventionally green for low risk, yellow for moderate risk, and red for high risk. 'Composite CRS' (CCRS) – or play risk – maps are made by compositing play component presence and effectiveness risk maps (and, ideally, uncertainty or data confidence maps). The only segments that can be characterised as low risk on the composite risk map are those that are characterised as low risk on all of the individual play component risk maps. Similarly, segments

characterised as moderate risk on the composite risk map are those that are characterised as no higher than moderate risk on all of the individual play component risk maps. Segments that are characterised as high risk on any of the individual play component risk maps are characterised as high risk on the composite risk map.

Drilling, petrophysical logging and testing technologies

Drilling technologies. Once a prospect has been matured, the technical and commercial risk of drilling evaluated, and the decision to drill approved, drilling can go ahead. The choice of rig will depend on a number of well-site and geological factors, including, for example, whether there is environmental legislation governing site planning or operations, and whether shallow gas pockets or over-pressured intervals are anticipated from seismic data, site investigation or basin modelling, necessitating fitting or upgrading of blow-out preventers. Incidentally, the sort of deep-water drilling currently being undertaken in the Gulf of Mexico and South Atlantic is close to the limits of available technology, and presents particular rig design, mobilisation and other logistical problems.

Put simply, drilling is effected by rotation of a diamond bit at the end of a drill string formed of lengths of pipe connected by collars, the weight of the apparatus providing the necessary downward force (Fig. 5.6).

Drilling can be either vertical, deviated or high-angle, or horizontal, with direction provided by various types of 'geosteering', including 'biosteering'. Periodic casing of the borehole minimises down-hole contamination or caving, and generally maintains the condition of the hole, reducing the risk of the drill-pipe sticking and having to be fished out, which can be a time-consuming and costly exercise. Importantly, casing can also be used to contain over-pressured zones, reducing the risk of potentially extremely dangerous leaks or blow-outs.

Drilling involves the use of a specially formulated drilling mud which is pumped down the drill-pipe to lubricate the bit and seal the walls of the bore, and which returns to the surface between the drill-pipe and the walls of the bore, bringing with it samples of rock broken off by the action of the bit.

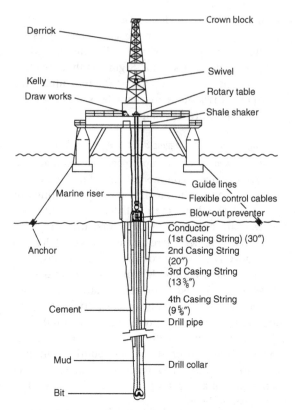

Fig. 5.6 **Components of a semi-submersible offshore drilling rig**. © Springer. Reproduced with permission from Copestake, in Jenkins (1993).

These samples, generally known as ditch-cuttings or simply cuttings, are collected in a trap or 'shale shaker' prior to recycling of the drilling mud. Cuttings samples are generally acquired every 3 m or 10′ for the purposes of lithological description, and for micropalaeontological and other analyses (see 1.2.2 above). On account of the costs involved, conventional – and side-wall – core samples are generally only acquired on an *ad hoc* basis to address specific reservoir sedimentological and related issues, such as depositional and diagenetic facies, and porosity and permeability characteristics, although they can also be used for micropalaeontological and other analyses (see 1.2.2 above).

Petrophysical logging technologies. A wide range of petrophysical or wireline logging technologies are available for formation evaluation. (Note in this context that formation evaluation techniques

practised by former eastern-bloc geologists differ somewhat from those practised by western geologists.) The most commonly run petrophysical or wireline logs are gamma, sonic and resistivity logs. Together, these provide an indication not only of lithology but also of porosity and pore fluid content. 'Logging-while-drilling' or LWD technologies have recently been developed that allow these logs to be run in real-time in the reservoir. Other logs that are becoming more widely used on account of their usefulness specifically in structural geological and sedimentological interpretation are dipmeter logs and image logs. Interestingly from the palaeontologist's point of view, individual fossils and trace fossils are occasionally identifiable on image logs.

Gamma logs measure radioactivity. Because in sedimentary rocks this is essentially a function of lithology, being high in fine clastics in which radioactive constituents tend to be concentrated, and low in coarse clastics and carbonates, gamma logs also provide a measure of this parameter. Note, though, that sandstones containing igneous clasts can yield anomalously high values, and non-radioactive clay minerals such as kaolinites anomalously low values. Note also, though, that the so-called natural gamma ray spectrometry tool can be used to identify individual clay minerals (incidentally, it is also useful in source-rock evaluation, since it can distinguish uranium, commonly associated with source-rocks, from other radioactive elements).

Sonic logs measure sonic transit time. Because this is essentially a function of density and porosity, sonic logs provide measures of these parameters. Density values are highest in dense carbonates, and lowest in uncompacted clastics. Porosity values are highest in porous, and lowest in tight, carbonates and clastics. Porosity can also be measured using neutron logs. Porosity and permeability can be measured using nuclear magnetic resonance logs.

Resistivity logs measure resistivity (the inverse of electrical conductivity). Because this is essentially a function of pore fluid content, resistivity logs provide a measure of this parameter (and of hydrocarbon saturation). Values are highest in rocks whose pore spaces are occupied by hydrocarbons, and lowest in rocks whose pore spaces are occupied by water or brine.

Testing technologies. Testing is a process initiated on indication of hydrocarbons, whereby the bottom-hole formation is made to flow some of its contents into the drill-pipe for recovery and analysis. It provides valuable information on potential flow rate, or reservoir deliverability.

Applications in petroleum exploration

Palaeontology plays a key role in petroleum exploration in the description and prediction of source-, reservoir- and cap-rock distributions, in time and in space, in a sequence stratigraphic framework (Schenck, 1940; Fleisher & Lane, in Beaumont & Foster, 1999; Gluyas & Swarbrick, 2004; Jones, 2006; Fig. 1.1b). Basic application calibrates cores, wells and seismic sections, and facilitates correlation. More advanced application constrains the timing of structural geological events and enables the establishment of a tectono-sequence stratigraphic framework, and the construction of chronostratigraphic diagrams, for the basin (see Section 4.7, above); and also provides important inputs into basin and petroleum systems analysis (see 5.1.1, and Box 5.1, above), and play fairway analysis or evaluation (see above). In play fairway analysis or evaluation, the most important application is in the constraint of sequence boundaries and maximum flooding surfaces and characterisation of systems tracts, and hence in the definition of meaningful time-slices for the purpose of GDE, and source-, reservoir- and cap-rock CRS, mapping, and in the population and annotation of GDE and CRS maps with pertinent data and interpretations. An example of play fairway map, for the Cenozoic of the Middle East, is given by Goff *et al.*, in Al-Husseini (1995) (see 5.2.1).

Quantitative approaches are discussed in Sections 2.7 and 3.7 above.

We now turn to specific case studies of applications in petroleum exploration.

5.2.1 Middle East

The following is an abridged account of the geology and petroleum geology of the Middle East (Fig. 5.7).

Readers interested in further details are referred to Sharland *et al.* (2001), Ruiz *et al.* (2004), Jones (2006), Jassim and Goff (2006), Aqrawi *et al.* (2010a), Leturmy and Robin (2010) and van Buchem *et al.* (2010). Those interested in the geology and petroleum geology of north Africa, contiguous with the Middle East until the opening of the Red Sea in the Oligocene, are referred to Said (1990), Salem and Belaid (1991), Askri *et al.* (1995), MacGregor *et al.* (1998), Sola and Worsley (2000), Pique (2001), Tawadros (2001), Hallett (2002), Arthur *et al.* (2003), Salem and Oun (2003), Sampsell (2003), Mejri *et al.* (2006), Schluter (2006) and Craig *et al.* (2009). Those interested in the Red Sea are referred to Coleman (1993) and Purser and Bosence (1998).

Geological setting

The geological history of the Middle East is long and complex (Fig. 5.8).

It essentially began with ?extensional tectonism and the formation of a series of salt basins, the evidence of which is most clearly seen in the present-day south Oman salt basin, in the area around the Strait of Hormuz in the southern Gulf, and in Bandar Abbas and Fars provinces in Iran, in the Precambrian or 'Infracambrian' (Megasequence AP1 of Sharland *et al.*, 2001). These basins underwent deformation, inversion, erosion and peneplanation at the 'Infracambrian'/Cambrian boundary. The entire Arabian plate then underwent passive margin subsidence in the Cambrian to Ordovician (Megasequence AP2). A glaciation took place in the latest Ordovician, Ashgillian (Hirnantian), and a deglaciation, resulting in the widespread deposition of transgressive marine sediments, in the Early Silurian, Llandoverian, and the subsequent deposition of regressive sediments in the Late Silurian–Devonian (Megasequence AP3). The plate then underwent ?compressional tectonism and uplift, ultimately resulting in extensive erosion associated with the so-called 'Hercynian unconformity' or, in Saudi Arabia, 'Pre-Unayzah unconformity', in the Early Carboniferous (Megasequence AP4). A second glaciation took place in the Late Carboniferous to Early Permian (Megasequence AP5).

Late Permian to Early Jurassic (Megasequence AP6) rifting and post-rift subsidence resulted in the break-up of the supercontinent of Gondwana, and, for the first time, extensive carbonate deposition. Middle to Late Jurassic (Megasequence AP7) rifting resulted in the break-up of Pangaea, and the formation of a series of intra-shelf basins, of which the most important were the Gotnia and Diyab

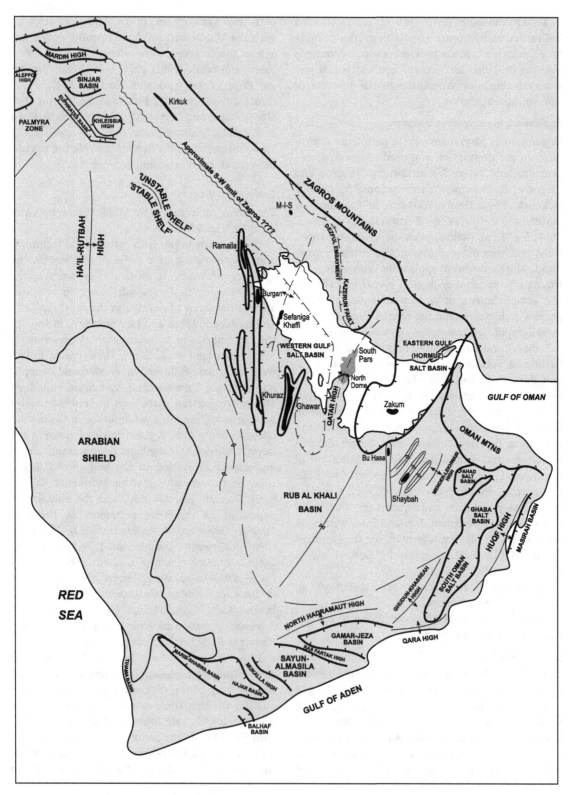

Fig. 5.7 **Location map, Middle East**. From Jones (2006).

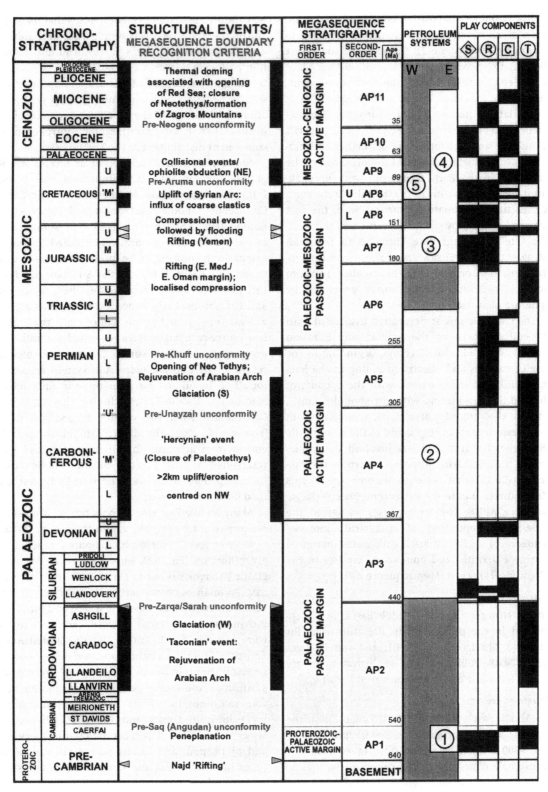

Fig. 5.8 **Stratigraphic framework and distribution of petroleum resource, Middle East.**
From Jones (2006). Petroleum systems: (1) Infracambrian; (2) Silurian; (3) Jurassic;
(4) Cretaceous–Tertiary; (5) 'Middle' Cretaceous.

basins. Carbonate deposition continued to dominate, with clastic deposition essentially confined to certain graben systems in the extreme southwest (Yemen). Extensive evaporite deposition took place at the end of the Jurassic. The Early Cretaceous, Neocomian to early Aptian, and 'Middle' Cretaceous, late Aptian to middle Turonian (Megasequence AP8) was characterised by post-rift subsidence and the formation of further of intrashelf basins, of which the most important were the Bab basin in the Early Cretaceous and the Kazhdumi and Shilaif basins in the 'Middle' Cretaceous. At this time, carbonate sedimentation in the east became balanced by clastic deposition in the west. The Late Cretaceous, late Turonian to Senonian (Megasequence AP9) saw structuration and extensive erosion associated with the so-called, in Saudi Arabia, 'Pre-Aruma unconformity', generated by ophiolite obduction in the east.

The Cenozoic saw structuration associated with the collision between the Arabian and Eurasian Plates and the closure of Tethys, beginning in the Late Cretaceous and effectively ending in the Miocene, and, to a lesser extent, with the opening of the Red Sea, beginning, with rift shoulder uplift, and the shedding of coarse clastic sediments from the 'Western Arabian Highlands' to the west, in the Oligocene. The initial phases involved the formation of a deep marine foredeep known as the Pabdeh basin in what were to become the Zagros Mountains in Iran in the Palaeocene–Eocene (Megasequence AP10). The later phases witnessed the widespread deposition of a typical 'molasse' sequence of shallow marine carbonates, marginal marine evaporites and non-marine clastics in the Oligocene–Holocene (Megasequence AP11).

Outcrop geology

The outcrop geology of the Middle East is well documented in the public-domain literature (see, for example, Sharland et al., 2001, and van Buchem et al. (2010). Published field geological maps are available for the entire area on various scales.

Petroleum geology

The Middle East is the site of 20% (103 out of 509) of the world's 'giant' oil and gas fields, that is, those containing >500 million barrels of oil – or equivalent – recoverable (Jones, 2006). The consensus view as to why the area is so rich in giant fields is that its large and on the whole relatively simple traps were able to drain large source kitchens by way of simple migration pathways (Fraser et al., 2003). This process was facilitated by a peculiarly fortuitous juxtaposition of source, reservoir and seal units in trap configurations, a function of the area's unique geological history. There is some stratigraphic control on play component distribution in the Middle East.

Source-rocks. Source-rocks occur at several stratigraphic horizons, the most important ones being Infracambrian, Silurian, Middle–Late Jurassic, Early Cretaceous, 'Middle' Cretaceous and Palaeocene–Eocene (Fig. 5.8). The Infracambrian source-rock is apparently of comparatively limited regional extent, only proven to be present in the south Oman salt basin. However, the Silurian, Middle–Late Jurassic, Early Cretaceous, 'Middle' Cretaceous and Palaeocene–Early Miocene source-rocks are of at least sub-regional extent. Collectively, these prolific source-rock basins have generated a significant proportion of the world's known petroleum reserves. The Silurian petroleum system accounts for all of the oil and gas reservoired in the Palaeozoic of the Middle East, including that in North Dome/South Pars, the world's largest gas field (Fraser et al., 2003). The Middle–Late Jurassic petroleum system accounts for all of the oil and gas reservoired in the Jurassic of the Middle East, including that in Ghawar, the world's largest oil field (Fraser et al., op. cit.).

Many Middle East source-rocks can be studied at outcrop (see, for example, van Buchem et al., 2010).

Reservoir-rocks. Reservoir-rocks occur at practically every horizon (Fig. 5.8). Important Early Permian clastic reservoirs occur in Oman and Saudi Arabia. Late Permian platform carbonate reservoirs occur across the Arabian Peninsula and Gulf and into Fars province in Iran. Jurassic platform carbonate reservoirs occur in Saudi Arabia and the United Arab Emirates. Cretaceous paralic and peri-deltaic clastic reservoirs occur in Kuwait and Iraq; and Cretaceous platformal, and locally peri-reefal, carbonate reservoirs in Kuwait, Iraq, Khuzestan and Lurestan provinces in Iran, Oman and the UAE. Oligocene–Miocene paralic and peri-deltaic clastic reservoirs and platformal, and locally peri-reefal, reservoirs occur in Kuwait, Iraq and Khuzestan and Lurestan provinces in Iran. The quality of the carbonate reservoirs is controlled partly by primary depositional factors and partly by secondary diagenetic factors (fracturing also locally plays an important role in enhancing permeability, as, for example, in the Zagros Mountains of Iran and Iraq). There is some stratigraphic control on carbonate reservoir development within the sequence stratigraphic framework, with preferential deposition during HSTs, and preferential diagenetic enhancement during LSTs in the case of dissolution, and TSTs in the case of dolomitisation.

Many regional reservoirs can be studied at outcrop (see, for example, Sharp *et al.*, in van Buchem *et al.*, 2010 (Cretaceous Khami and Bangestan groups, Zagros mountains, southwest Iran); Droste, in van Buchem *et al.*, 2010 (Early–'Middle' Cretaceous Shu'aiba and Natih formations, Oman mountains); Razin *et al.*, in van Buchem *et al.*, 2010 ('Middle' Cretaceous Sarvak formation, Zagros mountains, southwest Iran), and van Buchem *et al.*, in van Buchem *et al.*, 2010 (Oligo-Miocene Asmari formation, Dezful embayment, Zagros mountains, southwest Iran)).

Cap-rocks (seals) and traps. Cap-rocks (seals) occur at several horizons (Fig. 5.8). The most important are Infracambrian halites and anhydrites, Early Triassic shales, Late Jurassic halites and anhydrites, 'Middle' Cretaceous shales, Late Cretaceous shales, and Middle Miocene halites and anhydrites. Locally, as in parts of the Zagros Mountains, some of the seals have been breached by surface erosion, or rendered ineffective by fracturing. Generally speaking: the Infracambrian cap-rocks delineate the top of the Infracambrian petroleum system; the Early Triassic cap-rocks the top of the Silurian petroleum system; the Late Jurassic cap-rocks the top of the Middle–Late Jurassic petroleum systems; the 'Middle' Cretaceous cap-rocks the top of the Early Cretaceous petroleum systems; the Late Cretaceous cap-rocks the top of the 'Middle' Cretaceous petroleum systems; and the Middle Miocene cap-rocks the top of the Palaeocene–Early Miocene petroleum system.

Traps are predominantly structural, the style varying from inverted basement-cored anticlinal in the foreland to thrust anticlinal in the fold-belt of the Zagros Mountains of Iran and Iraq. Mobilisation of Infracambrian salt has played an important role in trap formation in Iran, the Gulf and the eastern part of the Arabian Peninsula. Combination or straight stratigraphic traps include that of the Fateh field in Dubai in the UAE.

Applications of palaeontology: biostratigraphy

A wide range of biostratigraphically and/or palaeobiologically useful fossil groups has been recorded in the surface and subsurface sections of the Middle East, including dinoflagellates, calcareous nannoplankton, calcareous Algae, acritarchs, foraminifera, radiolarians, calpionellids, plants, 'ediacarans', SSFs, stromatoporoids, corals, brachiopods, bryozoans, bivalves, gastropods, ammonoids, trilobites, ostracods, crinoids, echinoids, graptolites, chitinozoans, fish, reptiles and birds, and mammals (Jones, 1996, 2006; see also Sections 2.1 and 3.1 above; and 2.7 above for a discussion of quantitative approaches to biostratigraphy). A similarly wide range of biostratigraphic zonation schemes is applicable, the scale ranging from regional to reservoir, the resolution in inverse proportion to the scale. The resolution of some reservoir schemes is high enough to enable detection of intraformational unconformities, facies changes and faults, and an assessment of their implications for reservoir distribution. Non-biostratigraphic techniques have also been employed in dating and correlation, including chemostratigraphy (carbon and strontium isotope stratigraphy) and cyclostratigraphy.

Proterozoic. Acritarchs, 'ediacarans' and SSFs have proved of at least some use in biostratigraphy and palaeobiology, and/or in the characterisation of MFSs in the Proterozoic, Precambrian (Jones, 1996, 2000, 2006; Sharland *et al.*, 2001, 2004). For example, the occurrence of the 'ediacaran' *Spriggina*, with or without associated *Charnia*, *Dickinsonia* and unidentified medusioid forms, has enabled a correlation between the Hormuz salt formation of southwest Iran and the Rizu and Esfordi formations of northeast Iran, radiometrically dated to 595–715 Ma (Jones, 2000).

Palaeozoic. Acritarchs, plant micro- and macrofossils, calcareous Algae, LBFs, spores and pollen, SSFs, corals, brachiopods, ammonoids, trilobites, ostracods, graptolites, chitnozoans and conodonts have proved of use in biostratigraphy and palaeobiology, and/or in the characterisation of MFSs in the Palaeozoic (Jones, 1996; Al-Hajri & Owens, 2000; Sharland *et al.*, 2001, 2004; Al-Husseini, 2004; Miller & Melvin, 2005; Hughes, 2006; Jones, 2006; Molyneux *et al.*, 2006; Stephenson, 2006; Ghavidel-syooki & Vecoli, 2008; Penney *et al.*, 2008; Stephenson, 2008; Hughes, in Demchuk & Gary, 2009).

Mesozoic. Dinoflagellates, calcispheres, calcareous nannoplankton, calcareous Algae, calpionellids, LBFs, planktonic foraminifera, radiolarians, spores and pollen, stromatoporoids, bryozoans, bivalves, gastropods, ammonoids, ostracods, crinoids and echinoids have proved of use in biostratigraphy and palaeobiology, and/or in the characterisation of MFSs in the Mesozoic (Dunnington, 1955; Arkell, 1956; Adams *et al.*, 1967; Jones, 1996; Whittaker *et al.*, 1998; Sharland *et al.*, 2001, 2004; Hughes, 2004, 2006; Jones, 2006; Enay *et al.*, 2007; Hughes, 2009; Hughes & Naji, 2009; Hughes, in Demchuk & Gary, 2009; Hughes *et al.*, 2009). Modified global calcareous nannoplankton and planktonic foraminiferal zonation schemes, and regional calcisphere and LBF zonation schemes, are applicable in the appropriate facies. For example, LBFs have proved of use in the shallow marine platform carbonate facies of the Jurassic, as in the reservoir-prone Marrat, Dhruma, Tuwaiq Mountain, Hanifa and Arab formations of Saudi Arabia and their equivalents elsewhere, and in the Cretaceous; calcispheres and planktonic foraminifera in the deep marine pelagic carbonate or basinal clastic facies of the Cretaceous of Iran, Iraq and elsewhere (Fig. 5.9).

Planktonic foraminifera have proved of particular use in the deep marine pelagic carbonate or basinal clastic facies of the Cretaceous of Iraq, where the resolution of the local zonation scheme based on variations in the proportions of globotruncanid species groups is high enough to enable it to be used not only in the prediction of the thickness of eroded sediment, but also in the prediction of the thickness of unpenetrated sediment (Dunnington, 1955). Palynozonations are also locally applicable in the appropriate facies, for example in the marginal marine, peri-deltaic facies of the Cretaceous, as in the reservoir-prone Burgan formation of Kuwait and its equivalents elsewhere.

The chronostratigraphic ranges of a number of key Mesozoic species of LBFs have now been established either directly or indirectly by means of correlation, supplemented by graphic correlation (Whittaker *et al.*, 1998; Simmons *et al.*, in Hart *et al.*, 2000; Jones *et al.*, in Bubik & Kaminski, 2004; Jones, 2006). This has facilitated regional and reservoir scale-correlation (see 5.3.2 below).

Cenozoic. LBFs, planktonic foraminifera, calcareous nannoplankton, calcareous Algae, echinoids, reptiles and birds, and mammals have proved of use in biostratigraphy and palaeobiology, and/or in the characterisation of MFSs, in the Cenozoic (Jones & Racey, in Simmons, 1994; Jones, 1996; Sharland *et al.*, 2001, 2004; Jones, 2006). Modified global calcareous nannoplankton and planktonic foraminiferal zonation schemes, and regional larger benthic foraminiferal zonation schemes, are applicable in the appropriate facies.

Palaeobiological interpretation

Benthic foraminifera and Algae are perhaps the most important groups of fossils in palaeobiological, palaeoecological and palaeoenvironmental interpretation in the Middle East. Interpretation is based on analogy with their living counterparts, and on associated fossils and sedimentary environments (Banner & Simmons, in Simmons, 1994; Jones, 1996; Simmons *et al.*, in Hart *et al.*, 2000; Jones, 2006; Hughes *et al.*, 2009).

Palaeobathymetry. The palaeobathymetric ranges of a large number of benthic foraminifera have been established by means of calibration against photosynthetic Algae with known light requirements and hence water depth – or, more correctly, water clarity – preferences (Banner & Simmons, in Simmons, 1994; Jones, 1996; Jones *et al.*, in Bubik & Kaminski, 2004; Jones, 2006).

Using this information, high-resolution palaeobathymetric interpretations have been undertaken – and indeed detailed palaeobathymetric curves constructed – whenever and wherever sufficiently closely spaced samples are available for analysis. Integrated with sedimentology, this has facilitated detailed depositional modelling and reservoir characterisation (and identification of reservoir 'sweet spots'). For example, detailed palaeobathymetric curves have been constructed for the Early Cretaceous (Barremian) Kharaib and Shu'aiba formations (Thamama group) and the 'Middle'

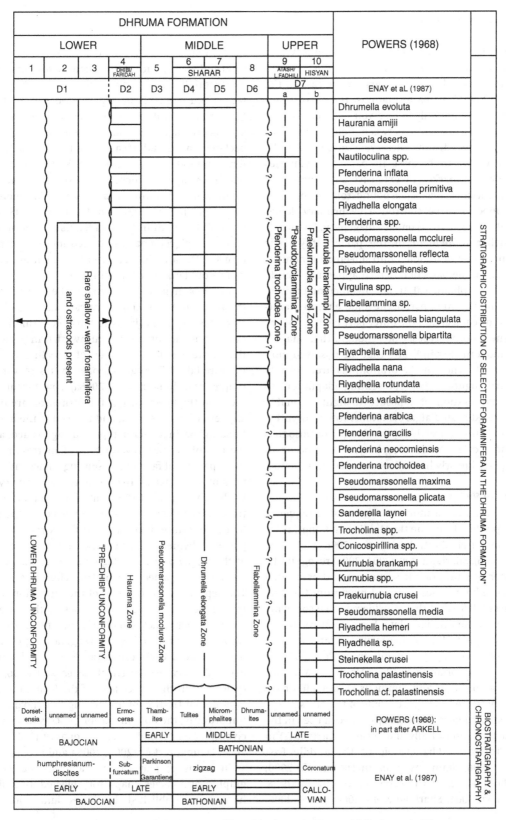

Fig. 5.9 **Stratigraphic distribution of benthic foraminifera, Middle Jurassic Dhruma formation, Saudi Arabia.** From Jones (1996).

Cretaceous Nahr Umr and Natih formations (Wasia group) of parts of Eastern Arabia – which are all important petroleum reservoirs. In the case of the Kharaib and Shu'aiba formations, the curves have proved useful not only in reservoir characterisation but also in correlaton between fields some tens of kilometres distant. Reservoir quality can be demonstrated to be best in palaeobathymetries of 10–30 m – marked by extensive development of porous *Bacinella/Lithocodium* boundstone facies – and thus appears to be controlled by primary depositional rather than secondary diagenetic factors.

Further refinement of the palaeobathymetric interpretation of middle ramp sub-environments is possible on the basis of observed apparently depth-related morphological trends in ancient orbitolinids, with the inner part of the middle ramp characterised by conical and the outer part by discoidal or flattened morphotypes. Note, though, that analogy with modern LBFs indicates that the observed morphological trends in the orbitolinids are probably primarily light-related, representing adaptations to the maximisation of exposure to light to enable interpreted hosted algal symbionts to photosynthesise, and only indirectly depth-related. Note also that discoidal or flattened morphotypes can occur in interpreted poorly lit shallow-water environments.

Integrated studies

There have been a number of integrated sequence stratigraphic studies on the Middle East (Jones, 1996; Sharland et al., 2001, 2004; Jones, 2006; Molyneux et al., 2006). Sharland et al. (2001, 2004) recognised a number of stratigraphic sequences and associated sequence boundaries (SBs) and maximum flooding surfaces (MFSs) that they considered to be correlatable across the Arabian plate. For ease of reference, they assigned alpha-numeric names and ages to each of their MFSs, ranging from Pc10, 570 Ma in the Precambrian to Ng40, 12 Ma in the Miocene. Al-Husseini (2007) and Simmons et al. (2007) have recently revised the ages of some of the MFSs.

A wide range of fossil groups has been used in the characterisation of SBs and systems tracts (STs) in the Proterozoic–Phanerozoic of the Middle East (Jones, 1996; Simmons et al., in Hart et al., 2000; Sharland et al., 2001; Jones et al., in Bubik & Kaminski, 2004; Sharland et al., 2004; Jones, 2006;

Al-Husseini, 2007). TSTs are characterised by orbitolinid and planktonic foraminiferal facies; early HSTs by essentially aggradational algal (*Lithocodium*) reef facies, and late HSTs by aggradational to progradational rudist–coral reef facies, orbitolinid–valvulinid–echinoderm and dasycladacean (*Hensonella*)–miliolid back-reef facies and, locally (Bab Basin), planktonic foraminiferal fore-reef facies. TSTs are also characterised by upward increases in the incidences of orbitolinid foraminifera, and the replacement of conical forms by flattened forms, and HSTs by upward decreases in the incidences of orbitolinids, and the replacement of flattened forms by conical forms. MFSs are characterised by maximum abundances of planktonic foraminifera.

Palaeobathymetric analysis of the benthic foraminifera and associated calcareous Algae and macrofossils of the Jurassic Surmeh formation of the Zagros mountains of southwest Iran indicates the existence of a number of shoaling-upward, interpreted HST progradational cycles of varying frequency (Lehmann et al., 2006a). Lower-frequency cycles in the lower part of the formation are typically tens of metres thick, and marked at the base by flooding basinal facies representing palaeobathymetries of at least several tens of metres, and at the top by shoal facies representing palaeobathymetries of the order of 10 m. Higher-frequency cycles in the upper part of the formation are typically of the order of 10 m thick, and are marked at the base by flooding fore-shoal or shoal facies representing palaeobathymetries of the order of ten metres, and at the top by back-shoal lagoonal or subtidal facies representing palaeobathymetries close to sea-level. The depositional lithofacies of the formation range from mudstones and wackestones at the bases of cycles to packstones and oolitic and oncolitic grainstones at the tops of cycles. Diagenesis appears to have involved early cementation of – subaerially exposed – grainstone shoal facies at the tops of cycles within the upper part of the formation, and later dolomitisation of underlying fore-shoal facies. Importantly, the dolomitisation process was enhancive of porosity and reservoir quality (Goff et al., 2004; Lehmann et al., 2004, 2006a and b).

In terms of play fairway analysis, palaeogeographic and facies maps for selected time-slices have been published by Jones and Racey, in

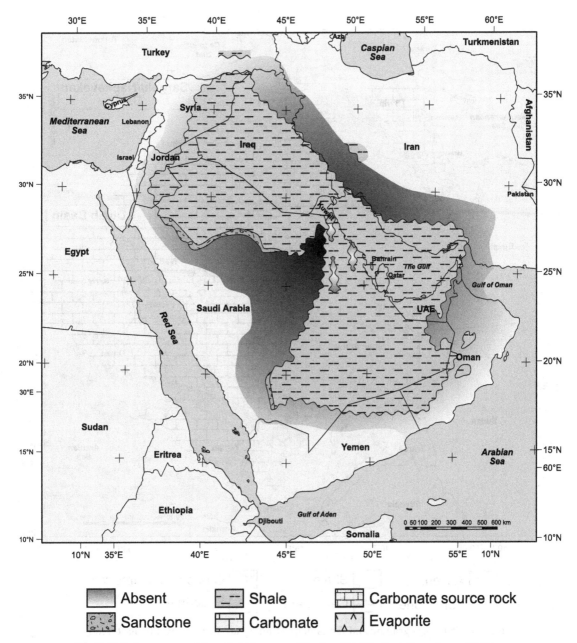

Fig. 5.10 **Palaeogeographic and facies map, Silurian, Middle East**. From Jones (2006). Showing maximum extent of source facies.

Simmons (1994), Goff *et al.*, in Al-Husseini (1995), Jones (2006), Konert *et al.* (2001) and Ziegler (2001). The time-slices have been identified in part on the basis of biostratigraphy; and the palaeogeography and facies on the basis of palaeobiology.

Palaeogeographic and facies maps showing the distributions of the Silurian, Middle–Late Jurassic, earliest Cretaceous, 'Middle' Cretaceous and Palaeocene–Eocene source facies are given in Figs. 5.10–5.14, respectively. It is evident that the Silurian source

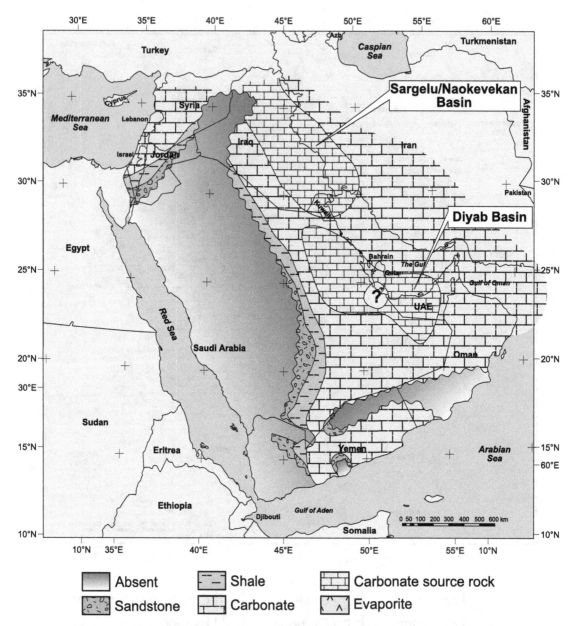

Fig. 5.11 **Palaeogeographic and facies map, Middle–Late Jurassic, Middle East.** From Jones (2006). Showing maximum extent of source facies (Sargelu/Naokelekan and Diyab basins).

facies extends throughout the Middle East, and indeed into north Africa. The Middle–Late Jurassic, earliest Cretaceous and 'Middle' Cretaceous source facies are restricted to intra-shelf basins. The Palaeocene–Eocene source facies is restricted to the Zagros foredeep basin.

Palaeogeographic and facies maps showing the distribution of proven Oligocene and Early Miocene

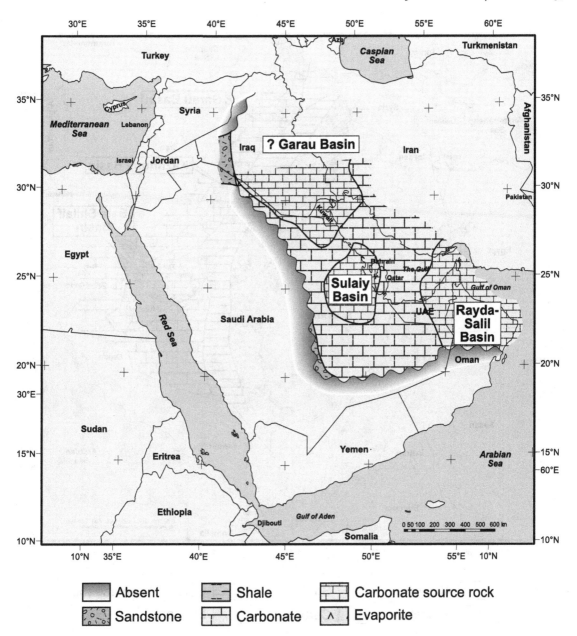

Fig. 5.12 **Palaeogeographic and facies map, earliest Cretaceous, Middle East**. From Jones (2006). Showing maximum extent of source facies (Garau, Sulaiy and Rayda/Salil basins).

paralic and peri-deltaic, and platformal and peri-reefal, reservoirs and associated facies are given in Figs. 5.15 and 5.16, respectively. The reservoirs are essentially restricted to the margins of the Zagros foredeep basin in Khuzestan and Lurestan Provinces in Iran, and in contiguous Iraq. Note, though,

that the clastic reservoirs, sourced from the 'Western Arabian Highlands', emerging in response to rifting in the Red Sea, extend further west.

A palaeogeographic and facies map showing the distribution of Middle Miocene evaporitic seal facies is given in Fig. 5.17. The seals

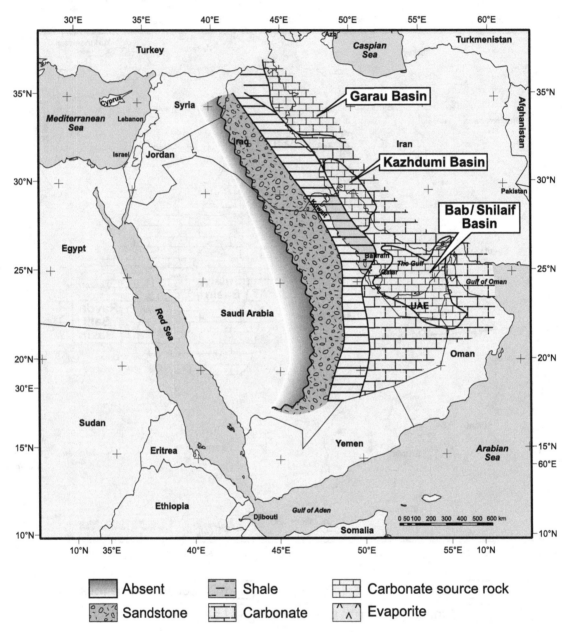

Fig. 5.13 **Palaeogeographic and facies map, 'Middle' Cretaceous, Middle East.** From Jones (2006). Showing maximum extent of source facies (Garau, Kazhdumi and Bab/Shilaif basins).

are essentially restricted to the Zagros foredeep basin.

A Cenozoic play fairway map is given in Fig. 5.18. The producing fields lie along the axis of the Zagros foredeep basin.

5.2.2 North Sea

The following is an abridged account of the geology and petroleum geology of the North Sea (Fig. 5.19).

Readers interested in further details are referred to Ziegler *et al.* (1997), Gluyas and Hichens (2003),

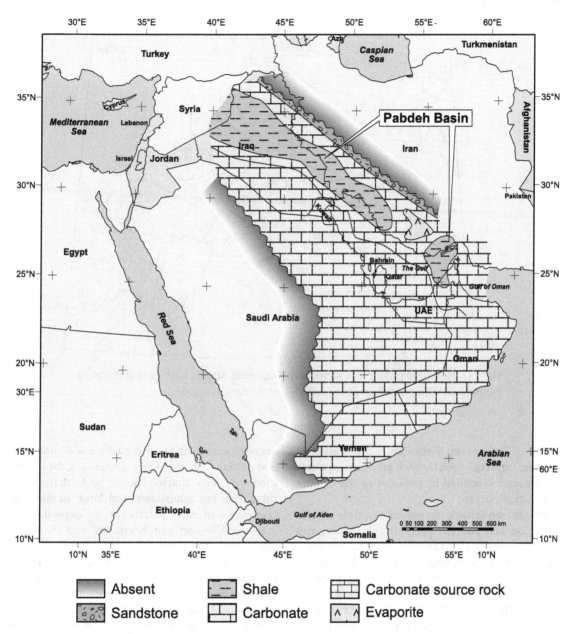

Fig. 5.14 **Palaeogeographic and facies map, Palaeocene-Eocene, Middle East**. From Jones (2006). Showing maximum extent of source facies (Pabdeh basin).

Evans *et al.* (2003), Dore & Vining (2005), Jones (2006) and Vining & Pickering (2010). Those interested in the geology and petroleum geology of the more or less analogous Wessex basin of the southwest of England are referred to Underhill (1998).

Geological setting

The North Sea basin essentially formed in response to extensional episodes associated with rifting in the North Atlantic in the Permo-Triassic and Jurassic. Regional evidence from the onshore United Kingdom also points towards an

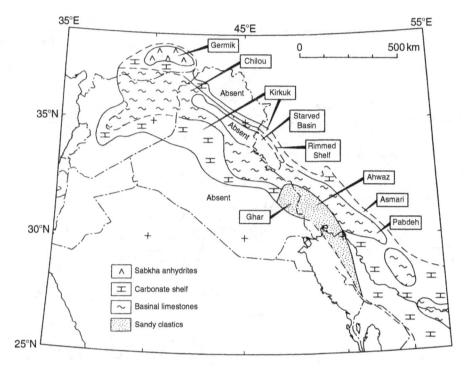

Fig. 5.15 **Palaeogeographic and facies map, Oligocene, Middle East.** © Gulf PetroLink.
Reproduced with permission from Goff *et al.*, in Al-Husseini (1995).

earlier – Variscan or Hercynian – compressional episode, during which back-arc rift systems formed and deformed in response to the closure of the Rheic Ocean.

The Permo-Triassic witnessed multiple phases of rifting and post-rift thermal subsidence, and was characterised by the deposition of continental clastics and evaporites. The Early–Middle Jurassic was a time of thermal doming prior to another extensional phase in the Late Jurassic, and was characterised by deposition of volcanics, volcaniclastics and marginal marine, peri-deltaic clastics, and by locally significant erosion at the 'mid-Cimmerian unconformity'. The – diachronous – Late Jurassic syn-rift sequence was characterised by deposition of deep marine shales and submarine fan sandstones in the basin centre, and of shallow marine, coastal plain sandstones at the basin margins. Significant erosion took place at or near the Jurassic–Cretaceous boundary 'Base Cretaceous Unconformity' ('BCU') or 'late Cimmerian

unconformity'. The post-rift package was characterised initially, in the Early Cretaceous, by deposition of deep marine marls and shales and submarine fan sandstones, and later, in the Late Cretaceous to Early Palaeocene, by deposition of marls in the northern North Sea and chalks in the central North Sea.

The Late Palaeocene saw the onset of sea-floor spreading in the North Atlantic, and attendant uplift of hinterland areas in Scotland and northern England. The amount of uplift has been estimated to have been as much as 300–400 m at the time of the 'Chron-25N Hiatus' or 'Forties lowstand'. Rejuvenation of sediment supply resulted in the progradation of successive sequences of deltaic and associated submarine fan clastics into the basin in the Late Palaeocene and Eocene. The morphology of the fan units was controlled by the evolving structural configuration of the basin. Inversion at the end of the Eocene resulted in the generation of the 'Pyrenean Unconformity'.

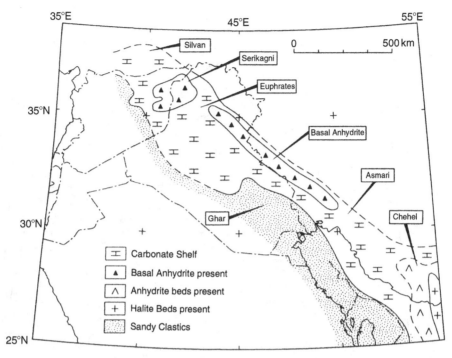

Fig. 5.16 **Palaeogeographic and facies map, Early Miocene, Middle East.** © Gulf PetroLink. Reproduced with permission from Goff *et al.*, in Al-Husseini (1995).

The succeeding Oligocene to Holocene was characterised by renewed basin fill.

Outcrop geology

The outcrop geology of the margins of the North Sea basin is extremely well documented in the public-domain literature (see, for example, the British Geological Survey's 'British Regional Geology' guides; and the Danish Geological Survey GEUS's publications, including Henriksen, 2008). Published field geological maps are available for the entire area on various scales (in some cases with accompanying 'sheet memoirs').

Petroleum geology

Source-rocks. The most important source-rock, and also incidentally one of the most important regional cap-rocks, in the North Sea is the latest Jurassic to earliest Cretaceous Kimmeridge Clay formation. The Kimmeridge clay is modelled as having been deposited in a basin characterised by at least intermittently anoxic bottom conditions brought about by salinity stratification, and

inimical to most life forms. These conditions favoured the low consumption and hence high preservation of organic material observed in palynological preparations.

The Kimmeridge Clay can be studied at outcrop in the 'natural laboratory' of the Wessex basin (Taylor, 1995; Underhill, 1998).

Reservoir-rocks. Reservoir-rocks are developed at several stratigraphic horizons (Fig. 5.20).

In the central North Sea, the most important reservoirs are Late Jurassic to Early Cretaceous marginal- and shallow-marine sandstones, Late Cretaceous to Early Palaeocene chalks, and Late Palaeocene to Middle Eocene submarine fan sandstones. In the northern North Sea, the most important reservoirs are Middle Jurassic peri-deltaic sandstones, and Late Jurassic, Late Palaeocene and Early Eocene submarine fan sandstones. The quality of the Late Cretaceous to Early Palaeocene chalk reservoirs in the central North Sea is controlled partly by primary porosity associated with the constituent calcareous dinoflagellates and nannofossils.

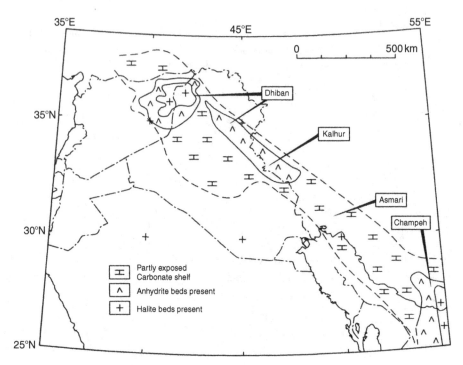

Fig. 5.17 **Palaeogeographic and facies map, Middle Miocene, Middle East.** © Gulf PetroLink. Reproduced with permission from Goff *et al.*, in Al-Husseini (1995).

Direct or indirect analogues for many of the producing reservoirs of the North Sea can be studied at outcrop in the British Isles and elsewhere. Analogues for the submarine fan sandstone reservoirs can be studied at outcrop in the Austrian and Swiss Alps, Polish Carpathians and Spanish Pyrenees, and in the Karoo basin of South Africa.

Cap-rocks (seals) and traps. Cap-rocks are predominantly 'intraformational'. Traps are predominantly structural, those of the Brent province in the northern North Sea being classical tilted fault-blocks. However, combination and stratigraphic traps are becoming increasingly important exploration targets. Mobilisation of Permian salt has locally played an important role in trap development.

Applications of palaeontology: biostratigraphy

A wide range of biostratigraphically and/or palaeobiologically useful fossil groups has been recorded in the surface and subsurface sections of the North Sea basin and its margins, including dinoflagellates, silicoflagellates, diatoms, calcareous nannoplankton,

Bolboforma, foraminifera, radiolarians, plants, sponges, brachiopods, bivalves, gastropods, ammonoids, belemnites, ostracods, crinoids, echinoids, and fish (Arkell, 1956; Jones, 1996, 2006; see also Sections 2.1 and 3.1 above; and Sections 2.7 and 3.7 above for discussions of quantitative approaches). A similarly wide range of biostratigraphic zonation schemes is applicable, the scale ranging from regional to reservoir, the resolution in inverse proportion to the scale. The resolution of some reservoir schemes is high enough to enable detection of intraformational unconformities, facies changes and faults, and an assessment of their implications for reservoir distribution. Unfortunately, on account of their commercial sensitivity, many such schemes remain either unpublished or published only in coded or otherwise unusable form.

Mesozoic. Palynostratigraphy is the most important tool in the zonation and correlation of Jurassic to Early Cretaceous of the North Sea basin and its margins (Mitchener *et al.*, in Morton, 1992; Powell, 1992; Whittaker *et al.*, in Morton, 1992; Partington

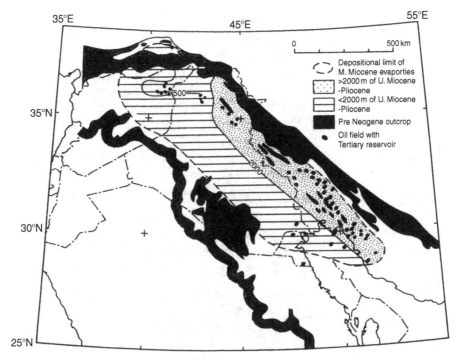

Fig. 5.18 **Play fairway map, Cenozoic, Middle East.** © Gulf PetroLink. Reproduced with permission from Goff *et al.*, in Al-Husseini (1995).

et al., in Parker, 1993; Jones, 1996; Ainsworth *et al.*, in Underhill, 1998; Charnock *et al.*, in Martinsen & Dreyer, 2001; Butler *et al.*, in Powell & Riding, 2005; Jones, 2006; Fig. 5.21).

A number of applicable palynozonation schemes incorporating ammonite control have now been published, including those of Mitchener *et al.*, in Morton (1992), Partington *et al.*, in Parker (1993) and Charnock *et al.*, in Martinsen and Dreyer (2001), which last-named, importantly, also incorporates independent micropalaeontological schemes based on benthic foraminifera and ostracods.

Global biostratigraphic zonation schemes based on calcareous plankton, that is, calcareous nanno-plankton and planktonic foraminifera, are essentially inapplicable in the prevalent proximal or otherwise unfavourable facies of the Jurassic to Early Cretaceous. Indeed, there is no published micro- or nanno-palaeontological zonation of any sort available for the Jurassic to Early Cretaceous of the offshore. However, the established onshore

ranges of benthic foraminiferal and ostracod species and the established onshore zonations of Ainsworth *et al.*, in Underhill (1998), can be applied with varying degrees of confidence. Moreover, off-shore ranges have been established for some species of benthic foraminifera, and for some species of radiolarians. The radiolarian datums, in particular, are of considerable chronostratigraphic and correlative value in the 'Central Graben', where Jurassic sediments are deeply buried, and where contained palynomorphs are often rendered unidentifiable by thermal alteration. Siliceous sponges are also of chronostratigraphic and correlative value in the 'Central Graben'. Modified global schemes based on calcareous plankton are applicable in the calcareous facies of the Late Cretaceous. There are also local micropalaeontological schemes not only for the onshore but also for the offshore (King *et al.*, in Jenkins & Murray, 1989).

Applications of micropalaeontological and paly-nological biostratigraphy to problem solving in the Mesozoic – and Cenozoic – of the North Sea are

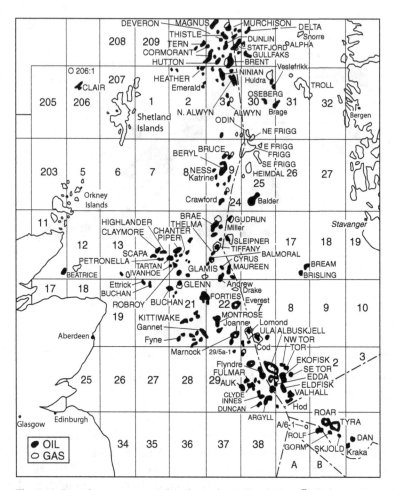

Fig. 5.19 **Location map, central and northern North Sea.** © Springer. Reproduced with permission from Copestake, in Jenkins (1993).

discussed by, among others, Copestake, in Jenkins (1993) and Jones (1996, 2006). Copestake, in Jenkins (1993), highlighted the importance of integration of micropalaeontological and palynological data both in exploration and in reservoir exploitation, with specific reference to the Jurassic reservoir of Don field, where models established on the basis of lithostratigraphic rather than biostratigraphic correlations proved seriously flawed.

Cenozoic. Micropalaeontological and palynological stratigraphy are equally important tools in the zonation and correlation of the Cenozoic of the North Sea (Jones, 1996, 2006). Global biostratigraphic zonation schemes based on calcareous plankton are essentially inapplicable, although

some modified schemes and many datums are locally applicable, especially in the Early Palaeocene part of the Chalk group.

There are established micropalaeontological zonation schemes not only for the onshore but also for the offshore (King *et al.*, in Jenkins & Murray, 1989).

The principal microfossil group utilised is the benthic foraminifera. Stratigraphically and environmentally important species of agglutinated benthic foraminifera are illustrated by Charnock and Jones, in Hemleben *et al.* (1990) and by Charnock and Jones (1997). Siliceous microfossils, including radiolarians, problematic microfossils, such as *Bolboforma*, and macrofossils, including fish otoliths, are also of some use.

(a)

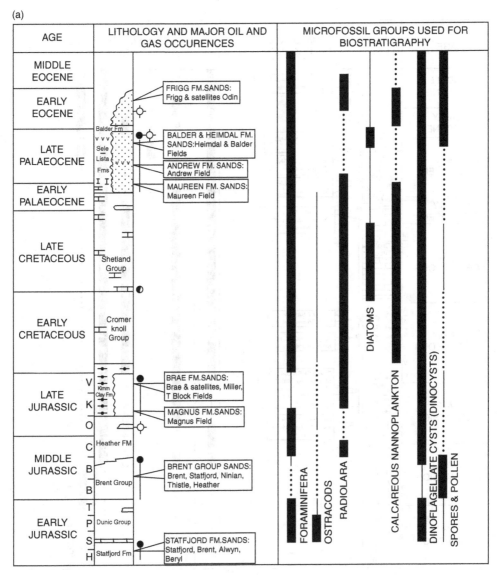

Fig. 5.20 **Stratigraphic distribution of reservoir-rocks, North Sea**. © Springer.
Reproduced with permission from Copestake, in Jenkins (1993). (a) Northern North
Sea; (b) central North Sea. Ranges of microfossil groups used in (reservoir) zonation
are also shown.

A number of applicable onshore-based palynozonation schemes, some incorporating calcareous nannoplankton control, have been established in academia and hence are available in the public domain (Powell, 1992). Unfortunately, though, most offshore-based schemes established in the petroleum industry remain in the proprietary domain,

notable exceptions including that of Schroder (1992) (Fig. 5.22).

Palaeobiological interpretation
Calcareous microfossils, especially benthic foraminifera, and organic-walled microfossils or palynomorphs, especially marine dinoflagellates and

(b)

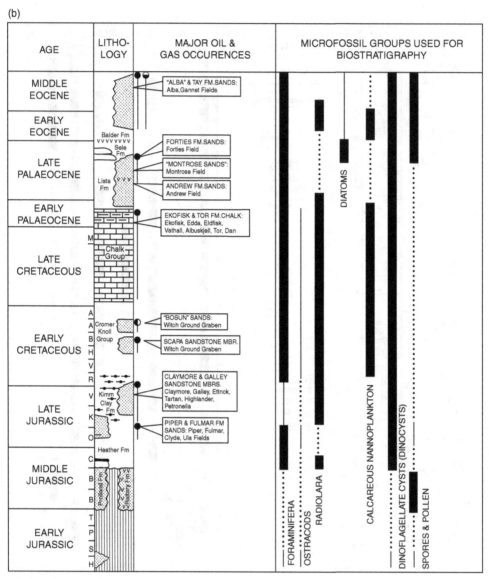

Fig. 5.20 (*cont.*)

terrestrially derived spores and pollen, are perhaps the most important groups of fossils in palaeobiological, palaeoecological and palaeoenvironmental interpretation in the North Sea. Interpretation is based on analogy with their living counterparts, on functional morphology, and on associated fossils and sedimentary environments (Jones, 1996; 2006).

Mesozoic. Microfacies and palynofacies analysis are equally important tools in the interpretation

of the predominantly marginal to shallow marine, peri-deltaic and paralic environments of the Jurassic–Cretaceous. Case studies include those of Nagy (1992) (microfacies) and Whittaker *et al.*, in Morton (1992), Williams, in Morton (1992), Mussard *et al.* (1994), Sawyer and Keegan (1996), Abbink (1998), Abbink *et al.*, (2004a) and Abbink *et al.* (2004b) (palynofacies). Nagy (1992) was able to discriminate between interdistributary bay, delta

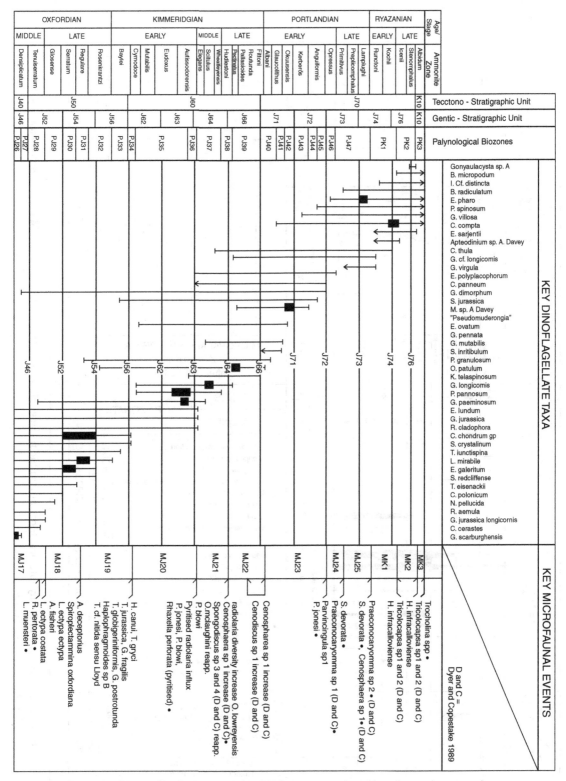

Fig. 5.21 **Micropalaeontological and palynological biostratigraphic framework, Late Jurassic, North Sea.** © The Geological Society, London. Reproduced with permission from Partington *et al.*, in Parker (1993).

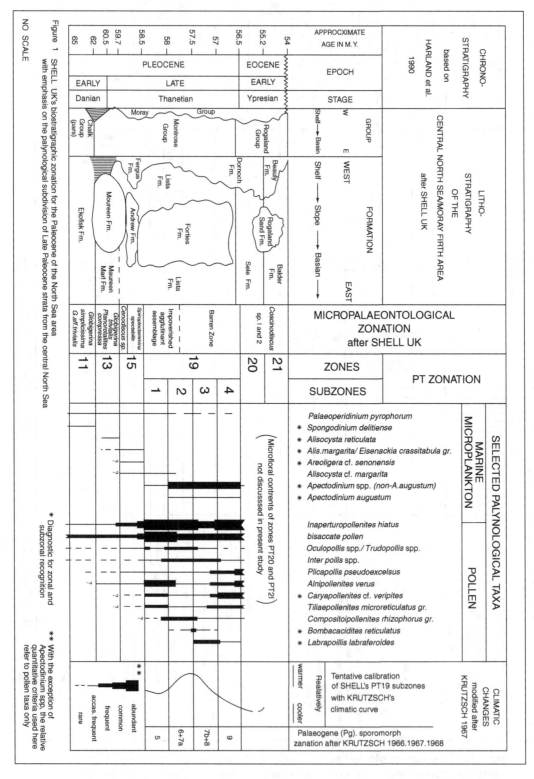

Fig. 5.22 **Micropalaeontological and palynological biostratigraphic framework, Palaeocene, North Sea.** © The Micropalaeontological Society. Reproduced with permission from Schröder (1992).

front and prodelta sub-environments of the Brent delta using microfacies analysis, specifically triangular cross-plots of epifaunal, infaunal and surficial 'morphogroups' (see also 5.3.2 below). (Note that his 'morphogroups' are essentially the same as those recognised earlier by Jones and Charnock, 1985.) Whittaker et al., in Morton (1992), Williams, in Morton (1992), Mussard et al. (1994) and Sawyer and Keegan (1996) were able to discriminate between sub-environments of the Brent delta using palynofacies analysis (see also 5.3.2 below). Abbink (1998) and Abbink et al., (2004a, b) were able to characterise systems tracts using 'sporomorph ecogroups' or SEGs.

Cenozoic. Microfacies is the most important important tool in the interpretation of the deep marine, submarine fan environments of the Palaeogene. Case studies include that of Charnock and Jones, in Hemleben et al. (1990). Charnock and Jones (*op. cit.*) were able to demonstrate on the basis of those species present in the Late Palaeocene of the North Sea and ranging through to the Recent that the sediments there are of essentially deep marine, submarine fan, rather than marginal to shallow marine, deltaic, aspect, as had previously been interpreted (Fig. 5.23).

Incidentally, microfacies and agglutinated foraminiferal 'morphogroup' analysis has also been used in reservoir characterisation, specifically in the identification of intra-reservoir mudstones constituting potential barriers and baffles to fluid flow, both in the North Sea and elsewhere (Holmes, in Jones & Simmons, 1999; Jones, in Jones & Simmons, 1999; Payne et al., in Jones & Simmons, 1999; Jones, 2001; Wakefield et al., 2001; Jones, 2003a, b; Jones et al., in Koutsoukos, 2005; Jones, 2006; see also 5.4.4 below).

Integrated studies

There have been a number of integrated sequence stratigraphic studies on the North Sea (Jones, 1996, 2006). Sequence stratigraphic schemes for the Jurassic have been published by, among others, Mitchener et al., in Morton (1992), Partington et al., in Parker (1993) and Charnock et al., in Martinsen and Dreyer (2001) (Fig. 5.24).

Sequence stratigraphic schemes for the Palaeogene have been published by Stewart, in Brooks and Glennie (1987) and Jones et al., in Evans et al. (2003) (Fig. 5.25).

Palaeogeographic and facies maps for selected time-slices have been published by Evans et al. (2003). The time-slices have been identified in part on the basis of biostratigraphy; and the palaeogeography and facies on the basis of palaeobiology. A palaeogeographic and facies map showing the distribution of the Late Jurassic Kimmeridge clay source facies is given by Fraser et al., in Evans et al. (2003) (Fig. 5.26).

'Common risk segment' (CRS) maps showing source presence and effectiveness have been published by Kubala et al., in Evans et al. (2003). Palaeogeographic and facies maps showing the distribution of proven and potential Eocene submarine fan reservoir and associated facies have been published by Jones et al., in Evans et al. (2003) (Figs. 5.27–5.34).

Micropalaeontological bathymetry has been integrated with structural restoration by Kjennerud and Gillmore (2003).

5.2.3 Northern South America and the Caribbean

The following is an abridged account of the geology and petroleum geology of northern South America and the Caribbean, particular emphasis on the eastern Venezuelan basin, including Trinidad (Fig. 5.35).

Readers interested in further details are referred to Saunders (1968), Gonzalez de Juana et al. (1980), Woodside (1981), Eva et al. (1989), Martinez (1989), Aymard, in Brooks (1990), James, in Brooks (1990), Stephan et al. (1990), Rohr (1991), Erlich and Barrett, in MacQueen and Leckie (1992), Donovan and Jackson (1994), Gallango and Parnaud, in Tankard et al. (1995), Hoorn et al. (1995), Parnaud, Gou, Pascual, Truskowski et al., in Tankard et al. (1995), Passalacqua et al., in Tankard et al. (1995), Higgins (1996), Jones, in Ali et al. (1998), Hsu (1999), Jones et al., in Jones and Simmons (1999), Wood (2000), Bartolini et al. (2003), Summa et al. (2003), Cobos (2005), MEEI (2009), James et al. (2009) and Mann and Escalona (2010). Those interested specifically in the geology and petroleum geology of western Venezuela and Colombia immediately to the west are referred to Cooper et al. (1995), Cooper et al., in Tankard et al. (1995), Lugo and Mann, in Tankard et al. (1995), Parnaud, Gou, Pascual, Capello et al., in Tankard

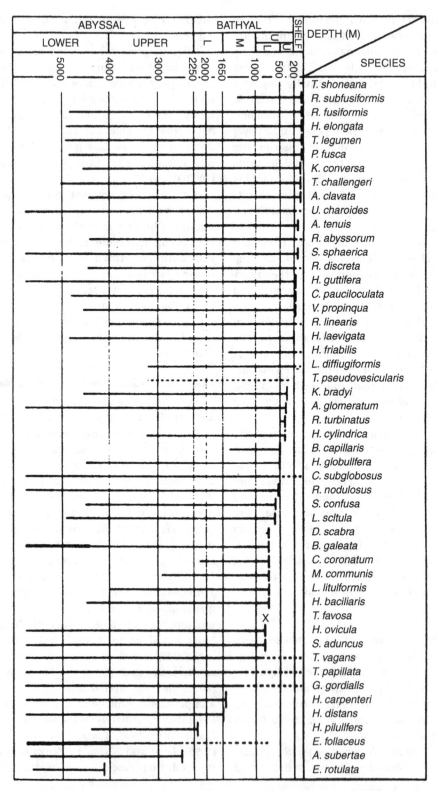

Fig. 5.23 **(Palaeo)bathymetric ranges of agglutinating foraminifera, Palaeogene, North Sea**. From Jones (1996).

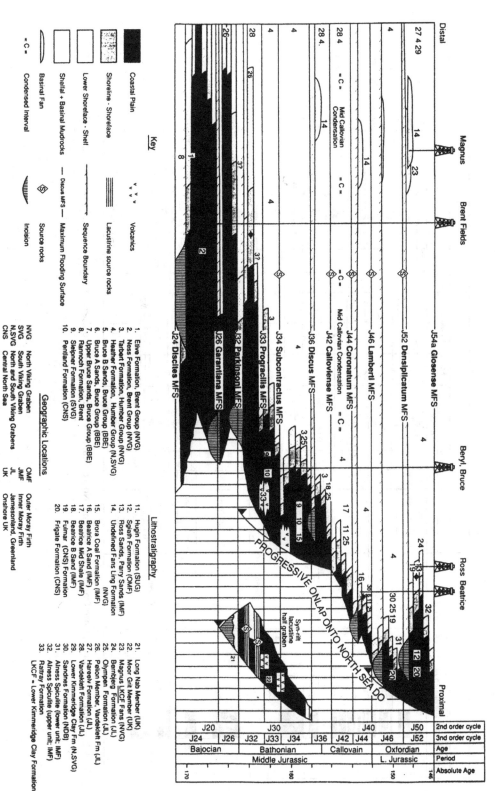

Fig. 5.24 **Sequence stratigraphic framework, Jurassic, North Sea.** © The Geological Society, London. Reproduced with permission from Partington *et al.*, in Parker (1993).

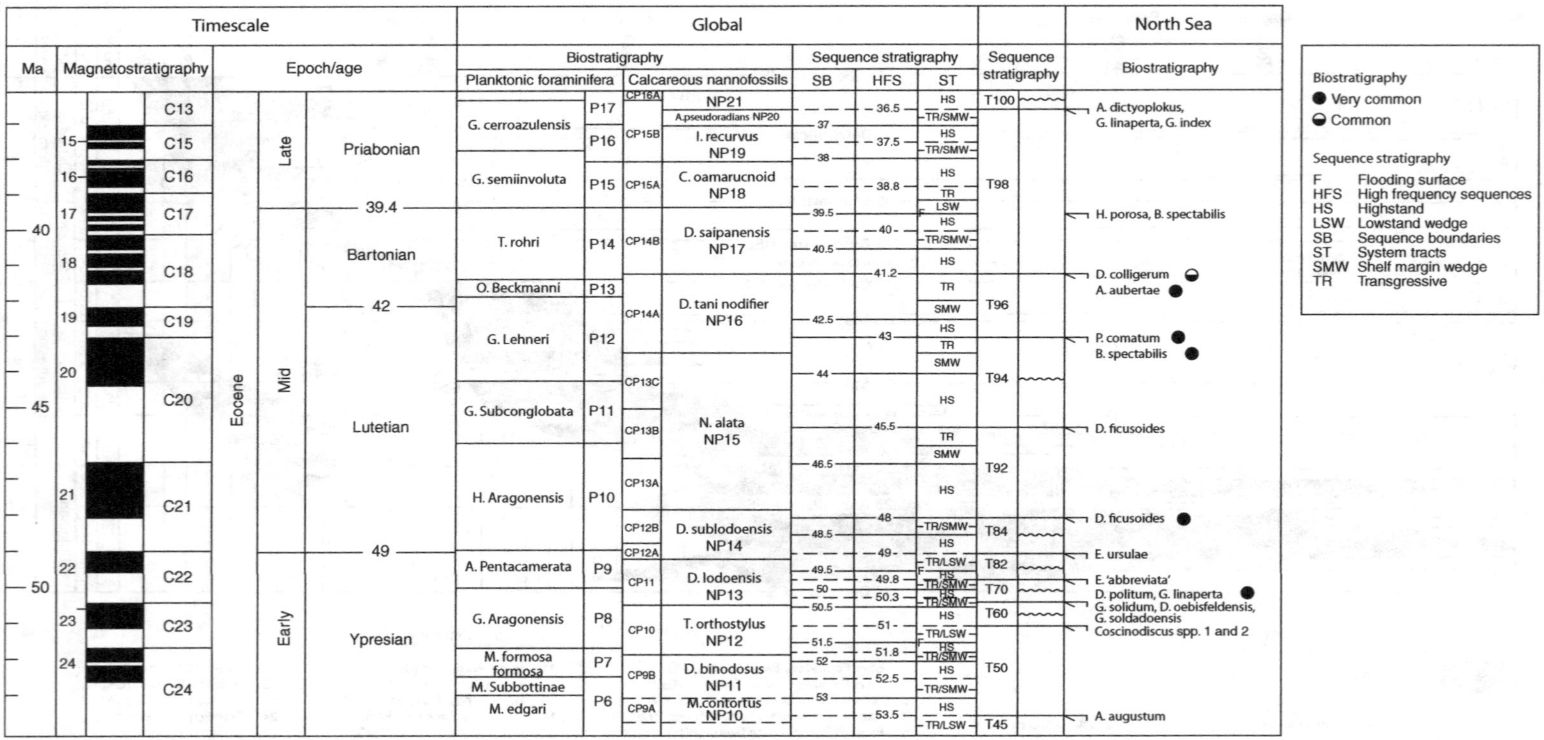

Fig. 5.25 **Sequence stratigraphic framework, Eocene, North Sea.** From Jones (2006), modified after Jones *et al.*, in Evans *et al.* (2003).

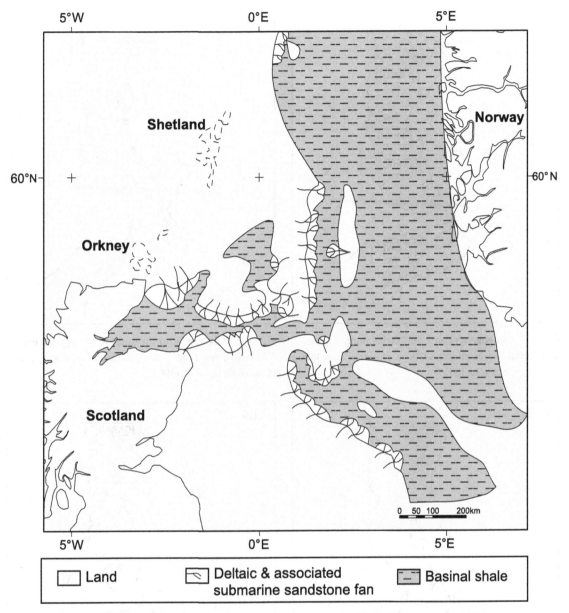

Fig. 5.26 **Palaeogeographic and facies map, Late Jurassic, North Sea.** From Jones (2006), modified after Jones *et al.*, in Evans *et al.* (2003). Showing maximum extent of Kimmeridge Clay formation source-rock.

et al. (1995), Erlich *et al.* (1999) and Jones *et al.*, in Koutsoukos (2005) (see also 5.4.3 below). Those interested in Mexico and the Gulf of Mexico are referred to Moran-Zenteno (1994), Jones (1996), O'Neill, in Jones and Simmons (1999), Bartolini *et al.* (2001) and Bartolini *et al.* (2003).

Geological setting

The geological history of northern South America and the Caribbean is long and complex, although much of the complexity is associated with the comparatively recent collision, in the Oligocene, between the Caribbean and South American plates, and the

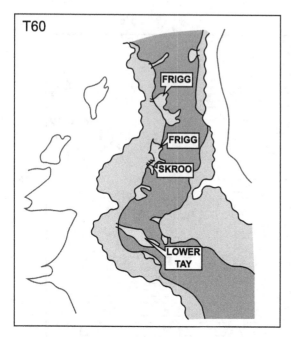

Fig. 5.27 **Palaeogeographic and facies map, Eocene, T60, North Sea.** From Jones (2006), modified after Jones *et al.*, in Evans *et al.* (2003).

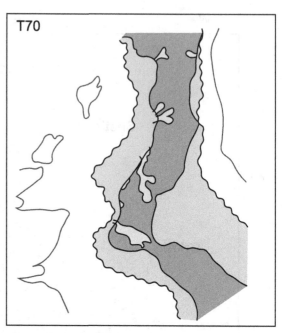

Fig. 5.28 **Palaeogeographic and facies map, Eocene, T70, North Sea.** From Jones (2006), modified after Jones *et al.*, in Evans *et al.* (2003).

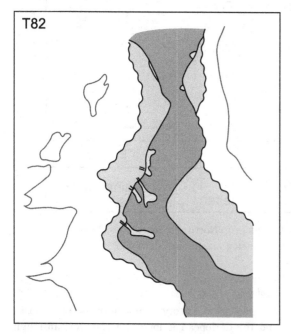

Fig. 5.29 **Palaeogeographic and facies map, Eocene, T82, North Sea.** From Jones (2006), modified after Jones *et al.*, in Evans *et al.* (2003).

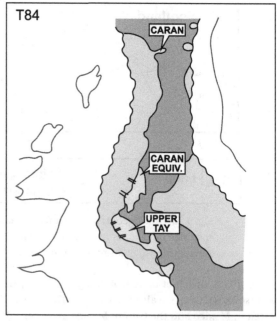

Fig. 5.30 **Palaeogeographic and facies map, Eocene, T84, North Sea.** From Jones (2006), modified after Jones *et al.*, in Evans *et al.* (2003).

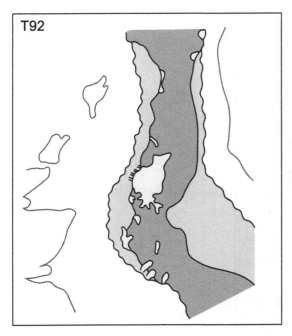

Fig. 5.31 **Palaeogeographic and facies map, Eocene, T92, North Sea.** From Jones (2006), modified after Jones *et al.*, in Evans *et al.* (2003).

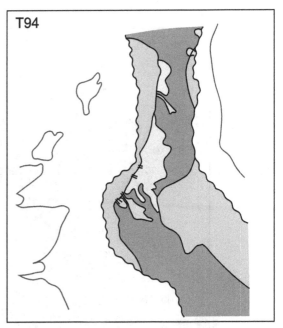

Fig. 5.32 **Palaeogeographic and facies map, Eocene, T94, North Sea.** From Jones (2006), modified after Jones *et al.*, in Evans *et al.* (2003).

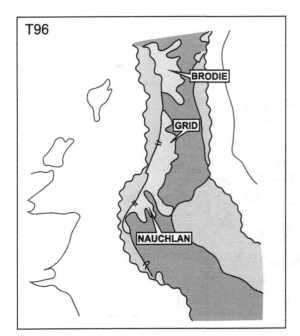

Fig. 5.33 **Palaeogeographic and facies map, Eocene, T96, North Sea.** From Jones (2006), modified after Jones *et al.*, in Evans *et al.* (2003).

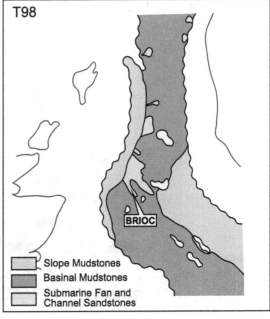

Fig. 5.34 **Palaeogeographic and facies map, Eocene, T98, North Sea.** From Jones (2006), modified after Jones *et al.*, in Evans *et al.* (2003).

(a)

(b)

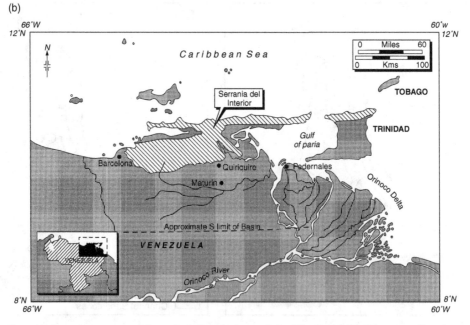

Fig. 5.35 **Location map, northern South America and Caribbean.** (a) Western part, showing location of Cusiana field in Colombia. Reproduced from Jones *et al.*, in Koutsoukos (2005). (b) Eastern part, showing location of Pedernales field in Venezuela. © The Geological Society, London. Reproduced with permission from Jones *et al.*, in Jones and Simmons (1999).

associated formation and uplift of the Colombian and Venezuelan Andes, and re-organisation of drainage systems. To (over-)simplify, before the collision, foreland basin sedimentation characterised the – only locally preserved – Palaeozoic, and rifting and passive margin sedimentation the Jurassic to the earlier part of the Oligocene. (The best evidence for rifting comes from surface outcrop and subsurface seismic and well data from the Perija and Merida Andes in western Venezuela, the Serrania del Interior and Espino graben in eastern Venezuela, and the Gulf of Paria between Venezuela and Trinidad.) After the collision, foreland basin sedimentation came to dominate the later part of the Oligocene to the present. The collision caused the uplift of the Andes and the diversion of the Orinoco to its present-day position, and the shoaling and closure of the former Strait of Panama and the creation of the Isthmus of Panama, some of the palaeobiogeographic evidence for which is outlined under 'Palaeobiogeography' below. Orinoco delta loading triggered extensive growth-faulting in the Columbus basin, offshore Trinidad, in the Pliocene–Pleistocene. The entire area remains extremely tectonically active. There is abundant evidence of extensional as transtensional as well as compressional and transpressional tectonism.

In the eastern Venezuelan basin, the passive margin phase is recorded in eastern Venezuela by the Barranquin, Garcia, El Cantil, Chimana, Querecual, San Antonio, San Juan, Vidono, Caratas, Tinajitas, Los Jabillos and Areo formations, and in Trinidad by the Naparima Hill, Guayaguayare, Lizard Springs and Navet (Fig. 5.36). Associated predominantly clastic sediments, derived from the Guyana Shield to the south, range from marginal marine sandstones through shallow marine sandstones to deep marine shales. Lithology varies with eustatically mediated sea-level, with the sandstones associated with regressions and the shales with transgressions. Carbonate sediments (platform and pelagic limestones) are only locally volumetrically significant.

The foreland basin phase is recorded in eastern Venezuela by the seismically demonstrably syntectonic Naricual, Carapita, La Pica, Las Piedras and Mesa/Paria formations, and in Trinidad by the Cipero, Lengua, Cruse, Forest and equivalents. Associated clastic sediments, derived not only from the Guyana Shield to the south but also from the rising Serrania del Interior and Northern Ranges to the north, range from alluvial and fluvial conglomerates and sandstones through peri-deltaic sandstones and coals and prodeltaic shales to turbiditic sandstones and basinal shales. On a basinal scale, lithology varies with tectonically mediated basin fill, with the turbiditic lithologies ('flysch') associated with early stages, and the prodeltaic, peri-deltaic, fluvial and alluvial lithologies ('molasse') associated with late stages. On a smaller scale, lithology varies with eustatically mediated sea-level, with the peri-deltaic sandstones associated with regressions and the prodeltaic shales with transgressions.

Outcrop geology

The outcrop geology of the eastern Venezuelan basin is reasonably well documented in the public-domain literature (see, for example, Hedberg, 1937; Renz, 1948, 1962; Blow, 1959; Stanley, 1960; Bermudez, 1962, 1966; Lamb, 1964; Lamb & Sulek, in Saunders, 1968; Metz, in Saunders, 1968; Stainforth, in Saunders, 1968; Bermudez & Stainforth, 1975; Bolli et al., 1994; Martinez, 1995; and Natural History Museum, Basel, 1996). Published field geological maps are available for the entire area on various scales (see, for example, Natural History Museum, Basel, 1996).

Petroleum geology

The principal petroliferous basins of northern South America and the Caribbean, with remaining reserves in excess of 500 million barrels of oil equivalent, are the Llanos basin of Colombia, the Maracaibo basin of western Venezuela, and the eastern Venezuelan basin of Venezuela and Trinidad (Kronman et al., in Tankard et al., 1995).

Source-rocks. The restricted deep marine La Luna formation of Colombia and western Venezuela, Querecual of eastern Venezuela, and Naparima Hill of Trinidad, all of essentially equivalent 'Middle' Cretaceous age, constitute the most important source-rock(s) in the region (Macellari & de Vries, 1987; Tribovillard et al., 1991; Rodrigues, 1993; Requejo et al., 1994; de Romero & Galea-Alvarez, 1995; Mann & Stein, 1997; Crespo de Cabrera et al., 1999; de Romero et al., 2003; Martinez, 2003;

(a)

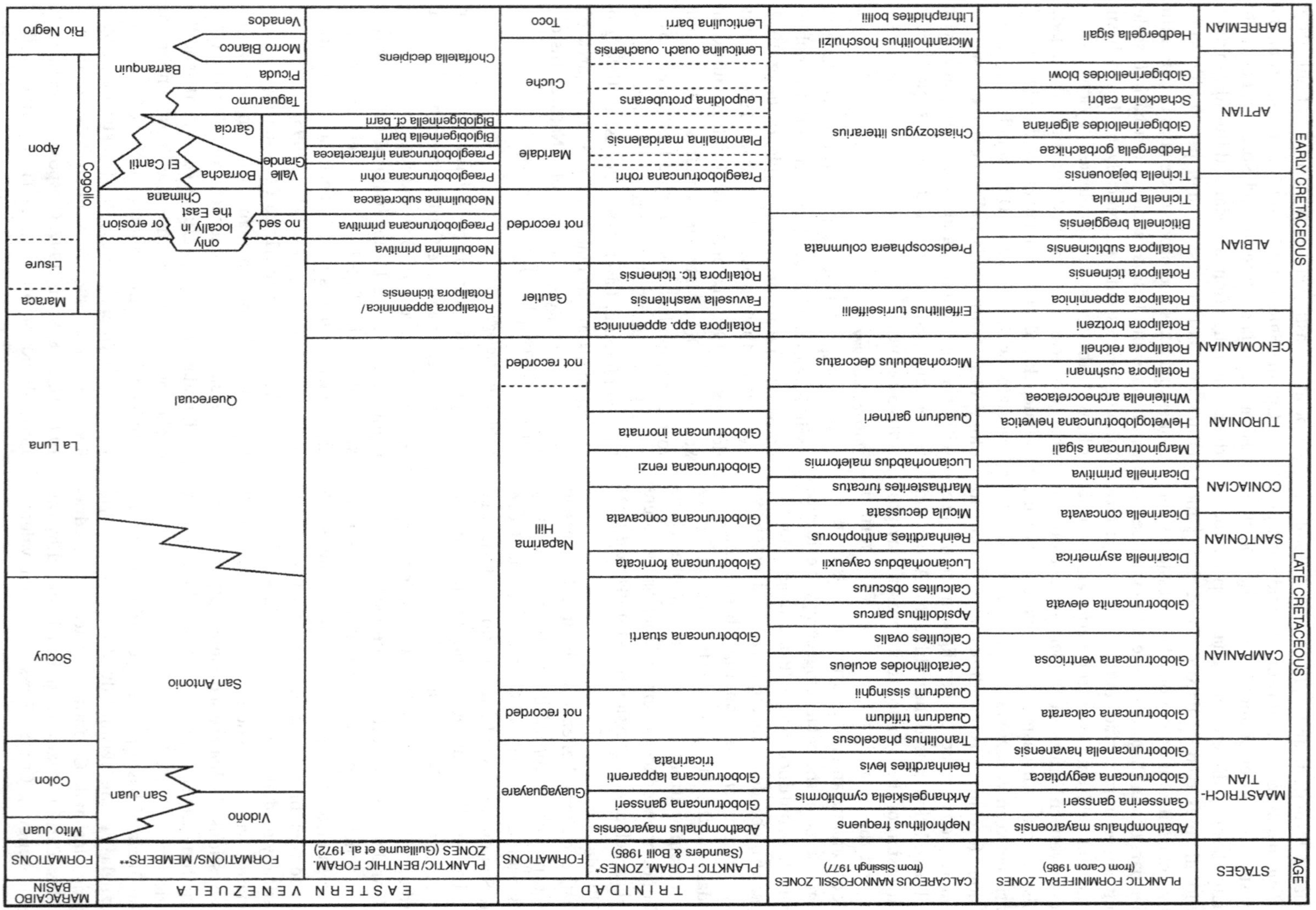

AGE	STAGES	PLANKTIC FORAMINIFERAL ZONES (from Caron 1985)	CALCAREOUS NANNOFOSSIL ZONES (from Sissingh 1977)	TRINIDAD — PLANKTIC FORAM. ZONES* (Saunders & Bolli 1985)	TRINIDAD — FORMATIONS	EASTERN VENEZUELA — PLANKTIC/BENTHIC FORAM. ZONES (Guillaume et al. 1972)	EASTERN VENEZUELA — FORMATIONS/MEMBERS**	MARACAIBO BASIN — FORMATIONS
LATE CRETACEOUS	MAASTRICHTIAN	Abathomphalus mayaroensis	Nephrolithus frequens	Abathomphalus mayaroensis	Guayaguayare	Abathomphalus mayaroensis	Vidoño	Mito Juan
		Gansserina gansseri	Arkhangelskiella cymbiformis	Globotruncana gansseri		Globotruncana gansseri	San Juan	Colon
	CAMPANIAN	Globotruncana aegyptiaca	Reinhardtites levis	Globotruncana lapparenti tricarinata	not recorded			
		Globotruncanella havanensis	Tranolithus phacelosus					
		Globotruncana calcarata	Quadrum trifidum	Globotruncana stuarti	Naparima Hill	San Antonio	Socuy	
		Globotruncanita elevata	Quadrum sissinghii					
		Globotruncana ventricosa	Ceratolithoides aculeus					
			Calculites ovalis					
	SANTONIAN	Dicarinella asymetrica	Calculites obscurus	Globotruncana concavata				
			Lucianorhabdus cayeuxii	Globotruncana fornicata				
	CONIACIAN	Dicarinella concavata	Reinhardtites anthophorus					
			Micula decussata	Globotruncana renzi			La Luna	
	TURONIAN	Dicarinella primitiva	Marthasterites furcatus					
		Marginotruncana sigali	Lucianorhabdus maleformis	Globotruncana inornata				
		Helvetoglobotruncana helvetica	Quadrum gartneri					
		Whiteinella archeocretacea		not recorded				
	CENOMANIAN	Rotalipora cushmani	Microrhabdulus decoratus			Neobulimina primitiva		
		Rotalipora reicheli		Rotalipora app. appenninica		Praeglobotruncana primitiva		
		Rotalipora brotzeni	Eiffellithus turriseiffelii	Favusella washitensis	Gautier	Rotalipora appenninica / Rotalipora ticinensis		
EARLY CRETACEOUS	ALBIAN	Rotalipora appenninica		Rotalipora tic. ticinensis		Neobulimina subcretacea	Querecual	Maraca
		Rotalipora ticinensis	Prediscosphaera columnata					Lisure
		Rotalipora subticinensis		Praeglobotruncana rohri	not recorded	Neobulimina primitiva	Chimana / Valle Grande	
		Biticinella breggiensis				Praeglobotruncana rohri	Borracha / El Cantil	
		Ticinella primula		Planomalina maridalensis	Maridale	Praeglobotruncana infracretacea	Garcia / Taguarumo	Apon
		Ticinella bejaouensis				Bigloglobigerinella barri		
	APTIAN	Hedbergella gorbachikae	Chiastozygus litterarius	Leupoldina protuberans	Cuche	Bigloglobigerinella cf. barri	Picuda	
		Globigerinelloides algeriana					Morro Blanco	Cogollo
		Schackoina cabri					Barranquin	
		Globigerinelloides blowi	Micrantholithus hoschulzii	Lenticulina ouach. ouachensis	Toco	Choffatella decipiens	Venados	
	BARREMIAN	Hedbergella sigali	Lithraphidites bollii	Lenticulina barri				Rio Negro

Fig. 5.36 Stratigraphic framework and zonation by means of planktonic foraminifera and calcareous nannoplankton, Cretaceous (a) and Palaeocene–Miocene (b), Trinidad and Venezuela. © Cambridge University Press. Reproduced with permission from Bolli et al. (1994).

(b)

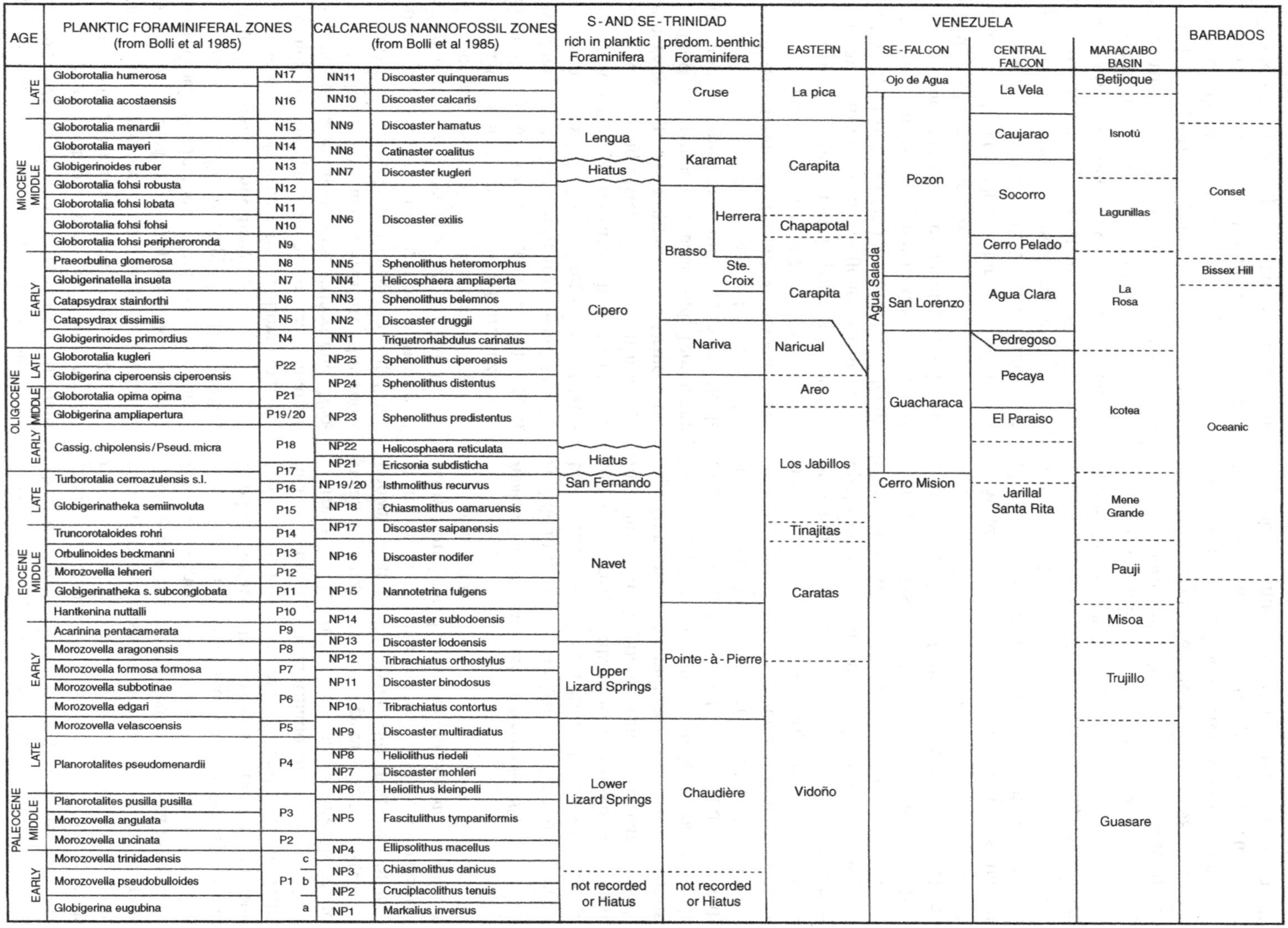

AGE

PALEOCENE			EOCENE			OLIGOCENE			MIOCENE		
EARLY	MIDDLE	LATE	EARLY	MIDDLE	LATE	EARLY	MIDDLE	LATE	EARLY	MIDDLE	LATE

PLANKTIC FORAMINIFERAL ZONES (from Bolli et al 1985)

Globigerina eugubina; Morozovella pseudobulloides; Morozovella trinidadensis; Morozovella uncinata; Morozovella angulata; Planorotalites pusilla pusilla; Planorotalites pseudomenardii; Morozovella velascoensis; Morozovella edgari; Morozovella subbotinae; Morozovella formosa formosa; Morozovella aragonensis; Acarinina pentacamerata; Hantkenina nuttalli; Globigerinatheka s. subconglobata; Morozovella lehneri; Orbulinoides beckmanni; Truncorotaloides rohri; Globigerinatheka semiinvoluta; Turborotalia cerroazulensis s.l.; Cassig. chipolensis/Pseud. micra; Globigerina ampliapertura; Globorotalia opima opima; Globigerina ciperoensis ciperoensis; Globigerinoides primordius; Globorotalia kugleri; Globorotalia tohsi fohsi; Globorotalia tohsi peripheroronda; Praeorbulina glomerosa; Globorotalia tohsi fohsi; Globorotalia tohsi lobata; Globorotalia tohsi robusta; Globigerinoides ruber; Globorotalia mayeri; Globorotalia menardii; Globorotalia acostaensis; Globorotalia humerosa

| | | P1a / P1b / P1c | P2 | P3 | P4 | P5 | P6 | P7 | P8 | P9 | P10 | P11 | P12 | P13 | P14 | P15 | P16 | P17 | P18 | P19/20 | P21 | P22 | N4 | N5 | N6 | N7 | N8 | N9 | N10 | N11 | N12 | N13 | N14 | N15 | N16 | N17 |

CALCAREOUS NANNOFOSSIL ZONES (from Bolli et al 1985)

| NP1 | NP2 | NP3 | NP4 | NP5 | NP6 | NP7 | NP8 | NP9 | NP10 | NP11 | NP12 | NP13 | NP14 | NP15 | NP16 | NP17 | NP18 | NP19/20 | NP21 | NP22 | NP23 | NP24 | NP25 | NN1 | NN2 | NN3 | NN4 | NN5 | NN6 | NN7 | NN8 | NN9 | NN10 | NN11 |

Markalius inversus; Cruciplacolithus tenuis; Chiasmolithus danicus; Ellipsolithus macellus; Fasciculithus tympaniformis; Heliolithus kleinpelli; Discoaster mohleri; Heliolithus riedeli; Discoaster multiradiatus; Tribrachiatus contortus; Discoaster binodosus; Tribrachiatus orthostylus; Discoaster lodoensis; Discoaster sublodoensis; Nannotetrina fulgens; Discoaster nodifer; Discoaster saipanensis; Chiasmolithus oamaruensis; Isthmolithus recurvus; Ericsonia subdisticha; Helicosphaera reticulata; Sphenolithus predistentus; Sphenolithus distentus; Sphenolithus ciperoensis; Triquetrorhabdulus carinatus; Discoaster druggii; Sphenolithus belemnos; Helicosphaera ampliaperta; Sphenolithus heteromorphus; Discoaster exilis; Discoaster kugleri; Catinaster coalitus; Discoaster hamatus; Discoaster calcaris; Discoaster quinqueramus

S - AND SE-TRINIDAD

predom. benthic Foraminifera (rich in planktic Foraminifera): not recorded or Hiatus; Lower Lizard Springs; Upper Lizard Springs; Navet; San Fernando; Hiatus; Cipero; Hiatus; Lengua; Lengua

VENEZUELA

EASTERN: not recorded or Hiatus; Chaudière; Pointe-à-Pierre; Caratas; Los Jabillos; Areo; Nariva; Naricual; Carapita; Chapapotal; Herrera; Ste. Croix; Brasso; Karamat; Cruse; La pica

Nariva; Carapita; Agua Salada; Guacharaca; Cerro Misión; San Lorenzo; Pozon; Ojo de Agua

SE-FALCON: Agua Salada

CENTRAL FALCON: Vidoño; Jarillal Santa Rita; Mene Grande; El Paraiso; Pecaya; Pedregoso; Agua Clara; Cerro Pelado; La Rosa; Socorro; Lagunillas; Caujarao; Isnotú; La Vela

MARACAIBO BASIN: Guasare; Trujillo; Misoa; Pauji; La Rosa; Lagunillas; Isnotú; Bettijoque

BARBADOS: Oceanic; Bissex Hill; Conset

Fig. 5.36 (cont.)

Summa *et al.*, 2003; Zapata *et al.*, 2003; Erlich & Keens-Dumas, 2007; Fig. 5.36). The preferred depositional model for the source-rock(s) invokes enhanced production of organic matter as a function of upwelling (Martinez, 2003; Jones, 2006).

The La Luna, Querecual and Naparima Hill formations can be studied at outcrop in Colombia and western Venezuela, in the Serrania del Interior of eastern Venezuela, and at Naparima Hill in San Fernando in Trinidad, respectively.

Reservoir-rocks. Marginal to shallow marine, including peri-deltaic, and deep marine, submarine fan clastics of Late Cretaceous, Palaeogene, Neogene and Pleistocene age constitute the most important reservoir-rocks in the region (Fig. 5.36). Marginal to shallow marine clastics constitute of Late Cretaceous to Palaeogene age constitute the most important reservoirs in Colombia and western Venezuela (Gonzalez-Guzman, 1985; Martinez, 1995; Jones *et al.*, in Koutsoukos, 2005; see also 5.4.3 below). Marginal to shallow marine, and, locally, deep marine, clastics of Late Cretaceous, Palaeogene and Neogene age constitute the most important reservoirs in the eastern Venezuelan basin in eastern Venezuela, for example in the Orinoco tar belt, along the Furrial, Oficina and Temblador trends, and in Quiriquire and Pedernales fields (Jones *et al.*, in Jones & Simmons, 1999; see also 5.3.3 below). Marginal to shallow marine, and, locally, deep marine, clastics of Neogene and Pleistogene constitute the most important reservoirs onshore and offshore Trinidad (Jones, in Ali *et al.*, 1998; Wood, 2000; MEEI, 2009). A reservoir depositional model for the marginal to shallow marine clastic reservoirs of the Columbus basin, offshore Trinidad is given by Wood (2000).

Direct or indirect analogues for many of the producing reservoirs of the region can be studied at outcrop. Outcrop analogues for the marginal to shallow marine, including peri-deltaic, clastic reservoirs are located in Colombia and western Venezuela, in the Serrania del Interior in eastern Venezuela, and on the Moruga and Mayaro coasts of Trinidad. Outcrop analogues for the deep marine, submarine fan, clastic reservoirs are located on Barbados.

Cap-rocks and traps. The ultimate cap-rock to the Late Cretaceous Guadalupe formation and Palaeogene Mirador and Misoa formation reservoirs in Colombia and western Venezuela is the Pauji. The cap-rock to the Late Cretaceous and Palaeogene 'Narical' formation *sensu lato* (*s.l.*) reservoirs along the Furrial trend in the eastern Venezuelan basin is the Areo or Carapita, a significant regional flooding surface at the base of the foreland basin megasequence; and that to the Neogene La Pica formation, Pedernales member reservoir in Pedernales field is the Cotorra mudstone, an intraformational flooding surface. The cap-rocks to the Neogene and Pleistogene reservoirs onshore and offshore Trinidad are similarly intraformational.

Traps in the region are predominantly structural, although a wide range of structural styles is known. That to the Pedernales field may have a stratigraphic component to it, although it is essentially a mud-cored anticline (Jones *et al.*, in Jones & Simmons, 1999).

Applications of palaeontology: biostratigraphy

A wide range of biostratigraphically and/or palaeobiologically useful fossil groups have been recorded in the surface and subsurface sections of the region, including dinoflagellates, calcareous nannoplankton, foraminifera, radiolarians, plants, corals, bivalves, gastropods, ammonoids, ostracods, fish, amphibians, reptiles and birds, and mammals (Jones, 2006; see also Sections 2.1 and 3.1 above; and Section 2.7. above for a discussion of quantitative approaches to biostratigraphy). A similarly wide range of biostratigraphic zonation schemes is applicable, the scale ranging from regional to reservoir, the resolution in inverse proportion to the scale (Saunders & Bolli, 1985; Lorente, 1986; Bolli *et al.*, 1994; Wood, 2000; Fig. 5.36; see also 5.3.3 and 5.4.3, below).

Mesozoic. Global standard calcareous nannoplankton and planktonic foraminiferal schemes are applicable in the Mesozoic (Saunders & Bolli, 1985; Bolli *et al.*, 1994; Carillo *et al.*, 1995; Carr-Brown, in MEEI, 2009; Fig. 5.36).

Local marine dinoflagellate and terrestrially derived spore and pollen schemes are also applicable (Muller *et al.*, 1987; Helenes *et al.*, 1998; Helenes & Somoza, 1999). And other groups have also proven to be of at least local biostratigraphic use, including benthic foraminifera (Bartenstein, 1985, 1987; Bolli *et al.*, 1994), and ammonoids (Arkell, 1956;

Renz, 1982; Bolli *et al.*, 1994; Villamil & Pindell, in Pindell & Drake, 1998).

Cenozoic. Global standard calcareous nannoplankton and planktonic foraminiferal schemes are applicable in the Cenozoic (Saunders & Bolli, 1985; Bolli *et al.*, 1994; Flor *et al.*, 1997; Wood, 2000; Carr-Brown, in MEEI, 2009; Jaramillo *et al.*, in Demchuk & Gary, 2009; Fig. 5.36).

Local spore and pollen schemes are also applicable (van der Hammen & Wymstra, 1964; Leonard, 1983; Gonzalez-Guzman, 1985; Lamy, 1986; Lorente, 1986; Muller *et al.*, 1987; Wood, 2000; Jaramillo *et al.*, in Demchuk & Gary, 2009), as are local ostracod schemes (van den Bold, 1960, 1963, 1974).

And other groups have also proven to be of at least local biostratigraphic use, including larger as well as smaller benthic foraminifera (Caudri, 1996; Jones, 2009), radiolarians (Jones, 2006), and bivalves (Hunter, 1978; Jones, 2006). Interestingly, larger benthic foraminifera, although of shallow marine origin, are sufficiently common – contemporaneously – reworked into the evidently deep marine deposits of the Scotland group in Barbados as to form the basis of a workable stratigraphy there (Jones, 2009).

Palaeobiological interpretation

Benthic foraminifera are perhaps the most important group of fossils in palaeobiological, palaeoecological and palaeoenvironmental interpretation in northern South America and the Caribbean, although bivalves, corals and trace fossils are also locally important, and foraminifera, plants, bivalves gastropods, fish, amphibians, reptiles and mammals are all important in the field of palaeobiogeography (see below). Interpretation is based on analogy with their living counterparts, and on associated fossils and sedimentary environments (van Andel & Postma, 1954; Todd & Bronnimann, 1957; Saunders, 1958; Bermudez, 1966; Batjes, in Saunders, 1968; Hofker, 1976; Hunter, 1976; Culver & Buzas, 1982; Koutsoukos & Merrick, 1986; Galea-Alvarez & Moreno-Vasquez, 1994; Carrillo *et al.*, 1995; Moreno-Vasquez, 1995; Jones, 1996; Jones, in Ali *et al.*, 1998; Jones *et al.*, in Jones & Simmons, 1999; Wilson, 2003, 2004, 2005; Jones, 2006, 2009; Smith *et al.*, 2010).

Palaeobathymetry. Benthic foraminifera are the principal group of use in palaeobathymetric, interpretation in the region. They are of particular use in the discrimination of sedimentary sub-environments in marginal to shallow marine, peri-deltaic, and also in deep marine, submarine fan environments and reservoirs (Batjes, in Saunders, 1968; Jones, in Ali *et al.*, 1998; Jones *et al.*, in Jones & Simmons, 1999; Fig. 5.37; see also 5.3.3, below).

Palaeobiogeography. Plants, bivalves, gastropods, fish, amphibians, reptiles and mammals record the uplift of the Andes and diversion of the Orinoco to its present-day position (Hooghiemstra, in Vrba *et al.*, 1995; Rasanen *et al.*, 1995; Kay *et al.*, 1997; Gregory-Wodzicki, 2000; Albert *et al.*, 2006; Hooghiemstra *et al.*, 2006; Jones, 2006; Hershkovitz *et al.*, 2006; Sanchez-Villagra & Aguilera, 2006; Wesselingh & Macsotay, 2006; Graham, 2009; Hoorn & Wesselingh, 2010; Montoya *et al.*, 2010; Mora *et al.*, in Hoorn & Wesselingh, 2010; Riff *et al.*, in Hoorn & Wesselingh, 2010). Specifically, plants record the uplift of the Andes; and bivalves, gastropods, fish, amphibians, reptiles and mammals, the diversion of the Orinoco to its present-day position, those of Orinoco basin affinity, including, for example, piranha, gharial, and megatheriid and mylodontid sloth, occurring in western Venezuela from the Oligocene to Early Miocene, and in eastern Venezuela from the Middle Miocene to Holocene, following the uplift of the (Merida) Andes (and of the Vaupes arch, forming the watershed between the Orinoco and Amazon).

Foraminifera, bivalves, gastropods, fish and mammals record the shoaling and closure of the former Strait of Panama and the creation of the Isthmus of Panama in response to the convergence of the Caribbean and South American plates (Keller *et al.*, 1989; Kirby & MacFadden, 2005; Jones, 2006; Schmidt, in Williams *et al.*, 2007a; Wang & Tedford, 2008; Campbell *et al.*, 2010). Specifically, fish record the existence of the Strait of Panama in the Middle Miocene to Early Pliocene, those from the Pacific and Caribbean being of identical palaeobiogeographic aspect at this time; foraminifera, the shoaling of the Strait of Panama throughout the Early Pliocene; marine bivalves and gastropods, the closure of the Strait of Panama in the Middle Pliocene, those from the Pacific and Caribbean becoming isolated and beginning to evolve separately at this time; and terrestrial mammals, the creation of the Isthmus of Panama also in the Middle Pliocene, those from North

(a)

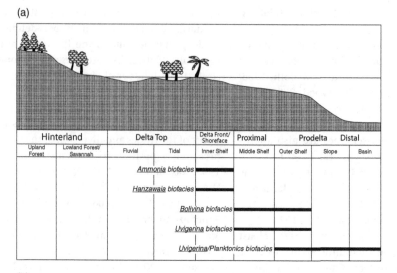

(b)

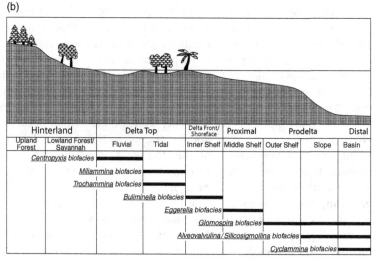

Fig. 5.37 **Palaeoenvironmental interpretation of micropalaeontological biofacies, late Cenozoic, eastern Venezuelan basin.** (a), 'Clear water'; (b), 'turbid water'. © The Geological Society, London. Reproduced with permission from Jones *et al.*, in Jones and Simmons (1999).

and South America beginning to interchange freely at this time.

Integrated studies

Integrated bio- and sequence-stratigraphic schemes for the region have been published by, among others, Galea-Alvarez and Moreno-Vasquez (1994), Carrillo *et al.* (1995), Crespo de Cabrera and di Gianni Canudas (1994), Moreno-Vasquez (1995),

Sams (1995), Flor *et al.* (1997), Erlich *et al.* (1999), Helenes and Somoza (1999), Jones *et al.*, in Jones and Simmons (1999) and Wood (2000).

Palaeogeographic and facies maps of sequences have been published by, among others, Rohr (1991), Sams (1995), Jones, in Ali *et al.* (1998), Jones *et al.*, in Jones and Simmons (1999; Fig. 5.38) and Hoorn and Wesselingh (2010). Source and/or reservoir distributions can be predicted from maps such as these.

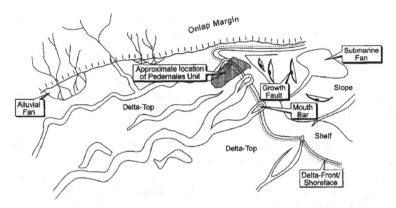

Fig. 5.38 **Palaeogeographic and facies map, Pedernales member, eastern Venezuelan basin**. © The Geological Society, London. Reproduced with permission from Jones *et al.*, in Jones and Simmons (1999).

5.2.4   South Atlantic salt basins

The following is an abridged account of the geology and petroleum geology of the South Atlantic salt basins, modified after Jones (1996). Readers interested in further details are referred to Nairn and Stehli (1973), Ojeda (1982), Magnavita and da Silva (1995), Cainelli and Mohriak (1999), Cameron *et al.* (1999), Davison, in Cameron *et al.* (1999), Karner and Driscoll, in Cameron *et al.* (1999), Karner, in Mello and Katz (2000), Mello and Katz (2000), Arthur *et al.* (2003) and Karner and Gamboa, in Schreiber *et al.* (2007). Those interested in the geology and petroleum geology of the Niger Delta immediately to the north are referred to Jones (1996), Armentrout *et al.*, in Jones and Simmons (1999), van der Zwan and Brugman, in Jones and Simmons (1999), Haack *et al.*, in Mello and Katz (2000), and Jermannaud *et al.* (2010).

Geological setting

The South Atlantic salt basins under consideration are those situated in eastern South America and west Africa between the Equatorial Fracture Zone to the north, and the Rio Grande Rise and Walvis Ridge to the south.

The South Atlantic salt basins owe their existence to South Atlantic rifting in the Early Cretaceous, Neocomian to 'early' Aptian, and drifting from the 'middle' Aptian to Recent, some of the palaeobiogeographic evidence for which is outlined under 'Palaeobiogeography' below. The nature of the pre-rift sequence is poorly understood. The Neocomian

to 'early' Aptian syn-rift sequence is restricted to grabens oriented sub-parallel to the modern coastlines of the formerly conjugate eastern South American and west African margins. It is characterised by predominantly non-marine facies, although also by marked lateral and vertical changes in facies. The boundary between the syn- and post-rift sequences is marked by a major – 'break-up' – unconformity. The 'middle' Aptian–Recent post-rift sequence is characterised by predominantly marine facies, and also by an upward transition from marginal through shallow to deep marine facies. It comprises, in ascending stratigraphic order, 'middle' Aptian marginal marine clastics and evaporites, 'late' Aptian–Albian shallow marine carbonates, and Late Cretaceous–Cenozoic deep marine clastics. The Aptian evaporites, including salts, were mobilised by sediment loading in the Albian to produce a variety of structures (initially salt wells and peripheral sinks, and subsequently diapirs and salt-withdrawal structures, including rim synclines). The deep marine clastics were predominantly fine-grained in the retrogradational Late Cretaceous–Eocene interval; and predominantly coarse-grained in the progradational Oligocene–Recent (following major tectono- or glacio-eustatically mediated sea-level falls in the Late Eocene and Early Oligocene).

Outcrop geology

The outcrop geology of the onshore portions of the South Atlantic salt basins is reasonably well documented in the public-domain literature. Pub-lished field geological maps are available.

Petroleum geology

The South Atlantic salt basins are interpreted as having remaining reserves in excess of 30 billion barrels of oil equivalent (Coward *et al.*, in Cameron *et al.*, 1999).

Source-rocks. Non-marine, lacustrine source-rocks are developed in the pre-salt; marine source-rocks at least locally at several horizons in the post-salt, including the Apto-Albian, Cenomanian–Turonian, Senonian, Palaeogene and Neogene (Koutsoukos *et al.*, 1990, 1991; Mello *et al.*, in Katz & Pratt, 1993; Burwood, in Katz, 1994; Trindade *et al.*, in Katz, 1994; Jones, 1996; Burwood, in Cameron *et al.*, 1999; Coward *et al.*, in Cameron *et al.*, 1999; Schiefelbein *et al.*, in Cameron *et al.*, 1999; Cole *et al.*, in Mello & Katz, 2000; Guardado *et al.*, in Mello & Katz, 2000; Harris, 2000; Katz & Mello, in Mello & Katz, 2000; Schoelkopf & Patterson, in Mello & Katz, 2000; Szatmari, in Mello & Katz, 2000; Harris *et al.*, 2004). Source-rock depositional models are given by Koutsoukos *et al.* (1990, 1991) and Mello *et al.*, in Katz & Pratt, 1993. The marine source-rocks in the Apto-Albian, Cenomanian–Turonian and Senonian appear to coincide with Cretaceous oceanic anoxic events or OAEs 1, 2 and 3, respectively.

Reservoir-rocks. Non-marine, lacustrine carbonate reservoir-rocks are developed in the pre-salt; marine carbonate and clastic reservoirs at least locally at several horizons in the post-salt, including the Albian (marginal to shallow marine, peri-deltaic clastics and shallow marine, platform carbonates), Late Cretaceous (deep marine, submarine fan clastics) and Cenozoic (again, deep marine, submarine fan clastics) (Bertani & Carozzi, 1984, 1985; Peres, 1993; Trindade *et al.*, in Katz, 1994; Jones, 1996; Coward *et al.*, in Cameron *et al.*, 1999; de Carvalho *et al.*, in Gierlowski-Kordesch & Kelts, 2000; Harris, in Mello & Katz, 2000; Liro & Dawson, in Mello & Katz, 2000; de Castro, 2006; Gomes *et al.*, 2009). Reservoir depositional models for the non-marine, lacustrine carbonate reservoirs of the pre-salt of the Campos basin in Brazil are given by Bertani and Carozzi (1984, 1985) and de Carvalho *et al.*, in Gierlowski-Kordesch and Kelts (2000). A reservoir depositional model for the deep marine, submarine fan clastic reservoirs of the post-salt of the Campos basin in Brazil is given by Peres (1993).

Analogues for the submarine fan sandstone reservoirs of the region can be studied at outcrop in the Austrian and Swiss Alps, Polish Carpathians and Spanish Pyrenees, and in the Karoo basin of South Africa.

Cap-rocks and traps. The ultimate cap-rock to the pre-salt reservoirs is the salt. The principal cap-rocks to the post-salt turbidite reservoirs are associated hemipelagites (see Fig. 5.5 above).

Traps are predominantly structural, although stratigraphic traps are developed in the Candeias field in the Reconcavo basin, and in the Lagoa Parda field in the Espirito Santo basin in Brazil (Jones, 1996).

Applications of palaeontology: biostratigraphy

A wide range of biostratigraphically and/or palaeobiologically useful fossil groups have been recorded in the South Atlantic salt basins, including Cyanobacteria, dinoflagellates, diatoms, calcareous nannoplankton, foraminifera, radiolarians, plants, fungi, bivalves, gastropods, ammonoids, ostracods, branchiopods, crinoids, fish, and reptiles and birds (Jones, 1996, 2006; see also Sections 2.1 and 3.1; and Sections 2.7 and 3.7 for discussions of quantitative approaches). A similarly wide range of biostratigraphic zonation schemes is applicable, the scale ranging from regional to reservoir, the resolution in inverse proportion to the scale.

Pre-salt Mesozoic. Local terrestrially derived spore and pollen, freshwater ostracod, and integrated biostratigraphic zonation schemes are applicable in the predominantly non-marine Neocomian to 'early' Aptian, 'Gondwana Wealden' sequences of the pre-salt (Grekoff, 1953; Krommelbein, 1962, 1963, 1964a, b, 1965a, b; Fonseca, in van Hinte, 1966; Krommelbein, in van Hinte, 1966; Muller, in van Hinte, 1966; Viana, in van Hinte, 1966; Grekoff & Krommelbein, 1967; Grosdidier, 1967; Krommelbein, 1968; Krommelbein & Weber, 1971; Viana *et al.*, 1971; Moura, 1972; Jardine *et al.*, 1974; Regali *et al.*, 1974a, b; Doyle *et al.*, 1977; Cunha & Moura, 1979; Petri & Campanha, 1981; Moura, in Hanai *et al.*, 1988; Colin *et al.*, 1992; Goodall *et al.*, 1992; Grosdidier *et al.*, 1996; Jones, 1996; Braccini *et al.*, 1997;

Bate, in Cameron *et al.*, 1999; Bate, 2001; Jones, 2006; Davison & Jones, 2005; Ramos *et al.*, 2006; do Carmo *et al.*, 2008).

Post-salt Mesozoic and Cenozoic. Global standard calcareous nannoplankton and planktonic foraminiferal schemes, or modifications thereof, are applicable in the predominantly marine 'late' Aptian–Recent sequences of the post-salt (Seiglie & Baker, in Watkins & Drake, 1982; Seiglie & Baker, 1983; Seiglie & Baker, in Schlee, 1984; Kogbe & Mehes, 1986; Jones, 1996; Grant *et al.*, 2000; Jones, 2006; Davison & Jones, 2005).

Other groups have also proved to be of at least local biostratigraphic use in the post-salt, including dinoflagellates, diatoms, benthic foraminifera, radiolarians, spores and pollen, ammonoids, ostracods and fish (Regali *et al.*, 1974a, b; Seiglie & Baker, 1983; Miller *et al.*, 2002; Jones, 2006; Lazarus *et al.*, 2006; Kender *et al.*, in Kaminski & Coccioni, 2008). Diatoms have proved of use in the stratigraphic subdivision and correlation of the Late Miocene part of the Malembo formation of the Pacassa and Veado wells in Block 3, offshore Angola (Jones, 2006). Here, two zones, a lower, *Thalassiosira sira* Zone and an upper, *Nitzschia porteri* Zone, have been recognised, as also at DSDP Site 362 on the Walvis Ridge to the south, that can be calibrated against the geochronological time-scale between 8.8 and 7.4 Ma. Benthic foraminifera have proved of use in the stratigraphic subdivision and correlation of the Oligocene of Angola (Kender *et al.*, in Kaminski & Coccioni, 2008; see also Section 2.7 above).

Palaeobiological interpretation

Calcareous microfossils, especially benthic foraminifera, ostracods and branchiopods, and organic-walled microfossils or palynomorphs, especially marine dinoflagellates and terrestrially derived spores and pollen, are perhaps the most important groups of fossils in palaeobiological, palaeoecological and palaeoenvironmental interpretation in the South Atlantic salt basins, although planktonic foraminifera, plants, crinoids, fish and reptiles are all important in the field of palaeobiogeography (see below). Interpretation is based on analogy with their living counterparts, on functional morphology, and on associated fossils and sedimentary environments (Brun *et al.*, 1984; Koutsoukos *et al.*, 1991;

Debenay & Basov, 1993; Jones, 1996; Bate, in Cameron *et al.*, 1999; do Carmo *et al.*, 1999; Preece *et al.*, in Cameron *et al.*, 1999; Jones, 2001; Lana *et al.*, 2002; Miller *et al.*, 2002; Jones, 2003a, b; Carvalho, 2004; Dilcher *et al.*, 2005; Jones, 2006; Ramos *et al.*, 2006; Batten, in Martill *et al.*, 2007; Heimhofer & Martill, in Martill *et al.*, 2007; Mohr *et al.*, in Martill *et al.*, 2007; Martinez *et al.*, 2008; Volkheimer *et al.*, 2008; Heimhofer & Hochuli, 2010; see also 3.7 above).

Incidentally, microfacies and agglutinated foraminiferal 'morphogroup' analysis has also been used in reservoir characterisation, specifically in the identification of intra-reservoir mudstones constituting potential barriers and baffles to fluid flow (Jones, 2001; Jones, 2003a, b).

Palaeobathymetry. The bathymetric distributions of modern benthic foraminifera around the margins of the South Atlantic can be used as proxies for the palaeobathymetric interpretation of the marine sequences of the post-salt.

Marginal, shallow and deep marine environments can be differentiated on the basis of benthic foraminifera. Shallow marine, inner, middle and outer neritic or shelf; and deep marine, upper, middle and lower bathyal or slope, and abyssal sub-environments can at least locally also be differentiated. Some measure of depth or distance from shoreline can also be derived from the ratio of planktonic to benthic foraminifera (although the trend to a higher ratio in deep marne environments is non-linear and indeed is locally reversed in upper slope sub-environments). The palaeobathymetric distributions of Cenozoic and post-salt Cretaceous benthic foraminifera in the Atlantic basins of Brazil have been established by Koutsoukos (1985) and Koutsoukos *et al.* (1991). Note that terrestrially derived plant spores and pollen, and fungal spores and hyphae, occur, often to the virtual exclusion of marine palynomorphs and other microfossils in some interpreted deep marine submarine fan reservoir sandstones in the Oligocene of Angola, interpreted as representing hyperpycnal flows of dense, sediment-laden, fresh water, originating in river mouths (Jones, 2006).

Ostracod diversity has been used as a proxy for the palaeobathymetric interpretation of the non-marine sequences of the pre-salt (Bate, in Cameron *et al.*, 1999).

Palaeobiogeography. The occurrence of essentially only land plants, land-plant-derived spores and pollen, and freshwater ostracods in eastern South America and west Africa on either side of the proto-South Atlantic Ocean in the Neocomian to 'early' Aptian pre-salt indicates that oceanic conditions had not yet become established, and by implication that drifting had not yet commenced, at this time (Jones, 2006). Plant macro- and microfossil distributions point toward the existence of a single West Africa/South America or WASA biogeographic province in the Gondwana realm in the Neocomian to 'early' Aptian (Traverse, 2008). The WASA province is characterised by *Classopollis* pollen and its parent plant, an extinct cheirolepidacean (Seward, in McCabe & Parrish, 1992). The local occurrence of the essentially Tethyan marine planktonic foraminifer *Lilliputianella globulifera [Hedbergella maslakovae]* and ostracod *Orthonotocythere* in the South Atlantic in the 'late' Barremian to 'early' Aptian indicates local marine influence, and a through-going marine connection to Tethys, by this end of this time period (Jones, 2006).

The occurrences of the riverine crocodilian *Araripesuchus* in the Araripe basin in Brazil in eastern South America and in Niger in west Africa, and of the shallow marine coelacanth fish *Mawsonia* also in Brazil in eastern South America and in Congo, Niger, Algeria, Morocco and Egypt in west, northwest and north Africa, in the 'late' Aptian–Cenomanian post-salt indicates either that the South Atlantic Ocean had still not become established, and by implication that drifting had still not commenced, even at this time, or, alternatively, that the ocean had become established, although not to such an extent as to constitute a significant barrier to the dispersal of non-marine or shallow marine animals from one side to the other (Jones, 2006; de Carvalho & Maisey, in Cavin *et al.*, 2008). The occurrence of essentially Tethyan marine stemless crinoids – saccocomids and rovearinids – in the South Atlantic in the Albian-Turonian indicates, in context, that a throughgoing ocean had become established by the end of this time period (Dias-Brito & Ferre, 2001; Ferre & Granier, 2001).

Integrated studies

Integrated bio- and sequence-stratigraphic schemes for the South Atlantic basins have been published

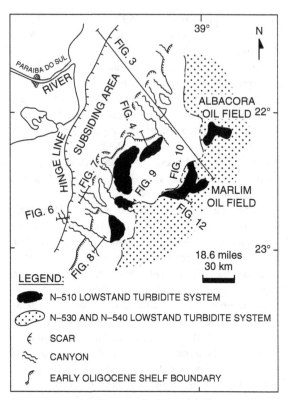

Fig. 5.39 Palaeogeographic and facies map, Oligocene, Campos basin, Brazil. © Springer. Reproduced with permission from Carminatti and Scarton, in Weimer and Link (1991).

by, among others, Braccini *et al.* (1997), Bate, in Cameron *et al.* (1999) and Moreira *et al.* (2007) for the pre-salt, and by, again among others, Seiglie and Baker, in Watkins and Drake (1982), Seiglie and Baker, in Schlee (1984), Brown *et al.* (1995), Carminatti and Scarton, in Weimer and Link (1991), Grant *et al.* (2000) and Moreira *et al.* (2007) for the post-salt. Palaeogeographic and facies maps of sequences have been published by, among others, Carminatti and Scarton, in Weimer and Link (1991) (Fig. 5.39). Source and/or reservoir distributions can be predicted from maps such as these.

Calcareous nannoplankton and planktonic foraminifera form the principal bases of the calibration of the regional sequence stratigraphic framework for the post-salt (and the basis of the identification of time-slices for facies mapping purposes). Benthic foraminiferal abundance and diversity trends, and bathymetric trends, form part of the basis of the characterisation of the maximum flooding surfaces that constitute the bounding surfaces of the sequences (Jones, 2001, 2003a and b, 2006). Calcareous nannoplankton abundance peaks in what benthic foraminiferal sedimentological data demonstrate to be autochthonous deposits also form part of the basis of the characterisation of MFSs.

5.2.5 Indian subcontinent

The following is an abridged account of the geology and petroleum geology of the Indian subcontinent, compiled from various sources. Readers interested in further details are referred to Raju (1968), Sastri *et al.* (1973), Brown and Dey (1975), Naini and Talwani, in Watkins and Drake (1982), Gombos *et al.* (1995), Veevers and Tewari (1995), Sheikh and Maseem, in Sahni (1996), Zutshi and Panwar (1997), Warwick *et al.* (1998), Biswas (1999), Nanda and Mathur (2000), Lakshminarayana (2002), Mehrotra *et al.* (2002), Segev (2002), Swamy and Kapoor (2002), Briggs (2003), Chakraborty and Ghosh (2005), Downing and Lindsay (2005), Mehrotra *et al.* (2005), Molnar (2005), Raju *et al.* (2005), Bastia (2006), Harris (2006), Rathore (2007), Raju (2008), Goswami (2008), Ramakrishnan and Vaidyanadhan (2008), Vaidyanadhan and Ramakrishnan (2008), Kassi *et al.* (2009), Lal *et al.* (2009), Melluso *et al.* (2009), Veevers and Saaed (2009) and Mehrotra *et al.* (2010). Those interested in further details of the unconventional – coal-bed methane (CBM) or coal-seam gas (CSG) – petroleum geology of the Gondwana basins of the Indian subcontinent are referred to Aravamudhan and Das, in Swamy and Kapoor (2002), Biswas *et al.*, in Swamy and Kapoor (2002) and Mishra *et al.*, in Swamy and Kapoor (2002) (see also 5.5.1 below). Readers interested in the geology and petroleum geology of southeast Asia to the east are referred to Jones (1996) and Fraser *et al.* (1997).

Geological setting

The geological history of the Indian subcontinent is long and complex, like that of the Middle East and of other formerly contiguous parts of the supercontinent of Gondwana, with which it shares certain common characteristics (Fig. 5.40; see also 5.2.1 above; and Fig. 5.8 above).

It essentially began with ?extensional tectonism and the formation of a series of basins, including salt basins, the best known of which is in the Salt Ranges in Pakistan, in the Precambrian-'Infracambrian' (Arabian Plate Megasequence AP1 of Sharland *et al.*, 2001). Nothing remains other than in parts of the Himalaya of the later Cambrian to Carboniferous stratigraphy (Megasequences AP2–AP4).

The preserved latest Carboniferous to Early Jurassic stratigraphy (Megasequences AP5–AP6) records a series of incipient rifting events related to the break-up of Gondwana, some of the palaeobiogeographic evidence for which is outlined under 'Palaeobiogeography' below. Continental clastic sedimentation tended to dominate, although there is also evidence of intermittent marine incursion onto the continent. Glacial sediments, including tillites such as the widely developed Talchir Boulder Bed and its correlatives, characterise the latest Carboniferous to earliest Permian, Gzelian to Asselian; and deglacial sediments, including marine mudstones, the 'early' Early Permian, Sakmarian. The so-called Barakar 'Coal Measures' and their correlatives characterise the 'mid' Early Permian, Artinskian; the Barren Measures the 'late' Early Permian, Kungurian; and the Raniganj 'Coal Measures' the Middle–Late Permian, Ufimian–Tatarian. Further continental clastics, including the Panchet and supra-Panchet and their correlatives, characterise the Triassic–Early Jurassic.

Greater India, including Madagascar and the Seychelles, and Antarctica and Australasia, drifted away from east Africa in the Middle Jurassic, approximately 160 Ma. In the Middle–Late Jurassic, shallow marine platform carbonate sedimentation commenced along the west coast of the subcontinent, although continental clastic sedimentation continued to dominate in the interior and on what was to become the east coast, which was still attached to Antarctica and Australia, and there was also widespread volcanism associated with the development of the Karoo–Ferrar Large Igneous Province. Antarctica and Australasia drifted away in the earliest Cretaceous, approximately 137 Ma, Madagascar and the Seychelles drifted away in the Late Cretaceous, approximately 90 Ma and 65 Ma, respectively. Continental to marine clastic and shallow to deep marine carbonate sedimentation were equally important in the Cretaceous, and again there was also widespread volcanism, this time

Fig. 5.40 **Location map, Indian subcontinent**.

associated with the Rajhamal hotspot event in the Early Cretaceous, the St Mary's Trap event in the Late Cretaceous and the Deccan Trap event and the development of the Deccan Large Igneous Province at the end of the Late Cretaceous.

India collided with Eurasia at the end of the Palaeocene, approximately 55 Ma. The initial elevation of the Himalaya and the associated initiation of southerly drainage was during the time the clastic Ghazij formation was deposited in Pakistan, in the Early Eocene, approximately 50 Ma. By the end of the Eocene, clastic sedimentation once more dominated over carbonate sedimentation across the entire subcontinent. By the end of the Miocene,

approximately 9–7 Ma, the Himalaya was sufficiently elevated to affect atmospheric circulation, and to cause the climate to become markedly seasonal, or monsoonal. Himalayan 'molasse' and associated continental clastic sedimentation commenced across the entire subcontinent, exemplified by the Siwalik group and its correlatives (see Box 3.8 above, and below).

Outcrop geology

The outcrop geology of the Indian subcontinent is well documented in the public-domain literature. Published field geological maps are available for the entire area on various scales.

Petroleum geology

The principal petroliferous basins of the Indian subcontinent, are, anti-clockwise from west to east, the Indus basin of Pakistan, the Cambay basin on the west coast of India, the Bombay or Mumbai basin also on west coast of India, the Cauvery basin on the east coast of India, the Krishna–Godavari basin also on the east coast of India, and the Assam basin of India and contiguous Surma basin of Bangladesh (Fig. 5.40).

Readers interested in the petroleum geology of the Indus basin are referred to Williams (1958), Shuaib (1982), Khan *et al.* (1986), Quadri and Shuaib (1986), Dolan, in Brooks (1990), Copestake *et al.*, in Caughey *et al.* (1996), Quadri and Quadri (1998), Zaigham and Mallick (2000), Daley and Alam, in Clift *et al.* (2002), Smewing *et al.*, in Clift *et al.* (2002), Siddiqui (2004) and Carmichael *et al.* (2009) (Fig. 5.41); those interested in the Cambay basin, to Thomas *et al.*, in Swamy and Kapoor (2002), Chowdhary (2004) and Negi *et al.* (2006); those interested in the Bombay or Mumbai basin, to Rao and Talukdar, in Halbouty (1980) and Roychoudhury and Deshpande (1982); those interested in the Cauvery basin, to Watkinson *et al.* (2007); those interested in the Krishna–Godavari basin, to Venkatachala and Sinha (1986), Mohinuddin *et al.* (1991), Rao *et al.* (1996), Das *et al.* (1999), Gupta *et al.* (2001), Rao (2001), Satyanarayana *et al.*, in Swamy and Kapoor (2002), Singh *et al.*, in Swamy and Kapoor (2002), Biswas (2003), Reddy *et al.* (2005), Bastia (2006), Bastia and Nayak (2006), Gupta (2006) and Bastia *et al.* (2007); and those interested in the Assam and Surma basins to Alam (1989), Imam (2005) and Mukherjee *et al.* (2009). Those interested in the emerging potential of the Barmer basin in Rajasthan are referred to Ellet *et al.* (2005) and Compton (2009).

Source-rocks. A range of peri-deltaic, coaly, prodeltaic and marine thermogenic – and terrestrially influenced marine biogenic – source-rocks are developed in the Indian subcontinent (Basu *et al.*, 1980; Khan *et al.*, 1986; Quadri & Shuaib, 1986; Dolan, in Brooks, 1990; Raza *et al.*, 1990; Biswas *et al.*, in Magoon & Dow, 1994; Gombos *et al.*, 1995; Peters *et al.*, 1995; Dhannawat & Mukherjee, 1997; Zutshi & Panwar, 1997; Quadri & Quadri, 1998; Robison *et al.*, 1999; Husain *et al.*, 2000; Pandey *et al.*, 2000; Prasad *et al.*, 2000; Raza Khan *et al.*, 2000; Swamy & Kapoor, 2000; Zaigham & Mallick, 2000; Rao, 2001; Aswal *et al.*, in Swamy & Kapoor, 2002; Banerjee *et al.*, 2002; Curiale *et al.*, 2002; Gourshetty & Bhattacharya, in Swamy & Kapoor, 2002; Imam & Hussain, 2002; Mehrotra *et al.*, 2002; Vyas *et al.*, in Swamy & Kapoor, 2002; Chowdhary, 2004; Wandrey, 2004; Wandrey *et al.*, 2004; Ahmad & Ahmad, 2005; Chandra *at al.*, in Raju *et al.*, 2005; Imam, 2005; Baillie *et al.*, 2006; Gupta, 2006; Bhatnagar *et al.*, 2007; Goswami *et al.*, 2007; Kapoor & Nanjundaswamy, 2007; Frielingsdorf *et al.*, 2008; Pande *et al.*, 2008; Singh *et al.*, 2008; Carmichael *et al.*, 2009; Fazeelat *et al.*, 2010; see also comments on 'palaeobiological controls on (biogenic) source-rock development', below).

Restricted marine source-rocks are developed in the Proterozoic–Palaeozoic, for example in the Salt Ranges in Pakistan, and the Bikaner–Nagaur basin in Rajasthan in neighbouring northwest India. Predominantly peri-deltaic, coaly source-rocks are developed in the Palaeozoic, for example in the Nilawahan and Zaluch formations in the Potwar basin in Pakistan; in the Barakar and Raniganj and equivalent formations in the Krishna–Godavari, Pranhita–Godavari, Dhamodar/Satpura and Bengal basins, and the Barren Measures in the Pranhita–Godavari basin, in eastern India; and in Bangladesh. Predominantly prodeltaic to marine source-rocks are developed in the Mesozoic, for example in the Mianwali and Datta formations in the Potwar basin, and the Sembar formation in the Indus basin, in Pakistan; in the Jhurio and Jumara formations in the Kutch basin, and the Baisakhi–Bedesir and Pariwar formations in the Jaisalmer basin, in western India; in the Mannar basin in Sri Lanka; and in the Andimadam and Sattapadi formations in the Cauvery basin, and the Mandapeta, Krishna, Gajulapadu and Raghavapuram formations in the Krishna–Godavari basin, in eastern India. Peri-deltaic, coaly, prodeltaic and marine source-rocks are developed in the Cenozoic, for example in the Lockhart, Patala and Sakesar formations in the Potwar basin, and in the Indus basin, in Pakistan; in the Olpad, Cambay Shale and Kalol formations in the Cambay basin, and the Panna formation in the Bombay basin, in western India; in the Palakollu, Vadaparru and Ravva formations in the

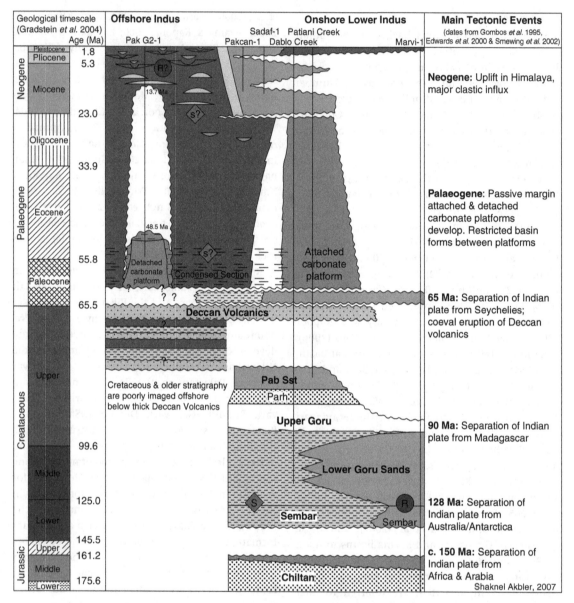

Fig. 5.41 **Chronostratigraphic diagram, Indus basin, Pakistan.** © The Geological Society, London. Modified with permission after Carmichael *et al.* (2009).

Krishna–Godavari basin, in the Mahanadi–Bengal basin, and in the Kopili and Jenam formations in Assam, in eastern India; and in the Tura, Kopili, Jenam and Bhuban formations in Bangladesh.

Some of the more important of these, which have been identified on the basis of typing as contributing to important working petroleum systems,

are the Cambay Shale formation in the Cambay basin; the Panna formation in the Bombay basin; the Andimadam formation in the Cauvery basin; the Kommugedem, Gajulapadu, Raghavapuram, Palakollu and Vadaparru formations in the Krishna–Godavari basin; and the Kopili and Jenam formations in Assam.

Direct or indirect analogues for the source-rocks of the Cambay basin can be studied at outcrop (Mahanti et al., 2006).

Reservoir-rocks. A range of continental to marginal marine, peri-deltaic to deep marine, submarine fan clastic, and shallow marine, platformal carbonate, reservoir-rocks are developed in the Indian subcontinent (Rao & Talukdar, in Halbouty, 1980; Dolan, in Brooks, 1990; Gombos et al., 1995; Zutshi & Panwar, 1997; Quadri & Quadri, 1998; Husain et al., 2000; Raza Khan et al., 2000; Zaigham & Mallick, 2000; Rao, 2001; Curiale et al., 2002; Gourshetty & Bhattacharya, in Swamy & Kapoor, 2002; Imam & Hussain, 2002; Chowdhary, 2004; Wandrey, 2004; Wandrey et al., 2004; Chandra at al., in Raju et al., 2005; Imam, 2005; Compton, 2009).

Continental clastic reservoir-rocks are developed in the Palaeozoic, for example in the Jhelum and Nilawahan formations in the Potwar basin in Pakistan, and the Mandapeta formation in the Krishna–Godavari basin on the east coast of India. Predominantly continental to marginal marine, peri-deltaic to deep marine, submarine fan clastic reservoir-rocks are developed in the Mesozoic, for example in the Datta, Shinawari, Samana Suk, Chichali and Lumshiwal formations in the Potwar basin, and the Lower Goru formation in the Indus basin, in Pakistan; in the Kutch basin in western India; and in the Bhuvanagiri and Nannilam formations in the Cauvery basin, and the Golapalli, Kanakollu and Tirupati formations in the Krishna–Godavari basin, in eastern India. Clastic and carbonate reservoir-rocks are developed in the Cenozoic, for example in the Potwar basin and in Kirthar and Sulaiman Ranges in Pakistan; in the Fatehgarh formation in the Barmer basin, the Anklesvar, Kadi and Kalol formations in the Cambay basin, and the Bombay formation in the Bombay basin, in western India; in the Niravi formation in the Cauvery basin, the Pasarlapudi and Ravva formations in the Krishna–Godavari basin, and the Tura, Sylhet and Kopili formations and Barail and Tipam groups in Assam, in eastern India; and the Jaintia, Barail and Surma groups in Bangladesh.

Direct or indirect analogues for the reservoir-rocks of the region can be studied at outcrop (Fitzsimmons et al., 2005; Mahanti et al., 2006).

Cap-rocks and traps. A range of continental to marine predominantly fine-grained clastic cap-rocks are developed in the Indian subcontinent (Rao & Talukdar, in Halbouty, 1980; Dolan, in Brooks, 1990; Gombos et al., 1995; Zutshi & Panwar, 1997; Quadri & Quadri, 1998; Husain et al., 2000; Raza Khan et al., 2000; Zaigham & Mallick, 2000; Rao, 2001; Curiale et al., 2002; Gourshetty & Bhattacharya, in Swamy & Kapoor, 2002; Imam & Hussain, 2002; Chowdhary, 2004; Wandrey, 2004; Wandrey et al., 2004; Chandra at al., in Raju et al., 2005; Imam, 2005; Compton, 2009). The Upper Goru formation seals the Lower Goru formation reservoir in the Indus basin; the Tarapur seals the Anklesvar, Kadi and Kalol reservoirs in the Cambay basin; the Bandra seals the Bombay reservoir in the Bombay basin; the Kudavasal seals the Bhuvanagiri reservoir, the Komarakshi, the Nannilam reservoir, and the Shiyali, the Niravi reservoir, in the Cauvery basin; the Raghavapuram seals the Kanakollu reservoir, the Razole, the Tirupati reservoir, the Vadaparru, the Pasarlapudi reservoir, and the Godavari, the Ravva reservoir, in the Krishna–Godavari basin in the east, in India; and the Upper Manna seals the Surma reservoir in Bangladesh.

Traps are predominantly structural, although a wide range of structural styles is known. Some recent discoveries in deep marine, submarine fan reservoirs appear to have a stratigraphic component.

Applied palaeontology: biostratigraphy

A wide range of biostratigraphically and/or palaeobiologically useful fossil groups have been recorded in the surface and subsurface sections of the Indian subcontinent, including dinoflagellates, diatoms, calcareous nannoplankton, calcareous Algae, acritarchs, foraminifera, plants, ammonoids, ostracods, branchiopods, fish, amphibians, reptiles and birds, and mammals (Jones, 2006; see also Sections 2.1 and 3.1 above; and 2.7 and 3.7 above for discussions of quantitative approaches). A similarly wide range of biostratigraphic zonation schemes is applicable, the scale ranging from regional to reservoir, the resolution in inverse proportion to the scale.

Proterozoic. A regional acritarch biostratigraphic zonation scheme is applicable to the marine Proterozoic of the Ganga basin in India (Prasad et al., in Swamy & Kapoor, 2002). Here, the Late Precambrian, Riphean is characterised by the *Tappania plana* and *Navifusa granulatus* Zones, and the Late

Precambrian, Vendian to Early Cambrian by the *Lophosphaeridium rarum* Zone.

Palaeozoic. Regional acritarch biostratigraphic zonation schemes are applicable to the marine Palaeozoic, and regional spore and pollen biozonation schemes to the non-marine to marine Palaeozoic, of the Indian subcontinent. Regional estheriid biostratigraphic zonation schemes are applicable to the marginal marine Palaeozoic. Larger benthic foraminifera and brachiopods are of local biostratigraphic use in marine environments.

A regional acritarch biostratigraphic zonation scheme is applicable to the marine Palaeozoic of the Ganga basin in India (Prasad *et al.*, in Swamy & Kapoor, 2002). Here, the Late Precambrian, Vendian to Early Cambrian is characterised by the *Lophosphaeridium rarum* Zone, the Cambrian by the *Micrhystridium tornatum*, *Cristallinum cambriense* and *Cymatiosphaera crameri* Zones, the Late Cambrian to Early Ordovician, Tremadoc by the *Eupoikilofusa squama* Zone, the Late Ordovician, Ashgill to Silurian, early Llandovery by the *Eupoikilofusa striatifera* Zone, the Silurian, late Llandovery to early Ludlow by the *Helosphaeridium citrinipeltatum*, *Dictyotidium stenodictyum* and *Helosphaeridium latispinosum* Zones, and the Silurian, late Ludlow to Early Devonian, Siegenian by the *Polydrixium multifrons* Zone.

Regional spore and pollen schemes are applicable in the non-marine to marine Palaeozoic, for example in the Gondwana basins on the east coast and in the interior of India, and in parts of Pakistan (Prasad *et al.*, 1995; Veevers & Tewari, 1995; Prasad, in Pandey *et al.*, 1996; Iqbal *et al.*, 1998; Prasad & Pundir, 1999; Prasad *et al.*, 2000; Mehrotra *et al.*, 2002; Archayya, in Raju *et al.*, 2005; Awatar, in Sinha, 2007; Jan *et al.*, 2009; Vijaya, 2009; Mehrotra *et al.*, 2010). In the Gondwana basins of India, the regional Talchir formation is dated as Late Carboniferous–Early Permian, early Sakmarian, the Karharbari as Early Permian, late Sakmarian, the Barakar Coal as Early Permian, Artinskian–early Kungurian, the Barren Measures as Early–Middle Permian, late Kungurian–Ufimian (early Roadian), and the Raniganj Coal as Middle-Late Permian, Kazanian–Tatarian (late Roadian–?Changsingian).

Regional estheriid biostratigraphic zonation schemes are applicable to the marginal marine Permian of the Gondwana basins of India (Ghosh, 1984).

Mesozoic. Global standard calcareous nannoplankton and planktonic foraminiferal zonation schemes are applicable to the marine Mesozoic of the Indian subcontinent. Regional dinoflagellate, larger benthic foraminiferal and ammonite zonation schemes are also applicable in marine environments; and regional spore and pollen biozonation schemes in non-marine to marine environments. Regional estheriid biostratigraphic zonation schemes are applicable in marginal marine environments. Diatoms, bivalves, belemnites and ostracods are of local biostratigraphic use in non-marine to marine environments.

Global standard calcareous nannoplankton and planktonic foraminiferal schemes, or minor regional modifications thereof, and regional larger benthic foraminiferal schemes, are applicable to the marine Mesozoic of the Indian subcontinent, for example in the Krishna–Godavari and Cauvery basins on the east coast of India, in the Kutch basin on the west coast of India, in the Jaisalmer basin in the interior of India adjacent to the border with Pakistan, and in parts of Pakistan (Bhalla, 1969; Kureshy, 1977a, b, 1978a; Govindan & Sastri, in Verdenius *et al.*, 1983; Kothe, 1988; Saxena & Misra, 1995; Abbas, in Pandey *et al.*, 1996; Govindan *et al.*, in Sahni, 1996; Jafar, in Sahni, 1996; Kumar & Saxena, 1996; Watkinson *et al.*, 2007).

Regional dinoflagellate schemes are applicable to the marine Mesozoic, and regional spore and pollen schemes to the non-marine to marine Mesozoic, for example in the Gondwana basins in the interior and on the east coast of India, in the Krishna–Godavari basin on the east coast, and in the Kutch basin on the west coast (Venkatachala & Sinha, 1986; Prasad *et al.*, 1995; Veevers & Tewari, 1995; Kumar & Saxena, 1996; Prasad *et al.*, in Sahni, 1996; Prasad, 1997; Prasad & Pundir, 1999; Vijaya, 1999; Aswal *et al.*, 2001; Aswal & Mehrotra, 2002; Aswal *et al.*, in Swamy & Kapoor, 2002; Prasad & Pundeer, 2002; Vijaya & Bhattacharji, 2002; Vijaya & Kumar, 2002; Mehrotra & Aswal, 2003; Mehrotra & Singh, 2003; Mehrotra, in Raju *et al.*, 2005; Mehrotra *et al.*, 2005; Khowaja-Ateequzzaman *et al.*, 2006; Tripathi & Ray, 2006; Prakash, 2008; Tripathi, 2008; Vijaya, 2009; Mehrotra *et al.*, 2010).

In the Gondwana basins, the regional Panchet formation is dated as Early Triassic, the supra-Panchet or Lower Dubrajpur as Late Triassic, Carnian, the Upper Dubrajpur as late Middle Jurassic to earliest Cretaceous, Callovian–Berrisian, and the Rajmahal as Early Cretaceous.

Regional ammonite schemes are applicable to the marine Mesozoic, for example in the Himalaya, in the interior of India, in the Krishna–Godavari and Cauvery basins on the east coast of India, in the Kutch basin on the west coast of India, in the Jaisalmer basin in the interior of India adjacent to the border with Pakistan, and in parts of Pakistan (Arkell, 1956; Fatmi, 1972; Gradstein et al., 1989; Jafar, in Sahni, 1996; Juyal et al., in Pandey et al., 1996; Vijaya & Kumar, 2002; Kennedy et al., 2003; Bhargava, in Raju et al., 2005; Krishna, in Raju et al., 2005; Krystyn et al., in Raju et al., 2005; Raju & Ramesh, in Raju et al., 2005; Roy et al., 2007; Vaidyanadhan & Ramakrishnan, 2008; Shome & Bardhan, 2009).

Regional estheriid biostratigraphic zonation schemes are applicable to the marginal marine Triassic of the Gondwana basins of India (Ghosh, 1984).

Cenozoic. Global standard calcareous nannoplankton and planktonic foraminiferal zonation schemes are applicable to the marine Cenozoic of the Indian subcontinent. Regional dinoflagellate and larger benthic foraminiferal schemes are also applicable in marine environments; regional spore and pollen biozonation schemes in non-marine to marine environments; and regional vertebrate schemes in non-marine environments. Ostracods and fish are of local biostratigraphic use in non-marine to marine environments.

For example, global standard calcareous nannoplankton and planktonic foraminiferal schemes, or minor regional modifications thereof, and regional larger benthic foraminiferal schemes are applicable in the Krishna–Godavari and Cauvery basins on the east coast of India, in the Bombay and Cambay basins on the west coast, and in parts of Pakistan (Eames, 1952; Haque, 1956; Samanta, 1973; Raju, 1974; Kureshy, 1978a, b; Kothe, 1988; Abbas, in Pandey et al., 1996; Mishra, 1996; Ramesh, in Pandey et al., 1996; Jones, 1997; Dave, 2000; Warraich et al., 2000; Warraich & Nishi, 2003; Govindan, 2004; Afzal et al., 2005; Mehrotra et al., 2005; Raju, in Raju et al., 2005; Raju et al., in Raju et al., 2005; Raju, Ramesh et al., in Raju et al., 2005; Raju & Uppal, in Raju et al., 2005; Vaidyanadhan & Ramakrishnan, 2008; Afzal et al., 2009; Keller et al., 2009; Fig. 5.42).

Importantly, LBFs are applicable to the characterisation of the Oligocene–Miocene carbonate reservoir in the Bombay High field in the offshore Bombay or Mumbai basin on the west coast of India (Raju, Ramesh et al., in Raju et al., 2005; Raju & Uppal, in Raju et al., 2005).

Regional dinoflagellate schemes are applicable to the marine Cenozoic, and regional spore and pollen schemes, calibrated against global calcareous nannoplankton and planktonic foraminiferal schemes, to the non-marine to marine Cenozoic, on the east and west coasts of India and in Pakistan (Koshal & Uniyal, 1984; Kothe, 1988; Juyal et al., in Pandey et al., 1996; Jones, 1997; Aswal et al., 2001; Mehrotra et al., 2002; Prasad & Pundeer, 2002; Singh et al., in Swamy & Kapoor, 2002; Mehrotra, in Raju et al., 2005; Khowaja-Ateequzzaman et al., 2006; Mehrotra et al., 2010).

Regional vertebrate schemes are applicable to the non-marine Cenozoic, for example in the Himalayan molasse (Nanda & Sehgal, in Raju et al., 2005; Jones, 2006; Vaidyanadhan & Ramakrishnan, 2008; Metais et al., 2009). Here, the Oligocene to earliest Miocene, Aquitanian Chitarwata formation of the Bugti hills in central Pakistan is divided into 12 mammal zones, namely Mein Zones MP21–MP30 (Oligocene) and MN1–MN2 (Miocene). The late Early Miocene, Burdigalian to Pliocene part of the Siwalik group of India and Pakistan is also divided into 12 mammal zones, namely MN3–MN14 (the lower Siwalik, represented by the Kamlial and Chinji formations, is divided into six zones, namely MN3–MN8; and the middle Siwalik, represented by the Nagri and Dhok Pathan, and the upper Siwalik, represented by the Tatrot and Pinjor, into a further six, namely MN9–MN14). And the partly overlapping Late Miocene to Pleistocene part of the Siwalik group of India and Nepal is divided into four mammal zones, namely the Hipparion s.l. Interval Zone (Late Miocene, 9.5–7.4 Ma), the Selenoportax lydekkeri Interval Zone (Late Miocene, 7.4–5.3 Ma), the Hexaprotodon sivalensis Interval Zone (Early–early Late Pliocene, 5.3–2.9 Ma), and the Elephas planifrons Interval Zone (latest Pliocene–earliest Pleistocene, 2.9–1.5 Ma).

Palaeobiological interpretation

Calcareous microfossils, especially benthic foraminifera, and to a lesser extent organic-walled microfossils or palynomorphs, especially marine

Chronostratigraphic and biostratigraphic chart showing the stratigraphic ranges of selected larger benthonic and planktonic foraminifera species across the Cretaceous through Holocene. The chart includes columns for Chronostratigraphy (Holocene, Pleistocene, Pliocene, Miocene, Oligocene, Eocene, Palaeocene, Cretaceous), Biostratigraphy (Planktonic Foraminifera zones N1–N23, P1A–P22; Calcareous Nannoplankton zones NN1–NN21, NP1–NP25), and ranges of selected fossil species including GLOBOROTALIA (MOROZOVELLA) ex gr. ANGULATA / CONICOTRUNCATA, GLOBOROTALIA (MOROZOVELLA) ex gr. SUBBOTINAE, GLOBOROTALIA (ACARININA) ex gr. PENTACAMERATA, "GLOBIGERINA" LOZANOI, ASSILINA MAIOR PUNCTULATA, ALVEOLINA ELLIPTICA, GLOBOROTALIA (ACARININA) ex gr. BULLBROOKI, ASSILINA EXPONENS, NUMMULITES BEAUMONT, ORBULINOIDES BECKMANN, TRUNCOROTALOIDES LIBYAENSIS, and GLOBIGERINA PRAEBULLOIDES.

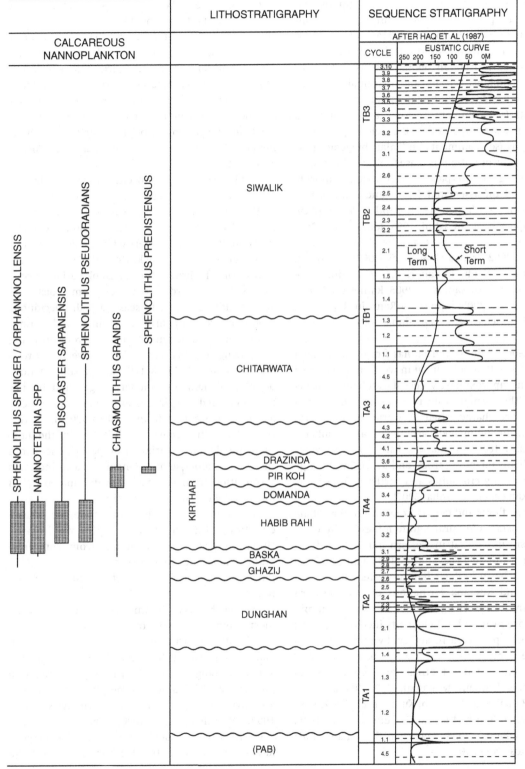

Fig. 5.42 **Stratigraphic distribution of selected benthic foraminifera, Cenozoic, Sulaiman ranges, Pakistan.** From Jones (1996).

dinoflagellates and terrestrially derived spores and pollen, are perhaps the most important groups of fossils in palaeobiological, palaeoecological and palaeoenvironmental interpretation in the Indian subcontinent, although plants, insects, ostracods, fish, amphibians, reptiles and mammals are also locally important in marginal- to non-marine environments (see comments below), and plants and vertebrates are important in the field of palaeobiogeography (see comments on 'Palaeobiogeography' below). Interpretation is based on analogy with their living counterparts, on functional morphology, and on associated fossils and sedimentary environments (Kumar, 1984; Pandey & Choubey, 1984; Spicer et al., in Sahni, 1996; Jones, 1997; Raju et al., 1999; Warraich et al., 2000; Wakefield & Monteil, 2002; Warraich & Nishi, 2003; Govindan, 2004; Raju & Naidu, in Raju et al., 2005; Raju, Satyanarayana et al., in Raju et al., 2005; Ramesh & Peters, in Raju et al., 2005; Ramesh & Raju, in Raju et al., 2005; Bhattacharjee et al., in Sinha, 2007; Afzal et al., 2009).

Palaeocene floras and faunas from the Mumbai Island formation of Mumbai in western India – and from the approximately coeval Fatehgarh formation of the Barmer basin in Rajasthan in western India – are characterised by warm, humid floral elements including podocarps, palms and bamboos; by freshwater and marine fish, by freshwater amphibians including frogs, and by freshwater reptiles including crocodiles and turtles (Cripps et al., 2005; Rana et al., in Sinha, 2007).

Eocene floras and faunas from the Cambay Shale of the Vastan and Surat lignite mines in the Cambay basin in Gujurat in western India – and from the approximately coeval Khuiala formation of the Jaisalmer basin in Rajasthan in western India, and Ghazij and Kuldana formations of northwestern Pakistan – are characterised by warm, humid mangrove-swamp floral elements including palms; by insects, preserved in amber; by freshwater and marine fish; by freshwater amphibians; by freshwater and marine reptiles; and by terrestrial mammals (Samant & Bajpai, 2001; Rana et al., 2004; Bajpai et al., 2005a, b; Kumar et al., 2007; Rose et al., 2007; Smith et al., 2007; Gunnell et al., 2008; Prasad & Bajpai, 2008; Rage et al., 2008; Rana et al., 2008; Rose et al., 2009).

Oligocene–Early Miocene floras and faunas from the Chitarwata formation (34–20 Ma) of the Bugti Hills in central Pakistan are characterised by moist deciduous, semi-evergreen tropical forest and rainforest floral elements including Terminalioxylon, hygrophilous ferns, pines, Amaranthaceae–Chenopodiceae–Caryophyllaceae and Palmae; by freshwater and marine fish; and by terrestrial mammals including elephants, hippos and rhinos (Welcomme et al., 2001; Adnet et al., 2007; de Franceschi et al., 2008; Metais et al., 2009).

Middle–early Late Miocene floras and faunas from the Siwalik group Kamlial formation (17–14 Ma), Chinji formation (14–10 Ma) and Nagri formation (10–9 Ma) of the Potwar plateau in northern Pakistan are also characterised by generally warm, humid elements (Retallack, 1991, 1995). In contrast, floras from the latest Miocene Dhok Pathan formation (9–5 Ma) record a transition from moist Gangetic deciduous forest to seasonally dry deciduous monsoon-forest to arid grassland, and faunas record an accompanying transition from browsing to grazing feeding strategies, between approximately 9 and 7 Ma (Retallack, op. cit.). Analogy between ancient and modern vegetation biomes and associated soils indicates a reduction in mean annual precipitation from 1200 to 600 mm at this time, that is, the time of the initiation of the monsoon. Incidentally, carbon isotope evidence from the Siwalik molasse and Bengal fan also indicates a transition from an ecosystem dominated by C3 forest to C4 grassland at this time.

Palaeoaltimetry. The palaeoaltitudes or palaeoelevations, and elevation or uplift rate, of the Himalaya has been calculated from contained plant macro- and microfossils (Hoorn et al., 2000; Jones, 2006; Xiao-Yan Song et al., 2010; see also Box 3.8).

Palaeobathymetry. Benthic foraminifera are the most important group of fossils in palaeobathymetric interpretation in the Indian subcontinent. Those from the Palaeogene formations of the Sulaiman Ranges of Pakistan represent a range of predominantly shallow marine platformal carbonate environments and palaeobathymetries (Jones, 1997; Warraich et al., 2000; Warraich & Nishi, 2003; Afzal et al., 2009). Those from the Neogene Ravva formation of the Krishna–Godavari basin off the east coast of India represent a range of predominantly deep marine clastic environments and palaeobathymetries (Govindan, 2004). Here, palaeoenvironmental and palaeobathymetric interpretation has been undertaken on the basis of analogy between contained fossil deep-water agglutinated foraminifera and their living counterparts (see Box 3.3 above). Three biofacies groupings have been recognised: A, interpreted as representing bathyal to abyssal bathymetries; B, interpreted as representing upper bathyal bathymetries; and C, interpreted as representing outer neritic to upper bathyal bathymetries. The observed succession from biofacies C–B–A is interpreted as representing shallowing from abyssal to lower to middle bathyal, through upper bathyal, to upper bathyal to outer neritic. The ancient oxygen minimum zone (OMZ) in upper bathyal bathymetries is characterised by infaunal agglutinated foraminifera. The modern OMZ in upper bathyal bathymetries is characterised by infaunal buliminide calcareous foraminifera (Bhattacharjee et al., in Sinha, 2007).

Palaeobathymetric interpretation is capable of being automated by the use of a proprietary software package that contains links to bathymetric distribution datasets and 'look-up' tables, as in the case of the Mesozoic and Cenozoic of the Indus basin of Pakistan (Wakefield & Monteil, 2002; see also 3.7.1 above).

Palaeobiogeography. In the Palaeozoic, the occurrence of the distinctive land plant *Glossopteris* in the Carboniferous–Permian not only of the Indian subcontinent but also of Africa, South America, Antarctica and Australia indicates that at this time these areas constituted part of a contiguous land-mass, Gondwana (Anderson et al., 1999). In the Mesozoic, the continuing occurrence of essentially identical suites of land plants in the Jurassic–Early Cretaceous of India and Australia indicates that at this time these areas continued to constitute part of Gondwana (McLoughlin & Pott, 2009). However, the increasing difference between Indian and Australian land plants from the 'Middle' Cretaceous onwards indicates that by this time Gondwana had started to rift apart and its former constituent parts to drift apart, and the resulting proto-eastern Indian Ocean had started to form a barrier to the dispersal of terrestrial organisms. In the Cenozoic, the occurrence of similar suites of land vertebrates in the Eocene of Europe, west Asia and India, and different suites in the rest of south and east Asia, indicates that even at this time the rising mountain range of the proto-Himalaya in between had started to form a barrier to the dispersal of terrestrial organisms.

Palaeobiological controls on (biogenic) source-rock development

As noted in Box 5.5 above, there are palaeobiological controls on (biogenic) source-rock development. Source-rock development is dependent on an excess of production over consumption of the organic matter (OM) that constitutes the feed-stock for the methanogenic 'Archaebacteria' that produce biogenic methane. Production of OM is controlled essentially by primary biological productivity in the terrestrial and marine realms. In the terrestrial realm, the production of OM, controlled by the productivity of land plants, is typically highest in places or at times of optimal climate. In the marine realm, the production of OM, controlled by the productivity of marine Algae, is highest in places or at times of optimal climate, or of enhanced nutrient supply associated with upwelling. Consumption and preservation of OM is controlled essentially by the availability of oxygen. Consumption is lowest, and preservation highest, in places or at times of dysoxia or anoxia.

In the case of the Indian subcontinent, the production of OM in the terrestrial realm is significantly higher in the Ganges–Brahmaputra and Godavari drainage basins to the east, within which the vegetation is characterised by deciduous monsoon-forest and, in estuarine areas, mangrove swamp or 'sundurban' elements, than in the Indus drainage basin to the west, within which the vegetation is characterised by montane forest and semi-desert elements (in the Hindu Kush mountains and Thar desert, respectively). Moreover, the flux of terrigenous OM from the Ganges–Brahmaputra and Godavari rivers into the marine realm is significantly higher than that from the Indus river, and the concentration of terrigenous and total OM on the Bengal fan, fed by the Ganges–Brahmaputra and Godavari, is higher than that on the Indus fan, fed by the Indus. Also in the case of the Indian subcontinent, consumption of OM is lowest, and preservation of OM is highest, in places of dysoxia or anoxia, such as the Indian Ocean, where there is a peculiarly well-developed OMZ (Ansari & Vink, 2007; Bhattacharjee et al., in Sinha, 2007).

Palaeobiological de-risking of (biogenic) source presence. In the case of the Krishna–Godavari basin off the east coast of India, where good-quality prior palaeobiological information is available, palynofacies indicates a significant flux of terrigenous OM into the marine realm (Satyanarayana et al., in Swamy & Kapoor, 2002); and microfacies indicates the development of an environment favourable for the preservation of OM, actual preservation of OM within the sediment, and even the possibility of active gas seepage, allowing a low risk to be assigned to the presence of a – biogenic – source-rock in the basin (author's unpublished observations). Benthic foraminiferal microfacies and palaeobathymetry based on modern analogy indicates an upper slope environment of deposition for the Miocene–Pliocene Ravva formation of the Krishna–Godavari basin (Jones, 1994). A significant proportion of the species encountered (38%) exhibit internal functional morphological adaptations to low-oxygen conditions, and indicate the development of an oxygen minimum zone, favourable for the preservation of OM (Charnock & Jones, 1997). In addition, a significant proportion of the species encountered (46%) exhibit external functional morphological adaptations to infaunal habitats and detritivorous feeding strategies, and indicate actual preservation of OM within the sediment (Jones & Charnock, 1985). One species, *Arenomeandrospira glomerata*, previously described from regions of shallow gas occurrence, even indicates the possibility of active gas seepage (Jones & Wonders, in Hart et al., 2000).

Integrated studies

Integrated bio- and sequence-stratigraphic schemes for the Indian subcontinent have been published by, among others, Copestake et al., in Caughey et al. (1996) (Indus basin), Jones (1997) (Indus basin), Das et al. (1999) (Krishna–Godavari basin), Aswal et al., in Swamy and Kapoor (2002) (Kutch basin), Mehrotra et al. (2002) (multiple basins), Satyanarayana et al., in Swamy and Kapoor (2002) (Krishna–Godavari basin), Singh et al., in Swamy and Kapoor (2002) (Krishna–Godavari basin),

Smewing et al., in Clift et al. (2002) (Indus basin), Reddy et al. (2005) (Krishna–Godavari basin), Singh et al. (2008) (Kutch basin), Carmichael et al. (2009) (Indus basin), Mehrotra et al. (2010) (multiple basins) and Nagendra et al., (2010) (Cauvery basin). Palaeogeographic and facies maps of sequences have been published by, among others, Smewing et al., in Clift et al. (2002) and Carmichael et al. (2009) (Indus basin). Source and reservoir distributions can be predicted from maps such as these.

Importantly, certain of these publications include information on palynofacies, which enables not only enhanced palaeoenvironmental interpretation, but also petroleum source-rock identification and characterisation, and thermal maturity assessment (Aswal et al., in Swamy & Kapoor, 2002 (Kutch basin); Satyanarayana et al., in Swamy & Kapoor, 2002 (Krishna–Godavari basin)). Palaeoenvironmental interpretation based on palynofacies enables the population of sequence palaeogeographic and facies maps, also known as gross depositional environment or GDE maps. Source-rock identification and characterisation and thermal maturity assessment based on palynofacies enables the population of source-rock GDE maps, and source-rock presence and effectiveness common risk segment or CRS maps.

5.3 APPLICATIONS AND CASE STUDIES IN RESERVOIR EXPLOITATION

In reservoir exploitation, applied palaeontology assists generally, in the description and prediction of reservoir-rock distributions in time and in space, in a high-resolution sequence-stratigraphic framework; and specifically, for example in integrated reservoir characterisation, and also in well-site operations, including 'biosteering' (see Fig. 1.1c; see also below, and Section 5.4, below).

Reservoir exploitation

Reservoir exploitation involves the processes of appraisal, development and production (Selley, 1998; Gluyas & Swarbrick, 2004; Jones, 2006). Appraisal is the process whereby the extent of the reservoir and its contained reserves of hydrocarbon is determined, and its economic viability evaluated. Development and production are the processes whereby the contained reserves of the reservoir are exploited as efficiently as possible, through the implementation of an integrated reservoir management strategy. (Development refers specifically to the installation of pipelines and associated infrastructure; production to the flow of hydrocarbons from the reservoir into the pipeline system.)

It also involves a number of geological, geophysical and integrated techniques including not only those common to petroleum exploration (see above), but also integrated reservoir characterisation or description, and reservoir modelling (see below). Well-site operations in reservoir exploitation involve a number of production technologies (see below, and Section 5.4).

Geological, geophysical and integrated techniques

Integrated reservoir characterisation (and reservoir modelling). Integrated reservoir characterisation is the process, implemented across appraisal, development and production, whereby general geological, geophysical, including borehole and 4D seismic (4D seismic being essentially 'time-lapse' 3D, illustrating the movement of reservoir fluids through time), and static and dynamic reservoir engineering data, and specialist sedimentological and palaeontological data, are integrated and iterated. The objective is to describe reservoir layering and architecture, permeability distribution and flow behaviour (for input into a computer-simulated reservoir model).

Failure to integrate palaeontological data can result in serious flaws in the reservoir model, and costly inefficiencies in the appraisal and development programmes.

Production technologies

Production from petroleum reservoirs can be enhanced by the application of so-called secondary and tertiary recovery technologies, including, importantly, biotechnologies such as microbially enhanced oil recovery or MEOR (see, for example, McKinerny et al., in Ollivier & Magot, 2005). MEOR involves the use of microbial surfactants, or biosurfactants, such as those produced by certain species of *Bacillus* and *Pseudomonas*, which serve to increase the solubility of petroleum in aqueous media, and hence to mobilise residual petroleum. It also involves the use of biopolymers such as those produced by certain other species of *Bacillus* (which, incidentally, occur naturally in petroleum reservoirs), which serve to

preferentially reduce the permeability of zones of anomalously high permeability within the reservoir, and hence to even out uneven permeability distribution, and to improve sweep efficiency.

Incidentally, biotechnologies can also be applied in the control of hydrogen sulphide production in petroleum reservoirs (Sunde & Torsvik, in Ollivier & Magot, 2005); and in the desulphurisation and denitrogenation of, and the removal of metals from, petroleum (Kilbane, in Ollivier & Magot, 2005).

Applications in reservoir exploitation

Palaeontology plays a key role in reservoir exploitation in the description and prediction of reservoir-rock distributions in time and in space, in a high-resolution sequence stratigraphic framework, or, to put it another way, in the establishment of reservoir layering and architecture; and through input into integrated reservoir characterisation (Fig. 1.1b). In integrated reservoir characterisation, the most important application is in the identification of stratigraphic heterogeneity and compartmentalisation, and of potential barriers and baffles to fluid flow, such as condensed sections or CSs (Slatt, 2006; McKie et al., in Jolley et al., 2010; Scott et al., in Jolley et al., 2010; Smalley & Muggeridge, in Jolley et al., 2010). Biofacies and agglutinated foraminiferal 'morphogroup' analysis has been extensively used in the identification of intra-reservoir mudstones constituting potential barriers and baffles to fluid flow in submarine fan reservoirs (Holmes, in Jones & Simmons, 1999; Jones, in Jones & Simmons, 1999; Payne et al., in Jones & Simmons, 1999; Jones, 2001; Wakefield et al., 2001; Jones, 2003a, b; Jones et al., in Koutsoukos, 2005; Jones, 2006; see 5.3.4, and 5.4.4, below).

Palaeontological data is acquired at a higher resolution in reservoir exploitation than would be the case in routine exploration. The fossil groups used will vary depending on the reservoir age and facies. Non-biostratigraphic technologies are applicable in non-fossiliferous reservoir ages and facies (Payne et al., in Powell & Riding, 2005).

Quantitative approaches are discussed in Sections 2.7 and 3.7 above.

Case studies of applications in reservoir exploitation follow.

5.3.1 Reservoir characterisation, shallow marine, peri-reefal carbonate reservoir – Al Huwaisah, Dhulaima, Lekhwair and Yibal fields, Oman, and Shaybah field, Saudi Arabia

Al Huwaisah, Dhulaima, Lekhwair and Yibal fields

The Al Huwaisah, Dhulaima, Lekhwair and Yibal fields are located in Oman in the Middle East (Jones et al., in Bubik & Kaminski, 2004; see Fig. 5.7 above). They all produce from Thamama group, Shu'aiba formation shallow marine, peri-reefal carbonate reservoirs (Figs. 5.43–5.45).

Reservoir biostratigraphy. In the case of the Al Huwaisah field of central Oman, the reservoir can be dated to Early Aptian on the basis of the presence of *Mesorbitolina lotzei*, while in the cases of the Yibal, Dhulaima and Lekhwair fields of northern Oman, the upper part of the reservoir can be dated to Late Aptian on the basis of the presence of *M. texana* (Figs. 5.43–5.44).

Palaeobiology/palaeoenvironmental interpretation. The observation that Shu'aiba formation reservoir deposition persisted into the Late Aptian in northern but not in central Oman is consistent with the emerging model of a relatively deep ('Bab') intra-shelf basin in the northern area and emergence – through eustatic sea-level fall – of the formerly shallow-marine platform of the central area at this time (Fig. 5.45).

Integration. This is important from the petroleum geological point of view because the porosity and hence the quality of the Shu'aiba formation reservoir was enhanced – through subaerial exposure and leaching – only on the margin of the basin.

Shaybah field

The Shaybah field is located in Saudi Arabia, on the border with the United Arab Emirates (Hughes, in Bubik & Kaminski, 2004; Hughes, in Powell & Riding, 2005; see Fig. 5.7 above). Like the Al Huwaisah, Dhulaima, Lekhwair and Yibal fields of neighbouring Oman (see above), it produces from a Shu'aiba formation shallow marine, peri-reefal carbonate reservoir, deposited on the margin of the 'Bab' basin. (The reefs are constructed chiefly of rudist bivalves, which are recognisable in cores, in CT scans of cores, and on image logs: see Box 5.3 above.)

Because the Shaybah field is located in the sand sea of the Rub' Al-Khali, romantically known as the 'Empty Quarter', it is extremely difficult to acquire seismic data over it. Reservoir characterisation is therefore more than usually reliant on other types of subsurface data such as core and log data. To assist with the process, some 17 000′ of conventional core has been cut from over 50 wells, and semi-quantitative micro- and macropalaeontological analysis has been undertaken on a substantial number of core samples.

Reservoir biostratigraphy. Reservoir biostratigraphy is based on calcareous nannoplankton, larger benthic foraminifera and rudist bivalves (Hughes, in Bubik & Kaminski, 2004; Hughes, in Powell & Riding, 2005). The entire reservoir section appears to be of Aptian age.

Palaeobiology/palaeoenvironmental interpretation. Palaeoenvironmental interpretation of the reservoir is based on calcareous Algae, larger benthic foraminifera and rudist bivalves (Hughes, in Bubik & Kaminski, 2004; Hughes, in Powell & Riding, 2005). A total of 10 micropalaeontological biofacies have been recognised, and named after their dominant components. These are, more or less in order from distal to proximal: *Hedbergella* biofacies (ShBf-1), interpreted as basinal; *Hedbergella–Palorbitolina* biofacies (ShBf-2), interpreted as fore-reefal to basinal; *Praechrysalidina* biofacies (ShBf-3), interpreted as distal fore-reefal (and deep back-reef lagoonal); *Rotalia* biofacies (ShBf-4), interpreted as proximal fore-reefal; *Offneria* biofacies (ShBf-5), interpreted as peri-reefal; *Offneria–Oedomyophorus* biofacies (ShBf-6), also interpreted as peri-reefal; *Glossomyophorus* biofacies (ShBf-7), interpreted as shallow back-reef lagoonal to peri-reefal; *Lithocodium* biofacies (ShBf-8), interpreted as shallow back-reef lagoonal; *Agriopleura* biofacies (ShBf-9), interpreted as deep back-reef lagoonal; and *Horiopleura* biofacies (ShBf-10), interpreted as fore-reefal.

Integration. Integration of core micro- and macropalaeontological data and interpretation with core petrographic and sedimentological data and interpretations and wireline log petrophysical data has proved critical in developing reservoir layering and architecture schemes (Hughes, in Bubik & Kaminski, 2004; Hughes, in Powell & Riding, 2005). Four stratigraphic subdivisions or layers of the reservoir have been recognised, namely, in ascending stratigraphic order: Lower (Sh-1), interpreted as 'early'–'middle' Aptian; Middle (Sh-2), also interpreted as 'early'– 'middle' Aptian; Upper (Sh-3), also

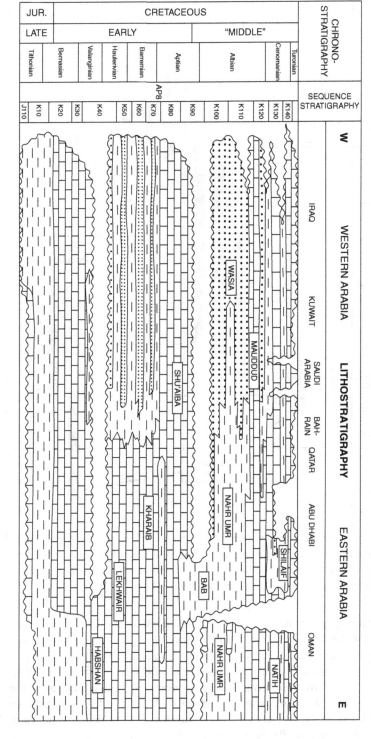

Fig. 5.43 **Chronostratigraphic diagram, Cretaceous, Middle East.** © Grzybowski Foundation. Reproduced with permission from Jones *et al.*, in Bubik and Kaminski (2004).

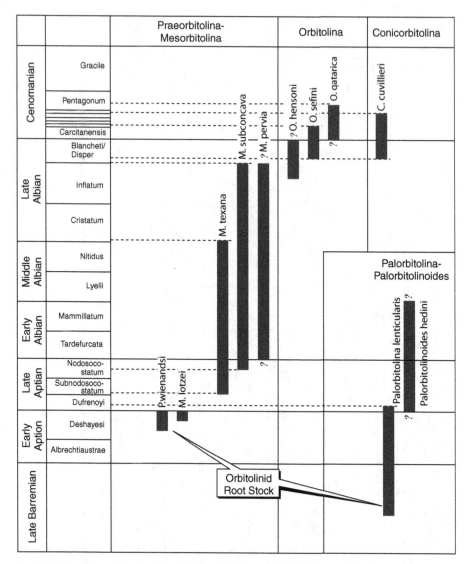

Fig. 5.44 **Stratigraphic distribution of selected orbitolinid larger benthic foraminifera, 'Middle' Cretaceous, Middle East.** © Grzybowski Foundation. Reproduced with permission from Jones *et al.*, in Bubik and Kaminski (2004).

interpreted as 'early'–'middle' Aptian; and, locally, onlapping the flanks of the field, and unrepresented on the crest, uppermost (Sh-4), interpreted as 'late' Aptian. Within each reservoir layer, a number of architectural elements have been recognised that appear to be related to the biofacies outlined above. Within the lower Shu'aiba reservoir layer (Sh-1), the interpreted distal *Hedbergella–Palorbitolina* biofacies (ShBf-2) and *Praechrysalidina*

biofacies (ShBf-3) are developed over essentially the entire field, indicating transgression. Within middle Shu'aiba layer Sh-2, the interpreted peri-reefal *Offneria* biofacies (ShBf-5), *Offneria–Oedomyophorus* biofacies (ShBf-6) and *Glossomyophorus* biofacies (ShBf-7) are widely developed especially over the central part of the field; the interpreted shallow back-reef lagoonal *Lithocodium* biofacies (ShBf-8), only over the southwestern part; and the interpreted

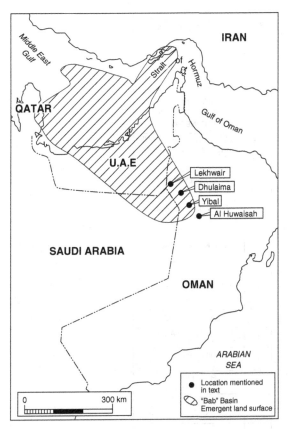

Fig. 5.45 **Palaeogeographic and facies map, 'Middle'**
Cretaceous, Middle East. © Grzybowski Foundation.
Reproduced with permission from Jones *et al.*, in Bubik
and Kaminski (2004).

fore-reefal to basinal *Hedbergella* biofacies (ShBf-1),
Hedbergella–Palorbitolina biofacies (ShBf-2) and *Rotalia*
biofacies (ShBf-4), only over the northeastern part;
indicating differentiation. Within upper Shu'aiba
layer Sh-3, the interpreted deep back-reef lagoonal
Agriopleura biofacies (ShBf-9) is developed over essentially the entire field; indicating regression.

5.3.2 Reservoir characterisation, marginal to shallow marine, peri-deltaic clastic reservoirs – Gullfaks, Snorre and Statfjord fields, Norwegian sector, and Ninian and Thistle fields, UK sector, North Sea

The Gullfaks, Snorre and Statfjord fields are located
in the Norwegian sector, and the Ninian and Thistle
fields in the UK sector, of the North Sea (see 5.2.2,

and Fig. 5.19, above). They all produce from Brent
group marginal to shallow marine, peri-deltaic clastic reservoirs (Morton, 1992).

Reservoir biostratigraphy. Reservoir biostratigraphy is based on spore and pollen – and dinoflagellate – palynostratigraphy (Helland-Hansen *et al.*, in
Morton, 1992; Williams, in Morton, 1992). The
entire reservoir section appears to be of late Early?
to Middle Jurassic age.

Palaeobiology/palaeoenvironmental interpretation. Palaeoenvironmental interpretation of the reservoir is
based on integration of biofacies, that is, palynofacies (Williams, in Morton, 1992; Mussard *et al.*, 1994;
Sawyer & Keegan, 1996) and microfacies (Nagy,
1992), and core, log and seismic facies. Palynofacies
analysis has led to the recognition of a total of 12
reservoir facies in the Brent group reservoir in the
Ninian and Thistle fields (Williams, in Morton, 1992;
Fig. 5.46).

Microfacies analysis has led to the distinction of
a range of peri-deltaic sub-environments in the
Gullfaks, Snorre and Statfjord fields, and sub-
regionally (Nagy, 1992; Nagy *et al.*, 2010). Vegetated
delta-top and to a lesser extent delta-front sub-
environments have been distinguished on cross-
plots of epiphytic, other epifaunal, and infaunal
'morphogroups' of agglutinating foraminifera on
the basis of significant incidences of the epiphytic
'morphogroup', and prodelta sub-environments by
significant incidences of the other epifaunal and
infaunal 'morphogroups' (see 3.1.3 and 3.2 above).

Integration. Integration of biostratigraphy and
palaeoenvironmental interpretation, with core, log
and seismic data, has enabled the construction of
time-slice palaeogeographic and facies maps that have
proved of considerable use in sub-regional-scale mapping of peri-deltaic reservoir fairways (Helland-Hansen
et al., in Morton, 1992; Nagy, 1992; Nagy *et al.*, 2010).

5.3.3 Reservoir characterisation, marginal to shallow marine, peri-deltaic clastic reservoir – Pedernales field, Venezuela, northern South America

Pedernales field is located in the wetlands of the
Orinoco delta some 100 km north-northeast of
Maturin in the northeasternmost part of the east-
ern Venezuelan basin, sometimes referred to as the
Maturin sub-basin (Jones *et al.*, in Jones & Simmons,
1999; see also 5.2.3, and Fig. 5.35, above). It produces

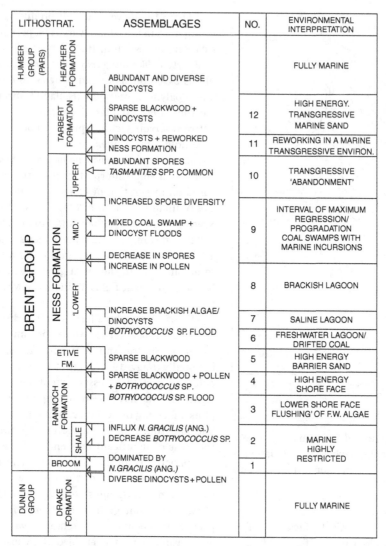

LITHOSTRAT.			ASSEMBLAGES	NO.	ENVIRONMENTAL INTERPRETATION
HUMBER GROUP (PARS)	HEATHER FORMATION				FULLY MARINE
BRENT GROUP			ABUNDANT AND DIVERSE DINOCYSTS		
	TARBERT FORMATION		SPARSE BLACKWOOD + DINOCYSTS	12	HIGH ENERGY. TRANSGRESSIVE MARINE SAND
			DINOCYSTS + REWORKED NESS FORMATION	11	REWORKING IN A MARINE TRANSGRESSIVE ENVIRON.
	NESS FORMATION	'UPPER'	ABUNDANT SPORES TASMANITES SPP. COMMON	10	TRANSGRESSIVE 'ABANDONMENT'
		'MID'	INCREASED SPORE DIVERSITY	9	INTERVAL OF MAXIMUM REGRESSION/ PROGRADATION COAL SWAMPS WITH MARINE INCURSIONS
			MIXED COAL SWAMP + DINOCYST FLOODS		
			DECREASE IN SPORES INCREASE IN POLLEN		
		'LOWER'		8	BRACKISH LAGOON
			INCREASE BRACKISH ALGAE/ DINOCYSTS	7	SALINE LAGOON
			BOTRYOCOCCUS SP. FLOOD	6	FRESHWATER LAGOON/ DRIFTED COAL
	ETIVE FM.		SPARSE BLACKWOOD	5	HIGH ENERGY BARRIER SAND
	RANNOCH FORMATION		SPARSE BLACKWOOD + POLLEN + BOTRYOCOCCUS SP.	4	HIGH ENERGY SHORE FACE
			BOTRYOCOCCUS SP. FLOOD	3	LOWER SHORE FACE FLUSHING' OF F.W. ALGAE
		SHALE	INFLUX N. GRACILIS (ANG.) DECREASE BOTRYOCOCCUS SP.	2	MARINE HIGHLY RESTRICTED
	BROOM		DOMINATED BY N.GRACILIS (ANG.)	1	
DUNLIN GROUP	DRAKE FORMATION		DIVERSE DINOCYSTS + POLLEN		FULLY MARINE

Fig. 5.46 **Palynofacies characterization of Brent group reservoir, Thistle field, northern North Sea.** © The Geological Society, London. Reproduced with permission from Williams, in Morton (1992).

principally from a La Pica formation, Pedernales member marginal to shallow marine, peri-deltaic clastic reservoir (Fig. 5.47).

Reservoir biostratigraphy. The stratigraphic layering of the reservoir is based on integration of biostratigraphy, and core, log and seismic stratigraphy, in a work-station environment. Historically, the reservoir biostratigraphy was based mainly on qualitative micropalaeontology (Jones *et al.*, in Jones & Simmons, 1999). More recently, it has come to be based more on integrated high-resolution quantitative micropalaeontology, nannopalaeontology, and spore and pollen, and to a lesser extent dinoflagellate, palynostratigraphy (although note that the palynostratigraphic data is of questionable usefulness on account of contamination of palynological samples by drilling fluid formulated from local river water and mud, and containing modern mangrove pollen and other palynomorphs effectively indistinguishable

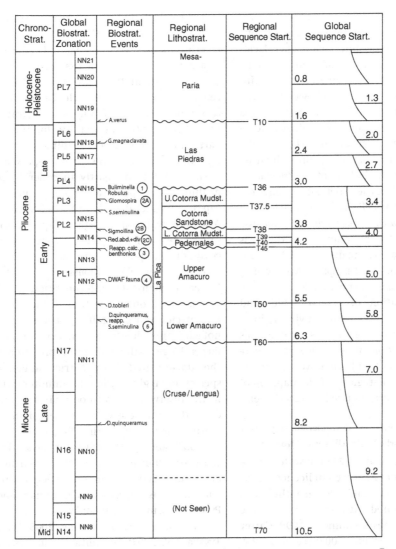

Fig. 5.47 **Stratigraphic framework, Pedernales field, eastern Venezuela.** © The Geological Society, London. Reproduced with permission from Jones *et al.*, in Jones and Simmons (1999).

from their ancient counterparts in the subsurface). The formal reservoir biozonation has had to be based essentially on ecologically controlled benthic foraminifera, with questionable regional chronostratigraphic and correlative significance, because of the rarity in the predominantly marginal marine facies of the reservoir of the open oceanic planktonic foraminifera and calcareous nannoplankton that form the principal bases of global standard biozonations (Fig. 5.47).

The zonation has been supplemented by an informal 'event stratigraphy' based on abundance and diversity peaks and troughs, biofacies and bathymetric trends etc. Abundance and diversity peaks have been used in the identification of sequence stratigraphically significant condensed sections in the same way that they have in Palo Seco field in southwestern Trinidad.

Palaeobiology/palaeoenvironmental interpretation. Palaeoenvironmental interpretation of the reservoir is based on integration of biofacies, that is, microfacies and to a lesser extent palynofacies, and ichnofacies, and core, log and seismic facies. Microfacies analysis has led to the distinction of a range of delta-top, delta-front and prodelta environments with modern and/or ancient analogues elsewhere in the Orinoco system (see Fig. 5.37 above). Within the area of influence of the Orinoco, fluvially dominated delta-top environments have been distinguished on the basis of the *Centropyxis* microfacies; tidally influenced delta-top environments on the basis of the *Miliammina* and *Trochammina* microfacies; delta-front environments on the basis of the *Buliminella* microfacies, and *Skolithos* ichnofacies with *Ophiomorpha* and *Thalassinoides*; proximal prodelta environments on the basis of the *Eggerella* microfacies; and distal prodelta environments on the basis of the *Glomospira, Alveovalvulina/ Silicosigmoilina* and *Cyclammina* microfacies.

Integration. Integration of reservoir bio- and sequence-stratigraphy and palaeoenvironmental interpretation, with core, log and seismic data, has enabled the construction of time-slice palaeogeographic and facies maps that have proved of considerable use in field- to sub-regional-scale mapping of peri-deltaic reservoir fairways (see Fig. 5.38 above).

5.3.4 Reservoir characterisation, deep marine, submarine fan reservoir – Forties field, UK sector, North Sea

Forties field is located in 104–128 m of water, some 180 km east-northeast of Aberdeen in licence blocks 22/10 (main part of field) and 22/6a (southeastern extension) in the United Kingdom sector of the North Sea (Jones, in Jones & Simmons, 1999; Carter & Heale, in Gluyas & Hichens, 2003; see also 5.2.2, and Fig. 5.19, above). It produces principally from a Montrose group, Forties formation deep marine, submarine fan reservoir.

Reservoir biostratigraphy. The stratigraphic layering of the reservoir is based on integration of biostratigraphy, and core, log and seismic stratigraphy, in a work-station environment. Historically, the reservoir biostratigraphy was based mainly on semi-quantitative diatom and benthic foraminiferal micropalaeontology and dinoflagellate palynostratigraphy. More recently, it has come to be based more on high-resolution semi-quantitative to quantitative micropalaeontology and spore and pollen palynostratigraphy (Fig. 5.48).

The resolution of the reservoir biostratigraphy is now high enough to demonstrate that historical layering models were wrong, and require righting as a matter of import, in view of the implications for the exploitation of the still substantial remaining reserves in the field (Payne *et al.*, in Jones & Simmons, 1999). Notably, palynostratigraphy is now being applied at well-site for the first time, partly in order to direct drilling up through the reservoir to access previously by-passed attic oil from beneath; and partly in order to terminate drilling before penetrating the – incompetent – cap-rock (Williams *et al.*, in Powell & Riding, 2005). Also notably, this palynostratigraphic work is being undertaken utilising green, non-acid well-site processing techniques (see Box 5.6 below).

Palaeobiology/palaeoenvironmental   interpretation. Palaeoenvironmental interpretation of the reservoir is based on integration of biofacies, that is, microfacies and generally to a lesser extent palynofacies, and ichnofacies, and core, log and seismic facies. Palaeobathymetric interpretation based on the documented bathymetric ranges of those species of agglutinating foraminifera that occur in the reservoir and also occur in the Recent indicates that the reservoir was deposited in a deep marine, hence submarine fan, environment (see Fig. 5.23 above). Microfacies analysis has led to the distinction of a range of submarine fan sub-environments with outcrop analogues in the Austrian and Swiss Alps, Polish Carpathians and Spanish Pyrenees (Grun *et al.*, 1964; Winkler, in Oertli, 1984; Jones *et al.*, in Powell & Riding, 2005; Lemanska, in Tyszka *et al.*, 2005). Turbiditic channel axis sub-environments have been distinguished on cross-plots of 'morphogroups' of epifaunal suspension-feeding, other epifaunal, and infaunal 'morphogroups' of agglutinating foraminifera on the basis of significant incidences of epifaunal suspension-feeding 'morphogroup' A; interturbiditic channel off-axis sub-environments on the basis of significant incidences of other epifaunal 'morphogroup' B; and hemipelagic channel levee and overbank to basin-plain sub-environments on the basis of significant incidences of infaunal 'morphogroup' C (Fig. 5.49; see also 3.1.3 and Section 3.2, and Fig. 3.13, above).

Interestingly, the palaeoenvironmental interpretation of the upper part of the reservoir, which

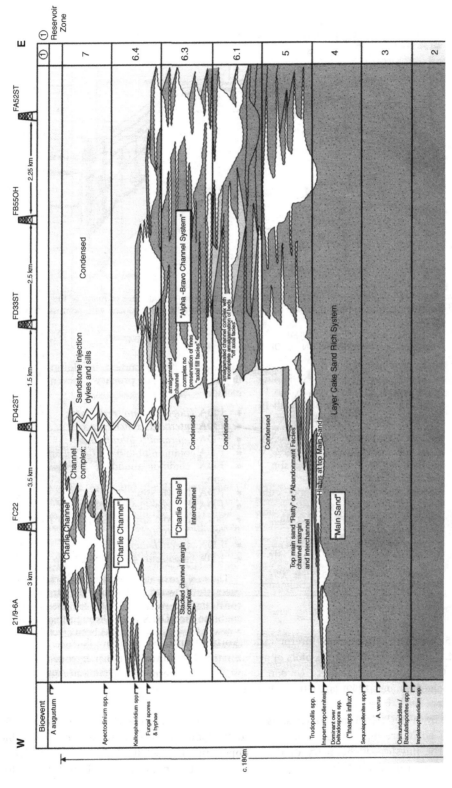

Fig. 5.48 **Revised reservoir zonation, Forties formation reservoir, Forties field, North Sea.** © The Geological Society, London. Reproduced with permission from Payne *et al.*, in Jones and Simmons (1999). Note biostratigraphy essentially based on spores and pollen.

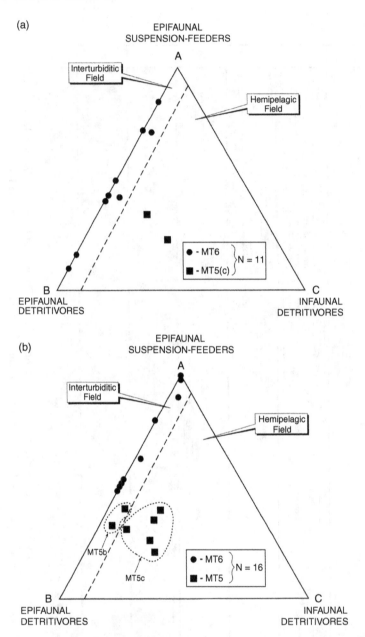

Fig. 5.49 **Palaeoenvironmental interpretation, Forties formation reservoir, Forties field, North Sea, based on cross-plots of agglutinating foraminiferal 'morphogroups' A–B–C.** © The Geological Society, London. Reproduced with permission from Jones, in Jones and Simmons (1999). (a) 21/10-FC22; (b) 22/6a-E9.

is characterised by a restricted microfacies, is based as much if not more on palynofacies than on microfacies. The physics or chemistry of the benthic environment represented by the upper part of the reservoir evidently deteriorated firstly to the point at which it was only habitable by opportunistic epifaunal foraminifera, and secondly to the point at which it was no longer habitable by foraminifera. Concomitantly, the pelagic environment locally deteriorated to the point at which it was only habitable by opportunistic dinoflagellates.

Integration. Integration of reservoir bio- and sequence-stratigraphy and palaeoenvironmental interpretation, with core, log and seismic data, has enabled the construction of time-slice palaeogeographic and facies maps that have proved of considerable use in field-scale mapping of the reservoir.

Time-lapse 3D, or 4D, seismic is currently being used to track production through time, to identify unswept oil, and to gear the future infill injection and production drilling programme to maximise sweep efficiency.

5.4 APPLICATIONS AND CASE STUDIES IN WELL-SITE OPERATIONS

Well-site operations in petroleum exploration involve a number of drilling, petrophysical and testing technologies; and well-site operations in reservoir exploitation involve a number of production technologies (see above; see also below).

Petroleum exploration

The main applications of palaeontology in well-site operations in petroleum exploration are in monitoring of stratigraphic position while drilling, determination of casing, coring and terminal depths, and post-well analysis and auditing (Jones, 2006). Note that successful application is dependent on appropriate sampling.

Monitoring of stratigraphic position while drilling. Monitoring while drilling exploration wells is seldom called for nowadays other than in frontier – 'wildcat' – areas. However, monitoring while drilling production wells has become a virtual necessity in many areas.

Determination of casing, coring and terminal depths. Casing, coring and terminal depths are normally determined prior to drilling, with appropriate input from the palaeontologist. In many cases the depths are derived from two-way time to a seismic horizon, and may be inaccurate owing to uncertainty as to which is the most appropriate velocity function to use to convert time to depth. Well-site palaeontology provides an accurate determination, effectively in real time (Lowe et al., 1988). It therefore enables decisions to be made also effectively in

real time, thereby saving significant amounts of time and money. It can also have important safety implications, as when casing off overpressured intervals.

Post-well analysis and auditing. Post-well analysis generally aims to provide a stratigraphic breakdown of the well within 6 weeks of the receipt of the final sample. Analytical data can be entered into software packages for the purposes of interpretation and integration. Auditing evaluates the technical and commercial success of the drilling. One of the most important parts involves the comparison of the prognosed well stratigraphy with the actual. This comparison is generally made by means of overlaying a synthetic seismogram of the well at the appropriate location, and transferring the actual well stratigraphy onto the seismic grid. This in turn helps to refine the predictive sequence-stratigraphic model.

Reservoir exploitation

The main application of palaeontology in well-site operations in reservoir exploitation is in 'biosteering' (Jones et al., in Koutsoukos, 2005; Jones, 2006). Note, though, that there are also other important applications, for example in managing high-pressure–high-temperature or HPHT wells (Mears & Cullum, in Demchuk & Gary, 2009).

Biosteering involves real-time monitoring of precise stratigraphic position relative to reservoir in a (deviated) well, through the well-site application of biostratigraphic technology. It also involves, as necessary, issuing instructions to redirect the well trajectory to ensure optimal reservoir penetration, for example when encountering a sub-seismic fault or a problem with seismic depth conversion or survey data. The high resolution of the technique (its 'window'), usually established by analyses of closely spaced samples from offset wells or a pilot hole, is often only a few feet or metres. The technique is, and will remain, critical to the exploitation of many petroleum reservoirs. The value added to date through the application of biosteering in BP runs into hundreds of millions of dollars. It is anticipated that this figure will further increase in the future, as the technology is transferred to fields in areas only now entering into production.

Case studies of applications in well-site operations follow.

5.4.1 'Biosteering', shallow marine carbonate reservoir – Sajaa field, United Arab Emirates, Middle East

The Sajaa field is situated in the frontal thrust-sheets of the Oman Mountains, some 25 miles east of Sharjah town in Sharjah in the UAE (Jones et al., in Koutsoukos, 2005; Jones, 2006; Marshall et al., 2007; Marshall, 2010; see also 5.2.1 and Fig. 5.7, above).

Reserves are estimated at between 100 million and 400 million barrels of condensate, and between 1.5 trillion and 6 trillion cubic feet of gas. The (Kharaib and Shu'aiba formation) reservoir comprises shallow marine, platform carbonates of Early Cretaceous, Barremian-Aptian, age. Production is through horizontal wells. Rates are of the order of 60 000 barrels per day.

A micropalaeontological biostratigraphic study enabled the reservoir interval to be divided into 21 zones, each a few tens of feet thick. Detailed palaeobathymetric interpretation of closely spaced samples has contributed significantly to the understanding of parasequence-scale reservoir facies and architecture, and to the identification of reservoir 'sweet spots'. Subsequent biosteering – by means of thin-section micropalaeontology and microfacies – has kept the well bore within optimal parts of the reservoir over distances of several thousand feet (biosteering of multilaterals has also been possible). It has effectively replaced 'geosteering' using coherency, which worked well in unfaulted but not in faulted sections, the approach taken on encountering a fault being to steer upwards to a known point in the stratigraphy and then back down again. This was not only time-consuming and expensive but potentially also dangerous, as it increased the risk of drilling out of the reservoir and into the seal, and causing the loss of control of the hole. Biosteering is currently being practised in conjunction with a combination of novel drilling technologies such as coiled tubing, slim-hole and underbalance (and also in conjunction with 'logging-while-drilling'). Even after the carbonate dissolution and sand-blasting effects associated with these drilling technologies, sufficient cuttings are still recoverable to allow the compositing of thin sections for micropalaeontological analysis. Note, though, that the drilling fluid has to be sufficiently buffered by the addition of alkaline substances to counter the acid dissolution (without damaging motors and seals).

Note that a similar 'biosteering' methodology has been applied in the exploitation of the Late Permian Khuff formation reservoirs of various fields in Saudi Arabia (Hughes et al., 2010a, b).

5.4.2 'Biosteering', deep marine carbonate reservoir – Valhall field, Norwegian sector, North Sea

The Valhall field is situated in the Norwegian sector of the North Sea (Jones et al., in Koutsoukos, 2005; Jones, 2006; see also 5.2.2, and Fig. 5.19, above).

Reserves are estimated at 705 million barrels. The (Chalk group, Tor formation) reservoir comprises deep marine, essentially pelagic carbonates, including allochthonous chalks and chalky turbidites, of Late Cretaceous, Maastrichtian, age. Production is through high-angle wells, of which some 50 have already been drilled. At the time of writing, rates are 105 000 barrels per day.

A micropalaeontological and nannopalaeontological biostratigraphic study enabled the reservoir interval to be divided into seven zones, each a few tens of feet thick. Subsequent biosteering – by means of micropalaeontology and nannopalaeontology – has targeted zones C and D – zones A and B possessing better reservoir properties in terms of porosity and permeability, but being unstable and prone to collapse under drawdown. One particularly successful well that was kept within zone D by biosteering produced 12 000 barrels per day. Other well-site applications of biostratigraphy include setting close to the base of the overburden without drilling overbalanced into the underpressured zone at the top of the reservoir, thereby causing formation damage; and terminating the well at the base of the reservoir, or 'biostopping'. Another application is identifying the origin of caved material and hence unstable zones in the tophole, impacting future well design. In terms of value added, 30% of the current production is attributed to optimal reservoir placement enabled by biosteering and associated technologies. Moreover, a cost saving of minimum of $1 million per well – 7 days drilling, at $150 000 per day – is achieved by being able to set casing in the correct place by means of biostratigraphy.

Note that a similar 'biosteering' methodology has been applied in the exploitation of the Chalk group reservoirs of other fields in the North Sea, for example Eldfisk (Yang-Logan et al., 1996).

5.4.3 'Biosteering', shallow marine clastic reservoir – Cusiana field, Colombia, northern South America

The Cusiana field is situated in the Llanos basin, more specifically in the frontal thrust-sheets of the Eastern Cordillera, some 150 miles northeast of Bogota, in Colombia (Jones *et al.*, in Koutsoukos, 2005; Jones, 2006; see also 5.2.3, and Fig. 5.35, above).

Recoverable reserves are estimated at 1.5 billion barrels of light oil and condensate and 3.4 trillion cubic feet of gas. The (Guadalupe formation) reservoir comprises shallow marine clastics of Late Cretaceous, Santonian–Campanian age. It is characterised by significant vertical variation in quality. Production is through horizontal wells. At the time of writing, rates are in excess of 300 000 barrels per day.

A palynological biostratigraphic study of cored offset wells enabled the reservoir interval to be divided into a number of zones, each a few tens of feet thick. Subsequent biosteering – by means of palynology – targetted Zones GR3–GR7, which are developed within the best quality reservoir (the 'upper Phosphate'). (Incidentally, note that the hazardous chemicals used in the palynology processing were contained and safely handled in a portable palynology processing unit designed by BP.) The production potential from the first biosteered or 'palynosteered' horizontal well was approximately 30 000 barrels per day, as against 12 000 barrels per day from the best conventional vertical well. Moreover, one biosteered well costing $26 million effectively does the work of three to four conventional wells costing $15–18 million each. The three that have been drilled to date have thus resulted in significant savings in drilling costs, and value addition.

Similar examples of palynosteering in western Venezuela are provided by Rull (2002).

5.4.4 'Biosteering', deep marine clastic reservoir – Andrew field, UK sector, North Sea

The Andrew field is situated in the UK sector of the North Sea (Jones *et al.*, in Koutsoukos, 2005; Jones, 2006; see also 5.2.2, and Fig. 5.19, above).

Reserves are estimated at 118 million barrels. The (Andrew formation) reservoir comprises deep marine, submarine fan clastics of Late Palaeocene age. The pay intervals are comparatively thin. Production is through horizontal wells. At the time of writing, rates are 64 000 barrels per day. Understanding of the reservoir facies and heterogeneities and consequences for fluid flow, and optimal placement of wells with respect to fluid contacts, are critical to maximisation of oil production prior to the inevitable early gas and/or water breakthough.

A micropalaeontological biostratigraphic study enabled the reservoir interval to be divided into seven zones, each a few tens of feet thick (Figs. 5.50–5.51).

Microfacies, integrated with core sedimentology, was used to identify facies and heterogeneities with each zone, thereby establishing the spatio-temporal distribution of reservoir and non-reservoir units, and potential consequences for fluid flow. Mudstones A3 and A1 were interpreted as essentially hemipelagic on the basis of contained high abundance and diversity, low dominance assemblages of agglutinated foraminifera further characterised by comparatively high incidences of complex infaunal 'morphogroup' C (Jones & Charnock, 1985). They were therefore also interpreted as potentially of regional or field-wide extent, and constituting barriers to fluid flow. This has subsequently been confirmed by pressure data. Mudstone A2 was interpreted as interturbiditic on the basis of contained low abundance and diversity, high dominance assemblages dominated by simple epifaunal 'morphogroups' A and B. It was therefore also interpreted as of local extent, and constituting only a baffle to fluid flow. Subsequent biosteering – by means of agglutinated foraminiferal micropalaeontology and microfacies – has targeted reservoir unit.

Fortuitously, the top of this unit is effectively coincident with the gas–oil contact over the crest of the field. Realisation of this fact has allowed the biosteered well-bore to be run more medially through the reservoir than in the initial well plan, with the overlying Mudstone A3 acting as a barrier to fluid flow and hence protecting it from gas invasion. It has been estimated that the optimal well placement enabled by the biosteering has added 10 million barrels of reserves to the books.

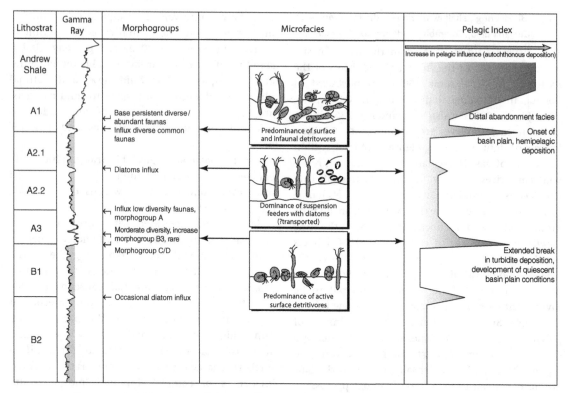

Fig. 5.50 **Biostratigraphically based reservoir zonation for Andrew formation reservoir, Andrew field, North Sea.** From Jones *et al.*, in Koutsoukos (2005).

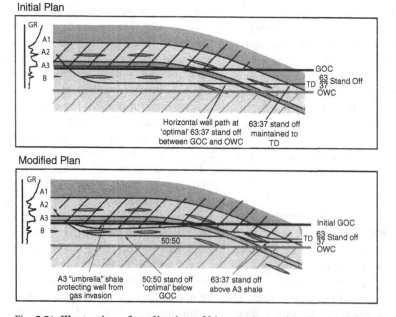

Fig. 5.51 **Illustration of application of biozonation to biosteering of Andrew formation reservoir. Andrew field, North Sea.** From Jones *et al.*, in Koutsoukos (2005).

BOX 5.6 PALAEONTOLOGY AND HEALTH, SAFETY AND ENVIRONMENTAL ISSUES IN THE PETROLEUM INDUSTRY (MODIFIED AFTER JONES, 2006)

Introduction

There are five principal areas in which palaeontology impacts on health, safety and environmental (HSE) issues – or on which HSE issues impact on palaeontology – in the petroleum industry. These are: site investigation; environmental impact assessment (EIA); pressure prediction; palynology processing, especially in the context of well-site operations; and environmental monitoring.

Site investigation

In my working experience in the petroleum industry, micropalaeontology has proved of use in site investigation in the Black Sea off Turkey, in the Caspian off Azerbaijan and in the Mediterranean off Egypt, specifically in helping to determine the safest sites for the locations of drilling rigs, pipelines and other facilities through the avoidance of geo-hazards (see also 8.1.4 below).

Environmental impact assessment (EIA)

In my working experience in the petroleum industry, the living biota has proved of use in EIA in baseline studies on the potential environmental impacts of industry projects, and the mitigation of these impacts, in and around BP's Prudhoe Bay and other oil fields on the North Slope of Alaska, Wytch Farm field under Poole Harbour in Dorset on the south coast of England, and in the Columbus basin off the east coast of Trinidad (Evans, 1997; Truett & Johnson, 2000; Gluyas & Swarbrick, 2004; author's unpublished observations; see also 9.1.1 and 9.1.2 below).

Pressure prediction

Micropalaeontology provides critical inputs into pressure prediction from petroleum systems analysis and basin modelling.

Well-site operations

There are a number of areas in which micropalaeontology impacts HSE in the field of well-site operations, including not only making real-time casing and terminal depth calls, thus avoiding pressure kicks, but also providing an independent assessment of the condition and stability of the borehole, from cavings.

There are also areas in which HSE impacts micropalaeontology. For example, there are particular HSE risks associated with the use of hazardous and potentially lethal chemicals such as hydrochloric acid (HCl), nitric acid (HNO_3) and hydrofluoric acid (HF) in conventional palynology processing. However, these risks can be mitigated by the appropriate training, equipment and supervision of palynology processing technicians, and the implementation of safety checks and audit processes. Importantly, the risks are capable of further mitigation through the containment of the hazardous chemicals in portable palynology processing units, developed by BP in the 1980s, which technique has since proved of use worldwide. Indeed, the risks can be mitigated altogether through the use of a non-conventional, 'green, non-acid well-site processing technique' (GNAWPT) developed by BP contractors in the 1990s (Williams et al., in Powell & Riding, 2005; Riding & Kyffin-Hughes, 2006). This technique has also proved of use almost worldwide, and of particular use in the Norwegian sector of the North Sea, where strict environmental legislation prohibits the use of the conventional, acid-processing technique.

Environmental monitoring

The living biota has proved of use in environmental monitoring in the petroleum industry, in baseline studies on industrial pollution (Mojtahid et al., 2006; Jorissen et al., 2009; Bicchi et al., 2010; Hess et al., 2010; see also 9.2.1 below).

5.5 UNCONVENTIONAL PETROLEUM GEOLOGY

5.5.1 Coal-bed methane or coal-seam gas

Coal constitutes not only an important economic resource in its own right (see Chapter 7), but also a source-rock producing petroleum in the form of oil or, more typically, gas (see 5.1.1 above), and a reservoir-rock capable of producing coal-bed methane (CBM) or coal-seam gas (CSG) (Gayer & Harris, 1996; Evans, 1997; Spitz & Trudinger, 2009; Scott, 2010).

Coal is characterised by surprisingly high microporosity and adsorbed gas and free water storage capacity, of up to $200-300\,m^2$ per gram (OSM, 2001). CSG is produced from coal reservoirs, from wells, by first producing the free water and reducing the hydrostatic pressure in the reservoir, and then separately producing the resulting desorbed gas (Evans, 1997; Spitz & Trudinger, 2009). The disposal of the produced water raises important environmental concerns, as it is commonly characterised by high concentrations of total dissolved solids, total suspended solids, halides, ammonia and metals (Spitz & Trudinger, 2009).

The quality of CSG reservoirs appear to be in part controlled by the composition of the coal, which is in turn in part controlled by its original depositional environment, as inferred from indicators such as the gelification index or GI, and the tissue preservation index or TPI (Diessel, 1986; Calder et al., 1991; Kalkreuth et al., 1991; Diessel, 1992; Crosdale et al., 1998; Kolcon & Sachsenhofer, 1999; Scott, 2000, 2002; Iordanidis & Georgakopoulos, 2003; Zdrakov & Kortenski, 2004; Scott et al., 2007; Potter et al., 2009; see also 7.1.1 below). Detailed studies on CSG reservoirs in the Lower Cretaceous, Mannville group, Medicine River formation in the Alberta basin in Canada have shown that the composition of the coal and the quality of the reservoir vary both laterally and vertically within individual seams (Potter et al., 2009; see also Holz et al., 2002).

Coal-seam gas in Great Britain

Interestingly, there is a proven CSG resource in Scotland, in the Bannockburn coal in the Late Carboniferous, Namurian (Serpukhovian to early Bashkirian) Limestone Coal formation of the Clackmannan group of the Clackmannan syncline, or precursor Kincardine basin, of the Airth gas-field, discovered in the 1990s (Pye & Brown, in Trewin, 2003; Rippon, in Trewin, 2003).

There is also a significant potential resource in the south Wales coalfield, as indicated by recent 3D modelling integrating surface geological data and historical subsurface mining records (Lockett & Andrews, 2010).

5.5.2 Shale gas

Shale, like coal, is characterised by high gas storage capacity (Ross & Bustin, 2009). The capacity of shale gas reservoirs appears to be in part controlled by the composition of the rock, and is highest in those rocks containing the highest total organic carbon or TOC, there being a positive correlation between TOC and gas sorption (Ross & Bustin, 2009). The quality of shale gas reservoirs also appears to be in part controlled by the composition of the rock, and in particular its pore structure, and its abiogenic and biogenic silica content and associated brittle fracture propensity. There is a positive correlation between biogenic silica content, in the form of diatoms, agglutinating or arenaceous foraminifera, radiolarians and sponge spicules, and TOC.

Gas is produced from shale-gas reservoirs by first hydraulically fracturing them in order to improve their producibility. The 'fraccing' process releases gas into the subsurface, and raises important environmental concerns surrounding contamination of the subsurface water supply.

5.5.3 Gas or methane hydrate

Gas or methane hydrate constitutes a potentially significant resource of methane in frozen form in shallow marine environments in polar regions, and, under pressure, in deep marine environments elsewhere (Ruffine et al., 2010). Its occurrence is indicated by bottom simulating reflectors or BSRs on seismic data.

Exploitation of the resource is currently unachievable for technological rather than technical reasons.

6 • Mineral exploration and exploitation

(Micro)palaeontology has proved of use in mineral exploration and exploitation (Hart, in Jenkins, 1993; Jones, 1996; Robbins & Burden, in Jansonius & MacGregor, 1996; Jones, 2006). Case studies of applications in mineral exploration and exploitation are given in Sections 6.1 and 6.2, respectively.

Readers interested in further details of mineral exploration and exploitation, and its environmental impact, are referred to Evans (1997), Spitz and Trudinger (2009), and Botkin and Keller (2010).

6.1 APPLICATION AND CASE STUDY IN MINERAL EXPLORATION

6.1.1 La Troya mine, Spain

Material and methods

The study was undertaken in and around the Early Cretaceous, Barremian–Aptian rudist-reef-hosted lead–zinc ore-body of the La Troya mine in the Basco-Cantabrian basin in the Basque country of Spain (Jones, 2006). The primary objective was to collect and analyse samples from the site of the ore-body itself and from – coeval – laterally equivalent sites varying distances away. This was to test the hypothesis that micropalaeontological assemblages from sites of ore-body deposition would be characterised by environmental stress, whereas those from laterally equivalent sites would not, enabling an 'environmental stress gradient', or, in three dimensions, a 'halo', to be defined, the likes of which might serve as pointers to the locations of hitherto undiscovered ore-bodies elsewhere (Fig. 6.1).

The secondary objective was to validate the established ore-body depositional model, specifically, to confirm the suspected syngenetic rather than epigenetic origin using micropalaeontological (palynological) indications of thermal maturity above and below.

Results and discussion

'Environmental stress gradient'. Unfortunately, in practice, it proved impossible to trace and sample the lateral equivalents of the ore-body, owing in part to

their extreme thinness (a function of the geochemically interpreted extremely short period of time over which the ore-body formed), and in part to fault complications. The hypothesis of the 'environmental stress gradient' thus remains effectively untested either in the Basco-Cantabrian basin or elsewhere.

Depositional model. Micropalaeontological analyses of the platform carbonates underlying the ore-body indicated an Urgonian (late Barremian to early Aptian) age. Palynostratigraphic analyses of the black pyritic shales ('margas negras') overlying it indicated a probable late Aptian age. Micropalaeontological analyses of sediments adjacent, although, in view of the problems outlined above, not necessarily exactly laterally equivalent to, the ore-body indicated a restricted marine environment, with only rare and stress-tolerant foraminifera and radiolarians present. Palynological analyses yielded 'spore colour indices' in the range 5–7, indicating over-maturity. No conclusions could be drawn on the basis of this evidence with regard to the syngenetic or epigenetic origin of the ore-body.

6.2 APPLICATIONS AND CASE STUDIES IN MINERAL EXPLOITATION

The appended case studies of applications in mineral exploitation are based on the work of Hart, in Jenkins (1993), and involve Late Cretaceous chalk sections in southern England. This reflects not only Hart's interest but also the peculiar problems faced in the detailed stratigraphic subdivision and correlation of macroscopically essentially featureless chalky lithotypes, especially the identification of lithological and rock property changes associated with unconformities, faults and facies changes.

6.2.1 Pitstone quarry, Hertfordshire, United Kingdom

The Pitstone quarry in Hertfordshire extracts chalk for use in the cement manufacturing industry. A balance of high- and low-calcimetry chalks provides the optimum properties. At Pitstone, calcimetry is controlled primarily by stratigraphic horizon, being low in the lower part of the lower Chalk and

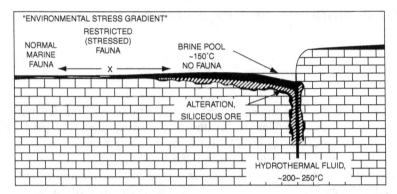

Fig. 6.1 **Anticipated faunal trends at time of ore genesis, La Troya mine.** From Jones (1996). Note 'environmental stress gradient'.

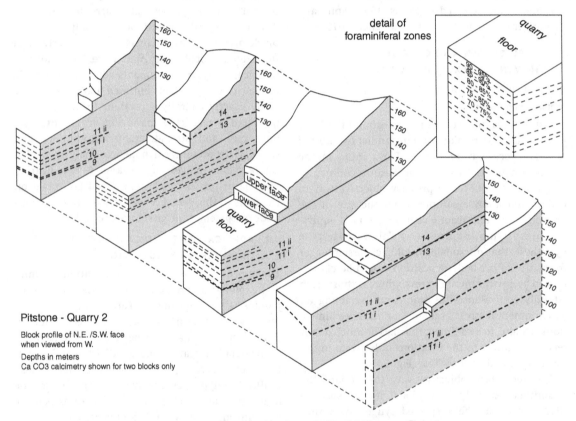

Fig. 6.2 **Pitstone quarry section showing foraminiferal biozones.** © Springer. Reproduced with permission from Hart, in Jenkins (1993).

Plenus marl, and high in the upper part of the lower Chalk and the lower part of the middle Chalk. A micropalaeontological study was therefore undertaken with the aim of identifying the local extent of these horizons. Closely spaced samples from a series of boreholes were analysed and the sections subdivided and correlated using a high-resolution biostratigraphic zonation scheme. The data thus obtained was presented in the form of a series of profiles projecting calcimetry (Fig. 6.2). These were utilised

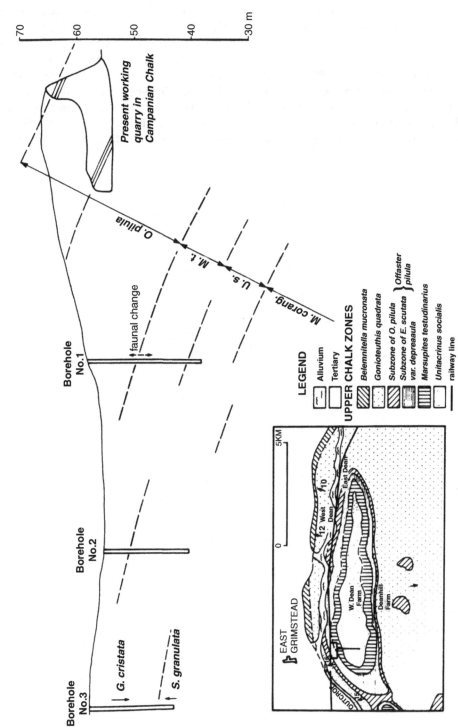

Fig. 6.3 East Grimstead quarry section showing foraminiferal and belemnite biozones. © Springer. Reproduced with permission from Hart, in Jenkins (1993).

in formulating a pit design plan optimising the existing infrastructure and providing the most ergonomic and economic extraction of future supplies.

6.2.2 East Grimstead quarry, Wiltshire, United Kingdom

The East Grimstead quarry in Wiltshire extracts chalk for use in the paper-whitener manufacturing industry. 'Bright' chalks provide the optimum properties. At East Grimstead, 'brightness' is controlled by strati-graphic horizon, being highest in the *Offaster pilula* Sub-Zone of the upper Chalk. A micropalaeontological study was therefore undertaken with the aim of identifying the local extent of this horizon. Closely spaced samples from a series of boreholes were analysed and the sections subdivided and correlated using a high-resolution biostratigraphic zonation scheme (Fig. 6.3). Results indicated that the zone of 'bright' chalks only extended for a short distance to the south of the existing quarry works.

7 • Coal geology and mining

The following is an abbreviated introduction to coal geology and mining. Readers interested in further details of coal geology and mining, and its environmental impact, are referred to Rahmani and Flores (1984), Scott (1987), Thomas (1992), Gayer and Harris (1996), Pickering and Owen (1994), Whateley and Spears (1995), Evans (1997), Miller (2005), Warwick (2005), Greb and diMichele (2006), Masters and Ela (2008), Catelin (2009), Spitz and Trudinger (2009), Tabak (2009) and Botkin and Keller (2010).

Studies of coal geology and mining are liable to become increasingly important in the future, as supplies of coal are set to outlast those of other fossil fuels, there being sufficient supplies of coal to last for approximately 130 years at current rates of production and consumption, but only sufficient gas and oil to last for approximately 60 years and 40 years, respectively (Catelin, 2009). The potential to sequester the greenhouse gas carbon dioxide in coal could also come to realisation, with both economic and environmental benefits (Catelin, 2009; Miranda *et al.*, 2009). Note that burning coal causes potentially damaging emissions of carbon dioxide, nitrogen oxides and sulphur oxides, particulates and trace metals, although it is possible to build or retro-fit coal-fired power stations with emission controls, albeit at some expense, and at the further expense of some efficiency loss.

7.1 INTRODUCTION TO COAL GEOLOGY AND MINING

7.1.1 Coal geology

By definition, coal is a rock composed of at least 50% by weight of carbonaceous matter. The carbonaceous matter is in turn itself composed of the physically compacted and thermally altered remains of land plants (Collinson & Scott, in Scott, 1987; DiMichele *et al.*, 2001; Jones, 2006).

There are three principal components, or maceral groups, namely: vitrinite; liptinite, formerly known as exinite; and inertinite (International Committee for Coal and Organic Petrology, 1998; Suarez-Ruiz & Crelling, 2008; Diessel, 2010). Vitrinite is derived from woody plant material that accumulated under dysoxic to anoxic surface- and ground-water conditions. Liptinite is derived from non-woody plant material such as cuticle, spores and pollen, and resins and waxes, that accumulated under dysoxic to anoxic conditions, together with algal material. Inertinite is derived from plant material that accumulated under oxic conditions.

Coal rank is determined by the nature and proportions of the constituent plant material and macerals, the physical conditions obtaining at the time of accumulation, and the complex of physical and thermal conditions acting during burial. It is low in the least compacted and altered type, that is, peat; moderate in moderately compacted and altered brown coal or lignite, and bituminous coal; and high in highly compacted and altered devolatilised bituminous coal and anthracite. High-rank anthracite is the most sought-after, as it has been compacted and altered to the extent that it retains little of its original volatile content, and consequently produces little smoke and ash on combustion.

The gelification index or GI of coal, based on the proportions of various maceral sub-groups, provides an indication of the ground-water level and oxygen content in its original depositional environment; and the tissue preservation index or TPI, an indication of the amount and type of plant input (Diessel, 1986; Calder *et al.*, 1991; Kalkreuth *et al.*, 1991; Diessel, 1992; Crosdale *et al.*, 1998; Kolcon & Sachsenhofer, 1999; Scott, 2000, 2002; Iordanidis & Georgakopoulos, 2003; Zdrakov & Kortenski, 2004; Scott *et al.*, 2007; Potter *et al.*, 2009). Cross-plots of the GI and TPI can be used to discriminate between a range of coal mire sub-environments, including: terrestrial; limnic, or, essentially, standing-water marsh; limno-telmatic, or fen, with encroaching forest vegetation; and telmatic, or forest-swamp; the last-named being particularly conducive to coal formation (Zdrakov & Kortenski, 2004; Potter *et al.*, 2009). Such cross-plots can also reveal 'drying-upwards' and other trends (Potter *et al.*, 2009). Other useful indices include the groundwater index or GWI and the vegetation index or VI (Calder *et al.*, 1991; Zdrakov & Kortenski, 2004); and the water-cover index or WCI (Jasper *et al.*, 2010). Cross-plots of the GWI and VI, like those of the GI and TPI, can be used to discriminate between a range of coal mire sub-environments, including not only marsh, fen and forest-swamp, but also various categories of bog (as recognised by McCabe, in Rahmani & Flores, 1984, McCabe, in Scott, 1987, Moore, in Scott, 1987, and Greb *et al.*, 2002).

7.1.2 Coal mining

Surface or 'open pit' mining

Mining of coal essentially at the surface takes two forms, namely, strip mining and contour mining (Kennedy, 1990; Evans, 1997). Strip mining is employed when the terrain is essentially flat, and involves the use of heavy machinery simply to remove a strip of overburden to reveal the coal beneath. Contour mining is employed when the terrain is hilly, and involves the removal of a section of hill-side, and the creation of a berm, to reveal the coal within. In both cases the spoil is at least initially simply heaped up elsewhere, although nowadays in 'developed' countries subsequently stabilised, spread with top-soil and sown with seed, in compliance with environmental legislation.

Underground mining

Mining of coal buried deeply underground involves the use of heavy machinery to sink shafts from the surface to enable access to the coal, and the use of light machinery and human labour to tunnel into and extract the coal from the individual seams, from within their often cramped confines (Evans, 1997). In 'room-and-pillar', or 'pillar-and-stall', mining operations, approximately half the coal has to be left behind in the form of pillars to support the roofs of the tunnels and to prevent them from collapsing. In contrast, in 'long-wall' mining, the roofs of the tunnels are allowed to collapse. The spoil from underground mining is carried to the surface and treated in the same way as that from surface mining (see above).

Underground mining was, and is, demanding and dangerous work, undertaken at the risk of sudden death caused by tunnel collapse or coal gas (methane) explosion, and of occupational disease caused by the inhalation of coal dust. As Mervyn Peake more poetically put it, in '*Rhondda Valley*':

> In the portentous dark
> For bread
> They make their stake,
> And gamble with the spark
> Of the non-dead.
> Their fear
> Is of the seeping gas; they pin
> Their faith in props...

It was also work that engendered a strong spirit among the miners, and in mining communities. Peake, again:

> And then, I heard them sing,
> And loose the Celtic bird that has no wing,
> No body, eye, nor feather,
> Only song,
> That indestructible, that golden thing.

7.2 APPLICATIONS AND CASE STUDIES IN COAL GEOLOGY AND MINING

Biostratigraphy

Coal occurs at virtually every stratigraphic level post-dating the evolution of land plants in the Ordovician, although it is especially characteristic of the generally 'greenhouse' climatic episodes of the Carboniferous, for example in Great Britain (see 7.2.1 below), the former Union of Soviet Socialist Republics, and the United States; of the Permian, for example in India (see 5.2.5 above), and South Africa (see 7.2.2 below); and of the Jurassic through Palaeogene (Collinson & Scott, in Scott, 1987; DiMichele *et al.*, 2001; Jones, 2006; Pfefferkorn *et al.*, in Fielding *et al.*, 2008; Tabor & Poulsen, 2008; Falcon-Lang & diMichele, 2010). The stratigraphically oldest occurrence of coal known to the author is in the Wabya formation of southern Shan state in Myanmar, whose age appears to be at least in part Silurian, Llandovery on graptolite evidence (Bender, 1983; Talent & Bhargawa, in Landing & Johnson, 2003). There is a major 'coal gap' in the Triassic, caused by a reduction in plant productivity following the end-Permian mass extinction (Retallack *et al.*, 1996).

Plant micro- and macrofossils are the most useful groups in the stratigraphic subdivision and correlation of the Carboniferous 'Coal Measures' of Euramerica and the Permian 'Coal Measures' of Gondwana (see above; see also below, and 7.2.1 and 7.2.2). Plant microfossils such as spores are especially useful, not least as they actually constitute a significant component of the coal (Peppers, 1996; Izart et al., 1998; see also 7.2.1 and 7.2.2 below). Indeed, high-resolution quantitative analysis of spore assemblages can even be used in the identification and correlation of individual coal seams (Marshall, in Bowden et al., 2005; see also 7.2.1 below).

In a recent application, spore analysis undertaken in the Carboniferous Pictou coal-field in Nova Scotia revealed (by idientifying additional areas where the seam was present) that the Acadia seam had not been fully worked in the Albion district, leading to the identification of an additional 22 million tonnes of coal reserves – and potentially also significant coal-bed methane or coal-seam gas (Hacquebard & Barss, 1966; MacGregor et al., in Jansonius & MacGregor, 1996; Gibling et al., in Pashin & Gastaldo, 2004; Jones, 2006). Spore analysis of samples from offshore oil and gas wells also revealed an extension of the Carboniferous Sydney coal-field under the Gulf of St Lawrence, making it the largest in eastern Canada (Barss et al., 1979; MacGregor et al., in Jansonius & MacGregor, 1996; Gibling et al., in Pashin & Gastaldo, 2004; Jones, 2006). Incidentally, the Sydney coal-field was discovered as long ago as 1672, when a reference to 'a mountain of very good coal' appeared in a book by the then Governor of Acadia, Nicholas Denys; and the Pictou coal-field was discovered in 1798 by Dr James MacGregor (Eyles & Miall, 2007).

In an earlier application, when speculative boreholes were sunk in central England in search of coal, the recovery of graptolites was enough to stop the drilling, and thus to save a significant amount of money – their presence indicating that the rocks being drilled were too old to contain coal (Jones, 2006). Still earlier, no less a personage than William Smith had attempted – albeit in vain – to stop mining for coal at what to him were also undoubtedly inappropriate stratigraphic levels.

Palaeobiology

Coal occurs in vegetated flood-basin, or fluvial flood-plain, in inter-distributary delta-plain and in coastal-plain, marsh or swamp environments in all other than the highest palaeolatitudes, although it is characteristic of tropical to temperate palaeolatitudes (Ziegler et al., in Scott, 1987; Jones, 2006). The highest palaeolatitude occurrences of coal known to the author are in the Permian of South Africa (MacRae, 1999; see 7.2.2 below); the Permian, Triassic and Jurassic of Australia (Brunnschweiler, 1984; White, 1986; Struckmeyer & Totterdell, 1990; White, 1990; Vickers-Rich & Rich, 1993; Johnson, 2004); the Cretaceous of Alaska (Graham, 1999; Skelton, 2003), of the former Union of Soviet Socialist Republics (Skelton, 2003), and of Greenland (Skelton, 2003; Henriksen, 2008); and the Palaeogene of Axel Heiberg and Ellesmere Islands in the Canadian Arctic (Eyles & Miall, 2007), and of Svalbard in the Norwegian Arctic (Worsley & Aga, 1986); all of which were located in palaeolatitudes of up to 70–80° (see also Parrish et al., 1982).

Plant micro- and macrofossil evidence indicates that the essentially tropical 'Coal Measure' flood-basin swamp-forests of the Carboniferous of Euramerica were vegetated by a range of pteridophytes, including giant and dwarf club-mosses and horsetails, and also newly evolved gymnosperms, including conifers and allied forms, such as tree-ferns, seed-ferns and cordaites (Collinson & Scott, in Scott, 1987; DiMichele et al., 2001; Eble, in Jansonius & MacGregor, 1996; Jones, 2006; see also 7.2.1 below). The club-mosses appear to have been characteristic of low-lying areas of the the flood-basin proper, or the fluvial flood-plain, inter-distributary delta-plain and coastal-plain; the horsetails, of the margins of the basin; and the gymnosperms, of higher, drier areas such as distributary channel-levees. The 'Coal Measure' flood-basin swamp-forests of the Carboniferous would have shared certain structural, though not taxonomic, similarities with those of the lower reaches of the Mississippi or the Orinoco of the modern era (and would have resounded to different insects).

Integrated studies

An integrated biostratigraphic and sequence stratigraphic framework for the Late Carboniferous,

Westphalian [Bashkirian–Moscovian] 'Coal Measures' of Great Britain is given by Cole *et al.*, in Collinson *et al.* (2005) (see also Quirk, in Ziegler *et al.*, 1997; and 7.2.1 below). Integrated sequence stratigraphic frameworks for the coal-bearing Carboniferous, Mississippian to Pennsylvanian of the Appalachian and Illinois basins in the northeastern United States are given by Martino, by Nadon and Kelly, by Kvale *et al.*, by Greb *et al.*, by Pashin, and by Gastaldo *et al.*, all in Pashin and Gastaldo (2004). An integrated framework for the Carboniferous of Atlantic Canada is given by Gibling *et al.*, in Pashin and Gastaldo (2004). Ones for the Early Permian Rio Bonito formation of the Parana basin in Brazil are given by Holz *et al.* (2002) and Holz and Kalkreuth, in Pashin and Gastaldo (2004); one for the Early Cretaceous Mannville group of the Alberta basin in Canada by Holz *et al.* (2002); one for the Palaeocene Fort Union formation of the Williston basin of the northwestern United States by Warwick *et al.*, in Pashin and Gastaldo (2004); and one for the Eocene Kapuni group of the Taranaki basin in New Zealand by Flores, in Pashin and Gastaldo (2004).

In general, in coaly sequences, interpreted transgressive systems tracts (TSTs) are characterised by coals; maximum flooding surfaces (MFSs), by marine bands or MBs; and high- or low-stand systems tracts (HSTs and LSTs) by sandstones (Cole *et al.*, in Collinson *et al.*, 2005; see also 7.2.1 below). TSTs are characterised palynologically by palynomorph assemblages indicating an upward transition from swamp, through fen, marsh and salt-marsh to marine; HSTs, by palynomorph assemblages indicating an upward transition from marine, through salt-marsh, marsh and fen to swamp (Holz *et al.*, 2002; see also Box 4.1 above).

7.2.1   Coal geology and mining in Great Britain

In Great Britain, coal is essentially restricted to the Late Carboniferous, Westphalian [late Bashkirian–Moscovian], although in Scotland there is also some in the Late Carboniferous, Namurian [Serpukhovian–early Bashkirian], and some in the Jurassic (Browne *et al.*, 1999; Read *et al.*, in Trewin, 2003; Rippon, in Trewin, 2003; Owens *et al.*, 2004; Cole *et al.*, in Collinson *et al.*, 2005; McLean *et al.*, in Collinson *et al.*, 2005; Jones, 2006, Waters & Davies, in Brenchley & Rawson, 2006; Waters *et al.*, 2007, 2009; Fig. 7.1; see also below).

It was deposited, in a series of cycles, in vegetated flood-basin swamp-forest environments, in the 'Pennine basin' on the northern margin, and in the South Wales, Forest of Dean, Somerset, Oxfordshire and Kent basins on the southern margin, of an emergent mass known as St George's Land, extending from mid-Wales in the west to Brabant massif in the east (Ramsbottom, in McKerrow, 1978; Cope *et al.*, in Cope *et al.*, 1992; Cleal & Thomas, 1994; Fraser & Gawthorpe, 2003; Read *et al.*, in Trewin, 2003; Rippon, in Trewin, 2003; Jones, 2006; Waters & Davies, in Brenchley & Rawson, 2006; see also below). Coal was used in Great Britain as long ago as the Bronze Age, for cremation; and later worked by the Romans, and through into the Middle Ages. Open-pit and underground mining operations upscaled considerably in the nineteenth century, and provided the power that drove the industrial revolution. By the beginning of the twentieth century, Great Britain was the largest producer of coal in the world, mining 300 million tonnes annually. By the end of the twentieth century, however, it was only the fifteenth largest producer, mining less than 50 million tonnes annually. At the time of writing, virtually all the coal mines in the country have been closed, and in former mining communities a long-established way of life has come to an end, with dire social consequences.

Biostratigraphy

The Westphalian [Bashkirian–Moscovian] 'Coal Measures' are zoned by means of plant micro- and macro-fossils and non-marine bivalves, and associated MBs by means of goniatites and conodonts (Trueman & Weir, 1946; Calver, 1956; Stubblefield & Trotter, 1957; Smith & Butterworth, 1967; Calver, 1968; Calver, 1969; Druce *et al.*, 1972; Ramsbottom *et al.*, 1978; Cleal, in Cleal, 1991; Francis, in Craig, 1991; Cleal & Thomas, 1994; Eagar, 1994; Vasey, 1994; McLean & Murray, in Gayer & Harris, 1996; McLean & Davies, in Jones & Simmons, 1999; Read *et al.*, in Trewin, 2003; Owens *et al.*, 2004; Cole *et al.*, in Collinson *et al.*, 2005; McLean *et al.*, in Collinson *et al.*, 2005; Jones, 2006; Waters & Davies, in Brenchley & Rawson, 2006; Fig. 7.2).

In detail, the 'Coal Measures' are marked by, in ascending stratigraphic order: the *Triquitrites sinanii–Cirratriradites saturnii*, *Radiizonates aligerens*,

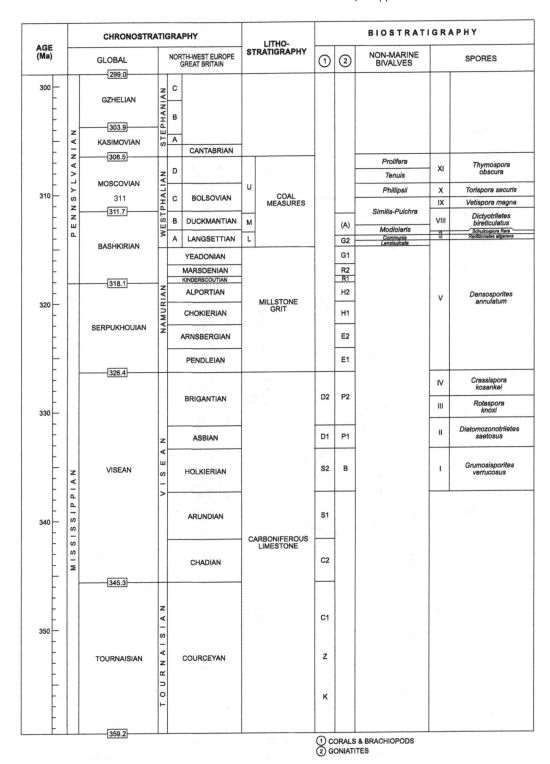

Fig. 7.1 **Carboniferous stratigraphy of Great Britain.** From Jones (2006).

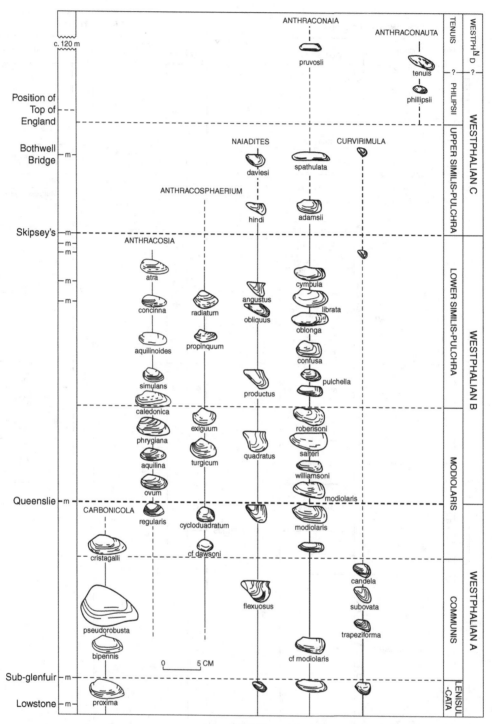

Fig. 7.2 Stratigraphic zonation of the Coal Measures of Scotland by means of non-marine bivalves. © The Geological Society, London. Reproduced with permission from Francis, in Craig (1991).

Microreticulatisporites nobilis–Florinites junior, Torispora securis and *Thymospora obscura*; or *Densosporites annulatus, Radiizonates aligerens, Schulzospora rara, Dictyotriletes bireticulatus, Vetispora magna, Torispora securis* and *Thymospora obscura*; or SS, RA, NJ, SL and OT, spore zones, and the *Lyringopteris hoenighausii, Lonchopteris rugosa, Paripteris linguaefolia, Linopteris bunburii* and *Lobatopteris* plant macrofossil zones; and the *Anthracomya* or *Carbonicola lenisulcata, Carbonicola communis, Anthraconaia modiolaris, Similis–Pulchra, Anthraconauta phillipsi, A. tenuis* and *A. prolifera* non-marine bivalve zones. Plant microfossils such as spores are especially useful in the stratigraphic subdivision and correlation of the 'Coal Measures'. Indeed, high-resolution quantitative analysis of spore assemblages can even be used in the identification and correlation of individual coal seams, a technique pioneered in the 1930s by the palynologist and polymath Arthur Raistrick (Marshall, in Bowden *et al.*, 2005). Successive MBs are marked by, in ascending stratigraphic order, the goniatites *Gastrioceras subcrenatum, G. listeri, Anthracoceras vanderbeckei, A. hindi, A. aegiranum* and *A. cambriense*. The oldest 'Coal Measures', and the youngest 'Barren Measures', known to generations of miners as the 'Farewell Rock', locally predate, and locally post-date, the *Gastrioceras subcrenatum* MB.

Palaeobiology

Plant micro- and macrofossil evidence indicates that the coal was deposited in flood-basin swamp-forests vegetated by a range of pteridophytes and newly evolved gymnosperms (Ramsbottom, in McKerrow, 1978; Cleal & Thomas, 1994). In general, lycopsids appear to have been characteristic of low-lying areas of the flood-basin proper; calamite horsetails, of the margins of the basin; and ferns, pteridosperms, cordaites, sphenopsids, and varieties of shrubs, herbs, ground-creepers and lianas, of higher, drier areas. However, the ecosystem appears to have been complex and dynamic, and, for example, specific lycopsids to have been characteristic of successive stages of colonisation of flood-basin habitats – *Paralycopodites* of pioneer colonisation, *Diaphorodendron* of climax colonisation; and *Synchysidendron, Lepidodendron* and *Lepidophloios* of opportunistic colonisation.

Integrated studies

An integrated – spore and non-marine bivalve – biostratigraphic and sequence stratigraphic framework for the Westphalian [Bashkirian–Moscovian] 'Coal Measures' of the southern North Sea and surrounding land areas is given by Cole *et al.*, in Collinson *et al.* (2005) (see also McLean & Murray, in Gayer & Harris, 1996, and McLean & Davies, in Jones & Simmons, 1999). The *Lenisulcata* non-marine bivalve zone, whose lower boundary is placed at the Westphalian A [Bashkirian], Langsettian *Subcrenatum* MB, a pronounced maximum flooding surface, is interpreted as representing a high-stand systems tract. The *Communis* non-marine bivalve zone, whose lower boundary is placed at the Burton Joyce MB, is interpreted as representing a low-stand systems tract. The lower *Modiolaris* non-marine bivalve zone, whose lower boundary is placed at the Tupton Coal, is interpreted as representing a transgressive systems tract. The upper *Modiolaris* non-marine bivalve zone, whose lower boundary is placed at the Westphalian B [Bashkirian], Duckmantian *Vanderbeckei* MB, another pronounced MFS, is interpreted as representing an HST/LST. The lower *Similis–Pulchra* non-marine bivalve zone, whose lower boundary is placed at the Top Hard (Barnsley) Coal, is interpreted as representing a TST. The upper *Similis–Pulchra* non-marine bivalve zone, whose lower boundary is placed at the Westphalian C [Moscovian], Bolsovian *Aegiranum* MB, yet another pronounced MFS, is interpreted as representing an HST. The *Phillipsi* non-marine bivalve zone, whose lower boundary is placed at the *Cambriense* MB, is interpreted as representing an LST. In general, interpreted TSTs are characterised by coals; MFSs, by MBs; and HSTs and LSTs by sandstones. MFSs/MBs may also mark stage boundaries.

7.2.2 Coal geology and mining in South Africa

South Africa contains significant reserves of principally Permian Gondwanan coal, and is a major coal producer, consumer and exporter (Hobday, in Scott, 1987; Evans, 1997; MacRae, 1999; Cairncross, 2001; Peatfield, 2003; Jeffrey, 2005; Schluter, 2006; McCarthy & Pretorius, 2009; Pinheiro & Hancox, 2009). Coal was first discovered in the country at Fransch-Hoek in the Western Cape as long ago as the 1690s, although the first actual coal mine was opened at Molteno in the Eastern Cape in the 1860s (Peatfield, 2003). Many more mines opened in the Witbank area in the 1890s following the discovery of the Kimberley diamond field in the 1870s and the Witwatersrand gold field in the 1880s, and the subsequent rapid growth of the diamond and gold mining industries. In 1927, South Africa produced 13 million tonnes of coal; in 1939, 18 million tonnes; in 1951, 29 million tonnes; in 2001, 221 million tonnes; in 2003, 236 million tonnes; and in 2006, 240 million tonnes (Peatfield, 2003; Jeffrey, 2005; Pinheiro & Hancox, 2009). In 1997, the country produced 218 million tonnes, of which it consumed 157 million tonnes, chiefly in the electricity generating, and synthetic fuel, petrochemicals and steel manufacturing, industries; and exported 59 million tonnes of bituminous coal and 2 million tonnes of anthracite (MacRae, 1999). That year, the value to the economy of South Africa of its coal exports was R8.8 billion. Approximately 40% of the country's coal production comes from the Witbank area (Schluter, 2006). Approximately 65% comes from underground mining, and 35% from surface mining. The environmental impact of coal mining in South Africa is discussed by McCarthy and Pretorius (2009), with particular emphasis on effects on water quality in the Vaal river system in the Highveld area.

Biostratigraphy and palaeobiology

In South Africa and neighbouring countries, there are palynostratigraphically dated 'Coal Measures' in the middle Early Permian, Artinskian part of the Ecca group, and in the Middle–Late Permian part of the Beaufort group (Falcon, 1975; Anderson, 1977; MacRae, 1988; MacRae, 1999; Stephenson & McLean, 1999; Bamford, 2004; Schneider & Marais, 2004; Cesari, 2007; Modie & le Herisse, 2009; Fig. 2.27). The Artinskian coals are age- and facies-equivalent to the Barakar and other coals of the Gondwanan basins of the east coast of the Indian subcontinent and in the continental interior, and indeed coals elsewhere in the former supercontinent of Gondwana, for example in South America and Madagascar, and similarly overlie latest Carboniferous to early Early Permian glacial diamictites and post-glacial lacustrine to marine shales (Veevers & Tewari, 1995; see also 5.2.5 above). The Middle–Late Permian coals are equivalent to the Raniganj and other coals of the Indian

subcontinent, and coals in Australia and Antarctica, and similarly overlie late Early Permian barren measures (Veevers & Tewari, 1995; see also 5.2.5 above). Note in this context that high-resolution quantitative analysis of spore assemblages has been used in the identification and correlation of individual coal seams in the Rampur coal-field in India (Brasier, 1980; Jones, 1996; see also above).

Interestingly, in South Africa, there are also, locally, palynostratigraphically dated 'Coal Measures' in the earliest to early Early Permian, Asselian to Sakmarian part of the Dwyka group, and in the early Late Triassic, Carnian part of the Stormberg group, which have either only local or essentially no equivalents elsewhere in Gondwana (Falcon, 1975; Anderson, 1977; MacRae, 1988; Anderson & Anderson, in Lucas & Morales, 1993; MacRae, 1999; Bamford, 2004; Cesari, 2007).

Earliest to early Early Permian, Asselian to Sakmarian part of Dwyka group. The locally developed earliest to early Early Permian, Asselian to Sakmarian coals of the Dwyka group of the Free State are intimately associated with the post-glacial lacustrine to marine shales of the latest Carboniferous to earliest Permian Gondwanan glaciations, and are interpreted as having been formed in a tundra- to taiga-type environment (MacRae, 1999). Plant microfossil evidence indicates that not only lower-order plants such as mosses, ferns and horsetails, but also some higher-order plants such as primitive gymnosperms were present in the landscape, possibly in close proximity to the edge of the retreating ice (Falcon, 1975; Anderson, 1977; MacRae, 1988; MacRae, 1999). The plants are representative of the so-called *Gangamopteris* or pre-*Glossopteris* flora (Wnuk, 1996).

Middle Early Permian, Artinskian part of Ecca group. The middle Early Permian, Artinskian coals of the Ecca group of the Witbank, Highveld and Ermelo areas constitute the bulk of South Africa's resource (MacRae, 1999; Peatfield, 2003; Jeffrey, 2005; McCarthy & Pretorius, 2009). These coals clearly developed during a warming phase following the Gondwanan glaciations, and are interpreted on various lines of palaeontological and other geological evidence as having accumulated in a complex of deltas and swamps fringing the epicontinental sea of the Karoo basin (Falcon, 1975; Anderson, 1977; MacRae, 1988; MacRae, 1999). Not only plant microfossils but also plant macrofossils, such as those described by Edna Plumstead, Heidi and John Anderson, and Eva Kovacs-Endrody, are extremely abundant and diverse in the Ecca coals and associated sediments (Plumstead, 1969; Anderson & Anderson, 1985; Kovacs-Endrody, 1991; MacRae, 1999). The plants are representative of the *Glossopteris* flora that came to colonise and characterise the entire supercontinent of Gondwana at this time (Wnuk, 1996; Anderson *et al.*, 1999).

Middle–Late Permian part of Beaufort group. The Middle–Late Permian coals of the Estcourt or Normandien formation of the Beaufort group, and of the equivalent Emakwezini formation, are atypical of the Beaufort group as a whole, which is otherwise dominated by continental sediments of hot, arid aspect, famous for, and dated by, their tetrapod vertebrate remains (Bordy & Prevec, 2008; Prevec *et al.*, 2009; see also 2.1.7 above). The plants of the coals and associated sediments continue to be representative of the *Glossopteris* flora (Wnuk, 1996). Insect feeding traces on plants, and inferred insect–plant behavioural interactions, have recently been described and discussed (Prevec *et al.*, 2009).

Early Late Triassic, Carnian part of Stormberg group. The locally developed early Late Triassic, Carnian coals and associated sediments of the Molteno formation of the Stormberg group are famous for their plant and insect remains, and have been intensively studied by palaeobotanists and palaeoentomologists (Anderson & Anderson, in Lucas & Morales, 1993; MacRae, 1999; Jones, 2006; Sues & Fraser, 2010). The plants are representative of the *Dicroidium* flora that succeeded the *Glossopteris* flora after the end-Permian mass extinction (Wnuk, 1996). The plants and insects have been interpreted as representing a range of habitats: including mature and immature *Dicroidium* riparian forest (bordering channels), with beetles, cockroaches, bugs and dragonflies; *Dicroidium* woodland (on the open flood-plain), with beetles, bugs and cockroaches; *Sphenobaiera* woodland (adjacent to lakes on the flood-plain), with beetles, cockroaches, bugs, dragonflies, crickets and moths; *Heidiphyllum* thicket (on flood-plains or channel sand-bars, or in areas of high water table), with beetles, cockroaches and bugs; *Equisetum* marsh (on flood-plains), with beetles and bugs; and fern/*Ginkgophytopsis* meadow (on sand-bars in braided rivers), with no insect faunas (Anderson & Anderson, in Lucas & Morales, 1993; MacRae, 1999; Jones, 2006). The insects are interpreted as herbivores, carnivores, omnivores, and, importantly, pollinators (of gymnosperms, excluding glossopterids, extinct by this time). Insect feeding traces on plants, and inferred insect–plant behavioural interactions, have recently been described and discussed (Scott *et al.*, 2004).

8 • Engineering geology

(Micro)palaeontology has proved of use in engineering geology, in the fields of site investigation and seismic hazard assessment (see below).

The living biota has also proved of use in environmental impact assessment (EIA) in civil engineering (see Section 9.1).

8.1 APPLICATIONS AND CASE STUDIES IN SITE INVESTIGATION

(Micro)palaeontology has proved of use in site investigation (Hart, in Jenkins, 1993; Jones, 1996; Hart, in Martin, 2000; Wright, 2001; Jones, 2006; author's unpublished observations; see also Box 5.5 above). Case studies are given in 8.1.1–8.1.4 below. Three of the four case studies (8.1.1–8.1.3) involve Late Cretaceous chalk sections in southern England. As with the case studies in mineral exploitation (6.2.1–6.2.2 above), this reflects the peculiar problems faced in the detailed stratigraphic subdivision and correlation of macroscopically essentially featureless chalky lithotypes, especially the identification of lithological and engineering property changes associated with unconformities, faults and facies changes. Note in this context that calcareous nannofossil chalks have peculiar engineering properties, controlled primarily by their 'nannofabrics'. For example, some are peculiarly dense on account of the close packing of the subcubical calcareous nannofossil *Micula*.

8.1.1 Channel tunnel, UK

The main application in this case was in the provision of precise and accurate stratigraphic control on the Albian Gault clay and Cenomanian lower Chalk (Hart, in Jenkins, 1993). Samples were taken with an average vertical spacing of as little as 1 m from the appropriate stratigraphic section from a number of sites along a cross-channel transect. A high-resolution foraminiferal biostratigraphic zonation scheme was erected, the resolution and correlative value of which was enhanced by quantitative measures of percentages of planktonic species and of planktonic 'morphogroups'

(epipelagic *Hedbergella*/*Whiteinella*, mesopelagic *Dicarinella*/*Praeglobotruncana* and bathypelagic *Rotalipora*). The resolution of the biozonation scheme was sufficient to allow recognition of local and regional onlap surfaces, as at the base of the lower Chalk, and erosional surfaces, as in the Middle Cenomanian part of the lower Chalk (Fig. 8.1). Rock property changes associated with the east–west loss through onlap of the particularly clay-rich basal Chalk have important engineering implications.

8.1.2 Thames barrier, UK

A number of boreholes were drilled to investigate the rock properties of the upper Chalk in the subsurface in order to select a suitable founding level for the 3300-tonne moveable steel gates of the Thames barrier and their supporting concrete piers (Hart, in Jenkins, 1993; Hart, in Martin, 2000). Most of the variation in engineering properties as revealed by tests was in the vertical (stratigraphic) rather than horizontal sense. There was also some overprint associated with weathering, frost-shattering and solifluction. The principal objective of the parallel micropalaeontological study was therefore to provide a detailed correlation of different stratigraphic horizons and any offset by faults. A subsidiary aim was to distinguish solifluction from *in situ* chalk using the presence of exotic components (Tertiary sand grains etc.) introduced during movement. The sample spacing was again extremely close (of the order of 1 m). Again, a high-resolution foraminiferal biozonation scheme was erected. The resolution was enhanced by separate plots, in the form of bar charts or kite diagrams, of selected superfamilies, and of the proportions of selected species or species groups within the superfamilies Cassidulinacea (*Globorotalites cushmani*, *Lingulogavelinella* aff. *vombensis*, other) and Globigerinacea (*Globotruncana bulloides*/*G. marginata*, *G. linneiana*/*G. pseudolinneiana*, *Hedbergella*). The resolution was sufficient to allow recognition not only of a widespread erosional unconformity of

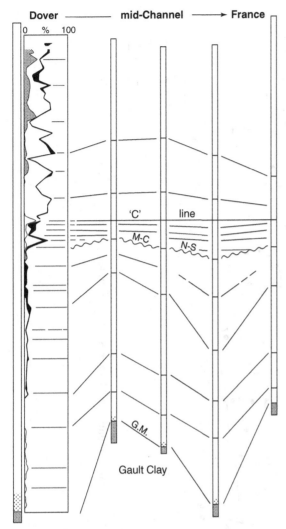

Dover —————— mid-Channel ————▶ France

'C' line

M-C   N-S

G.M.

Gault Clay

Fig. 8.1 **Channel tunnel section showing foraminiferal biozones**. © Springer. Reproduced with permission from Hart, in Jenkins (1993). C, Correlation line used as reference datum; M-C N-S, Mid-Cenomanian Non-Sequence; G.M., Glauconite marl. Note onlap associated with Mid-Cenomanian Non-Sequence.

varying extent at the top of the Chalk, but also of cross-cutting faults (Fig. 8.2). The engineering implications of rock property changes associated with faults at the foundation level were fully evaluated by engineering geologists prior to commencement of the construction operation.

8.1.3 'Project Orwell', UK

As part of the recent 'Project Orwell', completed in 1999, a 5.5-km-long tunnel was constructed to reduce flooding in and around Ipswich in Suffolk. The tunnel had to be drilled between two resistant flint bands within the Campanian *Gonioteuthis quadrata* Zone of the Chalk, since encountering either would have constituted a significant hazard. The trajectory of the tunnel was successfully kept within this narrow 'window' by means of high-resolution micropalaeontology (Wright, 2001).

8.1.4 Site investigation in the petroleum industry – Azerbaijan and Egypt

In my working experience in the petroleum industry, micropalaeontology has proved of use in site investigation in the Black Sea off Turkey, in the Caspian off Azerbaijan and in the Mediterranean off Egypt, specifically in helping to determine the safest sites for the locations of drilling rigs, pipelines and other facilities through the avoidance of geo-hazards (see also Box 5.6).

Azerbaijan

In the Caspian off Azerbaijan, high-resolution micro- and palynostratigraphy, essentially in the form of climatostratigraphy calibrated against the marine oxygen isotope record, have been used to date and correlate Quaternary sediments, and also to characterise potentially geo-hazardous sediments, in surface and shallow subsurface core samples (and high-resolution shallow seismic has been used to map the potentially geo-hazardous sediments away from the areas of core control).

Micro- and palynofacies has been used to characterise potentially geo-hazardous mud-volcano flows, on the basis of their non-contemporaneously reworked microfossil content.

Egypt

In the Mediterranean off Egypt, high-resolution micro- and palynostratigraphy, has been used, in conjunction with accelerator mass spectrometry (AMS), and optically stimulated luminescence (OSL), stratigraphy, to date and correlate Quaternary sediments, and also to characterise potentially geo-hazardous sediments, in surface and shallow subsurface core samples (and again high-resolution shallow seismic has been used to map

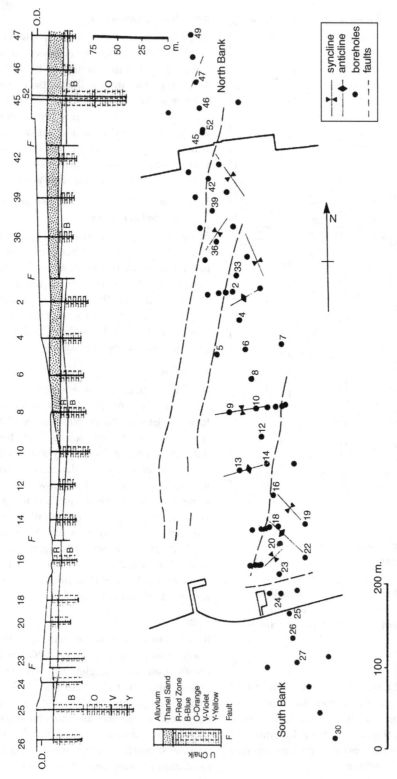

Fig. 8.2 **Thames barrier site investigation section showing foraminiferal biozones.** © Springer. Reproduced with permission from Hart, in Jenkins (1993).

the potentially geo-hazardous sediments away from the areas of core control).

Micro- and palynofacies has also been used to characterise potentially geo-hazardous land-slides on the basis of their contemporaneously transported microfossil content. It has also been used to demonstrate that land-slides significantly decreased in frequency between the 'early' and 'late' Holocene, and currently constitute a manageable geo-hazard. Palynofacies indicates that the 'early' Holocene was a time of humid climate and associated high sediment accumulation rate, and hence increased pore-pressure, leading to increased land-slide frequency. In contrast, the 'late' Holocene was a time of arid climate, low sediment accumulation rate, and decreased pore-pressure and land-slide frequency.

8.2 APPLICATIONS AND CASE STUDIES IN SEISMIC HAZARD ASSESSMENT

Palaeontology has proved to be of use in palaeoseismology, that is, the study of the historical and prehistoric record of earthquake ground effects, and hence in seismic hazard assessment. Case studies are given in 8.2.1–8.2.4 below.

8.2.1 Strait of Juan de Fuca, Vancouver Island, Canada, Iyo-nada Bay, Japan, and Hawke's Bay, New Zealand

Rapid changes in bathymetry – of as little as 0.05–1.5 m – recorded by diatom and foraminiferal assemblages in dated marginal marine environments in the Strait of Juan de Fuca off Vancouver Island in Canada, in Iyo-nada Bay in Japan, and in Hawke's Bay in New Zealand, have been used to infer uplift associated with past earthquake activity (Williams, 1999; Yasuhara et al., 2005; Hayward et al., 2006). In the case of Hawke's Bay in New Zealand, they have been used to infer uplift associated not only with the 1931 Napier earthquake (1.5 m), but also with previously unknown earlier earthquakes 600, 1600, 3000, 4200, 5800 and 7200 years before present, providing some measure of frequency probability if not predictability.

8.2.2 Cabo de Gata Lagoon, Almeria, Spain

Admixtures of autochthonous marginal marine and allochthonous shallow to deep marine foraminifera in dated marginal marine environments in the Cabo de Gata Lagoon in Spain have been used to identify tsunamites and infer earthquake activity associated with the known Almeria earthquake in 1522 (Reichterter & Becker-Heidman, in Reichterter et al., 2009).

Similar approaches have also been adopted elsewhere (Mamo et al., 2009; Ruiz et al., 2010).

8.2.3 Lake Sapanca, Turkey

Admixtures of autochthonous lacustrine and allochthonous terrestrial palynomorphs in dated lacustrine environments in Lake Sapanca in Turkey have been used to identify turbidites and infer earthquake activity associated with the known 1957 Abant, 1967 Mudurnu and 1999 Izmit earthquakes on the North Anatolian Fault (Leroy et al., 2009).

8.2.4 Western Crete

Assemblages of encrusting and boring corals, bryozoans and lithophagid bivalves on a dated former sea-bed surface now approximately 10 m above sea-level in western Crete have been used to infer uplift associated with the known M8.5 earthquake of 21 July AD 365, centred on Crete (Shaw et al., 2010).

Incidentally, the tsunami associated with this earthquake caused extensive loss of life and damage to property throughout the eastern Mediterranean (Guidoboni & Ebel, 2009); and indeed possibly even as far afield as the ancient Roman city of Baelo Claudia near Gibraltar in the western Mediterranean (Menanteau et al., 1983; Silva et al., 2005; Silva et al., in Reichterter et al., 2009). According to the eye-witness accounts of Ammianus Marcellinus (c. AD 330–400), John Cassian (c. AD 360–430) and others, there were numerous fatalities on what was to become known as the 'day of horror' in Alexandria in Egypt; the dune barrier on the seaward side of the Nile delta was breached, forming what was to become known as Lake Manzala; and the ancient city of Tenis was destroyed (Guidoboni & Ebel, 2009). As John Cassian wrote, of events in Panephysis, near present-day Lake Manzala:

The sea was disturbed by a sudden earthquake and broke its bounds … covering what used to be fertile lands with salt marshes … [T]owns perched on … hills were abandoned by their inhabitants, and the hills were turned into islands … thus providing the solitude desired by holy men seeking seclusion.

Repeat time estimates for such an event range from 800 to 5000 years.

9 • Environmental science

The living biota has proved of use in environmental science, specifically in environmental impact assessment (EIA), in environmental monitoring, in bioremediation, and in recording anthropogenically mediated global change. Applications and case studies in EIA, environmental monitoring, bioremediation and anthropogenically mediated global change are discussed in Sections 9.1–9.4, respectively, below.

9.1 APPLICATIONS AND CASE STUDIES IN ENVIRONMENTAL IMPACT ASSESSMENT

According to the UK Department of the Environment's definition, EIA is 'a process by which information about the environmental effects of a project is collected, both by the developer and other sources, and taken into account by the relevant decision making body before a decision is given on whether the development should go ahead' (Clift, in Harrison, 2001; Morris & Therivel, 2001; Wood, 2003; Jones, 2006; Spitz & Trudinger, 2009).

The living biota has proved of use in EIA in the civil engineering industry, in baseline studies on the potential environmental impacts of industry projects, and in the mitigation of these impacts, for example in the construction of the Rance barrage in France and the Severn barrage in Great Britain (Jones, 2006).

The living biota has also proved of use in EIA in the petroleum industry, in baseline studies on the potential environmental impacts of industry projects, and in the mitigation of these impacts, for example in and around BP's Prudhoe Bay and other oil fields on the North Slope of Alaska, in and around Wytch Farm field under Poole Harbour in Dorset on the south coast of England, and in the Columbus basin off the east coast of Trinidad (Evans, 1997; Truett & Johnson, 2000; Gluyas & Swarbrick, 2004; Jones, 2006; author's unpublished observations; see also Box 5.5, and below).

9.1.1 Environmental impact assessment, North Slope, Alaska

In the case of Prudhoe Bay and other oil fields, situated between the National Petroleum Reserve of Alaska to the west and the Arctic National Wildlife Refuge to the east, potential environmental impacts have been mitigated through the implementation of a number of measures approved by the appropriate regulatory bodies (Truett & Johnson, 2000). These measures include

… consolidation of facilities [into smaller areas than originally planned]; use of ice road technology to eliminate gravel roads adjacent to pipelines, and elevating … pipelines … to allow free movement of wildlife; directional drilling to reduce the number of gravel pads; [and] improved waste handling, and the elimination of … pits for surface storage of drilling muds and cuttings (these drilling by-products are now re-injected into confining geological formations).

9.1.2 Environmental impact assessment, Wytch Farm, Dorset, UK

In the case of Wytch Farm, situated in an Area of Outstanding Natural Beauty, potential impacts have been mitigated by targeting the offshore reservoir from an onshore rather than offshore base through 'extended reach' directional drilling, and by designing the base so as to blend in with the landscape as much as possible (Evans, 1997).

9.2 APPLICATIONS AND CASE STUDIES IN ENVIRONMENTAL MONITORING

The living biota has proved of use in environmental monitoring, in baseline studies on natural and artificial, anthropogenic effects on water quality, including domestic and industrial pollution (Pearson & Rosenberg, 1978; Jones, 1996; Martin, 2000; Clift, in Harrison, 2001; Harrison, in Harrison, 2001; Mason, in Harrison, 2001; Scott et al., 2001; Akimoto et al., 2004; Taylor & Morrissey, in Harris, 2004; Nigam, 2005; Ruiz et al., 2005; Scott *et al.*, 2005; Debenay & Bui Thi Luan, 2006; Ernst *et al.*, 2006; Jones, 2006; Mojtahid *et al.*, 2006; Wilson, 2006; Mojtahid *et al.*, 2008; Jorissen *et al.*, 2009; Alve & Dolven, 2010; Armynot du Chatelet & Debenay, 2010; Barras *et al.*, 2010; Bicchi *et al.*, 2010; Bouchet *et al.*, 2010; Hess *et al.*, 2010; Jorissen, 2010; see also Box 5.6 above, and 9.2.1 and 9.2.2 below).

Such studies will surely increase in importance in the future, not least in view of the impending European Water Framework Directive or EWFD, due to be implemented in 2015 (Jorissen, 2010). Preparatory work is already well under way on standardising foraminiferal sampling and preparation methods and taxonomy, on identifying 'sensitive', 'indifferent', 'tolerant', 'second-order opportunist' and 'first-order opportunist' indicator species, and on identifying 'high', 'good', 'average', 'poor' and 'bad' environmental quality categories on the basis of these indicator species. Other useful indices include AMBI (the AZTI Marine Biotic Index), the Benthic Quality Index or BQI, and the Infaunal Trophic Index or ITI.

9.2.1 Environmental monitoring of natural and anthropogenic effects on water quality, including domestic and industrial pollution

The living biota, including the foraminiferal microbiota, has been used to monitor natural effects on water quality in a number of studies, for example on salinity in the Everglades National Park in Florida.

The living biota, including the palynomorph and foraminiferal microbiotas, has also been used to monitor anthropogenic effects on water quality, including domestic and industrial pollution, and pollutant transport, in a number of studies.

The effects of pollution on the foraminiferal microbiota vary considerably, but can include any or all of the following:

modifications to the structure and composition of communities, including decreases in abundance and diversity, exclusion of certain species etc.;
modifications to the reproductive cycles of individual species, including increases in incidences of asexually reproducing individuals;
development of test deformities;
changes in shell chemistry, including increased absorption of metals, sulphur etc., and pyritisation;

all of which appear to become progressively more pronounced as the pollution continues;
and at least locally, death.

As far as can be ascertained from the available data, the effects appear to be more or less localised around the source of the pollution. Note, though, that the eutrophication associated with nitrogen-rich run-off from sugar cane plantations, shrimp farms and sewage outlets in eastern South America is experienced at least as far afield as Trinidad in the eastern Caribbean, owing to the effect of the Guyana current (Wilson, 2006).

The effects appear to disappear more or less as soon as the source of the pollution disappears (although this would probably not be so in the case of radioactive pollution). In a recent well-documented case involving the disposal of oil-contaminated cuttings samples from oilfield operations offshore Congo, concentric zones of comparatively strongly, moderately and weakly stressed environments were observed around the disposal site over a horizontal distance of 750 m initially, and a distance of 250 m four years later (Mojtahid *et al.*, 2006; compare also Jorissen *et al.*, 2009, Bicchi *et al.*, 2010, and Hess *et al.*, 2010). The strongly stressed environments, observed initially between 0 and 70 m from the disposal site, are characterised by extremely impoverished benthic foraminiferal assemblages; the moderately stressed environments, observed initially 70–250 m from the disposal site, by locally enriched benthic foraminiferal assemblages containing high proportions (>50%) of opportunistic – infaunal – species tolerant of eutrophication and/or dysoxia, including *Bolivina* spp., *Bulimina aculeata*, *Bulimina marginata*, *Textularia sagittula* and *Trifarina bradyi*; and the weakly stressed environments, observed initially 250–750 m from the disposal site, by impoverished assemblages containing lower proportions (<50%) of opportunistic species. The proportion of opportunistic species appears to be a useful measure of the extent of anthropogenic as well as natural eutrophication.

9.2.2 Environmental monitoring of coral reef vitality

The living biota, including the foraminiferal microbiota, has been used to monitor coral reef vitality in a number of studies, for example on Florida and

Hawaii in the United States (Hallock, in Martin, 2000), and on the Great Barrier Reef in Australia (Scheuth & Frank, 2008). An index based on foraminiferal assemblages, the so-called FORaminifera in Assessment and Monitoring, or FORAM, Index (FI), has been used to provide an indication of water quality and its ability to support healthy coral reefs (Hallock et al., 2003; Scheuth & Frank, 2008). The FI is given by the formula: $FI = (10 \times P_s) + (P_o) + (2 \times P_h)$; where P_s is the proportion of symbiont-bearing LBFs, such as are characteristic of natural, oligotrophic environments; P_o is the proportion of opportunistic foraminifera, such as are characteristic of naturally or artificially eutrophic environments, for example those affected by nutrient-rich run-off associated with agricultural activities; and P_h is the proportion of other foraminifera. The results of a recent study on the Low Isles Reef on the Great Barrier Reef in Australia revealed low FIs, and naturally or artificially eutrophic environments, unfavourable for coral reef vitality, only over a relatively small area of the reef, adjacent to a mangrove swamp providing a local, natural source of nutrient input in the form of leaf litter and bird guano from thousands of Torres Strait pigeons (Scheuth & Frank, 2008). Incidentally, this area had been essentially cleared of reef-forming corals by a particularly destructive cyclone in 1950, and subsequently recolonised by Algae and soft-bodied corals. Having visited the Low Isles Reef following Cyclone Joy in 1989, I can attest to the damage done to certain reef-forming corals, such as branching morphotypes of *Acropora*, by such natural events.

9.3 APPLICATIONS AND CASE STUDIES IN BIOREMEDIATION

Bioremediation is a natural or artificial process whereby elements of the living microbiota decontaminate surface water, ground-water or soil environments contaminated by pollutants (Admassu & Korus, in Crawford & Crawford, 1996; James, in Harrison, 2001; Masters & Ela, 2008).

Artificial bioremediation or decontamination of environments contaminated by hydrocarbon pollutants is achieved by means of the introduction of elements of the microbiota selected on the basis of their ability to adhere to, to degrade, and to desorb from, hydrocarbons; or of microbial surfactants

(Rosenberg & Ron, in Crawford & Crawford, 1996; Prince, in Ollivier & Magot, 2005; Masters & Ela, 2008). (Degradation is by means of oxidation or emulsification.) A surprisingly large number of Bacteria, Fungi and Algae fulfil these selection criteria.

9.3.1 Bioremediation of the Exxon Valdez oil spill

On 24 March 1989, the oil tanker *Exxon Valdez* ran aground near Bligh Island in Prince William Sound, Alaska, spilling 30–40 thousand tonnes of crude oil. On succeeding days, oil washed up onto some 100 miles of coastline around the sound. Some natural bioremediation was achieved by naturally occurring Bacteria. Artificial bioremediation was attempted by treating affected areas with Inipol EAP22, a fertiliser developed to stimulate the growth of Bacteria (Prince, in Fry et al., 1992; Rosenberg & Ron, in Crawford & Crawford, 1996; Taylor & Morrissey, in Harris, 2004). Subsequent residual oil measurements in Snug Harbor indicated an approximately 75% removal of oil in treated cobblestone plots, compared with 50% removal in untreated, control plots, ater 90 days (Pritchard et al., 1992).

Importantly, many bioremediation techniques were refined, and some new ones developed, during research stimulated by the *Exxon Valdez* spill.

9.3.2 Bioremediation in the aftermath of Operation Desert Storm

Operation Desert Storm was undertaken between 17 January and 28 February 1991, with the objective of forcing the Iraqi army out of Kuwait, which it had invaded on 2 August 1990. During the course of the operation, the retreating Iraqi army attempted to destroy the oil infrastructure in Kuwait by releasing oil from well-heads onto the land and into the sea, and locally by setting it alight. In the aftermath, some natural bioremediation of the oil spills at sea was achieved by naturally occurring oil-degrading Bacteria. Artificial bioremediation of the oil spills on land was attempted by treating affected areas with cultures of oil-degrading Bacteria isolated from the populations that had developed naturally on the surface of the oil spills at sea, including *Rhodococcus*, *Bacillus* and *Arthrobacter* (Sarkoh et al., 1995; Varnam & Evans, 2000). In this case, measurements indicated an approximately 50% removal of oil in treated areas over a

10–20 week period. During this period, the Bacteria *Pseudomonas*, *Streptomyces* and cf. *Thermoactinomyces*, and the moulds *Aspergillus* and *Penicillium* joined the microbial consortium alongside *Rhodococcus*, *Bacillus* and *Arthrobacter*.

9.4 APPLICATIONS AND CASE STUDIES IN ANTHROPOGENICALLY MEDIATED GLOBAL CHANGE

Fossils provide a long-term record of global environmental and climatic change, and a means of discriminating past natural from present anthropogenic causes and effects. Their study also provides pointers as to the ranges of rates of global change. It indicates, for example, that the past natural global warming associated with events such as the Palaeocene–Eocene Thermal Maximum was relatively rapid, implying that the present anthropogenic global warming could be, too, and with similarly widespread consequences (Dunkley Jones *et al.*, 2010; see also Box 3.7 above).

The Intergovernmental Panel on Climate Change (IPCC) concluded in 1996 that anthropogenic greenhouse gas emissions were responsible for the observed present, or rather recent, global warming and sea-level rise, as had long been suspected, and that action was required to limit such emission, the case and timetable for which was set out in the 'Kyoto protocol' in 1997 (Holman, in Harrison, 2001; Hulme, in Harris, 2004; Nunn, in Harris, 2004; Petford, in Harris, 2004; Houghton, 2004; Lovell, 2006; Masters & Ela, 2008; Botkin & Keller, 2010). Engineering interventions to counter the effects of recent global warming and sea-level rise are discussed by Launder and Thompson (2009).

Interestingly, anthropogenic greenhouse gas emissions and associated effects on the climate system have recently been shown to have begun in the pre-industrial era, with the clearance of forests for agriculture (Ruddiman, 2003, 2005; Nevle & Bird, 2008; Kaplan *et al.*, 2009; Xuefeng Yu *et al.*, 2010). Decreased anthropogenic greenhouse gas emissions associated with decreases in human populations during periods of pandemic in the pre-industrial era resulted in global cooling (Ruddiman, *op. cit.*); increased emissions associated with the burning of fossil fuels in the industrial era, in global warming.

9.4.1 Ocean acidification

The living calcareous nannoplankton, foraminiferal, coral, bivalve and pteropod biotas have proved of use in studies on increasing ocean acidification in response to increasing, anthropogenically mediated atmospheric carbon dioxide concentration, the results of which suggest that the abilities of at least these taxa to calcify will be adversely affected if the oceans continue to acidify (Gattuso *et al.*, 1998; Riebesell *et al.*, 2000; Orr *et al.*, 2005; Gazeau *et al.*, 2007; Iglesias-Rodriguez *et al.*, 2008; Kuroyanagi *et al.*, 2009; Dias *et al.*, 2010).

9.4.2 Carbon dioxide sequestration

The response of the petroleum industry to the challenge of reducing greenhouse gas emissions, through carbon capture and storage in underground reservoirs, is discussed by Lovell (2009).

The living foraminiferal biota has recently proved of use in a study on the potential environmental impact of carbon dioxide sequestration in a deep-sea reservoir in Monterey Bay in California (Ricketts *et al.*, 2009).

10.1 APPLICATIONS AND CASE STUDIES
IN ARCHAEOLOGY

In conventional palaeontology, fossils are typically used to provide information on the ages and environments of containing sediments.

In archaeology, the approach is subtly different, in that both sediments and fossils, including human remains and associated artefacts, and associated animal and plant remains, are used to provide information on the stratigraphic and environmental context of human evolution and dispersal, and of ancient human life and behaviour (Jones, 2006; see also comments on 'Archaeostratigraphy' and 'Environmental archaeology' below; Bobe et al., 2007; Dominguez-Rodrigo et al., 2007; Ungar, 2007; Grine et al., 2009; Hublin & Richards, 2009; Sinibaldi, 2010). The human remains provide information as to ancient human life and behaviour, and also as to disease and/or cause of death (Mays, 1998; Ortner, 2003; Roberts & Manchester, 2005). Tooth wear provides information as to diet; tooth composition, information as to the point of origination of the individual, and as to migration; DNA, as to the migration of populations (Renfrew & Boyle, 2000; Evans & Tatham, in Pye & Croft, 2004; Hillson, 2005; Weiner, 2010).

Palaeontology also provides information as to the provenance of materials used in the manufacture of pottery and of mosaics, and of building and decorative stone, flint, amber and other trade goods, and ship's ballast, and hence as to ancient trade links (MacGregor et al., in Jansonius & MacGregor, 1996; Jones, 2006; Quinn & Day, 2007; Tasker et al., 2009; Weiner, 2010). For example, fossil foraminifera have been used to pinpoint the precise source of the clay used in the manufacture of the Minoan pottery discovered in the recent excavation of Akrotiri on the island of Thera in the Aegean, which turned out to be most likely somewhere around Cape Loumaravi, only some 5 km to the northwest (Jones, 2006). (Thera was part of the much larger volcanic island of Santorini until the explosive volcanic eruption of 1640 BC, which, incidentally, is interpreted as having been partly responsible for the decline of the Minoan civilisation on Crete.) Fossil dinoflagellates have been used to identify the source of the shale used in the manufacture of a pre-Roman vase found at Harpenden, just north of London, in the southeast of England: it most probably came from Dorset, some 200 km away, in the southwest (MacGregor et al., in Jansonius & MacGregor, 1996). Calcareous nannofossils and foraminifera have been used to identify the source of the chalk used in the manufacture of the individual tesserae from the Roman mosaics of Calleva Atrenatum or modern Silchester in Hampshire, UK, which also turned out to be in Dorset, some 100 km to the southwest (Wilkinson et al., 2008; Tasker et al., 2009). Calcareous nannofossils have also been used to identify the source of the chalk used in medieval Norwegian art (Perch-Nielsen, 1973). Finally, foraminifera and ostracods have been used to pinpoint the precise source of the stone used in the construction of the English Civil War defensive earthworks at Wallingford Castle in Oxfordshire, which turned out to be an outcrop of the Glauconitic Marl member of the West Melbury Marly Chalk formation local to the castle site (Wilkinson et al., 2010). In this particular case the use of a local source of stone may have been born of necessity, the castle having been besieged by plucky Parliamentarian forces for the 65 days leading up to the Royalist surrender on 27 July 1646. Whether or not this was so, the stone, when compacted, proved to be eminently suited to the construction of defensive earthworks and even gun emplacements.

Incidentally, geology provides information as to the provenance of not only building stones, such as the standing stones of Stonehenge in the UK, but also worked metals. Assays of Bronze Age gold from all over Ireland have revealed a silver content suggestive of an origin in the Croagh Patrick area of

Connemara. And archaeometallurgy indicates what type of technology was available, and when.

Readers interested in further details are referred to Herz and Garrison (1998), Pollard (1999), Dincauze (2000), Garrison (2000), Renfrew and Bahn (2000), Sobolik (2003), Wilkinson and Stevens (2003) and Weiner (2010). Those interested specifically in further details of human evolution and dispersal are referred to Box 3.8. Those interested in the relationship between civilisation and climate are referred to Fagan (2004), Burroughs (2005) and Fagan (2009).

Archaeostratigraphy

A number of dating methods are employed in the Quaternary, and in the construction of the archaeological or archaeostratigraphic time-scale (Walker, 2005; Goldberg & MacPhail, 2006; Jones, 2006; Penkman *et al.*, 2010; Weiner, 2010; Fig. 10.1; see also 2.8.6 above).

These methods include conventional biostratigraphy, oxygen isotope stratigraphy, accelerator mass spectometry, amino-acid racemisation, cosmogenic chlorine-36 rock exposure, dendrochronology, electron spin resonance, magnetic susceptibility, molecular stratigraphy, optically stimulated luminescence, radiocarbon dating, thermoluminescence and uranium-series dating. Incidentally, not only dendrochronology but also temperature and climate can be derived from tree ring analysis (Pires *et al.*, 2005).

Note that in archaeostratigraphy, as against conventional biostratigraphy, the 'law of superposition of strata' does not always apply, as both burials and cave fills can introduce younger material below older.

Conventional biostratigraphy

Palynology. Palynostratigraphy provides a form of climatostratigraphy that can be calibrated directly against the radiocarbon record and indirectly against the deep-sea oxygen isotope record (Tzedakis, 2005; Jones, 2006; Weiner, 2010). In Great Britain and elsewhere in northern Europe, Late Glacial Older *Dryas* stadial Zone I(c) (Tundra) ranges to approximately 14 000 years before present or BP (Palaeolithic); Bolling interstadial (warmer), Older *Dryas* stadial (cooler) and Allerod interstadial (warmer) Zone II (= 'Windermere') from

14 000 to 12 000 years BP (Palaeolithic); Younger *Dryas* stadial Zone III (Arctic) from 12 000 to 11 000 years BP (Palaeolithic); Post-Glacial, Pre-Boreal Zone IV (warmer, Sub-Arctic) from 11 000 to 10 000 years BP (Mesolithic); Boreal Zones V–VI (warm, dry) from 10 000 to 8000 years BP (Mesolithic); Atlantic Zone VIIA (warm, wet, maritime) from 8000 to 6000 years BP (Mesolithic); Sub-Boreal Zone VIIB (dry, continental) from 6000 to 2000 years BP, or 4000 BC to 0 BC (Neolithic, Bronze Age and Iron Age); and Sub-Atlantic Zone VIII (cool, wet) from 2000 years BP, or AD 0, to the present day (including Middle Age) (Fig. 10.1). Incidentally, the vegetational and climatic trends indicated by the pollen are substantiated by other groups of organisms and by other techniques. For example, the carbon and nitrogen isotopic composition of red deer collagen from the Late Glacial of the northern Jura is indicative of a cool climate and of open vegetation; and that from the Post-Glacial, Pre-Boreal and Boreal, of a warmer climate and of forest vegetation. Also, the oxygen isotopic composition of (palaeoenvironmental water from) animal and human bone from the Post-Glacial of the Volga–Don steppe area of southern European Russia is indicative of a general climatic amelioration.

Palynology also provides an indication of anthropogenic disturbance of the environment. For example, the increased incidence of charcoal in a bog in the Cambara do Sul region of Brazil from approximately 7400 years BP is interpreted as indicating human habitation and the widespread use of fire at this time.

Vertebrate palaeontology. Vertebrate stratigraphy can also be calibrated directly against the radiocarbon record, or the magnetostratigraphic record. In Great Britain, a number of zones have been recognised on the basis of large and small mammals (Currant, 1989; Schreve, 2004; Jones, 2006; Fig. 10.2).

Some important mammal fossils from the British Isles are shown in Fig. 10.3.

In eastern Europe, a number of zones have been recognised on the basis of progressively more evolved mammal faunas; age dates have been assigned to these zones; and correlations to western Europe and elsewhere effected. Here, the Odessa fauna is characterised by the voles *Allophaiomys pliocaenicus* and *Prolagurus ternopolitanus [P. praepannonicus]* – and also by particular pollen, molluscs and ostracods – variously dated to approximately 1.7–1.8 Ma,

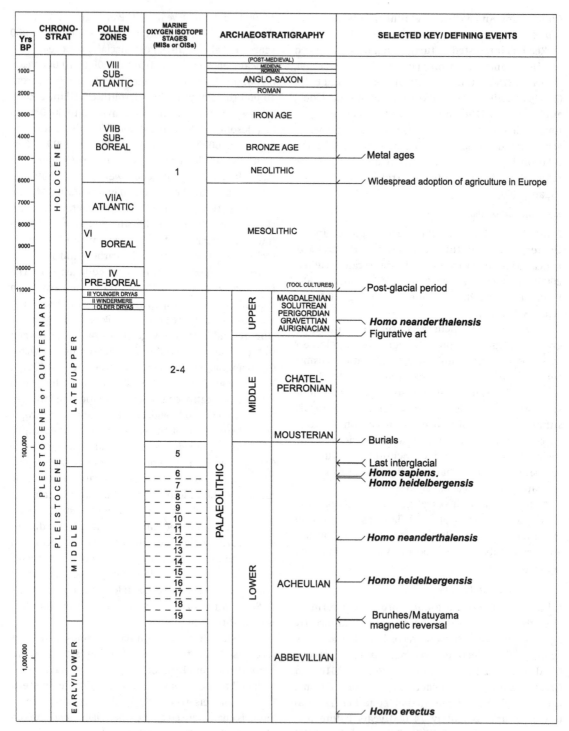

Fig. 10.1 **The archaeological time-scale.** From Jones (2006). Note change from linear time-scale in Holocene to logarithmic time-scale in Pleistocene.

INTERGLACIAL/ OXYGEN ISOTOPE STAGE	POST-CROMERIAN	HOXNIAN	PRE-IPSWICHIAN	PRE-IPSWICHIAN	IPSWICHIAN	RECENT
	13	11	9	7	5	1
Erinaceus europhaeus Hedgehog	*	*				*
Sorex araneus Common Shrew			*	*	*	*
Sorex minutus Pigmy Shrew	?	?	*	*	*	*
Neomys newtoni A water shrew	*	*				
Neomys fodiens Water Shrew			*	*		*
Crocidura sp. A white-toothed shrew				*		
Talpa europaea Mole	*	*		*		*
Macaca sylvanus Macaque	*		*	*		
Homo sapiens Man		*	*			*
Lepus timidus Mountain Hare	*	*	*		*	*
Sciurus whytei A squirrel	*	*				
Sciurus vulgaris Red Squirrel				*		*
Trogontherium cuvieri Giant Beaver	*	*	*			
Castor fiber Beaver	*	*	*	*		*
Muscardinus avellanarius Dormouse		*				*
Clethrionomys glareolus Bank Vole	*	*	*	*	*	*
Pliomys episcopalis An extinct vole	*	*				
Mimomys savini A water vole	*					
Arvicola cantiana A water Vole		*	*	*	*\	
Arvicola terrestris Water vole					*	*
Microtus agrestis Field Vole	?	(*)	*	*	*	*
Pitymys subterraneus Pine Vole	*	*	*			
Apodemus sylvaticus Wood Mouse	*	*	*	*	*	*
Canis mosbachensis A wolf	*	*\				
Canis lupus Wolf		*	*	*	*	*
Vulpes vulpes Red Fox					*	*
Ursus deningeri/spelaeus Cave Bear	*	*	*			
Ursus arctos Brown Bear				*	*	*
Martes martes Pine Marten	*	*	*			*

Fig. 10.2 **Stratigraphic zonation of the Quaternary of Great Britain by means of mammals.** Arrows indicate replacement on one species by another in evolutionary succession. From Jones, 2006 (reproduced with the permission of Poyser from Yalden, 1999).

Mustela nivalis Weasel	★	★				★
Mustela erminia Stoat		★			★	★
Meles meles Badger		★			★	★
Lutra lutra Otter	★		★	★		★
Felis silvestris Wild Cat	★		★		★	★
Equus ferus Wild Horse	★	★	★	★		
Dicerorhinus etruscus Etruscan Rhino	★	★				
Dicerorhinus kirchbergensis Merck's Rhino			★	★		
Dicerorhinus hemitoechus Narrow-nosed Rhino			★	★	★	
Hippopotamus amphibius					★	
Sus scrofa Wild Boar	★		★	★	★	★
Cervus elaphus Red Deer	★	★	★	★	★	★
Capreolus capreolus Roe Deer	★	★	★	★	★	★
Bos primigenius Aurochs			★	★	★	★
Temperature +						
Curve −						

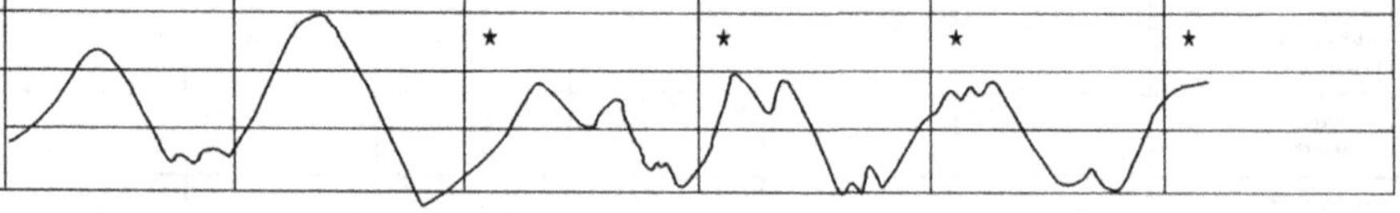

Fig. 10.2 (*cont.*)

1.2–1.4 Ma or 0.8–1.1 Ma. Importantly, the *Homo erectus* mandible from Dmanisi in Georgia was apparently found in association with this fauna.

Non-biostratigraphic technologies

Optically stimulated luminescence (OSL). OSL has been used to date the hill-wash overlying the figure of the 'Long Man of Wilmington' in the UK, carved into the chalk bedrock, as sixteenth century (Jones, 2006). Incidentally, molluscan stratigraphy yielded a similar age (one of the species recorded only being introduced into this country in the late medieval period). OSL has also been used to date the figure of the 'White Horse of Uffington', again in the UK, as Bronze Age.

Caption for Fig. 10.3 (*cont.*) Norfolk, ×0.33. Even-toed ungulates: (c) *Bos primigenius* (aurochs), upper and lower molars, Late Pleistocene, ×0.5; (d) *Hippopotamus amphibius* (hippopotamus), molar, Pleistocene, near Bedford, ×0.5; (e) *Cervus elephas* (red deer), antler, Late Pleistocene, Walthamstow, ×0.04; (f) *Dama clactoniana* (Clacton fallow deer), antler, Middle Pleistocene, Swanscombe, Kent, ×0.04; (g) *Megaloceros giganteus* (giant Irish deer), antler, Pleistocene, Ireland, ×0.04; (h) *Rangifer tarandus* (reindeer), antler, Late Pleistocene, Twickenham, Middlesex, ×0.04. Odd-toed ungulate: (i) *Coelodonta antiquitatis* (woolly rhinoceros), upper molar, Late Pleistocene, Kent's cavern, Torquay, Devon, ×0.5. Proboscideans: (j) *Anancus arvernensis* (mastodon), upper molar, Red Crag, Suffolk, ×0.25; (k) *Mammuthus primigenius* (woolly mammoth), upper molar, Late Pleistocene, Millbank, London, ×0.25; (l) *Palaeoloxodon antiquus* (straight-tusked elephant), upper molar, Pleistocene, Greenhithe, Kent, ×0.25.

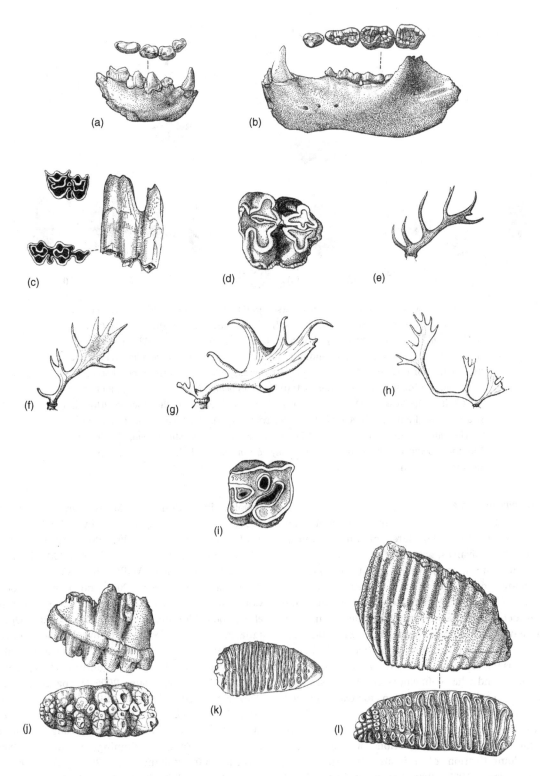

Fig. 10.3 **Some important Quaternary mammals**. From Jones (2006). Carnivores:
(a) *Crocuta crocuta spelaea* (cave hyena), right lower jaw, Late Pleistocene, Kent's cavern,
Torquay, Devon, ×0.25; (b) *Ursus deningeri* (bear), left lower jaw, Cromer forest-bed, Bacton,

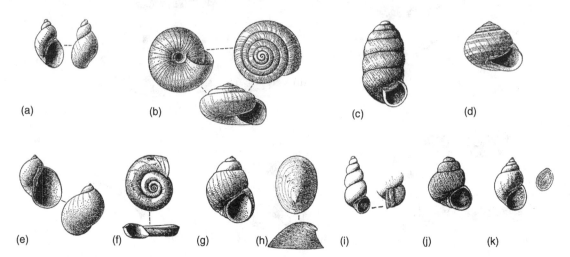

Fig. 10.4 **Some important Quaternary gastropods**. From Jones (2006). Terrestrial gastropods: (a) *Succinea oblonga* (amber snail), Pleistocene, Devensian, Tottenham, ×3; (b) *Hygromia hispida* (bristly snail), Holocene, Halling, Kent, ×3; (c) *Pupilla muscorum*, Pleistocene, Devensian, Ponders End, Middlesex, ×10; (d) *Cepaea nemoralis*, Pleistocene, ?Hoxnian, Ilford, Essex, ×1. Freshwater gastropods: (e) *Lymnaea peregra* (pond snail), Pleistocene, ?Hoxnian, Ilford, Essex, ×1; (f) *Planorbis planorbis* (trumpet snail), Pleistocene, West Wittering, Sussex, ×1.5; (g) *Viviparus diluvianus* (river snail), Pleistocene, ?Hoxnian, Swanscombe, Kent, ×1; (h) *Ancylus fluviatilis* (river limpet), Pleistocene, Clacton, Essex, ×2; (i) *Belgrandia marginata*, Pleistocene, Clacton, Essex, ×9; (j) *Valvata antiqua*, Pleistocene, ?Hoxnian, Swanscombe, Kent. ×2.5; (k) *Bithynia tentaculata* (shell and operculum), Pleistocene, Swanscombe, Kent, ×2.5.

Environmental archaeology

The exacting ecological requirements and tolerances of many dinoflagellates, diatoms, foraminifera, plants, bivalves, gastropods, ostracods, insects and mammals, and their rapid response to changing environmental and climatic conditions, render them useful in palaeoenvironmental and palaeoclimatic interpretation, and hence in environmental archaeology, that is, the interpretation of the environmental context of archaeological sites; the season(s) in which the sites were occupied by humans; and what behaviours and land uses were associated with the occupation, for example feeding behaviour, including scavenging and hunting/gathering, or farming behaviour, including the domestication and cultivation of plants, and the domestication of animals (Atkinson *et al.*, 1987; Brewer, 1987, 1989; Brewer, in Krzyzaniak & Kobusiewicz, 1989; Elias, 1994; Lyman, 1994; Bryant & Holloway, in Jansonius & MacGregor, 1996; Sobolik, in Jansonius & MacGregor, 1996; Elias, 1997; Claassen, 1998; Kerney, 1999; Sidell *et al.*, 2000; Brooks & Birks, 2001; Griffiths, 2001; Williamson & Stevens, 2001; Haslett, 2002; Holmes & Chivas, 2002; Sidell & Wilkinson, in Cotton & Field, 2004; Bar-Yosef Mayer, 2005; Branch *et al.*, 2005; Court-Picon *et al.*, 2005; Davies *et al.*, 2005; Legendre *et al.*, 2005; O'Connor & Evans, 2005; Verzi & Quintana, 2005; Bamford *et al.*, 2006; Brooks, 2006; Coope, 2006; Fernandez, 2006; Horton & Edwards, 2006; Jones, 2006; Maltby, 2006; Montuire *et al.*, 2006; Piperno, 2006; van Dam, 2006; Fernandez *et al.*, 2007; Horne, 2007; Joly *et al.*, 2007; Madella & Zurro, 2007; Blain *et al.*, 2008; Costeur & Legendre, 2008; Lyman, 2008; Marsh & Cohen, 2008; Wang & Tedford, 2008; Lopez-Saez & Dominguez-Rodrigo, 2009; Piperno, 2009; Rolland *et al.*, 2009; Holmes *et al.*, 2010; Horne, 2010; Horne *et al.*, 2010; Leroy *et al.*, 2010; Messager *et al.*, 2010; Weiner, 2010; Fernandez-Jalvo *et al.*, in press; Horsfield, 2010; Figs. 10.4–10.6).

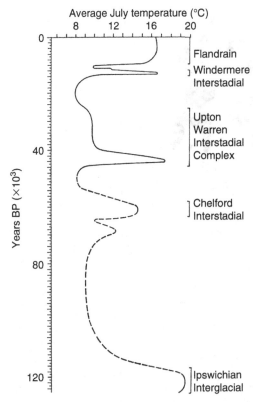

Average July temperature (°C)

Flandrain

] Windermere
Interstadial

Upton
Warren
Interstadial
Complex

] Chelford
] Interstadial

] Ipswichian
] Interglacial

Fig. 10.5 Palaeoclimatic interpretation of the Late Pleistocene to Holocene of Great Britain by means of insects. From Jones, 2006 (reproduced with the permission of the Smithsonian Institution Press from Elias, 1994).

Importantly, there is often a time lag between the onset of change of environmental or climatic condition and the response of the organism. This time lag is a function of the ability of the organism to respond to the change of condition, that is, its motility. It is shorter in motile animals than in sessile plants, such that animal and plant indications of condition can be out of accord with one another.

Environmental context of archaeological sites

In the case of diatoms, detailed palaeoenvironmental and palaeoclimatic interpretations have been made of the last glacial to post-glacial cycle (Sidell et al., 2000; Jones, 2006). Diatom evidence, specifically the occurrence of *Paralia sulcata*, *Cymatosira belgica*, *Thalassionema nitzschioides*, *Cyclotella striata* and *Cocconeis placentula* in sediments radiocarbon-dated

to 2800–2550 BC, indicates that the Thames of central London first began to be tidally influenced in the Neolithic (Sidell et al., 2000; Sidell & Wilkinson, in Cotton & Field, 2004).

In the case of mammals, the occurrence of the reindeer *Rangifer tarandus* has been used to infer an Arctic climate during glacials, and that of the hippopotamus *Hippopotamus amphibius*, an Ethiopian climate during interglacials, in the British Isles (Mitchell-Jones et al., 1999; Yalden, 1999; Hart-Davis, 2002; Jones, 2006). An Arctic climate during glacials has also been inferred from the occurrences of the extinct but interpreted cold-adapted woolly mammoth *Mammuthus primigenius* and woolly rhinoceros *Coelodonta antiquitatis* (Jones, 2006). Note in this context that the the woolly rhinoceros *Coelodonta antiquitatis* is recorded in the last glacial, Weichselian of Sourlie in western Scotland in association with plant and insect remains indicating mean annual temperatures of between −1 and 10 °C.

Note also that the small mammals from the Atelian of Cherni Yar in Astrakhan oblast in the Volga–Urals region occur in association with plant and insect remains indicating mean January temperatures of between −12 and −5 °C, mean July temperatures of between 21 and 23 °C, mean annual precipitation of 350–500 mm, and mean annual evaporation of 500–700 mm. Environmental conditions were evidently similar to those obtaining on the northern steppe of the Volga–Don interfluve at the present time.

Season(s) of occupation of archaeological sites

The season(s) of occupation of archaeological sites can be inferred from the remains of seasonal plant foods such as hazel nuts, which preserve well when charred (Weiner, 2010).

Interestingly, it can also be inferred from the remains of fish, since the season in which a fish was caught can be determined by the stage of development of its ear bones, or otoliths, or growth rings on its scales or spines (Brewer, 1987, 1989; Brewer, in Krzyzaniak & Kobusiewicz, 1989; Jones, 2006; Weiner, 2010).

Behaviours and land uses associated with occupation

In the case of plants, pollen provides information not only as to the environment and climate, but also as to the domestication and cultivation of plants, as

(a)

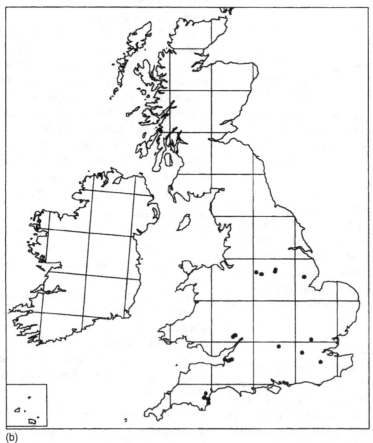

(b)

Fig. 10.6 **Modern (a) and ancient (b) distributions of the narrow-headed vole *Microtus (Stenocranius) gregalis*.** From Jones, 2006 (reproduced with the permission of Poyser from Yalden, 1999).

it is possible to distinguish the pollen of cultivated cereals from that of wild grasses by means of its greater size and annulus diameter (Jones, 2006; Joly *et al.*, 2007; Weiner, 2010). Pollen in coprolites provides information as to diet (Sobolik, in Jansonius & MacGregor, 1996). The pollen record can even inform as to the effects of earthquake activity on agriculture at archaeological sites, as in the case of Dead Sea sites during the Roman–Byzantine period (Leroy *et al.*, 2010).

Phytoliths provide information regarding not only the environment, but also the domestication and cultivation of plant species, and land use (Jones, 2006; Madella & Zurro, 2007; Piperno, 2009; Messager *et al.*, 2010; Weiner, 2010). Phytolith indices such as the Dicotyledonous/Poaceae or D/P ratio can even inform as to the extent of grassland cover and of water stress at archaeological sites, as in the case of the hominid site of Dmanisi in Georgia in the Lower Pleistocene (Messager *et al.*, 2010).

BOX 10.1   PALAEOENVIRONMENTAL INTERPRETATION OF THE PLEISTOCENE-HOLOCENE OF THE BRITISH ISLES, USING PROXY RECENT BENTHIC FORAMINIFERAL DISTRIBUTION DATA (MODIFIED AFTER JONES, 2006, AND JONES & WHITTAKER, IN WHITTAKER & HART, 2010)

Introduction

The palaeoenvironments of the Pleistocene to sub-Recent Holocene of the British Isles have been interpreted using proxy data on the environmental – in particular the biogeographic – distributions of benthic foraminifera from the Recent. The approach adopted has been qualitative and uniformitarian, that is to say, to assume that the empirically observed modern distributions are directly applicable to the interpretation of ancient environments.

The term 'Pleistocene' is used in the sense of the former and only recently overturned definition and ratification by the International Commission on Stratigraphy, with the base defined at the base of the Calabrian, dated to approximately 1.8 Ma, rather than at the base of the Gelasian, dated to approximately 2.6 Ma.

Biogeographic distribution data. The biogeographic provinces referred to are as follows:

The Arctic Province (which includes the Norwegian Sea), the southern boundary of which is approximately coincident with the 0 °C winter/ 5 °C summer isotherm;

The Subarctic Province (which includes the northern North Sea, Scandinavia and the Baltic), the southern boundary of which is approximately coincident with the 5 °C winter/10 °C summer isotherm;

The (Atlantic) Cool-Temperate Province (which includes the west of Scotland, the west of Ireland,

the Irish Sea, the English Channel and the southern North Sea), the southern boundary of which is approximately coincident with the 10 °C winter/ 15 °C summer isotherm;

The (Atlantic) Warm-Temperate Province (which includes the Bay of Biscay, Iberia and the Mediterranean), the southern boundary of which is approximately coincident with the 15 °C winter/ 20 °C summer isotherm.

The Arctic, Subarctic, (Atlantic) Cool-Temperate and (Atlantic) Warm-Temperate Provinces correspond to the Arctic, Norwegian, Celtic and Lusitanian/Mediterranean Provinces respectively, as recognised on semi-quantitative (latitudinal distribution) and quantitative (Q-mode cluster analysis) ostracod and physiographic evidence. Here, new names have been given so as to provide a climatic, rather than simply a geographic connotation.

In the context of the palaeoclimatic interpretation of the Pleistocene to sub-Recent Holocene of the British Isles, cold-water conditions have been inferred on the basis of the occurrence of the Subarctic–Arctic benthic foraminifera *Astacolus hyalacrulus, Astrononion gallowayi, Buccella tenerrima, Cassidulina norcrossi, C. reniformis s.l., C. teretis s.l., Cibicides grossus, Dentalina baggi, D. frobisherensis, D. ittai, Elphidiella arctica, E. groenlandica, E. nitida, Elphidium albiumbilicatum, E. bartletti, E. clavatum* (common-abundant), *E. hallandense, Esosyrinx curta, Fissurina serrata, Gordiospira arctica, Guttulina glacialis, Lagena parri, Laryngosigma hyalascidia, Miliolinella chukchiensis, Nonion orbicularis, Nonionellina labradorica, Oolina scalariformissulcata, Pseudopolymorphina novangliae, Pyrgo williamsoni, Quinqueloculina agglutinata,*

Q. arctica, *Q. stalkeri*, *Stainforthia feylingi*, *S. loeblichi*, *Trichohyalus bartletti*, *Trifarina fluens* and *Triloculina trihedra*, the southern limits of distribution of which at the present time are north of the British Isles on the warmer, western margin, or in the North Sea on the cooler, eastern margin.

Comparatively warm-water conditions have been inferred on the basis of the occurrence of the southern Cool-Temperate species *Aubignyna perlucida*, *Bulimina elongata*, *Lagena perlucida*, *Laryngosigma harrisi*, *Nonionella sp. A* and *Siphonina georgiana*, the northern limits of distribution of which are in the southern half of the British Isles at the present time. Comparatively warm-water conditions could also be inferred on the basis of the occurrence of the essentially southern Cool-Temperate species *Ammonia aberdoveyensis*, *A. falsobeccarii*, *A. flevensis*, *Bolivina variabilis*, *Cancris auricula*, *Cornuspira selseyensis*, *Elphidium crispum*, *E. incertum*, *Gaudryina rudis*, *Ophthalmidium balkwilli*, *Quinqueloculina bicornis*, *Q. cliarensis*, *Q. intricata*, *Q. lata*, *Q. oblonga*, *Rosalina anomala*, *R. milletti* and *Rotaliella chasteri*, the northern limits of distribution of which are in the northern half of the British Isles on the warmer, western margin, but in the southern half on the cooler, eastern margin.

Conditions warmer than those obtaining at the present time have been inferred on the occurrence of the Warm-Temperate species *Ammonia parkinsoniana*, *Elphidium cf. advenum* and *Elphidium fichtellianum*, the northern limits of distribution of which are south of the British Isles at the present time. (*Elphidium cf. advenum* and *E. fichtellianum* are Mediterranean species. *Ammonia parkinsoniana* occurs both in the Mediterranean and on the eastern seaboard of the United States from Long Island to North Carolina.) Conditions warmer than those obtaining at the present time have also been inferred on the occurrence of the southern Cool-Temperate species *Aubigyna perlucida* in the Saalian Equivalent, OIS9 of Norfolk and Lincolnshire and *Bulimina elongata* in the Saalian Equivalent, OIS7 of Co. Durham, as these species do not range this far north on the eastern margin of the British Isles at the present time.

Results and discussion

Note that no meaningful inferences can be drawn as to climatic influence on sites analysed but not discussed below.

Early Pleistocene, Pastonian. The Pastonian Paston member of the Cromer Forest-Bed formation of Paston in Norfolk (the stratotype locality) is developed in non-marine to marginal marine facies, and, to the best of our knowledge, contains no benthic foraminifera.

The slightly older Pastonian or Pre-Pastonian Sidestrand member of the Norwich Crag formation of Weybourne Hope in Norfolk (formerly known as the 'Weybourne Crag') is characterised by the benthic foraminifer *Elphidium bartletti*, indicating Subarctic–Arctic climatic conditions. A cold and/or arid environment is also inferred from palynological evidence, specifically a high incidence of non-arboreal pollen, and an absence of thermophilous trees.

Middle Pleistocene, Beestonian. The Beestonian Runton member of the Cromer Forest-Bed formation of West Runton in Norfolk (the stratotype locality) is developed in non-marine to marginal marine facies, and, to the best of our knowledge, contains no benthic foraminifera.

Cromerian Complex, Oxygen Isotope Stage 17. The Cromerian Cromer Forest-Bed formation of Pakefield (near Lowestoft) is characterised by *Elphidium incertum*, indicating marginal marine, southern? Cool-Temperate climatic conditions. The marine Subarctic–Arctic species *Elphidium bartletti* is also present, but rare, and interpreted as probably reworked, possibly from the equivalent of the Pastonian Sidestrand member of the Norwich Crag formation. It is worthy of note that the Pakefield site has yielded one of the earliest records of human activity in northern Europe, in the form of 32 worked flint artefacts (the nearby Happisburgh site has recently yielded *the* earliest record, dating to OIS21 or OIS25, again in the form of worked flint artefacts: Parfitt *et al.*, 2010). The coleopteran fauna recorded from the Pakefield site, albeit from a stratigraphically higher level, is of distinctly Warm-Temperate aspect.

The 'Cromer Forest-Bed formation' of Norton Subcourse Quarry (near Great Yarmouth), which, although not as yet positively dated, is possibly at least in part correlative with the Cromer Forest-Bed formation of Pakefield, is characterised by the Mediterranean species *Elphidium fichtellianum*, indicating Warm-Temperate conditions, warmer than those obtaining at the present time. The Subarctic–Arctic species *Elphidium bartletti* is also present, and interpreted as probably reworked. Note, incidentally, that the Norton Subcourse Quarry section also contains reworked foraminifera of mixed, including probable Jurassic, and Lower Cretaceous, rather than predominantly Upper Cretaceous ('Chalk with Flints') provenance, suggesting some sort of correlation with the comparatively flint-poor Kesgrave (or Bytham) Sands and Gravels.

Cromerian Complex, OIS13. The Cromerian, OIS13 Interglacial appears on benthic foraminiferal evidence to have been characterised on the south coast by southern Cool-Temperate to Warm-Temperate conditions, but in the North Sea by Subarctic conditions. Our preferred interpretation is that this was because warm waters were unable to flow eastwards into the North Sea from the English Channel, in turn because the Strait of Dover was closed by a land bridge, as has been proposed by numerous authors. Another interpretation is that it was because cool waters were able to flow westwards into the North Sea from the White Sea, because the 'Karelian Portal' was open, irrespective of whether the Strait of Dover was open or closed, as is thought to have been the case in the Ipswichian. Yet another is that it was because cool waters were able to flow southwards into the North Sea from the Norwegian Sea, again irrespective of whether the Strait of Dover was open or closed, as at the present time.

The Cromerian, OIS13 Steyne Wood member of the Solent formation of Bembridge is characterised by *Aubignyna perlucida* and *Lagena perlucida*, indicating southern Cool-Temperate conditions. The essentially Subarctic–Arctic species *Elphidium albiumbilicatum* is also present, but rare.

Cromerian, OIS13 or arguably Hoxnian (OIS11) Slindon Sand member of the West Sussex Coast formation of Boxgrove, Trumley Copse and Valdoe on the Goodwood–Slindon Raised Beach is characterised by *Ammonia parkinsoniana* and *Elphidium fichtellianum*, indicating Warm-Temperate conditions, warmer than those obtaining at the present time. The overlying Slindon Silt member (and associated palaeosol and local freshwater deposits) has yielded the oldest actual hominid remains in Britain, in the form of teeth and a leg-bone.

The interglacial identified in Foraminiferal Zone 34K in the 81/34 Borehole in the Devil's Hole area, underlain and overlain by units yielding isoleucine epimerisation ratios of 0.365 and 0.343 respectively, and interpreted as Cromerian, OIS13, is characterised by *Cassidulina reniformis s.l.*, *C. teretis*, *Elphidium albiumbilicatum*, *E. bartletti*, *E. clavatum*, *E. hallandense*, *Nonion orbicularis* and *Trifarina fluens*, indicating Subarctic–Arctic conditions, and by *Ammonia 'beccarii'* [?*A. batavus*], indicating Temperate to Subarctic conditions.

Anglian, OIS12. The Anglian, OIS12 Glacial appears to have been characterised on the east coast by Subarctic–Arctic conditions.

The stratotypical Anglian, as represented by the Lowestoft formation of Lowestoft, is herein equated to OIS12. This interpretation is partly based on the assumption that the overlying Hoxnian, as represented by the Hoxne formation of Hoxne, also in Suffolk, equates to OIS11. The Anglian Corton member of the North Sea Drift formation of Corton is characterised by reworked Jurassic, Cretaceous, Palaeogene and Neogene foraminifera, and by the possibly also reworked Quaternary foraminifera *Elphidiella arctica* and *Nonion orbicularis*, indicating Subarctic–Arctic conditions. The Anglian Woolpit Beds of Woolpit near Bury St Edmunds are characterised by *Cassidulina reniformis s.l.*, *C. teretis*, *Elphidium clavatum*, *Elphidiella nitida*, *Elphidium hallandense* and *Nonion orbicularis*, again indicating Subarctic-Arctic conditions.

The Anglian Leet Hill member of the North Sea Drift formation of Leet Hill is barren of *in situ* taxa, and contains only reworked Jurassic, Cretaceous and Quaternary taxa.

Hoxnian, OIS11. The Hoxnian, OIS11 Interglacial appears to have been characterised on the south coast by essentially Temperate conditions, but in the North Sea by Subarctic conditions. One interpretation is that this was because warm waters were unable to flow into the North Sea from the English Channel, because the Strait of Dover was closed, as is thought to have been the case in the Cromerian. Other interpretations are that it was because cool waters were able to flow into the North Sea either from the White Sea, because the 'Karelian Portal' was open, or from the Norwegian Sea, in either case irrespective of whether the Strait of Dover was open or closed.

The Hoxnian of Earnley, Bracklesham Bay, is characterised by *Ammonia 'beccarii'* [?*A. batavus*], *A. 'beccarii'* var., *Elphidium margaritaceum* and *E. williamsoni*, indicating either Temperate to Subarctic conditions, or Temperate conditions if *A. 'beccarii'* var. is a species other than *A. batavus*. The Cromerian or Hoxnian of Boxgrove, Trumley Copse and Valdoe discussed under Cromerian, OIS13 above. The Hoxnian or Saalian Equivalent of Aldingbourne and Tangmere is discussed under Saalian Equivalent, OIS7 below.

The interglacial identified in Foraminiferal Zone 34H in the 81/34 Borehole in the Devil's Hole area, marked by isoleucine epimerisation ratios ranging from 0.27 to 0.328 and averaging 0.285, and thus interpreted as Hoxnian, OIS11, is characterised by *Cassidulina norcrossi, C. reniformis s.l., Elphidium albiumbilicatum, E. clavatum* and *Nonion orbicularis*, indicating Subarctic–Arctic conditions, and by *Ammonia 'beccarii'* [?*A. batavus*] and *Spiroplectinella wrightii*, indicating Temperate to Subarctic conditions.

Saalian Equivalent, OISs10–6. The 'Saalian Equivalent' as herein defined equates to OISs10–6. It includes the Wolstonian *s.l.* of East Anglia, but not the Wolstonian *s.s.* of Wolston in the East Midlands, which is actually of Anglian, OIS12 age.

The interglacial identified Foraminiferal Zone 52–3 in the 81/52 Borehole in the Inner Silver Pit area, marked by isoleucine epimerisation ratios ranging from 0.185 to 0.321 and averaging 0.22, and thus interpreted as Saalian Equivalent, OIS9 and/or OIS7, is characterised by *Cassidulina reniformis s.l., E. clavatum* and *Quinqueloculina*

stalkeri, indicating Subarctic–Arctic conditions, and by *Ammonia 'beccarii'* [?*A. batavus*], indicating Temperate to Subarctic conditions. The southern Cool-Temperate species *Aubignyna perlucida* is also present, but rare.

OIS9 ('Purfleet' Interglacial). The Hoxnian, OIS11, or, more likely, Saalian Equivalent, OIS9 Nar Valley Clay of the Nar member of the Nar Valley formation of the East Winch No. 1 Borehole and of a now backfilled trench near Tottenhill, both close to the stratotype locality elsewhere in the Nar Valley, is characterised by *Aubignyna perlucida*, indicating southern Cool-Temperate conditions, slightly warmer than those obtaining at the present time (this species does not occur this far north at present). The Subarctic–Arctic species *Elphidium albiumbilicatum* and *E. hallandense* are also present, but rare. The Nar member of the Nar Valley formation of the actual stratotype locality has been dated as Saalian Equivalent, OIS9 on uranium-series evidence. It is datable between Hoxnian, OIS11 and Saalian Equivalent, OIS9, or even Saalian Equivalent, OIS7 or Ipswichian, OIS5 on amino-acid evidence, as isoleucine epimerisation ratios range from 0.109 to 0.289.

The Hoxnian, OIS11, or, more likely on the basis of correlation with the Nar Valley in Norfolk, Saalian Equivalent, OIS9 Kirmington formation of Kirmington is dominated by *Aubignyna perlucida*, indicating southern Cool-Temperate conditions, slightly warmer than those obtaining at the present time (this species does not occur this far north at the present time). The Subarctic–Arctic species *Dentalina baggi* and *Elphidium hallandense* are also present, but rare.

OIS7 ('Aveley' Interglacial). The Saalian Equivalent, OIS7 'Aveley' Interglacial appears to have been characterised on the south and east coasts by southern Cool-Temperate conditions (slightly warmer than those obtaining at the present time), implying that the Strait of Dover was open at this time, but in the North Sea by Subarctic–Arctic conditions. Our preferred interpretation is that this was because cool waters were able to flow into the North Sea from the White Sea or Norwegian Sea.

The Hoxnian, OIS11, or, more likely, Saalian Equivalent, OIS9 or – 'early' – OIS7 Aldingbourne member of the West Sussex Coast formation of Aldingbourne and Tangmere on the Aldingbourne Raised Beach is characterised by *Aubignyna perlucida*, *Bulimina elongata* and *Nonionella* sp. A, indicating southern Cool-Temperate conditions. The Saalian Equivalent, OIS7 Lifeboat Station member of the Lifeboat Station Channel, Selsey is characterised by *Lagena perlucida*, also indicating southern Cool-Temperate conditions. The Subarctic–Arctic species *Elphidiella arctica*, *Elphidium albiumbilicatum* and *E. hallandense* are also present, but rare.

The Saalian Equivalent, OIS7 Easington formation of Easington is characterised by *Bulimina elongata*, indicating southern Cool-Temperate conditions, slightly warmer than those obtaining at the present time (this species does not occur this far north at the present time).

The interglacial identified in Foraminiferal Zone E in the 81/34 Borehole in the Devil's Hole area, marked by an isoleucine epimerisation ratio of 0.189 and thus interpreted as Saalian Equivalent, OIS7, is characterised by *Cassidulina norcrossi*, *C. reniformis s.l.*, *C. teretis*, *Elphidium albiumbilicatum*, *E. bartletti*, *E. clavatum*, *Nonion orbicularis* and *Trifarina fluens*, indicating Subarctic–Arctic conditions.

'Late' OIS7. The Saalian Equivalent, 'late' OIS7 Interglacial appears to have been characterised on the south coast by Subarctic–Arctic conditions. The Saalian Equivalent, OIS7 or, more likely, 'late' OIS7 of the Black Rock and Norton members of the West Sussex Coast formation of numerous localities on the Brighton–Norton Raised Beach in Hampshire and Sussex is characterised by *Cassidulina reniformis s.l.*, *Elphidium albiumbilicatum*, *E. clavatum*, *E. hallandense* and *Oolina scalariformissulcata*, indicating Subarctic–Arctic conditions. The southern Cool-Temperate to Warm-Temperate species *Elphidium fichtellianum*, *Lagena perlucida* and *Rosalina milletti* are also present, but rare. As noted above, the fact that cold-water species are present in the older, shallow marine Norton Sand as well as the younger, marginal marine Norton Silt indicates that climatic cooling preceded rather than accompanied glacio-eustatic shallowing.

Late Pleistocene, Ipswichian, OIS5e. The Ipswichian, OIS5e Interglacial appears to have been characterised generally on the south coast by southern Cool-Temperate to Warm-Temperate conditions. However, it also appears to have been characterised locally, in Hampshire and Sussex, by Subarctic–Arctic conditions similar to those obtaining during the Saalian Equivalent, 'late' OIS7 Interglacial Glacial. Moreover, it appears to have been characterised in the North Sea by Subarctic conditions. Our preferred interpretation is that this was because cool waters were able to flow into the North Sea from the White Sea because the 'Karelian Portal' was open.

In terms of benthic foraminifera, the Ipswichian of Cardigan Bay is characterised by the Mediterranean species *Elphidium fichtellianum*, indicating Warm-Temperate conditions, warmer than those obtaining at the present time. The Subarctic–Arctic species *Astrononion gallowayi*, *Cassidulina reniformis s.l.*, *Elphidium albiumbilicatum*, *E. bartletti*, *Laryngosigma hyalascidia* and *Pseudopolymorphina novangliae* are also present, but rare.

The Ipswichian of Goldcliff on the Gwent Levels is characterised by *Rosalina milletti*, indicating southern Cool-Temperate conditions.

The Ipswichian of the Middlezoy member of the Burtle formation of Greylake on the Somerset Levels is characterised by the Mediterranean species *Elphidium cf. advenum*, indicating Warm-Temperate conditions, warmer than those obtaining at the present time.

The Ipswichian 'blown sand' of the Thatcher Stone member of the Torbay formation of Hope's Nose and North Thatcher is characterised by the Mediterranean species *Elphidium fichtellianum*, indicating Warm-Temperate conditions, warmer than those obtaining at the present time.

The Ipswichian Pagham member of the West Sussex Coast formation of numerous localities on the Pagham Raised Beach in Hampshire and Sussex is characterised by *Cassidulina reniformis s.l.* and *Elphidium albiumbilicatum*, indicating Subarctic–Arctic conditions.

The Ipswichian Equivalent Eemian 'Troll Interglacial' of Core 5.1/5.2 in the Troll area of the Norwegian sector is characterised by a high incidence of interpreted 'Boreal' species, including *Bulimina marginata* and *Cassidulina laevigata s.l.*, which are Temperate to Arctic species, and *Spiroplectinella wrightii*, which is Temperate to Subarctic. Interpreted 'Arctic' species from the older Saalian and younger Devensian Equivalent Weichselian include *Cassidulina reniformis s.l.* and *Elphidium clavatum*, which are indeed so, or at least Subarctic to Arctic. Samples from Core 5.1/5.2 are marked by isoleucine epimerisation ratios in the range 0.099–0.225 and averaging 0.137, confirming the Ipswichian Equivalent age. The Ipswichian Equivalent Eemian of the Roar, Skjold and Dan Boreholes in the Danish sector are characterised by interpreted 'Boreal–Lusitanian' faunas. Samples from the Roar Borehole are marked by isoleucine epimerisation ratios averaging approximately 0.1, confirming the Ipswichian Equivalent age. Similarly interpreted 'Boreal–Lusitanian' faunas from the Ipswichian Equivalent Eemian of the Apholm Borehole in Jutland in Denmark include *Bulimina marginata* and *Cassidulina laevigata s.l.*, which are Temperate to Arctic species. Interpreted 'Arctic' faunas from the older Saalian and younger Devensian Equivalent Weichselian of the Apholm Borehole again include *Cassidulina reniformis s.l.* and *Elphidium clavatum*, which are indeed essentially Arctic. Samples from the Apholm Borehole are marked by isoleucine epimerisation ratios in the range 0.079–0.099, confirming the Ipswichian Equivalent age.

Devensian, OIS5d to 2. The Devensian, OIS5d–2 Glacial appears to have been characterised generally by Subarctic–Arctic conditions. However, it also appears to have been characterised locally, in the Windermere Stadial, by Temperate influence.

OIS3. The Devensian, OIS3 Sistrakeel formation of Sistrakeel is characterised by dominant (90%) *Elphidium clavatum*, indicating Subarctic–Arctic conditions.

OIS2, Late Glacial, including Last Glacial Maximum. The radiocarbon-dated Late Glacial of Glenulra is characterised by *Buccella tenerrima, Cassidulina reniformis s.l., C. teretis, Dentalina baggi, Elphidiella groenlandica, Elphidium bartletti, E. clavatum, E. hallandense, Miliolinella chukchiensis, Nonion orbicularis, Nonionellina labradorica* and *Quinqueloculina arctica*, indicating

Subarctic–Arctic conditions. The southern Cool-Temperate species *Bulimina elongata* is also present. The Glenulra succession has been interpreted as representing the locally developed Glenavy Stadial. The radiocarbon-dated Belderg formation of Belderg is characterised by *Cassidulina reniformis s.l., Elphidium bartletti, E. hallandense, Miliolinella chukchiensis, Nonion orbicularis, Nonionellina labradorica, Pyrgo wiulliamsoni, Quinqueloculina agglutinata* and *Q. arctica*, indicating Subarctic–Arctic conditions. The Belderg formation represents the locally developed Belderg Stadial.

The AMS-dated Devensian Cooley Farm member of the Louth formation of the margins of Dundalk Bay, Co. Louth and Derryogue member of the Mourne formation of the Mourne region are dominated by *Elphidium clavatum*, indicating Subarctic–Arctic conditions. The Cooley Farm and Derryogue members of the Louth formation represent the locally developed Cooley Point Interstadial.

The undifferentiated Late Glacial Dog Mills member of the Orrisdale formation of Dog Mills is characterised by *Elphidiella hannai* (possibly reworked from an older glacial) and *Elphidium clavatum*, indicating Subarctic–Arctic conditions. Warmer-water species are also present.

The undifferentiated Late Glacial of the Western Irish Sea formation of the Irish Sea is characterised by *Cassidulina reniformis s.l., Elphidiella nitida, Elphidium bartletti, E. clavatum, E. hallandense, Nonion orbicularis, Nonionellina labradorica, Quinqueloculina arctica* and *Q. stalkeri*, indicating Subarctic–Arctic conditions. The Warm-Temperate species *Elphidium fichtellianum* is also present.

The undifferentiated Late Glacial of Cardigan Bay is characterised by *Buccella tenerrima, Cassidulina reniformis s.l., Elphidium albiumbilicatum, E. clavatum* and *Nonion orbicularis*, indicating Subarctic–Arctic conditions. The southern Cool-Temperate species *Bulimina elongata* and *Ophthalmidium balkwilli* are also present.

The undifferentiated Late Glacial of Broughton Bay is characterised by *Astrononion gallowayi, Cassidulina reniformis s.l., Elphidium albiumbilicatum* and *E. hallandense*, indicating Subarctic–Arctic conditions. Cool-Temperate species are also present.

The Last Glacial Maximum, Dimlington Stadial (?or Late Glacial Stadial) Bridlington member of the Holderness formation of Dimlington near Bridlington is characterised by *Astacolus hyalacrulus, Astrononion gallowayi, Buccella tenerrima, Cassidulina norcrossi, C. reniformis s.l., C. teretis, Cibicides grossa, Dentalina baggi, D. frobisherensis, Elphidiella arctica, E. groenlandica, E. nitida, Elphidium bartletti, E. clavatum, E. hallandense, Guttulina glacialis, Nonion orbicularis, Nonionellina labradorica, Oolina scalariformissulcata, Quinqueloculina agglutinata, Q. stalkeri, Pyrgo williamsoni, Trichohyalus bartletti* and *Trifarina fluens*, indicating Subarctic–Arctic conditions. The southern Cool-Temperate species *Bulimina elongata, Cancris auricula* and *Siphonina georgiana* are also present. Reworking from various stratigraphic horizons – including Jurassic, Cretaceous, Tertiary and possibly also older Quaternary – is a conspicuous feature of the Bridlington Crag. Essentially tropical to subtropical Tertiary larger benthic species (*Operculina* sp.) and planktonic species (*Hantkenina alabamensis, Morozovella* sp.) are present in some collections, and perhaps best interpreted as contaminants.

Late Glacial [Oldest Dryas Equivalent] Stadial. The radiocarbon-dated Oldest *Dryas* Equivalent Stadial of the Hebridean Shelf is characterised by *Cassidulina reniformis s.l., Elphidium albiumbilicatum, E. bartletti, E. clavatum, Nonion orbicularis, Nonionellina labradorica* and *Quinqueloculina stalkeri*, indicating Subarctic–Arctic conditions.

The interpreted Oldest *Dryas* Equivalent Stadial Killellan member of the Ardyne formation of Ardyne and a biostratigraphically equivalent unit in the '100' Beach Clay' of Benderloch are characterised by *Astacolus hyalacrulus, Buccella tenerrima, Elphidium bartletti, E. clavatum, E. hallandense, Laryngosigma hyalascidia, Nonion orbicularis, Pyrgo williamsoni* and *Quinqueloculina agglutinata*, indicating Subarctic–Arctic conditions.

The radiocarbon-dated Late Glacial Stadial of Corvish is characterised by *Cassidulina reniformis s.l., Elphidium clavatum, Pseudopolymorphina*

novangliae, Pyrgo williamsoni and *Quinqueloculina stalkeri*, indicating Subarctic–Arctic conditions. The southern Cool-Temperate species *Bulimina elongata* is also present.

The AMS-dated Devensian Killard Point formation of the drumlin fields around Killard Point in Co. Down is dominated by *Elphidium clavatum*, indicating Subarctic–Arctic conditions. The Killard Point formation represents the Late Glacial Killard Point Stadial (Heinrich Event 1).

Windermere Interstadial = Bolling–Allerod or Greenland Interstadial 1 Equivalent. The radiocarbon-dated Bolling and Allerod Equivalent Interstadials of the Hebridean Shelf are characterised by mixed, interpreted time-averaged assemblages including the essentially Temperate species *Ammonia batavus* (?including the southern? Cool-Temperate species *Ammonia falsobeccarii*), and the Subarctic–Arctic species *Cassidulina reniformis s.l., Elphidium albiumbilicatum, E. bartletti, Elphidium clavatum, Nonion orbicularis, Nonionellina labradorica* and *Quinqueloculina stalkeri*. The intervening Older *Dryas* Equivalent Stadial of the Hebridean Shelf (Wester Ross Re-advance) is characterised by *Cassidulina reniformis s.l., Elphidium albiumbilicatum, E. bartletti, E. clavatum, Nonion orbicularis, Nonionellina labradorica* and *Quinqueloculina stalkeri*, indicating Subarctic–Arctic conditions.

The radiocarbon-dated Toward member of the Ardyne formation of Ardyne and a biostratigraphically equivalent unit in the '100' Beach Clay' of Benderloch are characterised by mixed, interpreted time-averaged assemblages of including the southern? Cool-Temperate species *Ammonia falsobeccarii*, and the Subarctic–Arctic species *Astacolus hyalacrulus, Astrononion gallowayi, Dentalina frobisherensis, D. ittai, Elphidiella arctica, Elphidium albiumbilicatum, E. bartletti, E. clavatum, E. hallandense, Gordiospira arctica, Lagena parri, Laryngosigma hyalascidia, Nonion orbicularis, Nonionelllina labradorica, Pseudopolymorphina novangliae, Pyrgo williamsoni, Quinqueloculina agglutinata, Q. stalkeri, Trifarina fluens* and *Triloculina trihedra.*

Loch Lomond [Younger Dryas Equivalent] Stadial. The radiocarbon-dated Younger Dryas Stadial of the

Hebridean Shelf is characterised by *Cassidulina reniformis s.l., Elphidium albiumbilicatum, E. bartletti, E. clavatum, Nonion orbicularis, Nonionellina labradorica* and *Quinqueloculina stalkeri*, indicating Subarctic–Arctic conditions.

The radiocarbon-dated Ardyne Point member of the Ardyne formation of Ardyne and a biostratigraphically equivalent unit in the '100' Beach Clay' of Benderloch are characterised by *Astacolus hyalacrulus, Buccella tenerrima, Dentalina ittai, Elphidiella arctica, Elphidium albiumbilicatum, E. bartletti, E. clavatum, E. hallandense, Esosyrinx curta, Gordiospira arctica, Guttulina glacialis, Laryngosigma hyalascidia, Nonion orbicularis, Pseudopolymorphina novangliae, Pyrgo williamsoni, Quinqueloculina agglutinata, Trifarina fluens* and *Triloculina trihedra*, indicating Subarctic–Arctic conditions.

Holocene, Flandrian, OIS1. The Flandrian appears to have been characterised generally by Temperate conditions. However, it also appears to have been characterised locally, in the Pre-Boreal and Boreal Stages, by Subarctic–Arctic influence.

The radiocarbon-dated Flandrian of the Hebridean Shelf is characterised by the Subarctic–Arctic species *Cassidulina reniformis s.l., Elphidium albiumbilicatum, E. bartletti, E. clavatum, Nonion orbicularis, Nonionellina labradorica* and *Quinqueloculina stalkeri*, and the essentially Temperate species *Ammonia batavus.* Subarctic–Arctic species are commonest in the Pre-Boreal Stage and Boreal Stages, and Temperate species in the Atlantic Stage.

The Flandrian of the Dovey Marshes and of Cardigan Bay is characterised by the Subarctic–Arctic species *Cassidulina reniformis s.l., Elphidium albiumbilicatum, E. bartletti, E. clavatum, Nonion orbicularis, Nonionellina labradorica, Quinqueloculina agglutinata* and *Triloculina trihedra,* and the southern Cool-Temperate species *Aubignyna perlucida, Bulimina elongata, Lagena perlucida, Ophthalmidium balkwilli* and *Rosalina milletti.*

The Flandrian of Start Bay is characterised by *Lagena perlucida*, indicating southern Cool-Temperate conditions.

10.1.1 Westbury Cave, Somerset (Early Palaeolithic, 600 000 years BP)

The ancient cave exposed in the quarry above the village of Westbury-sub-Mendip in Somerset in the West Country is renowned for having yielded evidence, in the form of flints and cut marks on animal bone, of archaic human occupation of Britain, dating back to the early Middle Pleistocene, Cromerian interglacial, approximately 600 000 years BP, to the Early Palaeolithic (Andrews *et al.*, 1999; Barton, 2005; Jones, 2006; Stringer, 2006; Barton, in Hunter & Ralston, 2009). Incidentally, the immediately surrounding area is steeped in archaeology. There is a later Palaeolithic cave site at 'Hyaena Den', near Wookey Hole, and an important Neolithic cave site in Cheddar Gorge, where the skeleton of 'Cheddar Man' was discovered. Neolithic artefacts, trackways, settlements, burial barrows and henges abound in the Somerset Levels around Glastonbury, where there are also the remains of a Bronze Age lake village.

Recent work has indicated that the flints from Westbury Cave might in fact be ecofacts – or 'incerto-facts' – rather than worked artefacts, and might have been introduced naturally.

Associated animal remains are also for the most part thought to have been introduced into the cave by natural processes such as water transport, mud flow or surface collapse. Note, though, that certain of the animal remains appear to have been introduced by bears actually living in the cave, or, as droppings, by birds of prey. The remains include those of amphibians and reptiles, small mammals, large ruminant herbivores, and large carnivores.

Importantly, evidence of butchery by archaic humans has been found in the cave. The evidence is in the form of cut marks, although only on a single bone of the more than 5000 examined, a metacarpal of the red deer *Cervus elaphus.*

Interestingly, the unit containing the evidence of butchery by archaic humans, Unit 19, is characterised by an interpreted boreal fauna containing the Norway lemming *Lemmus lemmus,* the common vole *Microtus arvalis,* and the ancestral narrow-headed or Siberian vole *M. [Stenocranius] gregaloides* (while, in contrast, most of the older units in the cave are characterised by warmer, temperate faunas). *Lemmus lemmus* is essentially restricted to Fennoscandia at the present time (Mitchell-Jones *et al.*, 1999; Yalden, 1999; Shenbrot & Krasnov, 2005). *Microtus arvalis* ranges throughout much of northern – although not northernmost – and central Europe, and into central Asia. Its only British occurrence is in the Orkneys.

It is probable that Unit 19 represents a late part of the Cromerian interglacial (Currant, in Andrews *et al.*, 1999). Note in this context that the youngest occurrence of *Microtus [Stenocranius] gregaloides* has been correlated with the Cromerian III, OIS15 in northwest Europe. Note also, though, that the youngest occurrence of *M. [Stenocranius] gregaloides* has been tentatively correlated with the pre-Cromerian, OIS22, and that the oldest occurrence of *M. arvalis* has been tentatively correlated with the Cromerian IV, OIS13, at the Gran Dolina site in Atapuerca in Spain, although these tentative correlations may not be tenable (S. A. Parfitt, The Natural History Museum, personal communication).

10.1.2 Boxgrove, Sussex, UK (Early Palaeolithic, 500 000 years BP)

The raised beach at Boxgrove on the West Sussex coastal plain has also yielded evidence of archaic human occupation of Britain dating back to the early Middle Pleistocene, or Early Palaeolithic, although apparently to a younger stratigraphic level than that represented by the Westbury site (Roberts & Parfitt, 1999; Pope, in Rudling, 2003; Wilkinson & Stevens, 2003; Barton, 2005; Goldberg & MacPhail, 2006; Jones, 2006; Stringer, 2006; Barton, in Hunter & Ralston, 2009). The evidence is in the form of *Homo heidelbergensis* remains – a tibia and a tooth – and associated Acheulian flint hand-axes and other artefacts.

The human remains and associated artefacts are concentrated on an Early Palaeolithic foreshore occupation surface in the Slindon silt member of the Slindon formation of the West Sussex Coast group. Note, though, that artefacts also occur in the underlying Slindon sand member of the Slindon formation, and in the overlying Eartham lower and upper gravel members of the Eartham formation.

A range of terrestrial and freshwater animal remains have been found in association with the human remains and artefacts, principally in the Slindon silt. These include molluscs, ostracods, fish, amphibians and reptiles, birds, and mammals. The mammalian fauna includes insectivores, chiropterans, carnivores, proboscideans, perissodactyls, artiodactyls, rodents and lagomorphs as well as *Homo heidelbergensis*. Importantly, the molluscan fauna is indicative of a land surface for the most part covered with vegetation, and dotted with freshwater pools. The key environmental indicators are *Pupilla muscorum* and *Vallonia* spp., which indicate open country; *Lymnaea truncatula* and *Succinea oblonga*, which indicate some vegetation; *Aegopinella* and clausiliids, which indicate vegetation; *Spermodea lamellata* and *Acanthinula aculeata*, which indicate deeply shaded and moist hollows; and *Anisus leucostoma*, *Lymnaea peregra* and *Pisidium* spp., which indicate fresh water (Kerney, 1999). Some of these and some other palaeoenvironmentally important terrestrial and freshwater gastropods from the British Isles are shown on Fig. 10.4 (see also Kerney, 1999).

The remains of marine as well as terrestrial and freshwater organisms have also been found at Boxgrove, principally in the Slindon sand (Whittaker *et al.*, 2003; Wilkinson & Stevens, 2003; Jones, 2006). These include foraminifera, ostracods, dinoflagellates, molluscs and fish.

Clear evidence of butchery by humans of the extinct cave bear *Ursus deningeri*, the red deer *Cervus elaphus*, the extinct giant deer *Megaloceros* sp., the wild horse or tarpan *Equus ferus*, the extinct rhinoceros *Stephanorhinus hundsheimensis*, and *Bison* sp. has been found at Boxgrove (Jones, 2006). From impact as well as cut marks, it would appear that these animals were butchered for their marrowbone fat as well as their meat and skins. Significantly, in at least three cases, cut marks associated with human butchery clearly pre-date gnaw marks associated with non-human carnivory, indicating that humans got to the animal carcasses before hyenas and other carnivores. Whether the humans actually hunted and killed animals, or simply scavenged them, remains unknown. Note, though, that the arguable existence of a puncture wound in the scapula of a wild horse provides possible evidence of hunting. Interestingly, the technology employed in the manufacture of the flint hand-axes and other tools used in the butchery has been demonstrated by experimental archaeological techniques to have involved knapping with soft – bone and antler – as well as hard hammers.

The age of the Slindon formation of Boxgrove has been estimated by a variety of methods, including uranium-series dating, luminescence dating of brick-earth, OSL, electron spin resonance (ESR) and coupled ESR–uranium-series dating, amino acid geochronology, and magnetostratigraphy, as well as conventional biostratigraphy. Conventional biostratigraphic dating was in turn undertaken employing calcareous nannoplankton and mammals. Overall, age estimations range from approximately 175 000 to 525 000 years BP, or oxygen isotope stages OIS6–13. Those that are regarded as most reliable, namely those derived from amino acid geochronology and conventional biostratigraphy, range from approximately 350 000 to 525 000 years BP, or OIS11–13. Amino acid geochronology and calcareous nannoplankton biostratigraphy indicate approximately 350 000–425 000 years BP (OIS11, or Hoxnian), while mammalian biostratigraphy indicates approximately 475 000–525 000 years BP (OIS13, or Cromerian). The mammalian age estimation of approximately 525 000 years BP (OIS13, or Cromerian) is preferred, because the amino acid and calcareous nannoplankton age estimations are difficult to reconcile with regional observations, and because the latter is based not on positive but on negative evidence. Evidence of a Cromerian age is provided by the presence of the extinct shrews *Sorex runtonensis* and *S. (Drepanosorex) savini*, the vole *Pliomys episcopalis*, the cave bear *Ursus deningeri*, the rhinoceros *Stephanorhinus hundsheimensis*, and the giant deer *Megaloceros dawkinsi* and *M. verticornis*. All of the aforementioned species are recorded from the Cromerian, none from the Hoxnian. Evidence of a Cromerian IV, OIS13 rather than Cromerian III, OIS15 age is provided by the presence of the narrow-headed or Siberian vole *Microtus (Stenocranius) gregalis* (S. A. Parfitt, The Natural History Museum, personal communication). Incidentally, this species survives today in the Arctic, and also in the steppes of central Asia (Yalden, 1999; Shenbrot & Krasnov, 2005; Jones, 2006; Fig. 10.6).

Palaeoclimatic evidence from the contained fossils supports the interpretation that the Slindon formation of Boxgrove probably represents the latest part of the Cromerian interglacial (and the Eartham formation the Anglian glacial). Ostracod evidence from the Slindon silt, produced using the mutual climate range (MCR) and mutual ostracod temperature range (MOTR) methods, indicates similar summer temperatures but slightly cooler winter temperatures than those of today (Holmes *et al.*, 2010). Plant, mollusc, amphibian, reptile and mammal evidence also generally indicates cooler temperatures than those of today, with maritime Scandinavia cited as providing the best analogue (Horsfield, 2010). However, mammal evidence produced using the 'Taxonomic Habitat Index' has been interpreted as indicating warmer temperatures than those of today, with the Mediterranean providing the best analogue (Fernandez-Jalvo *et al.*, in press). Foraminiferal evidence from the Slindon *sand* rather than the Slindon *silt* has also been interpreted as indicating warmer temperatures than those of today (Jones, 2006; Jones & Whittaker, in Whittaker & Hart, 2010; Box 10.1).

10.1.3 Massawa, Eritrea (Middle Palaeolithic, 125 000 years BP)

Outcrops of Abdur reef limestone on raised beaches near the port of Massawa on the Red Sea coast of Eritrea in the Horn of Africa have yielded evidence of – by inference – 'anatomically modern' human occupation of the area, dating back to the last interglacial, OIS5e, 125 000 years BP, to the Middle Palaeolithic, or, as it is known in Africa, the 'Middle Stone Age' or MSA (Jones, 2006). The evidence is in the form of Acheulian flint bifacial hand-axes and cores, and MSA obsidian blades and flakes. The Acheulian artefacts are associated with transgressive gravels and high-stand reefal limestones, and oyster beds and shell middens. The MSA artefacts are associated with slightly later nearshore and beach facies, and both shell-fish and large mammal bones. The association of both the Acheulian and MSA artefacts with shell-fish is an indication of the exploitation of marine food resources as well as those to be found around the lakes and landscapes of east Africa. The exploitation of marine food resources has been interpreted as necessitated by the drying out of the inland watering holes, and the dying out of the game on the plains, during the penultimate glacial between 170 000 and 130 000 years BP. Incidentally, it may have been similar environmental stress associated with the last glacial that finally drove the successful dispersal of *Homo sapiens* out of Africa, and along the 'beachcomber route' to Eurasia.

10.1.4   Goat's Hole, Paviland, Gower, UK (Late Palaeolithic, 28 000 years BP)

Famously, in 1822 or 1823, the Reverend William Buckland discovered an apparently ceremonially buried 'anatomically modern' *Homo sapiens* skeleton in Goat's Hole in Paviland in the Gower Peninsula in south Wales, dating back to the Late Palaeolithic (Aldhouse-Green, 2000; Barton, 2005; Jones, 2006; Stringer, 2006; Barton, in Hunter & Ralston, 2009). The skeleton came to be popularly known as the 'Red Lady', although it eventually turned out to be male. A recent archaeological survey undertaken in the late 1990s revealed that excavations in the nineteenth and early twentieth centuries had essentially removed the unit that yielded the Late Palaeolithic skeleton, leaving only older Palaeolithic units, confirmed as such by uranium-series and thermoluminescence dating. As an adjunct to the recent survey, excavations were undertaken in adjacent Hound's Hole and Foxhole Cave. The excavation at Hound's Hole yielded a late glacial fauna and Palaeolithic finds. The partial excavation at Foxhole Cave yielded Neolithic human remains and finds in 'modern topsoil layer 1'; a post-glacial fauna, a human tooth dated to the Mesolithic, Mesolithic artefacts, and evidence of Mesolithic occupation, in the form of a hearth and burnt bone, in 'humic scree layer 2'; and a late glacial fauna in 'soliflucted scree layer 3'. The Foxhole Cave excavation did not reach the Palaeolithic.

The most recent radiocarbon evidence indicates that the Late Palaeolithic 'Red Lady' dates to 28 000 years BP rather than 26 000 years BP as previously reported, and other bones and bone artefacts from Goat's Hole to 37 800–15 250 years BP; and lithic artefacts to the Mousterian, Aurignacian, Gravettian, Creswellian, final Late Palaeolithic and Mesolithic. The non-human bones are a mixture of carnivores, including the spotted hyena *Crocuta crocuta*, the wolf *Canis lupus*, the fox *Vulpes vulpes*, ?the Arctic fox *Alopex lagopus*, and the brown bear *Ursus arctos*; and herbivores, including the woolly mammoth *Mammuthus primigenius*, the wild horse or

tarpan *Equus ferus*, the woolly rhinoceros *Coelodonta antiquitatis*, the wild boar *Sus scrofa*, the reindeer *Rangifer tarandus*, the red deer *Cervus elephas*, ?the giant deer *Megaloceros giganteus*, the steppe bison *Bison priscus*, and the wild sheep *Ovis aries*. The nonhuman carnivores are thought to have been principally responsible for the bone accumulations. There is no evidence of butchery by ancient humans. Analysis of the carbon and nitrogen isotopic composition of collagen indicates that the 'Red Lady' subsisted at certain times of year on sea fish and shell-fish, implying seasonal migration to the coast, at the time located some 100 km to the west; and/or to the 'palaeo-Severn', 20 km to the south, to exploit salmon running up-river to spawn. Interestingly, in contrast, analysis of the collagen from Mesolithic and Neolithic human remains from Foxhole Cave does not indicate any such dietary intake of sea fish or shell-fish. Note, though, that the data on the Mesolithic individual is equivocal, as he or she was an infant, and arguably still breast-feeding.

Importantly, the Late Palaeolithic 'Red Lady' has yielded a mitochondrial DNA sequence corresponding to the commonest extant lineage in Europe. This discovery does not support the so-called 'diffusion model' for the colonisation of Europe, which argues for a late, post-agricultural revolution, Neolithic replacement of the early, Palaeolithic population by migrants from the Near East. Instead, it supports the so-called 'continuity model'.

Analysis of the body proportions of the 'Red Lady' indicates that he was slightly more warm-adapted than most modern Europeans, and slightly more cold-adapted than most modern sub-Saharan Africans, though these observations do not permit any unequivocal inferences to be drawn with regard to population movement or gene flow.

10.1.5   'Doggerland', North Sea (Mesolithic, 11 600–7000 years BP)

In 1931, the fishing vessel *Colinda*, under the captaincy of the magnificently monickered Pilgrim E. Lockwood, trawled up some peat from the bottom of the North Sea some 25 miles off the coast of East Anglia in the UK that proved to contain a barbed spear-point fashioned from antler (Jones, 2006; Gaffney *et al.*, 2009). Subsequently, worked bone, stone and flint artefacts, and actual human and associated animal bones have been recovered from adjacent localities (Gaffney et al., 2009). The human bones date to 11 600–9200 years BP, to the immediately Post-Glacial period, to the earliest Mesolithic (van Kolfschoten & Laban, 1995; Coles, 1998; Glimmerveen et al., in Flemming, 2004; Ward et al., 2006; Gaffney et al., 2009).

Palynological analysis of peat samples from off-shore and onshore localities points toward the development initially of birch and subsequently of hazel woodland ('Noah's Woods') in Britain, in mainland Europe, and in between, in the immedi-ately Post-Glacial period, between 11 000 and 7000 years BP (IV–VI 'Zone' of Godwin, 1940; FlI–FlIIa–c 'Chronozone' of West, 1980; Pre-Boreal to Boreal Blytt–Sernander Periods of European usage); with the whole of what is now the North Sea then evi-dently a land bridge connecting Britain and main-land Europe, which has come to be known as 'Doggerland' (Reid, 1913; Verart, in Fischer, 1996; Waddington et al., in Waddington, 2007; Gaffney et al., 2009; Milner & Mithen, in Hunter & Ralston, 2009; Rivals et al., 2010). (It would also have been land in the preceding Late Glacial, although at this time, other than in short-lived interstadials, it would have been covered only with tundra vege-tation.) Palynological analysis also points toward the development ultimately of mixed woodland in Britain and in mainland Europe, but not in between, from 7000 years BP (VII–VIII 'Zone'; FlIId–FlIII 'Chronozone'; Atlantic to Sub-Atlantic Periods); with 'Doggerland' under water (Verart, in Fischer, 1996; Behre, 2007; Waddington et al., in Waddington, 2007; Gaffney et al., 2009; Milner & Mithen, in Hunter & Ralston, 2009). Interestingly, the tsunami associated with the earthquake-triggered 'Storegga slide' may have played a part in its inundation.

While it was in existence, 'Doggerland' formed not only a land bridge but also a landscape, where humans and other animals lived, and where humans hunted for other animals, initially rein-deer, and subsequently wild horse, aurochs, deer, elk, boar, otter, beaver and waterfowl (Jones, 2006; Gaffney et al., 2009; Rivals et al., 2010).

10.1.6 Mount Sandel, Coleraine, Co. Derry, Northern Ireland (Mesolithic, 9000 years BP)

Archaeological excavations in the bluffs alongside the River Bann at Mount Sandel near Coleraine in Co. Derry have yielded evidence of the oldest human occupation of Ireland, dating back to the Post-Glacial, Littletonian (Conneller & Warren, 2006; Jones, 2006; Milner & Mithen, in Hunter & Ralston, 2009). The evidence is in the form of flint axes and small blades or points known as micro-liths, in securely dated stratigraphic contexts, and of associated animal and plant remains. (Inciden-tally, the flint tool assemblage is comparable to that of the Mesolithic of Star Carr in Yorkshire in England, although it lacks the chisel-like so-called burin, interpreted as having been used to work bone, antler and possibly wood, all perfectly pre-served there below an unusually high water table.)

Conditions in Ireland during the Late Glacial, Nahanagan stadial – which preceded the Post-Glacial, Littletonian – were evidently sufficiently inhospitable as to have made it uninhabitable by humans. The pollen record indicates that the cli-mate at this time was Arctic, and that the vegetation was dominated by tundra and alpine scrub with Artemisia. The record of the northern European beetles Boreophilus henningianus and Diacheila arctica confirms an Arctic climate. The vertebrate palae-ontological record indicates that the fauna was dom-inated by reindeer, giant deer, wolf and brown bear.

Conditions during the Post-Glacial, Littletonian, ameliorated and made the land amenable to habi-tation. The pollen record indicates that the climate at this time became warmer, and that in response the vegetation became dominated initially by open scrub, subsequently by open woodland, and ultim-ately by juniper and birch woodland. The vertebrate palaeontological record indicates that at least ini-tially the fauna was still characterised by some essentially Arctic elements, such as the lemming. (Incidentally, the Post-Glacial return of the flora and fauna to Ireland from refugia to the south and east has been described as a 'steeplechase', complete with 'water jumps' over the English Channel and the Irish Sea, which (re-)formed 7500 years BP. Interestingly, snakes appear to have refused the 'water jump' over the Irish Sea. Alternatively, as legend has it, they were banished from Ireland by the early Christian mission-ary St Patrick in the fifth century AD.)

The early humans that came on to this scene approximately 9000 years BP – the Sandelians – evidently used their flint weapons and tools to hunt game such as wild boar, possibly with the assistance of domesticated dogs; and game birds such as capercaillie, grouse, duck and pigeon, possibly with the assistance of tamed and trained goshawk (also found at the Neolithic to Bronze Age site at Newgrange in Co. Meath). They evidently also fished for salmon, trout and eels in fresh water, and bass and flounder in salt water, and foraged for seasonal shell-fish, nuts, berries, white water-lily (*Nymphaea alba*) and other seeds, and roots. They are best thought of as hunter–gatherers.

10.1.7 Fayum, Egypt ('Epipalaeolithic'–Neolithic, 8000–6500 years BP)

The 'Epipalaeolithic'–Neolithic of the Fayum area on the shores of an ancient lake on the flood-plain of the Nile in Middle Egypt records the – 'pastroforaging' – transition from a nomadic, huntinggathering, to a more settled, farming culture, during the 'early Holocene wet phase' between approximately 14 000 and 6000 years BP (Butzer, 1959; Midant-Reynes, 2000; Sampsell, 2003; Holl, 2004; Linstadter & Kropelin, 2004; Kuper & Kropelin, 2006; Wengrow, 2006). Surrounding parts of Egypt are remarkably rich in archaeological remains, the most renowned and indeed iconic of which are the pyramids of Giza (or Gizeh) on the west bank of the Nile on the outskirts of Cairo, dating to the Dynastic period, approximately 5000 years BP. The oldest incontrovertible evidence of the human occupation of the country dates back to the time of the Acheulian stone tool technology, to the Early Palaeolithic, approximately 300 000 years BP. The Middle–Late Palaeolithic was marked by the diversification of hunter–gatherer cultures, and by 'Nilotic adaptation'.

The 'Epipalaeolithic' Fayum B or Qarunian culture has been dated to approximately 8000 years BP (Caton-Thompson & Gardner, 1934; Arkell & Ucko, 1965; Wendorf & Schild, 1976; Midant-Reynes, 2000; Kuper & Kropelin, 2006; Wengrow, 2006). The economy of that culture appears to have been based primarily on nomadic hunting and, particularly, 'Nilotically adapted' fishing, for shallow-water catfish (*Clarias*), which were probably harpooned, and for deep-water Nile perch (*Lates niloticus*), which were probably netted from boats (Brewer, 1987, 1989; Brewer, in Krzyzaniak & Kobusiewicz, 1989; Gautier, 1990; Midant-Reynes, 2000). Analysis of

seasonal growth rings on the pectoral spines of the catfish indicates that they were mostly caught during the Nilotic floods of the late summer, when they would have congregated in large numbers in order to breed, and would have been comparatively easy to catch. The Qarunian culture evidently made extensive use of bone tools, including harpoons crafted from the jaws of catfish; and stone tools, including a preponderance of bladelets or microliths, for which the principal raw material was silicified wood from a petrified forest of Oligocene age on Gebel Qatrani. The Qarunian people, judging from the skeleton of a 40-year-old woman found buried at site E29-G1, were of more recognisably modern aspect than the 'Mechtoids' of the Early Kubbaniyan or Fakhurian culture of the Palaeolithic, especially in terms of their more robust jaws and teeth (Henneberg et al., in Krzyzaniak & Kobusiewicz, 1989; Midant-Reynes, 2000). (Interestingly, the 'Mechtoids', also known as the 'African Cromagnoids', are believed to have descended from 'proto-Cro Magnons', such as those encountered at Qafzeh in Israel: Vandermeesch, 1981; Tillier, 1992; Midant-Reynes, 2000.)

The succeeding Neolithic Fayum A or Fayumian culture has been dated to approximately 7500–6000 years BP (Midant-Reynes, 2000; Kuper & Kropelin, 2006; Wengrow, 2006). The economy of that culture appears to have been based primarily on more settled farming, and only secondarily on hunting and fishing (Brewer, 1987, 1989; Brewer, in Krzyzaniak & Kobusiewicz, 1989; Gautier, 1990; Midant-Reynes, 2000; Wengrow, 2006). Thus far, only evidence of temporary settlement in the form of post-holes and hearths has been uncovered in the Fayum area, although evidence of permanent settlement in the form of actual dwellings been uncovered at a site in the nearby Merimda Beni Salama area, dated to approximately 6500–6000 years BP (Eiwanger, 1982; Hassan, 1988; Wengrow, 2006). The Fayumian culture evidently continued to make use of bone and stone tools, although the assemblages are more diverse than those of the Qarunian (Midant-Reynes, 2000; Wengrow, 2006). The bone tool assemblage includes not only harpoons, but also borers, needles and pins. The stone tool assemblage includes not only bladelets or microliths, but also precursors of the 'fishtail

lances' of the Pre-Dynastic period, and technologic-
ally advanced leaf-shaped points, arrowheads, axes,
and, importantly, sickle blades, implying cultivation of crops. (Cultivation of crops was evidently
also practised by the Natufian civilisation of the
Middle and Near East, dated to approximately
10 000 years BP, and apparently the first to have
become entirely sedentary and pastoral: Garrod,
1932; Cauvin, 1994; Midant-Reynes, 2000). Domesticated strains of cereals, including six-, four- and
two-row barley (*Hordeum hexastichum, H. vulgare* and
H. distichum, respectively) and wheat (*Triticum dicoc-
cum*) have been discovered in the Fayumian,
together with flax (*Linum usitatissimum*), and the
remains of domesticated dogs, sheep, goats, cattle
and pigs, and of wild hippopotamuses, elephants,
turtles, crocodiles, lizards, snakes, waterfowl,
fish, freshwater bivalves and the terrestrial
gastropod *Helix desertorum* (Wenke & Casini, in
Krzyzaniak & Kobusiewicz, 1989; Wetterstrom, in
Shaw *et al.*, 1993; Midant-Reynes, 2000; Wengrow,
2006). Curved or forked tamarisk sticks thought
to represent flails for separating wheat from
chaff have also been discovered, together with gran-
aries and silos for storing grain, and pestles and
mortars for grinding it. Pottery made from local
Tertiary clay is common in the Fayumian, with
some of the pots containing sea-shells interpreted
as having been used as spoons. Pierced shells,
ostrich egg-shells, stone discs and amazonite beads,
presumably used for personal adornment, are also
present. The presence of amazonite arguably indicates trade with cultures as far away as the Tibesti
region of southern Libya and northern Chad, possibly along the long-lost but recently rediscovered
'Trans-Saharan Highway' (Midant-Reynes, 2000;
Linstadter & Kropelin, 2004; Wengrow, 2006;
Young, 2007).

The Fayumian culture came to an end at the
time of the 'mid-Holocene arid phase', which began
approximately 6000 years BP, causing widespread
desertification and resulting in a mass population
migration to the fertile refuge of the Nile Valley,
and it was at this time and in this place that
the Dynastic or Pharaonic culture was founded
(Butzer, 1959; Midant-Reynes, 2000; Sampsell,
2003; Holl, 2004; Linstadter & Kropelin, 2004; Kuper
& Kropelin, 2006; Wengrow, 2006).

Note: While Western Europe was still in the
Mesolithic 8000 years BP, the Middle and Near East
had already entered into the Neolithic, and north
Africa was undergoing a process of 'Neolithicisa-
tion'. Civilisation clearly spread from the east to
the west!

10.1.8 Littleton, Co. Tipperary, Ireland (Neolithic to Medieval, 6200–800 years BP, 4200 BC to AD 1200)

An 8 m core from a raised bog near Littleton in Co.
Tipperary has yielded evidence of human occupation and activity in this part of Ireland from the
Neolithic to the present time (Malone, 2001; Jones,
2006). The evidence is in the form of the pollen
record of anthropogenic disturbance of the environment, specifically a decrease in the proportion of
native woodland pollen, and an increase in the
proportion of plantain, grass, and cultivated cereal
and herb pollen, interpreted as indicating forest
clearance for farming.

The pollen record from the Neolithic indicates a
generally warm, wet, maritime climate at this time,
and also indicates, for the first time, subsistence
farming – and, by inference, permanent settlement –
from 4200 to 2000 BC. The observed decline in elm
from 3900 to 2800 BC could also indicate human
activity, or disease, or both. Note in this context that
Stone Age – and Bronze Age – people are thought to
have used elm for firewood and for fodder for their
livestock, and to have pollarded it specifically for
these purposes, in the process predisposing it to disease. Note also that remains of the beetle (*Scolytus
scolytus*) that carries the fungus (*Ceratocystis ulmae*) that
causes Dutch Elm disease have been found in deposits
on Hampstead Heath in London that pre-date the
decline in elm there.

The pollen record from the Bronze Age indicates
a generally dry, continental climate in the Bronze
Age. It also indicates early Bronze Age farming
between 2000 and 1200 BC, and evidently more
extensive or efficient late Bronze Age farming
between 1200 and 300 BC. The increase in efficiency
in the late Bronze Age has been interpreted as associated with the use of the plough or 'ard'. Incidentally, land management practices that are now
thought of as characteristically Celtic were apparently also first implemented in the late Bronze Age.

Interestingly in this context, the Welsh word for plough or 'ard' is *aradr*.

The pollen record from the Iron Age to the present time indicates a generally cool, wet climate. That from the Iron Age indicates a reduction in farming between 300 BC and AD 400. That from the early Medieval period indicates a resumption associated with monasticism, followed by a reduction, both between AD 400 and 1200. Incidentally, that from the Anglo-Norman period to the present time indicates another resumption, from AD 1200 (and a further increase in efficiency, associated with the 'agricultural revolution', from AD 1700).

10.1.9 Skara Brae, Orkney (Neolithic, 5000 years BP, 3000 BC)

Archaeological excavation at Skara Brae in the Orkney Islands off the north coast of Scotland has yielded evidence of domestic and community life dating back to the late Neolithic (Jones, 2006; Bradley, 2007; Whittle, in Hunter & Ralston, 2009). The evidence is in the form of an entire village of stone houses, which would once have been home to 50–60 people, together with stone and bone artefacts, and associated animal and plant remains. The village apparently lay perfectly preserved for thousands of years under a protective blanket of dune sand, until a storm in the mid nineteenth century revealed it once more to the world.

Conditions obtaining in the Neolithic were evidently somewhat warmer than they are today, although there were still but a few trees to soften the landscape. The Neolithic inhabitants of the island – the Orcadians – therefore built their houses out of local flag-stones, which they were able to cleave from the country rock with antler picks. Like the maritime Sandelians of the Mesolithic, they fished, for red bream and corkfin wrasse, and foraged, for mussels and oysters (they may also have hunted game, with dogs). Unlike the Sandelians, though, they also farmed. They farmed cattle, for meat and milk, and they grew crops such as barley and wheat, where none will grow today (the land being suitable for pastoral but not for arable purposes).

10.1.10 Tyrolean Alps (Chalcolithic or Copper Age, 5000 years BP, 3000 BC)

The frozen and mummified body of an 'ice man' – now named 'Ötzi' – was discovered in the Tyrolean Alps in 1991 (Jones, 2006). Radiocarbon dating of plant and animal remains associated with the body has yielded a consistent age of around 5300 years BP.

Detailed archaeological examination of the plant and animal remains associated with Ötzi's body has yielded fascinating insights into his life (Bortenschlager & Oeggl, 2000; Jones, 2006). Remarkably, some 200 species of plants have been found in, on or around his body. Many are from lower altitudes than where he was discovered, and thus clearly introduced by him. Some of the plant species, such as hop hornbeam (*Ostrya carpinifolia*) and *Neckera*, indicate a southerly provenance, as, incidentally, does the flint in Ötzi's flint tool and weapons, which has been typed back to the locality of Monte Lessini in Italy by means of the siliceous microfossil assemblages that it contains. *Ostrya carpinifolia* is represented by incidentally ingested pollen in his gut, interestingly, with its cell content preserved, as it only is in late spring or early summer, indicating that this was the time of year at which he died. Traces of einkorn wheat (*Triticum monococcum*) has also been found in his gut, together with traces of red deer (*Cervus elaphus*) and ibex (*Capra ibex*), representing the remains of his last meals. A single species of *Sphagnum* moss has also been found in his gut, and interpreted as having been used as a food wrapping. Another species of *Sphagnum* found outside the body has been interpreted as having been used as a wound dressing. Bracket fungus also found outside the body is interpreted as having been used as a medicine (it has antiseptic properties). A pouch containing tinder fungus together with flints and pyrite has been found, and interpreted as a fire-lighting kit. Norway maple leaves have also been found, and interpreted as having been used to carry glowing embers. Seventeen types of wood or charcoal have been found. Ötzi evidently used wood from the wayfaring tree to make arrows; yew, to make a bow, and the haft of a copper-headed axe; ash, the handle of a dagger; and birch bark to make a bucket in which to carry a 'trail mix' of dried sloes. He also evidently used bark, together with grass, to make a serviceable cape; and goat-skin to make underclothing and shoes.

Detailed archaeological examination of the plant remains associated with Ötzi's body has also yielded insights into his death (Oeggl *et al.*, 2007). Several species are represented by incidentally ingested pollen in his gut in addition to the already mentioned hop hornbeam (*Ostrya carpinifolia*). Analysis of the pollen associated with the remains of his last three meals, in his sigmoid colon, transverse colon and ileum, indicates that he ate the ante-penultimate at an altitude of 2500 m; the penultimate at 1200 m; and the ultimate at 3000 m. Incidentally, it has been speculated that Ötzi ate his ultimate meal at altitude in an attempt to escape from an earlier conflict that had left him mortally wounded by an arrow under his left shoulder.

10.1.11   City of London (Medieval, 800–650 years BP, AD 1200? to AD 1350)

Archaeological excavation at a site in Tudor Street in the City of London in 1978, preparatory to site redevelopment, has yielded interesting evidence of domestic and industrial activity in the Middle Ages (Boyd, in Neale & Brasier, 1981; Jones, 2006). The evidence is essentially in the form of the record of ancient organismal remains in the medieval deposits of the River Fleet, a tributary of the Thames, and of the influence on it of human activity. Incidentally, the word 'fleet' is Anglo-Saxon in origin, and refers to the ability of the river to float boats on the rising tide. The post-Anglo-Saxon history of the River Fleet has been described as 'a decline from a river to a brook, from a brook to a ditch, and from a ditch to a drain'.

The medieval deposits of the Fleet are approximately 1 m thick, between −0.4 and +0.6 mOD. They unconformably overlie a bedrock of London clay and a layer of interpreted pre-Roman occupation deposits, and underlie a layer of post-medieval industrial waste, post-dating the period of land reclamation between AD 1350 and 1360. They have been provisionally dated to the thirteenth and fourteenth centuries on the basis of pot-sherds.

The medieval deposits of the Fleet altogether contain over 140 species of organisms, including diatoms, silicoflagellates, organic-walled microplankton, charophytes, angiosperms, testate *Amoebae*, foraminifera, sponges, thecate hydroids, polychaete worms, parasitic nematode worms (from human faeces!), freshwater bryozoans, bivalve and gastropod molluscs, cladocerans, ostracods and echinoids. The principal elements of the biota are essentially freshwater to only slightly brackish, and it is thought that the minor, marine elements are allochthonous. It is also thought that the salinity of the medieval Fleet may have been artificially low on account of the effects of the old London bridge, built in 1209 and not removed until as recently as 1832.

They also contain abundant industrial and domestic refuse such as wood chips (waste from wood-working), horn cores (waste from horn-working), hide (waste from butchery or tannery), leather shoes, meat- and domestic-animal bones, and animal and human dung (see also Thomas, 2002). The refuse and organic waste dumped into the river evidently resulted initially in eutrophication, and the proliferation of some, opportunistic, species, including freshwater molluscs and the macrophytes *Zannicellia* and *Groenlandia* (in Bed X). However, it also resulted untimately in hypertrophication, and the effective elimination of all species (in Bed Y), as recorded in archive records for 1343. Incidentally, the pollution became so bad that it was ordered in 1357 that 'no man shall take ... any manner of rubbish ... or dung, from out of his stable or elsewhere, to throw ... the same into the rivers of Thames and Fleet ... And if any one should be found doing the contrary thereof, let him have the prison for his body, and other heavy punishment as well, at the discretion of the Mayor and the Aldermen'!

10.2   APPLICATIONS AND CASE STUDIES IN FORENSIC SCIENCE

Calcareous nannofossils, fossil and living diatoms, living plant spore and pollen, and living insects have all proved of use in forensic science (Jones, 2006).

10.2.1   Use of calcareous nannofossils in forensic science

Calcareous nannofossils have recently proved of use in forensic science (Bailey & Gallagher, 2005). In a high-profile double murder case in the United Kingdom, calcareous nannofossils from chalk recovered from the front suspension and footwell of a suspect's car were typed back to an individual bed within the Upper Cretaceous, Cenomanian that was only exposed along the road beside which the victims' bodies were found. The suspect's denial that he had ever been along the road was therefore demonstrated to be false.

10.2.2   Use of diatoms in forensic science

Fossil diatoms have proved of use in identifying diatomaceous earth used in safe construction, and thus the scene where a safe was broken open, or the suspect, from the presence of diatomaceous earth at the scene or on the suspect's skin or clothing (Cameron, in Pye & Croft, 2004).

Living diatoms have proved of use in diagnosing death by drowning, from the diatom flora in the water in the lungs (Cameron, in Pye & Croft, 2004; Sugita & Marumo, in Pye & Croft, 2004). They have also proved of use in testing alibis, by comparing the diatom content in environmental samples taken from the suspect against that of samples taken from the scene of the crime and from the suspect's alleged whereabouts at the time of the crime (Cameron, in Pye & Croft, 2004; Sugita & Marumo, in Pye & Croft, 2004).

10.2.3   Use of spores and pollen in forensic science (forensic palynology)

Spores and pollen have proved of use in forensic science, for example in determining whether corpses have been moved after death, by comparing the spore and pollen content in environmental samples taken from the corpse against that in samples from the scene of recovery; and in testing alibis, by comparing the spore and pollen content in environmental samples taken from the suspect against that of samples taken from the scene of the crime and from the suspect's alleged whereabouts at the time of the crime (Bryant et al., in Jansonius & MacGregor, 1996; Dawson et al., in Pye & Croft, 2004; Pye, in Pye & Croft, 2004). The first reported use was in Austria in 1959, in which case a palynologist was able to identify the source of the mud on the soles of the shoes of a murder suspect as an area of the Danube valley north of Vienna, close to where the body of the victim was eventually found.

Spores and pollen have also proved of use in establishing the provenance and travel history of illicit drugs. In a recent case, cocaine seized by the New York Police Department turned out to contain not only Lycopodium spores and pollen from northern South America, probably introduced during initial processing of the coca leaves, but also Tsuga canadensis (hemlock) and Pinus banksiana (jack pine) pollen from New England, probably introduced during 'cutting', and a variety of pollen found in New York itself, probably introduced during storage.

10.2.4   Use of insects in forensic science (forensic entomology)

Living insects have proved of use in forensic science, principally in determining the time of death, from an assessment of the stage of advancement of the 'ecological succession' that results from changes in the attractiveness of the corpse to different groups through time (complicated by changes to the extent of exposure of the corpse, movement of the corpse, and so on). Blow-flies and house-flies are typically the first insects to visit a corpse, and oviposit eggs or drop live larvae onto it, which develop into maggots over a fixed period of time determined essentially by ambient temperature. Certain types of dermestid beetle larvae and adults only appear later, followed by cheese-skipper larvae. Fruit-flies and hover-flies appear later still, after the body has started to desiccate, and tineid moths latest of all, only after the body has completely desiccated.

10.3   MISCELLANEOUS OTHER APPLICATIONS

10.3.1   Medical palynology

Medical palynology aims to identify those airborne spores and pollen that can cause allergic reactions, and to alert particularly susceptible individuals to their presence through daily publication of a 'pollen count' (O'Rourke, in Jansonius & MacGregor, 1996).

10.3.2   Entomopalynology

Entomopalynology is the study of pollen associated with insects, and provides important information as to their feeding and migratory activities (Pendleton et al., in Jansonius & MacGregor, 1996).

10.3.3   Melissopalynology

Melissopalynology is the study of pollen associated with honey, and provides important information as to the source (Jones & Bryant, in Jansonius & MacGregor, 1996). This is by no means as esoteric as it might sound, as some sources result in honey commanding a higher price than others, and obviously beekeepers are keen to site their hives as close as possible to these sources!

References

Aaron, M. E., Farnsworth, E. J. & Merkt, R. E., 1999. Origins of mangrove ecosystems and the mangrove biodiversity anomaly. *Global Ecology and Biogeography*, **8**: 95–115.

Abbink, O. A., 1998. *Palynological Investigations in the Jurassic of the North Sea Region*. University of Utrecht (Laboratory of Palaeobotany and Palynology Contributions Series, No. 8).

Abbink, O. A., van Konijenburg-van Cittert, J. H. A. & Visscher, H., 2004a. A sporomorph ecogroup model for the northwest European Jurassic–Lower Cretaceous I: concepts and framework. *Geologie en Mijnbouw*, **83**: 17–31.

2004b. A sporomorph ecogroup model for the northwest European Jurassic–Lower Cretaceous II: application to an exploration well from the Dutch North Sea. *Geologie en Mijnbouw*, **83**: 81–91.

Adams, E. W., Grotzinger, J. P., Watters, W. A. *et al.*, 2005. Digital characterization of thrombolite-stromatolite reef distribution in a carbonate ramp system (terminal Proterozoic, Nama group, Namibia). *American Association of Petroleum Geologists Bulletin*, **89**(10): 1293–1318.

Adams, T. D., Khalili, M. & Said, A. K., 1967. Stratigraphic significance of some oligosteginid assemblages from Lurestan province, northwest Iran. *Micropaleontology*, **13**(1): 55–67.

Adnet, S., Antoine, P.-O., Hassan Baqri, S. R. *et al.*, 2007. New tropical carcarinids (chondrichthyes, Carchariformes) from the Late Eocene–Early Oligocene of Balochistan, Pakistan: paleoenvironmental and paleogeographic implications. *Journal of Asian Earth Sciences*, **30**: 303–23.

Afzal, J., Khan, F. R., Khan, S. N. *et al.*, 2005. Foraminiferal biostratigraphy and paleoenvironments of the Paleocene Lockhart limestone from Kotal pass, Kohat, northern Pakistan. *Pakistan Journal of Hydrocarbon Research*, **15**: 9–23.

Afzal, J., Williams, M. & Aldridge, R. J., 2009. Revised stratigraphy of the lower Cenozoic succession of the greater Indus basin in Pakistan. *Journal of Micropalaeontology*, **28**: 7–23.

Agusti, J., Cabrera, L., Garces, M. *et al.*, 2001. A calibrated mammal scale for the Neogene of western Europe: state of the art. *Earth-Science Reviews*, **52**: 247–60.

Agusti, J., Garces, M. & Krijgsman, W., 2006. Evidence for Africa–Iberian exchanges during the Messinian in the Spanish mammalian record. *Palaeogeography, Palaeoclimatology, Palaeoecology*, **238**: 5–14.

Agusti, J., Rook, L. & Andrews, P. (eds.), 1999. *Climatic and Environmental Change in the Neogene of Europe*. Volume 1 of *Hominoid Evolution and Climatic Change in Eurasia*. Cambridge University Press.

Ahmad, A. & Ahmad, N., 2005. Paleocene petroleum system and its significance for exploration in the southwest Lower Indus basin and nearby offshore of Pakistan. *Proceedings of the Society of Petroleum Engineers/Pakistan Association of Petroleum Geologists Annual Technical Conference, Islamabad*: 1–22.

Aigner, T., 1983. Facies and origin of nummulitic build-ups: an example from the Giza Pyramids plateau (Middle Eocene, Egypt). *Neues Jahrbuch für Geologie und Palaontologie, Abhandlungen*, **166**: 347–68.

1985. Biofabrics as dynamic indicators in nummulite accumulations. *Journal of Sedimentary Petrology*, **55**: 131–4.

Ainsworth, N. R., Burnett, R. D. & Kontrovitz, M., 1990. Ostracod colour change by thermal alteration, offshore Ireland and Western UK. *Marine and Petroleum Geology*, **7**: 288–97.

Akimoto, K., Nakahara, K., Kondo, H., Ishiga, H. & Dozen, K., 2004. Environmental reconstruction based on heavy metals, diatoms and benthic foraminifers in the Isahaya reclamation area, Nagasaki, Japan. *Journal of Environmental Micropaleontology, Microbiology, Meiobenthology*, **1**: 83–104.

Alam, M., 1989. Geology and depositional history of Cenozoic sediments of the Bengal basin of Bangladesh. *Palaeogeography, Palaeoclimatology, Palaeoecology*, **69**: 125–39.

Al-Ameri, T. K. & Lawa, F., 1986. Palaeoecological model and faunal interaction within Aqra Limestone formation, north Iraq. *Journal of the Geological Society of Iraq*, **19**(3): 7–27.

Albert, J. S., Lovejoy, N. R. & Crampton, W. G. R., 2006. Miocene tectonism and the separation of cis- and trans-Andean river basins: evidence from neotropical fishes. *Journal of South American Earth Sciences*, **21**: 1–17.

Aldhouse-Green, S., 2000. *Paviland Cave and the 'Red Lady': A Definitive Report*. Western Academic & Specialist Press.

Alemseged, Z., Spoor, F., Kimbel, W. H. *et al.*, 2006. A juvenile early hominin skeleton from Dikika, Ethiopia. *Nature*, **443**: 296–301.

Algeo, T. J., Berner, R. A., Maynard, J. B. & Scheckler, S. E., 1995. Late Devonian oceanic anoxic events and biotic crises: 'rooted' in the evolution of vascular land plants? *GSA Today*, **5**(45): 64–6.

Algeo, T. J. & Ingall, E., 2007. Sedimentary C_{org}:P ratios, paleocean ventilation, and Phanerozoic atmospheric pO_2. *Palaeogeography, Palaeoclimatology, Palaeoecology*, **256**: 130–155.

Algeo, T. J. & Scheckler, S. E., 1998. Terrestrial-marine teleconnections in the Devonian: links between the evolution of land plants, weathering processes, and marine anoxic events. *Philosophical Transactions of the Royal Society, London*, **B353**: 113–30.

Al-Hajri, S. & Owens, B. (eds.), 2000. *Stratigraphic Palynology of the Palaeozoic of Saudi Arabia*. Gulf PetroLink (GeoArabia Special Publication No. 1).

Al-Husseini, M. I. (ed.), 1995. *GEO'94: Selected Middle East Papers from the Middle East Geoscience Conference, April 25-27, 1994, Bahrain*. Gulf PetroLink.

2004. *Carboniferous, Permian and Early Triassic Arabian Stratigraphy*. Gulf PetroLink (GeoArabia Special Publication No. 3).

Al-Husseini, M. I., 2007. Stratigraphic note: revised ages (Ma) and accuracy of Arabian Plate maximum flooding surfaces. *GeoArabia*, **12**(4): 167-70.

Al-Husseini, M. & Matthews, R. K., 2005a. Arabian orbital stratigraphy: periodic second-order sequence boundaries. *GeoArabia*, **10**(2): 165-84.

2005b. Tectono-stratigraphic note: time calibration of Late Carboniferous, Permian and Early Triassic stratigraphy to orbital-forcing predictions. *GeoArabia*, **10**(2): 189-92.

2005c. Stratigraphic note: orbital-forcing calibration of the Late Cretaceous and Paleocene Aruma formation, Saudi Arabia. *GeoArabia*, **10**(3): 173-6.

2006a. Devonian Jauf formation, Saudi Arabia: orbital second-order depositional sequence 28. *GeoArabia*, **11**(2): 53-70.

2006b. Stratigraphic note: orbital calibration of the Arabian Jurassic second-order sequence stratigraphy. *GeoArabia*, **11**(3): 161-9.

Al-Husseini, M., Matthews, R. K. & Mattner, J., 2006. Stratigraphic note: orbital-forcing calibration of the Late Jurassic (Oxfordian-early Kimmeridgian) Hanifa formation, Saudi Arabia. *GeoArabia*, **11**(1): 145-9.

Ali, W., Paul, W. & Young On, V. (eds.), 1998. *Transactions of the 3rd Geological Conference of the Geological Society of Trinidad and Tobago and the 14th Caribbean Geological Conference*. The Geological Society of Trinidad and Tobago.

Allison, P. A., Hesselbo, S. P. & Brett, C. E., 2008. Methane seeps on an Early Jurassic sea floor. *Palaeogeography, Palaeoclimatology, Palaeoecology*, **270**: 230-8.

Allwood, A. C., Walter, M. R., Burch, I. W. & Kamber, B. S., 2007. 3.43 billion-year-old stromatolite reef from the Pilbara craton of Western Australia: ecosystem-scale insights into early life on Earth. *Precambrian Research*, **158**: 198-227.

Alroy, J., 1998. Diachrony of mammalian appearance events: implications for biochronology. *Geology*, **26**(1): 23-6.

Alsen, P. & Mutterlose, J., 2009. The Early Cretaceous of north-east Greenland: a crossroads of belemnite migration. *Palaeogeography, Palaeoclimatology, Palaeoecology*, **280**: 168-82.

Alve, E. & Dolven, J. K. L., 2010. Defining 'reference' conditions: monitoring inner Oslofjord, Norway. *Abstracts, Forams 2010 International Symposium on Foraminifera, Bonn*: **46**. Rheinische Friedrich-Wilhelms-Universitat.

Alvin, K. L., 1982. Cheirolepidaceae: biology, structure and paleoecology. *Review of Palaeobotany and Palynology*, **37**: 71-98.

1983. Reconstruction of a Lower Cretaceous conifer. *Botanical Journal of the Linnaean Society*, **86**: 169-76.

Amano, K. & Little, C. T. S., 2005. Miocene whale-fall community from Hokkaido, northern Japan. *Palaeogeography, Palaeoclimatology, Palaeoecology*, **215**: 345-56.

Amano, K., Little, C. T. S. & Inoue, K., 2007. A new Miocene whale-fall community from Japan. *Palaeogeography, Palaeoclimatology, Palaeoecology*, **247**: 236-42.

Ambwani, K., Sahni, A., Kar, R. K. & Dutta, D., 2003. Oldest known non-marine diatoms (*Aulacoseira*) from the uppermost Cretaceous Deccan Intertrappean beds and Lameta formation of India. *Revue de Micropaléontologie*, **46**: 67-71.

Anderson, J. M, 1977. *The Biostratigraphy of the Permian and Triassic, Part 3*. Botanical Survey of South Africa Memoir No. 41.

Anderson, J. M. & Anderson, H. M., 1985. *Palaeoflora of Southern Africa*. Balkema.

Anderson, J. M., Anderson, H. M., Archangelsky, S. *et al.*, 1999. Paterns of Gondwana plant colonisation and diversification. *Journal of African Earth Sciences*, **28**(1): 145-67.

Anderson, J. S. & Sues, H.-D. (eds.), 2007. *Major Transitions in Vertebrate Evolution*. Indiana University Press.

Andrews, P., Cook, J., Currant, A. & Stringer, C. 1999. *Westbury Cave: The Natural History Museum Excavations, 1976-1984*. Western Academic & Specialist Press.

Anketell, J. M. & Mriheel, I. Y., 2000. Depositional environment and diagenesis of the Eocene Jdeir formation, Gabes-Tripoli basin, western offshore Libya. *Journal of Petroleum Geology*, **23**(4): 425-47.

Ansari, M. H. & Vink, A., 2007. Vegetation history and palaeoclimate of the past 30 kyr in Pakistan as inferred from the palynology of continental margin sediments of the Indus delta. *Review of Palaeobotany and Palynology*, **145**: 201-16.

Anz, J. H. & Ellouz, M., 1985. Development and operation of the El Gueria reservoir, Ashtart field, offshore Tunisia. *Journal of Petroleum Technology, Tunis*, **37**: 481-7.

Aqrawi, A. A. M., Goff, J. C., Horbury, A. D. & Sadooni, F. N., 2010a. *The Petroleum Geology of Iraq*. Scientific Press.

Aqrawi, A. A. M., Mahdi, T. A., Sherwani, G. H. & Horbury, A. D., 2010b. Characterisation of the mid-Cretaceous Mishrif reservoir of the southern Mesopotamian basin, Iraq. *Search and Discovery Article* No. 50264. http://www.searchanddiscovery.com

Arkell, A. J. & Ucko, P. J., 1965. Reviews of Predynastic development in the Nile Valley. *Current Anthropology*, **6**: 145-66.

Arkell, W. J., 1956. *Jurassic Geology of the World*. Oliver & Boyd.

Armynot du Chatelet, E. & Debenay, J.-P., 2010. The anthropogenic impact on the western French coast as revealed by foraminifera: a review. *Revue de Micropaléontologie*, **53**: 129-37.

Aronson, R. B. (ed.), 2007. *Geological Approaches to Coral Reef Ecology*. Springer.

Arthur, T. J., MacGregor, D. S. & Cameron, N. R. (eds.), 2003. *Petroleum Geology of Africa: New Themes and Developing Technologies*. The Geological Society, London (Special Publication No. 207).

Ashkcenazi-Polivoda, S., Edelman-Furstenburg, Y., Almogi-Lavin, A. & Benjamini, C., 2010. Characteristics of lowest oxygen environments within ancient upwelling environments: benthic foraminiferal evidence. *Palaeogeography, Palaeoclimatology, Palaeoecology*, **289**: 134–44.

Ashworth, A.C., Buckland, P.C. & Sadler, J.P. (eds.), 1997. *Studies in Quaternary Entomology*. John Wiley & Sons Ltd.

Askri, H., Belmecheri, A., Benrabah, B. *et al.*, 1995. *Geologie de l'Algeria (Geology of Algeria)*. Sonatrach.

Aswal, H.S. & Mehrotra, N.C., 2002. Stratigraphic significance of Triassic–Jurassic dinoflagellate cysts in Krishna-Godavari basin, India. *Indian Journal of Petroleum Geology*, **11**(1): 9–35.

Aswal, H.S., Singh, K. & Mehrotra, N.C., 2001. Mesozoic–Cenozoic dinoflagellate cyst bio-chronostratigraphy of Krishna–Godavari basin, India – a summary. *Bulletin of the Oil and Gas Corporation*, **38**(2): 91–100.

Atkinson, T.C., Briffa, K.R. & Cooper, G.R., 1987. Seasonal temperatures in Britain during the past 22,000 years, reconstructed using beetle remains. *Nature*, **325**: 587–92.

Aubry, M, Ouda, K. Dupuis, C. *et al.*, 2007. The Global Standard Stratotype-section and Point (GSSP) for the base of the Eocene series in the Dababiya section (Egypt). *Episodes*, **30**(4): 271–86.

Babcock, L.E. & Shanchi Peng, 2007. Cambrian chronostratigraphy: current state and future plans. *Palaeogeography, Palaeoclimatology, Palaeoecology*, **254**: 62–6.

Baceta, J.I., Pujalte, V. & Bernaola, G., 2005. Paleocene coralgal reefs of the western Pyrenean basin, northern Spain: new evidence supporting an earliest Paleogene recovery of reefal ecosystems. *Palaeogeography, Palaeoclimatology, Palaeoecology*, **224**: 117–43.

Baco, A.R., Rowden, A.R., Levin, L.A. *et al.*, 2010. Initial characterization of cold seep faunal communities on the New Zealand Hikurangi margin. *Marine Geology*, **272**: 251–9.

Bailey, H. & Gallagher, L., 2005. Coccoliths and other microfossils in forensic palaeontology. *Abstracts, The Micropalaeontological Society Annual General Meeting Presentations, 2005*.

Baillie, P., Barber, P.M., Deighton, I. *et al.*, 2006. Petroleum systems of the deepwater Mannar basin, offshore Sri Lanka. *Proceedings of Indonesian Petroleum Association Deepwater and Frontier Exploration in Asia & Australia Symposium*, DFE04-PO-015.

Bajpai, S., Kapur, V.V., Thewissen, J.G.M. *et al.*, 2005a. First fossil marsupials from India: Early Eocene *Indodelphis* n. gen. and *Jaegeria* n. gen. from Vastan lignite mine, Surat, Gujurat. *Journal of the Palaeontological Society of India*, **50**(1): 147–51.

2005b. Early Eocene primates from Vastan lignite mine, Gujurat, western India. *Journal of the Palaeontological Society of India*, **50**(2): 43–54.

Bambach, R.K., Knoll, A.H. & Sepkoski, J.J. Jr, 2002. Anatomical and ecological constraints on Phanerozoic animal diversity in the marine realm. *Proceedings of the National Academy of Sciences USA*, **99**: 6854–9.

Bamford, M.K., 2004. Diversity of woody vegetation of Gondwanan southern Africa. *Gondwana Research*, **7**(1): 153–64.

Bamford, M.K., Albert, R.M. & Cabanes, D., 2006. Plio-Pleistocene macroplant fossil remains and phytoliths from Lowermost Bed II in the eastern palaeolake margin of Olduvai Gorge, Tanzania. *Quaternary International*, **148**: 95–112.

Banerjee, A., Pahari, S., Jha, M. *et al.*, 2002. The effective source rocks in the Cambay basin, India. *American Association of Petroleum Geologists Bulletin*, **86**(3): 433–56.

Baraboshkin, E.J. & Mutterlose, J., 2004. Correlation of the Barremian belemnite successions of northwest Europe and the Ulyanovsk–Saratov area (Russian Platform). *Acta Geological Polonica*, **54**(4): 499–510.

Barham, L. & Mitchell, P., 2008. *The First Africans: African Archaeology from the Earliest Toolmakers to Most Recent Foragers*. Cambridge World Archaeology Series. Cambridge University Press.

Barras, C., Geslin, E., Jorissen, F. *et al.*, 2010. Benthic foraminifera as bio-indicators of coastal water quality in the Mediterranean Sea in relation to the implementation of the water framework directive. *Abstracts, Forams 2010 Internationaxifera, Bonn*: **53**. Rheinische Friedrich-Wilhelms-Universitat.

Barss, M.S., Bujak, J.P. & Williams, G.L., 1979. Palynological zonation and correlation of sixty-seven wells, eastern Canada. *Geological Survey of Canada* Paper **78**-24: 1–118.

Bartenstein, H., 1985. Stratigraphic pattern of index foraminifera in the Lower Cretaceous of Trinidad. *Newsletters on Stratigraphy*, **14**(2): 110–7.

1987. Micropalaeontological synthesis of the Lower Cretaceous in Trinidad, West Indies: remarks on the Aptian/Albian boundary. *Newsletters on Stratigraphy*, **17**(3): 143–54.

Bartolini, C., Buffler, R.T. & Blickwede, J.F. (eds.), 2003. *The Circum-Gulf of Mexico and the Caribbean: Hydrocarbon Habitats, Basin Formation and Plate Tectonics*. American Association of Petroleum Geologists Memoir No. 79.

Bartolini, C., Buffler, R.T. & Cantu-Chapa, A. (eds.), 2001. *The Western Gulf of Mexico Basin: Tectonics, Sedimentary Basins and Petroleum Systems*. American Association of Petroleum Geologists Memoir No. 75.

Barton, N., 2005. *Ice Age Britain*, revised edition. B.T. Batsford/English Heritage.

Bar-Yosef Mayer, D.E. (ed.), 2005. *Archaeomalacology*. Oxbow Books.

Bastia, R., 2006. An overview of Indian sedimentary basins with special focus on emerging east coast deepwater frontiers. *The Leading Edge*, July 2006: 818–29.

Bastia, R. & Nayak, P.K., 2006. Tectonostratigraphy and depositional patterns in Krishna offshore basin, Bay of Bengal. *The Leading Edge*, July 2006: 839–45.

Bastia, R., Nayak, P. & Singh, 2007. Shelf delta to deepwater basin: a depositional model for Krishna–Godavari basin. *Search and Discovery* Article No. 40231. http://www.search-anddiscovery.com

Basu, D. N., Banerjee, A. & Tamhane, D. M., 1980. Source areas and migration trends of oil and gas in Bombay offshore basin, India. *American Association of Petroleum Geologists Bulletin*, **64**(2): 209–20.

Bate, R. H., 1975. Ostracods from Callovian to Tithonian sediments of Tanzania, east Africa. *Bulletin of the British Museum (Natural History), Geology*, **26**(5): 163–223.

1977. Upper Jurassic Ostracoda from Tanzania, east Africa. *Annales des Mines et de la Geologie*, **28**: 163–83.

2001. The Lower Cretaceous (pre-salt) stratigraphy of the Kwanza basin, Angola. *Newsletters on Stratigraphy*, **38**(2/3): 117–27.

Battarbee, R. W., Gasse, F. & Stickley, C. E. (eds.), 2004. *Past Climate Variability through Europe and Africa*. Springer.

Batten, D. J., 1975. Wealden palaeoecology from the distribution of plant fossils. *Proceedings of the Geologists' Association*, **85**: 433–58.

Beaudoin, A. B. & Head, M. J. (eds.), 2004. *The Palynology and Micropalaeontology of Boundaries*. The Geological Society, London (Special Publication No. 230).

Beaumont, E. A. & Foster, N. H. (eds.), 1999. *Treatise of Petroleum Geology: Exploring for Oil and Gas Traps*. American Association of Petroleum Geologists.

Beavington Penney, S. J., Wright, V. P. & Racey, A., 2005. Sediment production and dispersal on foraminifera-dominated early Tertiary ramps: the Eocene El Garia formation, Tunisia. *Sedimentology*, **52**: 537–69.

Becker, R. T. & Kirchgasser, W. T. (eds.), 2007. *Devonian Events and Correlations*. The Geological Society, London (Special Publication No. 278).

Beerling, D., 2007. *The Emerald Planet: How Plants Changed Earth's History*. Oxford University Press.

Beerling, D. J. & Woodward, F. I., 2001. *Vegetation and the Terrestrial Carbon Cycle*. Cambridge University Press.

Behre, K.-E., 2007. A new Holocene sea-level curve for the southern North Sea. *Boreas*, **36**: 82–102.

Behrensmeyer, A. K., Damuth, J. D., diMichele, W. A. *et al.*, 1992. *Terrestrial Ecosystems Through Time: Evolutionary Paleobiology of Terrestrial Plants and Animals*. University of Chicago Press.

Bekker-Migdisova, E. E., 1967. Tertiary Homoptera of Stavropol and a method of reconstruction of continental palaeobiocoenoses. *Palaeontology*, **10**(4): 542–53.

Bender, F., 1983. *Geology of Burma*. Gebruder Borntraeger.

Benson, R. B. J., Barrett, P. M., Powell, H. P. & Norman, D. B., 2008. The taxonomic status of *Megalosaurus bucklandii* (Dinosauria, Theropoda) from the Middle Jurassic of Oxfordshire, UK. *Palaeontology*, **51**(2): 419–24.

Benton, M. J., 1985. Patterns in the diversification of Mesozoic non-marine tetrapods and problems in historical diversity analysis. *Special Papers in Palaeontology*, **33**: 185–202.

Benton, M. J. & Emerson, B. C., 2007. How did life become so diverse? The dynamics of diversification according to the fossil record and molecular phylogenetics. *Palaeontology*, **50**(1): 23–40.

Benton, M. J. & Harper, D. A. T., 1997 *Basic Palaeontology*. Longman.

2009. *Introduction to Paleobiology and the Fossil Record*. Wiley-Blackwell.

Berger, L. R., de Ruiter, D. J., Churchill, S. E. *et al.*, 2010. *Australopithecus sediba*: a new species of *Homo*-like australopith from South Africa. *Science*, **328**(5975): 195–204.

Berggren, W. A., Kent, D. V., Aubry, M.-P. & Hardenbol, J. (eds.), 1995. *Geochronology, Time Scales and Global Stratigraphic Correlation*. Society of Economic Paleontologists and Mineralogists (Special Publication No. 54).

Berkowski, B., 2004. Monospecific rugosan assemblage from the Emsian hydrothermal vents of Morocco. *Acta Palaeontologica Polonica*, **49**(1): 75–84.

Bermudez, P. J., 1962. Foraminiferos de los lutitas de Punta Tolete, Terriotorio Delta Amacuro (Venezuela). *Geos, Caracas*, **8**: 35–8.

1966. Consideraciones sobre los sedimentos del Mioceno Medio al Reciente de los costas central; y oriental de Venezuela, primera parte. *Boletin de Geologia*, **14**: 333–412.

Bermudez, P. J. & Stainforth, R. M., 1975. Aplicaciones de foraminiferos planctonicos a la bioestratigrafia del Terziario en Venezuela. *Revista Espanola de Micropaleontologia*, **7**(3): 373–89.

Bernard, E. L., Ruta, M., Tarver, J. E. & Benton, M. J., 2010. The fossil record of early tetrapods: worker effort and the end-Permian mass extinction. *Acta Palaeontologica Polonica*, **55**(2): 229–39.

Berner, R. A., 2001. Modeling atmospheric O_2 over Phanerozoic time. *Geochimica et Cosmochimica Acta*, **65**: 685–94.

2006. GEOCARBSULF: a combined model for Phanerozoic atmospheric O_2 and CO_2. *Geochimica et Cosmochimica Acta*, **70**: 5653–64.

Berner, R. A. & Canfield, D. E., 1989. A new model of atmospheric oxygen over Phanerozoic time. *American Journal of Science*, **289**: 333–61.

Berner, R. A. & Kothavala, Z., 2001. GEOCARB III: a revised model of atmospheric CO_2 over Phanerozoic time. *American Journal of Science*, **301**: 182–204.

Bernor, R. L., 2007. New apes fill the gap. *Proceedings of the National Academy of Sciences*, **104**(50): 19661–2.

Bertani, R. T. & Carozzi, A. V., 1984. Lagoa Feia formation (Lower Cretaceous), Campos basin, offshore Brazil: rift valley stage lacustrine carbonate reservoirs. I. *Journal of Petroleum Geology*, **8**(1): 37–58.

1985. Lagoa Feia formation (Lower Cretaceous), Campos basin, offshore Brazil: rift valley stage lacustrine carbonate reservoirs. II. *Journal of Petroleum Geology*, **8**(2): 199–220.

Berthelin, M., Stolle, E., Kerp, H. & Broutin, J., 2006. *Glossopteris anatolica* Archangelsky and Wagner 1983 in a mixed middle Permian flora from the Sultanate of Oman: comments on the geographical and stratigraphical distribution. *Review of Palaeobotany and Palynology*, **141**: 313–7.

Bertini, A., 2006. The northern Apennines palynological record as a contribute for the reconstruction of the Messinian palaeoenvironments. *Sedimentary Geology*, **188–189**: 235–58.

Bertrand-Sarfati, J. & Monty, C. (eds.), 1994. *Phanerozoic stromatolites II.* Kluwer.

Besairie, H., 1971. Geologie de Madagascar. *Annales Geologiques de Madagascar*, **35**.

Bethoux, O., Papier, F. & Nel, A., 2005. The Triassic radiation of the entomofauna. *Comptes Rendus Palevol*, **4**: 541–53.

Bhalla, S. N., 1969. Foraminifera from the type Raghavapuram shales, east coast Gondwanas, India. *Micropalaeontology*, **15**(1): 61–84.

Bhatnagar, A. K., Bisht, R. S., Goswami, B. G. & Singh, R. R., 2007. Geochemical characterization, oil–oil and oil–source correlation studies in eastern margin fields of Ahmedabad block of Cambay basin, India. *Association of Petroleum Geologists Bulletin*, **1**: 161–77.

Bicchi, E., Barras, C., Denoyelle, M. *et al.*, 2010. The impact of offshore drilling activities monitored by recent and sub-fossil assemblages of benthic foraminifera. *Abstracts, Forams 2010 International Symposium on Foraminifera*, Bonn: **57**. Rheinische Friedrich-Wilhelms-Universitat.

Bison, K.-M., Versteegh, G. J. M., Orszag-Sperber, F. *et al.*, 2009. Palaeoenvironmental changes of the early Pliocene (Zanclean) in the eastern Mediterranean Pissouri Basin (Cyprus) evidenced from calcareous dinoflagellate cyst assemblages. *Marine Micropaleontology*, **73**: 49–56.

Biswas, S. K., 1999. A review of the evolution of the rift basins of India during Gondwana with special reference to western Indian basins and their hydrocarbon prospects. *PINSA*, **65A**(3): 261–83.

2003. Regional tectonic framework of the Pranhita-Godavari basin, India. *Journal of Asian Earth Sciences*, **21**: 543–51.

Bjerager, M. & Surlyk, F., 2007 Benthic palaeoecology of Danian deep-shelf bryozoan mounds in the Danish basin. *Palaeogeography, Palaeoclimatology, Palaeoecology*, **250**: 184–215.

Bjorlykke, K., 2010. *Petroleum Geosciences: From Sedimentary Environments to Rock Physics.* Springer.

Blain, H.-A., Bailon, S. & Cuenca-Bascos, G., 2008. The Early-Middle Pleistocene palaeoenvironmental change based on the squamate reptile and amphibian proxies at the Gran Dolina site, Atapuerca, Spain. *Palaeogeography, Palaeoclimatology, Palaeoecology*, **261**: 177–92.

Blow, W. H., 1959. Age, correlation and biostratigraphy of the upper Tocuyo (San Lorenzo) and Pozon formations, eastern Falcon, Venezuela. *Bulletins of American Paleontology*, **39**(178): 67–251.

Bobe, R., Alemgesed, Z. & Behrensmeyer, A. K. (eds.), 2007. *Hominin Environments in the East African Pliocene.* Vertebrate Paleobiology and Paleoanthropology Series. Springer.

Boisvert, C. A., 2005. The pelvic fin and girdle of *Panderichthys* and the origin of tetrapod locomotion. *Nature*, **438**: 1145–7.

Bolli, H. M., Beckmann, J.-P. & Saunders, J. B., 1994. *Benthic Foraminiferal Biostratigraphy of the South Caribbean Region.* Cambridge University Press.

Bolli, H. M., Saunders, J. B. & Perch-Nielsen, K. (eds.), 1985. *Plankton Stratigraphy.* Cambridge University Press.

Bond, D. P. G., Hilton, J., Wignall, P. B. *et al.*, 2010. The Middle Permian (Captianian) mass extinction on land and in the oceans. *Earth-Science Reviews*, **102**: 100–16.

Bond, D. P. G. & Wignall, P. B., 2008. The role of sea-level change and marine anoxia in the Frasnian–Famennian (Late Devonian) mass extinction. *Palaeogeography, Palaeoclimatology, Palaeoecology*, **263**: 107–18.

2009. Latitudinal selectivity of foraminifer extinctions during the late Guadalupian crisis. *Paleobiology*, **35**(4): 465–83.

Bonnischen, R., Lepper, B. T., Stanford, D. & Waters, M. R. (eds.), 2006. *Paleoamerican Origins: Beyond Clovis.* Texas A & M University Press.

Bordy, E. M. & Prevec, R., 2008. Sedimentology, palaeoentology and palaeo-environments of the Middle (?) to Upper Permian Emakwezini formation (Karoo supergroup, South Africa). *South African Journal of Geology*, **111**: 429–58.

Bornemann, A., Norris, R. A., Friedrich, O. *et al.*, 2008. Isotopic evidence for glaciation during the Cretaceous super-greenhouse. *Science*, **319**: 189–92.

Borowski, C. & Thiel, H., 1998. Deep-sea macrofaunal impacts of a large-scale physical disturbance experiment in the southeast Pacific. *Deep-Sea Research II*, **45**: 55–81.

Bortenschlager, S. & Oeggl, K. (eds.), 2000. *The Ice Man and his Natural Environment: Palaeobotanical Results.* Springer.

Bosellini, A., 2002. Dinosaurs re-write the geodynamics of the eastern Mediterranean and the paleogeography of the Apulia platform. *Earth-Science Reviews*, **59**: 211–34.

Bosellini, A., Russo, A., Aruch, M. A. & Cabdulqadir, M. M., 1987. The Oligo-Miocene of Eil (NE Somalia): a prograding coral–*Lepidocyclina* system. *Journal of African Earth Sciences*, **6**(4): 583–93.

Bosellini, F. R. & Perrin, C., 2008. Estimating Mediterranean Oligocene–Miocene sea-surface temperatures: an approach based on coral taxonomic richness. *Palaeogeography, Palaeoclimatology, Palaeoecology*, **258**: 71–88.

Bosence, D. W. J. & Allison, P. A., 1995. *Marine Palaeoenvironmental Analysis from Fossils.* The Geological Society, London (Special Publication No. 83).

Botha, J. & Smith, R. M. H., 2006. Rapid vertebrate recuperation in the Karoo basin of South Africa following the end-Permian extinction. *Journal of African Earth Sciences*, **45**: 502–14.

Botkin, D. B. & Keller, E. A., 2010. *Environmental Science: Earth as a Living Planet*, 7th edition. John Wiley & Sons.

Botquelen, A., Gourvennec, R., Loi, A. *et al.*, 2006. Replacements of benthic associations in a sequence stratigraphic framework: examples from Upper Ordovician of Sardinia and Lower Devonian of the Massif Armoicain. *Palaeogeography, Palaeoclimatology, Palaeoecology*, **239**: 286–310.

Bott, M.H.P., Saxov, S., Talwani, M. & Thiede, J. (eds.), 1983. *Structure and development of the Greenland–Scotland ridge: new methods and concepts*. NATO IV Marine Sciences, 8. Plenum Press.

Bouchet, V.M.P., Alve, E., Rygg, B. *et al.*, 2010. Determining ecological quality status of coastal waters using benthic foraminifera: calibration with benthic macrofauna. *Abstracts, Forams 2010 International Symposium on Foraminifera, Bonn*: **61**. Rheinische Friedrich-Wilhelms-Universitat.

Boucot, A.J. & Xu Chen, 2009. Fossil plankton depth zones. *Palaeoworld*, **18**: 213–34.

BouDagher-Fadel, M., 2008. *Evolution and Geological Significance of Larger Benthic Foraminifera*. Elsevier.

BouDagher-Fadel, M.K. & Price, G.D., 2010. Evolution and palaeogeographic distribution of the lepidocyclinids. *Journal of Foraminiferal Research*, **40**(1): 79–108.

Boudreau, R.E.A., Patterson, R.T., Dalby, A.P. & McKillop, W.B., 2001. Non-marine occurrence of the foraminifer *Cribroelphidium gunteri* in northern Lake Winnipegosis, Manitoba, Canada. *Journal of Foraminiferal Research*, **31**(2): 108–19.

Bourahrouh, A., Paris, F. & Elaouad-Debbaj, Z., 2004. Biostratigraphy, biodiversity and palaeoenvironments of the Chitinozoans and associated palynomorphs from the Upper Ordovician of the central Anti-Atlas, Morocco. *Review of Palaeobotany and Palynology*, **130**: 17–40.

Boureau, E., Lejal-Nicol, A. & Massa, D., 1978. A propos du Silurien et du Devonien en Libye. *Comptes Rendus des Séances de l'Académie des Sciences, Paris, Serie D*, **286**: 1567–71.

Bowden, A.J., Burek, C.V. & Wilding, R. (eds.), 2005. *History of Palaeobotany: Selected Essays*. London: The Geological Society (Special Publication No. 241).

Boyle, P. & Rodhouse, P., 2005. *Cephalopods*. Blackwell.

Braccini, E., Denison, C.N., Scheevel, J.R. *et al.*, 1997. A revised chrono-lithostratigraphic framework for the pre-salt (Lower Cretaceous) in Cabinda, Angola. *Bulletin du Centre de Recherches Exploration-Production Elf-Aquitaine*, **21**(1): 125–51.

Brack, P.H., Rieber, A., Nicora, A. *et al.*, 2005. The Global Boundary Stratotype Section and Point (GSSP) of the Ladinian stage (Middle Triassic) at Bagolino (southern lps, northern Italy) and its implications for the Triassic timescale. *Episodes*, **28**: 233–44.

Bradley, R., 2007. *The Prehistory of Britain and Ireland*. Cambridge University Press.

Branch, N., Clark, P., Green, C. & Turney, C., 2005. *Environmental Archaeology*. Hodder Arnold.

Brand, U., Legrand-Blain, M. & Streel, M., 2004. Biogeochemostratigraphy of the Devonian-Carboniferous boundary global stratotype section and point, Griotte formation, La Serre, Montagne Noire, France. *Palaeogeography, Palaeoclimatology, Palaeoecology*, **205**: 337–57.

Brandao, S.N., 2008. First record of a living Platycopida (Crustacea, Ostracoda) from Antarctic waters and a discussion of *Cytherella serratula* (Brady, 1880). *Zootaxa*, **1866**: 349–72.

Brandao, S.N. & Horne, D.J., 2009. The platycopid signal of oxygen depletion in the ocean: a critical evaluation of the evidence from modern ostracod biology, ecology and depth disribution. *Palaeogeography, Palaeoclimatology, Palaeoecology*, **286**: 126–33.

Brasier, M.D., 1975. An outline history of seagrass communities. *Palaeontology*, **18**(4): 681–702.

1980. *Microfossils*. George Allen & Unwin.

2009. *Darwin's Lost World: The Hidden History of Animal Life*. Oxford University Press.

Brasier, M.D., Green, O.R., Lindsay, J.F. *et al.*, 2005. Critical testing of Earth's oldest putative fossil assemblage from the ~3.5 Ga Apex chert, Chinaman Creek, Western Australia. *Precambrian Research*, **140**: 55–102.

Brayard, A., Heran, M.-A., Costeur, L. & Escarguel, C., 2004. Triassic and Cenozoic palaeobiogeography: two case studies in quantitative modelling using IDL [Interactive Data Language]. *Palaeontologia Electronica*, **7**(2).

Brenchley, P.J. & Harper, D.A.T., 1998. *Palaeoecology*. Chapman & Hall.

Brenchley, P.J. & Rawson, P.F. (eds.), 2006. *The Geology of England and Wales*, 2nd edition. The Geological Society, London.

Brewer, D.J., 1987. Seasonality in the prehistoric Fayum based on the incremental growth structures of the Nile catfish (Pisces: Clarias). *Journal of Archaeological Science*, **14**: 459–72.

1989. *Fishermen, Hunters and Herders: Zooarchaeology in the Fayum (ca. 8000–5000 Bp)*. British Archaeological Reports (International Series No. 478).

Briggs, D.E.G. (ed.), 2005. *Evolving Form and Function: Fossils and Development*. Yale University Press.

Briggs, D.E.G. & Crowther, P.R., 1990. *Palaeobiology*. Blackwell. (eds.), 2001. *Palaeobiology II*. Blackwell.

Briggs, J.C., 2003. The biogeographic and tectonic history of India. *Journal of Biogeography*, **30**: 381–88.

Briguglio, A. & Hohenegger, J., 2009. Nummulitids hydrodynamics: an example using *Nummulites globosus* Leymerie, 1846. *Bollettino della Societa Paleontologica Italiana*, **48**(2): 105–11.

Brodacki, M., 2006. Functional anatomy and mode of life of the latest Jurassic crinoid *Saccocoma*. *Acta Palaeontologica Polonica*, **51**(2): 261–70.

Bromage, T.G. & Schrenk, F., 1999. *African Biogeography, Climate Change and Human Evolution*. Oxford University Press.

Brooks, J. (ed.), 1990. *Classic Petroleum Provinces*. The Geological Society, London (Special Publication No. 50).

Brooks, J. & Glennie, K. (eds.), 1987. *Petroleum Geology of Northwest Europe*. Graham & Trotman.

Brooks, S.J., 2006. Fossil midges (Diptera: Chironomidae) as palaeoclimatic indicators for the Eurasian region. *Quaternary Science Reviews*, **25**: 1894–1919.

Brooks, S.J. & Birks, H.J.B., 2001. Chironomid-inferred air temperatures from Lateglacial and Holocene sites in north-west Europe: progress and problems. *Quaternary Science Reviews*, **20**: 1723–41.

Brown, J.C. & Dey, A.K., 1975. *The Mineral Fields of the Indian Subcontinent and Burma*. Oxford University Press.

Brown, L.F., Jr., Benson, J.M., Brink, G.J. *et al.*, 1995. *Sequence Stratigraphy on Offshore South African Divergent Basins*. American Association of Petroleum Geologists (Studies in Geology No. 41).

Browne, M.A.E., Dean, M.T., Hall, I.H.S. *et al.*, 1999. A lithostratigraphical framework for the Carboniferous rocks of the Midland Valley of Scotland. *British Geological Survey Research Report*, RR/99/07.

Bruch, A.A. & Zhilin, S.G., 2007. Early Miocene climate of central Eurasia – evidence from Aquitanian floras of Kazakhstan. *Palaeogeography, Palaeoclimatology, Palaeoecology*, **248**: 32–48.

Brun, L., Cherchi, M.A., Meijer, M. & Monteil, L., 1984. Repartition stratigraphique et paleoecologique des principales especes du Bolivinitidae (Foraminiferes) du Tertiare du Golfe de Guinee. *Bulletin du Centre de Recherches Exploration-Production Elf-Aquitaine, Mémoire*, **6**: 91–104.

Brunet, M., 2009. Origine et evolution des hominids: Toumai, une confirmation eclatante de la prediction de Darwin. *Comptes Rendus Palevol*, **8**: 311–9.

2010. Short note: the track of a new cradle of mankind in Sahelo-Saharan Africa (Chad, Libya, Egypt, Cameroon). *Journal of African Earth Sciences*, **58**: 680–3.

Brunnschweiler, R.O., 1984. *Ancient Australia*. Angus & Robertson.

Buatois, L.A., Mangano, M.G., Brussa, E.D. *et al.*, 2009. The changing face of the deep: colonization of the Early Ordovician deep-sea floor, Puna, northwest Argentina. *Palaeogeography, Palaeoclimatology, Palaeoecology*, **280**: 291–9.

Bubik, M. & Kaminski, M.A. (eds.), 2004. *Proceedings of the Sixth International Workshop on Agglutinated Foraminifera*. The Grzybowski Foundation (Special Publication No. 8).

Bulman, O.M.B., 1964. Lower Palaeozoic plankton. *Quarterly Journal of the Geological Society, London*, **119**: 401–18.

Burdige, D.J., 2006. *Geochemistry of Marine Sediments*. Princeton University Press.

Burney, D.A. & Flannery, T.F., 2005. Fifty millennia of catastrophic extinctions after human contact. *Trends in Ecology and Evolution*, **20**(7): 395–401.

Burroughs, W.J., 2005. *Climate Change in Prehistory: The End of the Reign of Chaos*. Cambridge University Press.

Buttler, C.J., Cherns, L. & Massa, D., 2007. Bryozoan mudmounds from the Upper Ordovician Jifarah (Djeffara) formation of Tripolitania, north-west Libya. *Palaeontology*, **50**(2): 479–94.

Butts, S.H., 2005. Latest Chesterian (Carboniferous) initiation of Gondwanan glaciation recorded in facies stacking patterns and brachiopod paleocommunities of the Antler foreland basin, Idaho. *Palaeogeography, Palaeoclimatology, Palaeoecology*, **223**: 275–89.

Butzer, K.W., 1959. Environment and human ecology in Egypt during pre-dynastic and early dynastic times. *Bulletin de la Société Géographique de l'Egypte*, **32**: 43–88.

Cagatay, M.N., Gorur, N., Flecker, R. *et al.*, 2006. Paratethyan connectivity in the Sea of Marmara region (NW Turkey) during the Messinian. *Sedimentary Geology*, **188–189**: 171–8.

Cainelli, C. & Mohriak, W.U., 1999. Some remarks on the evolution of sedimentary basins along the eastern Brazilian continental margin. *Episodes*, **22**(3): 206–16.

Cairncross, B., 2001. An overview of the Permian (Karoo) coal deposits of South Africa. *Journal of African Earth Sciences*, **33**: 529–62.

Calder, J., Gibling, M. & Mukopadhyay, P., 1991. Peat formation in a Westphalian B piedmont setting, Cumberland basin, Nova Scotia: implications for the maceral-based interpretation of rheotrophic and raised paleomires. *Bulletin de la Société Géologique de la France*, **162**(2): 283–98.

Callaway, J.M. & Nicholls, E.I. (eds.), 1997. *Ancient Marine Reptiles*. Academic Press.

Calver, M.A., 1956. Die stratigraphische Verbreitung der nicht-marinen Muscheln in den penninschen Kolenfeldern Englands. *Zeitschrift der Deutschen Geologischen Gesellschaft*, **107**: 26–39.

1968. Distribution of Westphalian marine faunas in northern England and adjoining areas. *Proceedings of the Yorkshire Geological Society*, **37**: 1–72.

1969. Westphalian of Britain. *Compte Rendu 6e Congres Internationale Stratigraphie et Géologie Carbonifere, Sheffield, 1967*, **I**: 233–54.

Calvin, W.H., 2002. *A Brain for All Seasons*. University of Chicago Press.

Cameron, D.W. & Groves, C.P., 2004. *Bones, Stones and Molecules: 'Out of Africa' and Human Origins*. Elsevier.

Cameron, N.R., Bate, R.H. & Clure, V.S. (eds.), 1999. *The Oil and Gas Habitats of the South Atlantic*. The Geological Society, London (Special Publication No. 153).

Campbell, K.A., 2006. Hydrocarbon seeps and hydrothermal vent paleoenvironments and paleontology: past developments and future research directions. *Palaeogeography, Palaeoclimatology, Palaeoecology*, **232**: 362–407.

Campbell, K.A. & Bottjer, D.J., 1995. Brachiopods and chemosymbiotic bivalves in Phanerozoic hydrothermal vent and 'cold seep' environments. *Geology*, **23**: 321–4.

Campbell, K.E., Prothero, D.R., Romero-Pittman, L. *et al.*, 2010. Amazonian magnetostratigraphy: dating the first pulse of the Great American Faunal Interchange. *Journal of South American Earth Sciences*, **29**: 619–26.

Cane, M.A. & Molnar, P., 2001. Closing of the Indonesian seaway as a precursor to east African aridification around 3–4 million years ago. *Nature*, **411**: 157–62.

Cantalamessa, G., di Celma, C. & Ragaini, L., 2005. Sequence stratigraphy of the Punta Ballena member of the Jama formation (Early Pleistocene, Ecuador): insights from integrated sedimentologic, taphonomic and paleoecologic analysis of molluscan shell concentrations. *Palaeogeography, Palaeoclimatology, Palaeoecology*, **216**: 1–25.

Canudo, J.I., Barco, J.L., Pereda-Suberbiola, X. *et al.*, 2009. What Iberian dinosaurs reveal about the bridge said to exist between Gondwana and Laurasia in the Early Cretaceous. *Bulletin de la Société Géologique de la France*, **180**(1): 5–11.

Capetta, H., Pfeil, F. & Schmidt-Kittler, N., 2000. New biostratigraphical data on the Upper Cretaceous and Palaeogene of Jordan. *Newsletters on Stratigraphy*, **38**(1): 81–95.

Caplan, M.L. & Bustin, R.M., 1999. Devonian-Carboniferous Hangenburg mass extinction event, widespread organic-rich mudrock and anoxia: causes and consequences. *Palaeogeography, Palaeoclimatology, Palaeoecology*, **148**: 187–207.

Carmichael, S.M., Akhter, S., Bennett, J.K. *et al.*, 2009. Geology and hydrocarbon potential of the offshore Indus basin, Pakistan. *Petroleum Geoscience*, **15**: 107–16.

Carrillat, A., Caline, B. & Davaud, E., 1999. Prediction of poroperm properties from rock fabric in a nummulite carbonate reservoir (El Garia fm., offshore Libya). *Journal of Conference Abstracts*, **4**(2): 907.

Carrillo, M., Paredes, I., Crux, J.A. & de Cabrera, S., 1995. Aptian to Maastrichtian paleobathymetric reconstruction of the eastern Venezuelan basin. *Marine Micropaleontology*, **26**: 405–18.

Carroll, R., 2009. *The Rise of Amphibians*. The Johns Hopkins University Press

Carvalho, M. de A., 2004. Palynological assemblage from Aptian/Albian of the Sergipe basin: paleoenvironmental interpretation. *Revista Brasiliera de Paleontologia*, **7**(2): 159–68.

Catelin, M., 2009. Coal and the environment. *3rd Symposium on Gondwana Coals, Pontifical Catholic University of Rio Grande do Sul (PUCRS)*, Porto Alegre, Brazil.

Caton-Thompson, G. & Gardner, E.W., 1934. *The Desert Fayum*. Royal Anthropological Institute of Great Britain and Ireland.

Catuneanu, O., 2006. Sequence stratigraphy of clastic systems. *Geological Society of Canada Short Course Notes*, **16**: 1–248.

Catuneanu, O., Abreu, V., Bhattacharya, J. *et al.*, 2009. Towards the standardization of sequence stratigraphy. *Earth-Science Reviews*, **92**: 1–33.

Caudri, C.M.B., 1996. The larger foraminifera of Trinidad (West Indies). *Eclogae Geologicae Helvetiae*, **89**(3): 1137–1309.

Caughey, C.A., Carter, D.C. & Clure, J. (eds.), 1996. *Proceedings of the International Symposium on Sequence Stratigraphy in SE Asia, May 1995*. Indonesian Petroleum Association.

Cauvin, J., 1994. *Naissance des divinités. Naissance de l'agriculture. La revolution des symboles au Neolithique*. Editions du Centre National de la Recherche Scientifique.

Cavin, L., Longbottom, A. & Richter, M. (eds.), 2008. *Fishes and the Break-Up of Pangaea*. The Geological Society London (Special Publication No. 295).

Cecca, F., 2002. *Palaeobiogeography of Marine Fossil Invertebrates*. Taylor & Francis.

Cesari, S.N., 2007. Palynological biozones and radiometric data at the Carboniferous–Permian boundary in western Gondwana. *Gondwana Research*, **11**: 529–36.

Chaix, C. & Cahuzac, B., 2005. The faunas of scleractinian corals in the faluns of the eastern Atlantic Middle Miocene (Loire and Aquitaine basins): paleobiogeography and climatic evolution. *Annales de Paleontologie*, **91**: 33–72.

Chakraborty, C. & Ghosh, S.K., 2005. Pull-apart origin of the Satpura Gondwana basin, central India. *Journal of Earth System Science*, **114**(3): 259–73.

Challinor, A.B. & Hikuroa, D.C.H., 2007. New Middle and Upper Jurassic belemnite assemblages from west Antarctica (Latady group, Ellsworth Land): taxonomy and paleobiogeography. *Palaeontologia Electronia*, **10**(1).

Charman, D.J., Gehrels, W.R., Manning, C. & Sharma, C., 2010. Reconstruction of sea-level change using testate amoebae. *Quaternary Research*, **73**: 208–19.

Charnock, M.A. & Jones, R.W., 1997. North Sea lituolid foraminifera with complex inner structures: taxonomy, stratigraphy and evolutionary relationships. *Annales Societatis Geologorum Poloniae*, **67**: 183–96.

Chatterjee, S. & Templin, R.J., 2004. *Posture, Locomotion and Paleoecology of Pterosaurs*. Geological Society of America.

Cheng Quan, Chunlin Sun, Yuewu Sun & Ge Sun, 2009. High resolution estimates of paleo-CO_2 levels through the Campanian (Late Cretaceous) based on *Ginkgo* cuticles. *Cretaceous Research*, **30**: 424–8.

Cherns, L. & Wheeley, J.R., 2007. A pre-Hirnantian (Late Ordovician) interval of global cooling – the Boda event re-assessed. *Palaeogeography, Palaeoclimatology, Palaeoecology*, **251**: 449–60.

Chowdhary, L.R., 2004. *Petroleum Geology of the Cambay Basin, Gujurat, India*. Indian Petroleum Publishers.

Claassen, C., 1998. *Shells*. Cambridge University Press.

Clack, J., 2006. The emergence of early tetrapods. *Palaeogeography, Palaeoclimatology, Palaeoecology*, **232**: 167–89.

Clack, J.A., 2007. Devonian climate change, breathing, and the origin of the tetrapod stem group. *Integrative and Comparative Biology*: 1–14 (published online 13/8/2007, DOI: 10.1093/icb/icm055).

Clarkson, E.N.K., 1998. *Invertebrate Palaeontology and Evolution*, 4th edition. Blackwell.

Cleal, C.J., 1991. *Plant Fossils in Geological Investigation: The Palaeozoic*. Ellis Horwood.

Cleal, C.J. & Thomas, B.A., 1994. *Plant Fossils of the British Coal Measures*. The Palaeontological Association (Field Guide to Fossils No. 6).

2005. Palaeozoic tropical rainforests and their effect on global climates: is the past the key to the present? *Geobiology*, **3**: 13–31.

2009. *Introduction to Plant Fossils*. Cambridge University Press.

Clement, G. & Letenneur, C., 2009. L'émergence des tétrapodes – une revue des récentes découvertes et hypothèses. *Comptes Rendus Palevol*, **8**: 221–32.

Clift, P. D., Kroon, D., Gaedicke, C. & Craig, J. (eds.), 2002. *The Tectonic and Climatic Evolution of the Arabian Sea Region*. The Geological Society, London (Special Publication No. 195).

Clift, P. D., Tada, R. & Zheng, H. (eds.), 2010. *Monsoon Evolution and Tectonics–Climate Linkage in Asia*. The Geological Society, London (Special Publication No. 340).

Cloudsley-Thompson, J. L., 2005. *Ecology and Behaviour of Mesozoic Reptiles*. Springer.

Coates, M. I., Ruta, M. & Friedman, M., 2008. Ever since Owen: changing perspectives on the early evolution of tetrapods. *Annual Review of Ecology and Evolutionary Systematics*, **39**: 571–92 (published online 10/10/2008, DOI: 10.1146/annurev.ecolsys.38.091206.095546).

Cobos, S., 2005. Structural interpretation of the Monagas foreland thrust belt. *Search and Discovery* Article No. 30031. http://www.searchanddiscovery.com

Coccioni, R., Luciani, V. & Marsili, A., 2006. Cretaceous anoxic events and radially elongated chambered planktonic foraminifera: paleoecological and paleoceanographic implications. *Palaeogeography, Palaeoclimatology, Palaeoecology*, **235**: 66–92.

Coccioni, R. & Marsili, A. (eds.), 2007. *Proceedings of the Giornate di Paleontologia 2005*. Grzybowski Foundation (Special Publication No. 12).

Cockell, C. S. (ed.), 2007. *An Introduction to the Earth-Life System*. The Open University/Cambridge University Press.

Cockell, C. S. & Bland, P. A., 2005. The evolutionary and ecological benefits of asteroid and comet impacts. *Trends in Ecology and Evolution*, **20**(4): 176–9.

Cocks, L. R. M., Fortey, R. A. & Rushton, A. W. A., 2010. Correlation for the Lower Palaeozoic. *Geological Magazine*, **147**(2): 171–80.

Cocks, L. R. M. & Torsvik, T. H., 2005. Baltica from the late Precambrian to mid-Palaeozoic times: the gain and loss of a terrane's identity. *Earth-Science Reviews*, **72**: 39–66.

Coe, A. L. (ed.), 2003. *The Sedimentary Record of Sea-Level Change*. The Open University/Cambridge University Press.

2010. *Geological Field Techniques*. Wiley-Blackwell/The Open University.

Coggin, H. C., Eschmeyer, E. N., Paxton, J. R., Zweifel, R. G. & Kirshner, D., 2003. *Encyclopaedia of Animals: Reptiles, Amphibians and Fishes*, 3rd edition. Fog City Press.

Cohen, A. S., 2003. *Paleolimnology: The History and Evolution of Lake Systems*. Oxford University Press.

Coleman, R. G., 1993. *Geologic Evolution of the Red Sea*. Oxford University Press.

Coles, B. J., 1998. Doggerland: a speculative survey. *Journal of the Prehistoric Society*, **64**: 45–81.

Coles, G. P., Ainsworth, N. R., Whatley, R. C. & Jones, R. W., 1996. Foraminifera and Ostracoda from Quaternary carbonate mounds associated with gas seepage in the Porcupine Basin, offshore western Ireland. *Revista Espanola de Micropaleontologia*, **XXVII**(2): 113–51.

Colin, J.-P., Brunet, M., Congleton, J. D. *et al.* 1992. Ostracodes lacustres des bassins d'age Cretace Inferieur du nord Cameroun: Hama-Koussou, Koum et Babouri-Figuil. *Revue de Paléobiologie*, **11**(2): 357–72.

Collinson, J. D., Evans, D. J., Holliday, D. W. & Jones, N. S. (eds.), 2005. *Carboniferous Hydrocarbon Geology: The Southern North Sea and Surrounding Onshore Areas*. Yorkshire Geological Society (Occasional Publications Series No. 7).

Compton, P. M., 2009. The geology of the Barmer basin, Rajasthan, India, and the origins of its major oil reservoir, the Fatehgarh formation. *Petroleum Geoscience*, **15**: 117–30.

Conneller, C. & Warren, G. (eds.), 2006. *Mesolithic Britain and Ireland*. Tempus.

Conway Morris, S., 2003. *Life's Solution: Inevitable Humans in a Lonely Universe*. Cambridge University Press.

Coope, G. R., 2006. Insect faunas associated with Palaeolithic industries from five sites of pre-Anglian age in central England. *Quaternary Science Reviews*, **25**: 1738–54.

Cooper, M. A., Addison, F. T., Alvarez, R. *et al.*, 1995. Basin development and tectonic history of the Llanos basin, eastern Cordillera and Middle Magdalena valley, Colombia. *American Association of Petroleum Geologists Bulletin*, **79**(10): 1421–43.

Cooper, R. A., Crampton, J. S., Raine, I. *et al.*, 2001. Quantitative biostratigraphy of the Taranaki basin, New Zealand: a deterministic and probabilistic approach. *American Association of Petroleum Geologists Bulletin*, **85**(8): 1469–98.

Cope, J. C. W., Ingham, J. K. & Rawson, P. F. (eds.), 1992. *Atlas of Palaeogeography and Lithofacies*. The Geological Society, London (Memoir No. 13).

Correge, T., 2006. Sea surface temperature and salinity reconstruction from coral geochemical tracers. *Palaeogeography, Palaeoclimatology, Palaeoecology*, **232**: 408–28.

Corsetti, F. A., Olcott, A. N. & Bakermans, C., 2006. The biotic response to Neoproterozoic snowball Earth. *Palaeogeography, Palaeoclimatology, Palaeoecology*, **232**: 114–30.

Cortese, G., Dolven, J. K., Bjorklund, K. R. & Malmgren, B. A., 2005. Late Pleistocene-Holocene radiolarian paleotemperatures in the Norwegian Sea based on artificial neural networks. *Palaeogeography, Palaeoclimatology, Palaeoecology*, **224**: 311–32.

Costeur, L. & Legendre, S., 2008. Spatial and temporal variation in European large mammals diversity. *Palaeogeography, Palaeoclimatology, Palaeoecology*, **261**: 127–44.

Cotton, J. & Field, D. (eds.), 2004. *Towards a New Stone Age*. Council for British Archaeology (Research Report No. 137).

Courtillot, V. (English translation), 1999. *Evolutionary Catastrophes*. Cambridge University Press.

Court-Picon, M., Buttler, A. & de Beaulieu, J.-L., 2005. Modern pollen-vegetation relationships in the Champsaur valley (French Alps) and their potential in the interpretation of fossil pollen records of past cultural landscapes. *Review of Palaeobotany and Palynology*, **135**: 13–39.

Craig, G.Y., 1991. *The Geology of Scotland*, 3rd edition. The Geological Society, London.

Craig, J., Thurow, J., Thusu, B. *et al.* (eds.), 2009. *Global Neoproterozoic Petroleum Systems: The Emerging Potential in North Africa*. The Geological Society, London (Special Publication No. 326).

Crame, J.A. & Owen, A.W. (eds.), 2002. *Palaeobiogeography and Biodiversity Change: The Ordovician and Mesozoic–Cenozoic Radiations*. The Geological Society. London (Special Publication No. 194).

Crawford, R.L. & Crawford, D.L. (eds.), 1996. *Bioremediation – Principles and Applications*. Cambridge University Press (Biotechnology Research Series No. 6).

Creaney, S. & Passey, Q.R., 1993. Recurring patterns of Total Organic Carbon and source rock quality within a sequence stratigraphic framework. *American Association of Petroleum Geologists Bulletin*, **77**(3): 386–401.

Crespo de Cabrera, S. & di Gianni Canudas, N., 1994. Biostratigraphy and paleogeography of the eastern Venezuelan basin during the Oligo/Miocene. *V Simposio Bolivariano (Exploracion Petrolera en las Cuencas Subandinas)*: 231–3.

Crespo de Cabrera, S., Sliter, W.V. & Jarvis, I., 1999. Integrated foraminiferal biostratigraphy and chemostratigraphy of the Querecual formation (Cretaceous), eastern Venezuela. *Journal of Foraminiferal Research*, **29**(4): 484–99.

Cripps, J.A., Widdowson, M., Spicer, R.A. & Jolley, D.W., 2005. Coastal ecosystem responses to late stage Deccan Trap volcanism: the post K–T boundary (Danian) palynofacies of Mumbai (Bombay), west India. *Palaeogeography, Palaeoclimatology, Palaeoecology*, **216**: 305–32.

Cronin, T.M., 2010. *Paleoclimates: Understanding Climate Change Past and Present*. Columbia University Press.

Cronin, T.M., Dowsett, H.J., Dwyer, G.S., Baker, P.A. & Chandler, M.A., 2005. Mid-Pliocene deep-sea bottom-water temperatures based on ostracode Mg/Ca ratios. *Marine Micropaleontology*, **54**: 249–61.

Crosdale, P.J., Beamish, B.B. & Valix, M., 1998. Coalbed methane sorption related to coal composition. *International Journal of Coal Geology*, **35**: 147–58.

Cross, N., Goodall, I., Hollis, C. *et al.*, 2010. Reservoir description of a mid-Cretaceous siliciclastic-carbonate ramp reservoir: Mauddud formation in the Raudhatain and Sabiriyah fields, north Kuwait. *GeoArabia*, **15**(2): 17–50.

Culver, S.J. & Buzas, M.A., 1982. Disribution of Recent benthic foraminiera in the Caribbean region. *Smithsonian Contributions to the Marine Sciences*, **14**: 1–382.

Cunha, M.C.C. & Moura, J.A., 1979. Especies novas de ostracodes nao-marinhos de serie do Reconcavo: paleontologia e bioestragrafia. *Boletim Tecnico de Petrobras*, **22**(2): 87–100.

Curiale, J.A., Covington, G.H., Shamsuddin, A.H.M. *et al.*, 2002. Origin of petroleum in Bangladesh. *American Association of Petroleum Geologists Bulletin*, **86**(4): 625–52.

Currant, A., 1989. The Quaternary origins of the modern British mammal fauna. *Biological Journal of the Linnean Society*, **38**: 23–30.

Dadson, S., 2008. Sediment delivery to ocean basins from steep mountain islands. *Abstracts, Geological Society of London/Society of Economic Paleontologists and Mineralogists Palaeogeography Conference*, Cambridge: 17.

Daizhao Chen & Tucker, M.E., 2004. Palaeokarst and its implication for the extinction event at the Frasnian–Famennian boundary (Guilin, south China). *Journal of the Geological Society, London*, **161**: 895–8.

Damste, J.S.S., Muyzer, G., Abbas, B. *et al.*, 2004. The rise of the rhizosolenid diatoms. *Science*, **304**: 584–7.

Danielian, T. & Moreira, D., 2004. Palaeontological and molecular arguments for the origin of silica-secreting marine organisms. *Comptes Rendus Palevol*, **3**: 229–36.

d'Argenio, B., Fischer, A.G., Premoli Silva, I *et al.*, 2004. *Cyclostratigraphy: Approaches and Case Histories*. Society of Economic Paleontologists and Mineralogists (Special Publication No. 81).

Darwin, C., 1859. *On the Origin of Species by Means of Natural Selection*. John Murray.

Das, A.K., Gupta, R.P., Hussam, R. & Maurya, S.N., 1999. Cenozoic sequence stratigraphy of Krishna mouth area and beyond in eastern offshore, India, with special reference to depositional setting and hydrocarbon habitat in Miocene sediments. *Geohorizons*, **4**(2): 1–10.

Dave, A., 2000. Biostratigraphy of the Neogene sequences in Andaman-Nicobar, Krishna–Godavari and western offshore basins of India. *Geological Survey of India Miscellaneous Publication*, **64**: 65–9.

Davies, E.H. & Bujak, J.P., 1987. Petroleum exploration applications of palynological assemblage sucessions in the Flexure Trend, Gulf of Mexico. *Proceedings of the Eighth Annual Research Conference, Gulf Coast Section, Society of Economic Paleontologists and Mineralogists*, Houston, Texas ('Innovative Biostratigraphic Approaches to Sequence Analysis – New Exploration Opportunities'): 47–51.

Davies, J., Fabis, M., Mainland, I. *et al.* (eds.), 2005. *Diet and Health in Past Animal Populations*. Oxbow Books.

Davies, N.S. & Gibling, M.R., 2010. Cambrian to Devonian evolution of alluvial systems: the sedimentological impact of the earliest land plants. *Earth-Science Reviews*, **98**: 171–200.

Davies, N.S., Sansom, I.J., Albanesi, G.L. & Cespedes, R., 2007. Ichnology, palaeoecology and taphonomy of a Gondwanan early vertebrate habitat: Insights from the Ordovician Anzaldo formation, Bolivia. *Palaeogeography, Palaeoclimatology, Palaeoecology*, **249**: 18–35.

Davies, P., 2008. *Snails, Archaeology and Landscape Change*. Oxbow Books.

Davison, I. & Jones, R.W., 2005. Ages and tectonics of the South Atlantic salt basins. *Abstracts, AAPG International Meeting*, Paris, 2005.

Dean, D.R., 1993. Gideon Mantell and the discovery of *Iguanodon*. *Modern Geology*, **18**: 209–19.

1999. *Gideon Mantell and the Discovery of Dinosaurs*. Cambridge University Press.

Debenay, J.-P. & Basov, I., 1993. Distribution of Recent benthic foraminifera on the west African shelf and slope: a synthesis. *Revue de Paléobiologie*, **12**(1): 265–300.

Debenay, J.-P. & Bui Thi Luan, 2006. Foraminiferal assemblages and the confinement indes as tools for assessment of saline intrusion and human impact in the Mekong Delta and neighbouring areas (Vietnam). *Revue de Micropaléontologie*, **49**(2): 74–85.

Debrenne, F., 2007. Lower Cambrian archaeocyathan bioconstructions. *Comptes Rendus Palevol*, **6**: 5–19.

de Castro, J.C., 2006. Evolucao dos conhecimentos sobre as coquinas-reservatorio da formacao Lagoa Feia no trend Badejo-Linguado-Pampo, bacia do Campos. *Geociencias, UNESP, Sao Paulo*, **25**(2): 175–86.

de Franceschi, D., Hoorn, C., Antoine, P.-O. *et al.*, 2008. Floral data from the mid-Cenozoic of central Pakistan. *Review of Palaeobotany and Palynology*, **150**: 115–29.

de Graciansky, P.-C., Hardenbol, J., Jacquin, T. & Vail, P.R. (eds.), 1998. *Mesozoic and Cenozoic Sequence Stratigraphy of European Basins*. Society of Economic Paleontologists and Mineralogists (Special Publication No. 60).

de Gregorio, B.T., Sharp, T.G., Flynn, G.J. *et al.*, 2009. Biogenic origin for Earth's oldest putative microfossils. *Geology*, **37**(7): 631–4.

de Man, E. & van Simaeys, S., 2004. Late Oligocene warming event in the southern North Sea Basin: benthic foraminifera as paleotemperature proxies. *Netherlands Journal of Geosciences/Geologie en Mijnbouw*, **83**(3): 227–39.

Demchuk, T.D. & Gary, A.C. (eds.), 2009. *Geologic Problem Solving with Microfossils*. Society of Economic Paleontologists and Mineralogists (Special Publication No. 93).

Demeter, F., Manni, F. & Coppens, Y., 2003. Late Upper Pleistocene human peopling of the Far East: multivariate analysis and geographic patterns of variation. *Comptes Rendus Palevol*, **2**: 625–38.

Dennell, R., 2009. *The Palaeolithic Settlement of Asia*. Cambridge University Press.

Denys, L., 1992. *A Check List of the Diatoms in the Holocene Deposits of the Belgian Coastal Plain with a Survey of their Apparent Ecological Requirements, 1: Introduction, Ecological Code and Complete List*. Service Geologique Belgique.

de Queiroz, A., 2005. The resurrection of oceanic dispersal in historical biogeography. *Trends in Ecology and Evolution*, **20**(2): 68–73.

de Romero, L.M. & Galea-Alvarez, F.A., 1995. Campanian *Bolivinoides* and microfacies from the La Luna formation, western Venezuela. *Marine Micropaleontology*, **26**: 383–404.

de Romero, L.M., Truskowski, I.E., Bralower, T.J. *et al.*, 2003. An integrated calcareous microfossil biostratigraphic and carbon-isotope stratigraphic framework for the La Luna formation, western Venezuela. *Palaios*, **18**: 349–66.

Dhannawat, B.S. & Mukherjee, M.K., 1997. Source rock studies in Jaisalmer basin, India. *Indian Journal of Petroleum Geology*, **6**(1): 25–42.

Dias, B.B., Hart, M.B., Smart, C.W. & Hall-Spencer, J.M., 2010. Modern seawater acidification: the response of foraminifers to high-CO_2 conditions in the Mediterranean Sea. *Journal of the Geological Society, London*, **167**: 1–4.

Dias-Brito, D. & Ferre, B., 2001. Roveacrinoids (stemless crinoids) in the Albian carbonates of the offshore Santos basin, southeastern Brazil: stratigraphic, palaeobiogeographic and palaeoceanographic significance. *Journal of South American Earth Sciences*, **14**: 203–18.

Diessel, C.F.K., 1986. On the correlation between coal facies and depositional environments. *20th Newcastle Symposium, 'Advances in the Study of the Sydney Basin'*, Department of Geology, University of Newcastle, Australia: 19–22.

1992. *Coal-Bearing Depositional Systems*. Springer Verlag.

2010. The stratigraphic distribution of inertinite. *International Journal of Coal Geology*, **81**: 251–68.

Dilcher, D.L., Bernardes de Oliveira, M.E., Pons, D. & Lott, T.A., 2005. Welwitschaceae from the Lower Cretaceous of northeastern Brazil. *American Journal of Botany*, **92**: 1294–310.

DiMichele, W.A., Rfefferkorn, H.W. & Gastaldo, R.A., 2001. Response of Late Carboniferous and Early Permian plant communities to climate change. *Annual Reviews of Earth and Planetary Science*, **29**: 461–87.

Dincauze, D.F., 2000. *Environmental Archaeology: Principles and Practice*. Cambridge University Press.

Dirks, P.H.G.M., Kibil, J.M., Kuhn, B.F. *et al.*, 2010. Geological setting and age of *Australopithecus sediba* from southern Africa. *Science*, **328**(5975): 205–8.

Dixon, J., Dietrich, J., McNeil, D.H. *et al.*, 1985. *Geology, Biostratigraphy and Organic Geochemistry of Jurassic to Pleistocene Strata, Beaufort–Mackenzie Area, Northwest Canada*. Canadian Society for Petroleum Geology.

Do Carmo, D.A., Sanguinetti, Y.T., Coimbra, J.C. & Guimaraes, E.M. 1999. Paleoecologia dos ostracodes nao-marinhos do Cretaceo Inferior da bacia Potiguar, RN, Brasil. *Simposio sobre o Cretaceo do Brasil*, Boletim: 383–391.

Do Carmo, D.A., Whatley, R, Queiroz Neto, J.V. & Coimbra, J.C., 2008. On the validity of two Lower Cretaceous non-marine ostracode genera: biostratigraphic and paleogeographic implications. *Journal of Paleontology*, **82**(4): 790–9.

Dominguez-Rodrigo, M., Barba, R. & Egeland, C.P. (eds.), 2007. *Deconstructing Olduvai: A Taphonomic Study of the Bed I Sites*. Springer (Vertebrate Paleobiology and Paleoanthropology Series).

Dominici, S., Cioppi, E., Danise, S. *et al.*, 2009. Mediterranean fossil whale falls and the adaptation of mollusks to extreme habitats. *Geology*, **37**(9): 815–8.

Donovan, S.K. & Jackson, T.A. (eds.), 1994. *Caribbean Geology: An Introduction*. University of the West Indies Publishers' Association.

Dore, A.G. & Sinding-Larsen, R. (eds.), 1996. *Quantification and Prediction of Petroleum Resources*. Norwegian Petroleum Society (Special Publication No. 6).

Dore, A. G. & Vining, B. A. 2005. *Petroleum Geology: North-West Europe and Global Perspectives*. The Geological Society, London (Proceedings of the Sixth Petroleum Geology Conference, 2003).

Dornbos, S. Q., 2006. Evolutionary palaeoecology of early epifaunal echinoderms: response to increasing bioturbation levels during the Cambrian radiation. *Palaeogeography, Palaeoclimatology, Palaeoecology*, 237(2–4): 225–39.

Dorschel, B., Hebbeln, D., Foubert, A. *et al.*, 2007. Hydrodynamics and cold-water coral facies distribution related to recent sedimentary processes at Galway Mound west of Ireland. *Marine Geology*, 244: 184–95.

Douglas, J. G. & Lejal-Nicol, A., 1981. Sure les premieres flores vasculaires terrestres datees du Silurien. *Comptes Rendus des Seances de l'Academie des Sciences, Paris, Serie II*, 292: 685–8.

Dov Por, F. & Dimentman, C., 2006. *Mare nostrum: Neogene and Anthropic Natural History of the Mediterranean Basin, with Emphasis on the Levant*. Pensoft.

Downing, K. F. & Lindsay, E. H., 2005. Relationship of Chitarwata formation paleodrainage and paleoenvironments to Himalayan tectonics and Indus River paleogeography. *Palaeontologia Electronica*, 8(1): 12pp.

Doyle, J. A., Biens, P., Doerenkamp, A. & Jardine, S., 1977. Angiosperm pollen from the pre-Albian Lower Cretaceous of equatorial Africa. *Bulletin du Centre de Recherches Exploration-Production Elf-Aquitaine*, 1(2): 451–73.

Doyle, P., 1996. *Understanding Fossils*. Wiley.

Drinia, H., Antonarakou, A., Tsaparas, N. & Dermitzakis, M. D., 2007. Foraminiferal stratigraphy and palaeoecological implications in turbidite-like deposits from the early Tortonian (Late Miocene) of Greece. *Journal of Micropalaeontology*, 26: 145–58.

Druce, E. C., Rhodes, F. H. T. & Austin, R. L., 1972. Statistical analysis of British Carboniferous conodont faunas. *Journal of the Geological Society*, 128: 53–70.

Dunay, R. E. & Hailwood, E. A. (eds.), 1995. *Non-biostatigraphical Methods of Dating and Correlation*. The Geological Society, London (Special Publication No. 89).

Dunkley Jones, T., Lunt, D. J., Schmidt, D. N. *et al.*, 2010. A review of the Paleocene–Eocene Thermal Maximum temperature anomaly. *Abstracts, International Palaeontological Congress*, London: 153.

Dunnington, H. V., 1955. Close zonation of Upper Cretaceous globigerinal sediments by abundance ratios of *Globotruncana* species groups. *Micropaleontology*, 1(3): 207–19.

Eagar, R. M. C., 1994. Discussion of 'classification of Carboniferous non-marine bivalves: systematic versus stratigraphy' by G. M. Vasey. *Journal of the Geological Society, London*, 151: 1030–3.

Eames, F. E., 1952. A contribution to the study of the Eocene in western Pakistan and western India. *Quarterly Journal of the Geological Society, London*, 107: 159–200.

Edgell, H. S., 2003. Upper Devonian Charophyta of Western Australia. *Micropaleontology*, 49(4): 359–74.

Einsele, G. & Seilacher, A. (eds.), 1982. *Cyclic and Event Stratification*. Springer.

Eiwanger, J., 1982. Die neolithische Siedlung von Merimde-Benisalame. *Mitteilungen des Deutschen Archaologischen Instituts, Abteilungen Kairo*, 38: 67–82.

El-Albani, A., Bengtson, S., Canfield, D. E. *et al.*, 2010. Large colonial organisms with co-ordinated growth in oxygenated environments 2.1 Gyr ago. *Nature*, 466: 100–4.

Elewa, A. M. T. (ed.), 2008. *Mass Extinctions*. Springer.

Elias, S. A., 1994. *Quaternary Insects and their Environments*. Smithsonian Institution Press.

   1997. The mutual climate range method of palaeoclimate reconstruction based on insect fossils: new applications and interhemispheric correlations. *Quaternary Science Reviews*, 16(10): 1217–25.

Elkins-Tanton, L. T. & Hager, B. H., 2005. Giant meteoroid impacts can cause volcanism. *Earth and Planetary Science Letters*, 239: 219–32.

Ellet, P., Heaton, R. & Watts, M., 2005. The Barmer basin, Rajastha, India – the ingredients which led to success. *Abstracts, American Association of Petroleum Geologists Annual Convention*.

El-Ouahibi, F. Z., Saint-Martin, S., Saint-Martin, J.-P. *et al.*, 2007. Messinian diatom assemblages from Boudinar basin (northeastern Rif, Morocco). *Revue de Micropaléontologie*, 50: 149–67.

Elrick, M., Berkyova, S., Klapper, G. *et al.*, 2009. Stratigraphic and oxygen isotope evidence for My-scale glaciations driving eustasy in the Early-Middle Devonian greenhouse world. *Palaeogeography, Palaeoclimatology, Palaeoecology*, 276: 170–81.

Emerson, S. R. & Hedges, J. I., 2008. *Chemical Oceanography and the Marine Carbon Cycle*. Cambridge University Press.

Emery, D. & Myers, K. J. (eds.), 1996. *Sequence Stratigraphy*. Oxford; Blackwell.

Enay, R., Mangold, C. & Almeras, Y., 2007. Le 'delta' du Wadi ad Dawisir, Arabie Saoudite centrale: une baisse du niveau marin relatif du Bathonien inferieur. *Revue de Paléobiologie*, 26(1): 307–34.

Epstein, A. G., Epstein, J. B. & Harris, L. D., 1977. Conodont color alteration – an index to organic metamorphism. *United States Geological Survey Professional Paper* 995.

Erba, E., 2006. The first 150 million years history of calcareous nannoplankton: biosphere–geosphere interaction. *Palaeogeography, Palaeoclimatology, Palaeoecology*, 232: 237–50.

Erickson, J., 2001. *Lost Creatures of the Earth: Mass Extinctions in the History of Life*. Checkmark Books/Facts on File.

Eriksson, M. E., Nilsson, E. K. & Jeppson, L., 2009. Vertebrate extinctions and reorganizations during the Late Silurian Lau event. *Geology*, 37(8): 739–42.

Erlich, R. N. & Keens-Dumas, J., 2007. Late Cretaceous palaeogeography of north-eastern South America: implications for source and reservoir development. *Transactions of the Fourth Geological Conference of the Geological Society of Trinidad and Tobago, Port of Spain*: 1–34.

Erlich, R. N., Macsotay, O., Nederbragt, A. J. & Lorente, M. A., 1999. Palaeoecology, palaeogeography and depositional environments of Upper Cretaceous rocks of western Venezuela. *Palaeogeography, Palaeoclimatology, Palaeoecology*, **153**: 203–38.

Ernst, S. R., Morvan, J., Geslin, E. *et al.*, 2006. Benthic foraminiferal response to experimentally induced *Erika* oil pollution. *Marine Micropaleontology*, **61**: 76–93.

Erwin, D. H., 2006. *Extinction: How Life on Earth Nearly Ended 250 Million Years Ago*. Princeton University Press.

Etienne, J. L., Allen, P., le Guerroue, E. & Rieu, R. (eds.), 2006. *Snowball Earth*. Monte Verita.

Eva, A. N., 1980. Pre-Miocene seagrass communities in the Caribbean. *Palaeontology*, **23**: 231–6.

Eva, A. N., Burke, K., Mann, P. & Wadge, G., 1989. Four-phase tectonostratigraphic development of the southern Caribbean. *Marine and Petroleum Geology*, **6**: 9–21.

Evans, A. M., 1997. *An Introduction to Economic Geology and its Environmental Impact*. Blackwell.

Evans, D., Graham, C., Armour, A. & Bathurst, P. (eds and co-ordinators), 2003. *The Millennium Atlas: Petroleum Geology of the Central and Northern North Sea*. The Geological Society, London/Norwegian Petroleum Society/Geological Survey of Denmark and Greenland.

Eyles, N. & Miall, A., 2007. *Canada Rocks*. Fitzhenry & Whiteside.

Fagan, B., 2004. *The Long Summer: How Climate Changed Civilization*. Granta.

Fagan, B. (ed.), 2009. *The Complete Ice Age: How Climate Change Shaped the World*. Thames & Hudso.

Fairbanks, R. G. & Matthews, R. K., 1978. The marine oxygen isotope record in Pleistocene coral, Barbados, West Indies. *Quaternary Research*, **10**: 181–196.

Falcon, R. M., 1975. Palynostratigraphy of the Lower Karoo sequence in the central Sebungwe district, mid-Zambezi, Rhodesia. *Palaeontologia Africana*, **18**: 1–29.

Falcon-Long, H. J., Benton, M. J. & Stimson, M., 2008. Ecology of earliest reptiles from basal Pennsylvanian trackways. *Journal of the Geological Society, London*, **164**: 1113–8.

Falcon-Lang, H. J. & DiMichele, W. A., 2010. What happened to the coal forest during Pennsylvanian glacial phases? *Palaios*, **25**: 611–7.

Falkowski, P. G. & Knoll, A. H. (eds.), 2007. *Evolution of Primary Producers in the Sea*. Academic Press (Elsevier).

Farooqui, A., Kumar, A., Jha, N. *et al.*, 2010. A thecamoebian assemblage from the Manjir formation (Early Permian) of northwest Himalaya, India. *Earth Science India*, **3**(3): 146–53.

Fatmi, A. N., 1972. Stratigraphy of the Jurassic and Lower Cretaceous rocks and Jurassic ammonites from northern areas of West Pakistan. *Bulletin of the British Museum (Natural History), Geology*, **20**(7): 299–380.

Fauquette, S., Clauzon, G., Suc, J.-P. & Zhuo Zheng, 1999. A new approach for palaeoaltitude estimates based on pollen records: example from the Mercantour Massif (southeastern France) at the earliest Pliocene. *Earth and Planetary Science Letters*, **170**: 35–47.

Favre, E., Francois, L., Fluteau, F. *et al.*, 2007. Messinian vegetation maps of the Mediterranean region using models and interpolated pollen data. *Geobios*, **40**: 433–43.

Fazeelat, T., Jalees, M. I. & Bianchi, T. S., 2010. Source rock potential of Eocene, Paleocene and Jurassic deposits in the subsurface of the Potwar basin, northern Pakistan. *Journal of Petroleum Geology*, **33**(1): 87–96.

Fenerci-Masse, M., Masse, J.-P. & Pernarcic, E., 2005. Quantitative stratigraphy of rudist limestones and its bearing on spatial organisation of rudist communities: the late Barremian, Urgonian, sequences of Provence (S.E. France). *Palaeogeography, Palaeoclimatology, Palaeoecology*, **215**: 265–84.

Feng-Sheng Xia, Sen-Gui Zhang & Zong-Zhe Wang, 2007. The oldest bryozoans: new evidence from the late Tremadocian (Early Ordovician) of east Yangtze gorges. *Journal of Paleontology*, **81**(6): 1308–26.

Fernandez, M. H., 2006. Rodent paleofaunas as indicators of climatic change in Europe during the last 125,000 years. *Quaternary Research*, **65**: 308–23.

Fernandez, M. H., Sierra, M. A. A. & Pelaez-Campomanes, P., 2007. Bioclimatic analysis of rodent palaeofaunas reveals severe climatic changes in southwestern Europe during the Plio-Pleistocene. *Palaeogeography, Palaeoclimatology, Palaeoecology*, **251**: 500–26.

Fernandez-Jalvo, Y., Scott, L. & Andrews, P., in press. Taphonomy in palaeoecological interpretations. *Quaternary Science Reviews*.

Ferre, B. & Granier, B., 2001. Albian roveacrinoids from the southern Congo basin off Angola. *Journal of South American Earth Sciences*, **14**: 219–35.

Fielding, C. R., Frank, T. D. & Isbell, J. L. (eds.), 2008. Resolving the Late Paleozoic ice age in time and space. Geological Society of America (Special Paper No. 441).

Fildani, A., Hanson, A. D., Chen, Z. Z. *et al.*, 2005. Geochemical characteristics of oil and source rocks and implications for petroleum systems, Talara basin, northwest Peru. *American Association of Petroleum Geologists Bulletin*, **89**(11): 1519–45.

Finney, S. C. & Berry, W. B. N., 1997. New perspectives on graptolite distributions and their use as indicators of platform magin dynamics. *Geology*, **25**: 919–22.

Fiorini, F., Scott, D. B. & Wach, G. D., 2009. Thecamoebians from the Early Cretaceous of the Scotian shelf. *Micropalaeontology*: **53**: 511–6.

2010. Characterization of paralic paleoenvironments using benthic foraminifera from Lower Cretaceous deposits (Scotian shelf, Canada). *Marine Micropaleontology*, **76**: 11–22.

Fischer, A. (ed.), 1996. *Man and the Sea in the Mesolithic*. Oxbow Monographs (No. 53).

Fischman, J., 2005. Family ties. *National Geographic Magazine*, April 2005: 16–27.

Fitzsimmons, R., Buchanan, J. & Izatt, C., 2005. The role of outcrop geology in predicting reservoir presence in the Cretaceous and Paleogene successions of the Sulaiman range, Pakistan. *American Association of Petroleum Geologists Bulletin*, **89**(2): 231–54.

Flecker, R. & Ellam, R.M., 2006. Identifying Late Miocene episodes of connection and isolation in the Mediterranean-Paratethyan realm using Sr isotopes. *Sedimentary Geology*, **188–189**: 189–203.

Fleet, A.J. & Boldy, S.A.R. (eds.), 1999. *Petroleum Geology of Northwest Europe: Proceedings of the 5th conference*. The Geological Society, London.

Flemming, I. (ed.), 2004. *Submarine Prehistoric Archaeology of the North Sea*. Council for British Archaeology (CBA/English Heritage Research Report No. 141).

Flor, L., Pilloud, A., Crux, J. & Arelis, F., 1997. Bioestratigrafia con nanofosiles calcareos del Paleogeno en las secciones Rio Aragua y Quebrada La Gallina, oriente de Venezuela. *I Congreso Latinamericano de Sedimentologio*.

Fontanier, C., Jorissen, F.J., Chaillou, G. *et al.*, 2005. Live foraminiferal faunas from a 2800 m deep lower canyon station from the Bay of Biscay: faunal response to focusing of refractory organic matter. *Deep-Sea Research, I*, **52**: 1189–227.

Foote, M. & Miller, A.I., 2007. *Principles of Paleontology*, 3rd edition. (Revised and updated edition of the classic text by David M. Raup and Steven M. Stanley). W.H. Freeman & Co.

Fortelius, M., Eronen, J., Liu, L. *et al.*, 2006. Late Miocene and Pliocene large land mammals and climatic changes in Eurasia. *Palaeogeography, Palaeoclimatology, Palaeoecology*, **238**(1–4): 219–27.

Fortelius, M., Kappelman, J., Sen, S. & Bernor, R.L. (eds.), 2003. *Geology and Paleontology of the Miocene Sinap Formation, Turkey*. Columbia University Press.

Fortey, R., 2009. *Fossils: The Key to the Past*. Revised and updated edition. The Natural History Museum, London.

Fox-Dobbs, K., Leonard, J.A. & Koch, P.L., 2008. Pleistocene megafauna from eastern Berinigia: paleoecological and paleoenvironmental interpretations of stable carbon and nitrogen isotope and radiocarbon records. *Palaeogeography, Palaeoclimatology, Palaeoecology*, **261**: 30–46.

Fraiser, M.L., Twitchett, R.J. & Bottjer, D.J., 2005. Unique microgastropod biofacies in the Early Triassic: indicator of long-term biotic stress and the pattern of biotic recovery after the end-Permian mass extinction. *Comptes Rendus Palevol*, **4**: 475–84.

Fraser, A., Goff, J., Simpson, I. *et al.*, 2003. A regional overview of the petroleum systems of the Middle East. *Abstracts, Petroleum Geology of the Middle East Conference*. The Geological Society, London.

Fraser, A., Matthews, S.J. & Murphy, R.W. (eds.), 1997. *Petroleum Geology of Southeast Asia*. The Geological Society, London (Special Publication No. 126).

Fraser, A.J. & Gawthorpe, R.L., 2003. *An Atlas of Carboniferous Basin Evolution in Northern England*. The Geological Society, London (Memoir No. 28).

Fraser, N., 2006. *Dawn of the Dinosaurs: Life in the Triassic*. Indiana University Press.

Fraser, R.H. & Currie, D.J., 1996. The species richness-energy hypothesis in a system where historical factors are thought to prevail: coral reefs. *American Naturalist*, **148**: 138–59.

Freiwald, A. & Roberts, J.M. (eds.), 2005. *Cold-Water Corals and Ecosystems*. Springer.

Frenzel, P. & Boomer, I., 2005. The use of ostracods from marginal marine, brackish waters as bioindicators of modern and Quaternary environmental change. *Palaeogeography, Palaeoclimatology, Palaeoecology*, **225**: 68–92.

Frey, R.W. & Pemberton, S.G., 1985. Biogenic structures in outcrops and cores. I. Introduction to ichnology. *Bulletin of Canadian Petroleum Geology*, **33**: 72–115.

Fricke, H.C., Rogers, R.R., Backlund, R. *et al.*, 2008. Preservation of primary stable isotope signals in dinosaur remains, and environmental gradients of the Late Cretaceous of Montana and Alberta. *Palaeogeography, Palaeoclimatology, Palaeoecology*, **266**: 13–27.

Friedrich, O., 2010. Benthic foraminifera and their role to decipher paleoenvironment during mid-Cretaceous Oceanic Anoxic Events – the 'anoxic benthic foraminifera' paradox. *Revue de Micropaléontologie*, **53**: 175–92.

Friedrich, O. & Erbacher, J., 2006. Benthic foraminiferal assemblages from Demerara Rise (ODP Leg 207, western tropical Atlantic): possible evidence for a progressive opening of the Equatorial Atlantic Gateway. *Cretaceous Research*, **27**: 377–97.

Friedrich, O. & Meier, K.J.S., 2006. Suitability of stable oxygen and carbon isotopes of calcareous dinoflagellate cysts for paleoclimatic studies: evidence from the Campanian/Maastrichtian cooling phase. *Palaeogeography, Palaeoclimatology, Palaeoecology*, **239**: 456–69.

Frielingsdorf, J., Islam, S.A., Block, M. *et al.*, 2008. Tectonic subsidence modelling and Gondwana source rock hydrocarbon potential, northwest Bangladesh: modelling of Kuchma, Singra and Hazipur wells. *Marine and Petroleum Geology*, **25**: 553–64.

Friis, E.M., Raunsgaard Pedersen, K. & Crane, R., 2006. Cretaceous angiosperm flowers: innovation and evolution in plant reproduction. *Palaeogeography, Palaeoclimatology, Palaeoecology*, **232**: 251–93.

Frostick, L.E., Renaut, R.W., Reid, I. & Tiercelin, J.J. (eds.), 1986. *Sedimentation in the African Rifts*. The Geological Society, London (Special Publication No. 25).

Fry, J.C., Gadd, G.M., Herbert, R.A. *et al.* (eds.), 1992. *Microbial Control of Pollution*. Cambridge University Press.

Fursich, F.T. & Pandey, D.K., 2003. Sequence stratigraphic significance of sedimentary cycles and shell concentrations in the Upper Jurassic-Lower Cretaceous of Kachchh, western India. *Palaeogeography, Palaeoclimatology, Palaeoecology*, **193**: 285–309.

Gaffney, V., Fitch, S. & Smith, D., 2009. *Europe's Lost World – The Rediscovery of Doggerland*. Council for British Archaeology (Research Report No. 160).

Galea-Alvarez, F.A. & Moreno-Vasquez, J., 1994. Foraminiferal biofacies of the Carapita formation and their relationship with tectonic events (Early–Middle Miocene). *V Simposio Bolivariano (Exploracion Petrolera en las Cuencas Subandinas)*: 225–7.

Galloway, W.E., 1989. Genetic stratigraphic sequences in basin analysis, I: architecture and genesis of flooding-surface bounded depositional units. *American Association of Petroleum Geologists Bulletin*, **73**(2): 125–42.

Galster, F., Guex, J. & Hammer, O., 2010. Neogene biochronology of Antarctic diatoms: a comparision between two quantitative approaches, CONOP and UAgraph. *Palaeogeography, Palaeoclimatology, Palaeoecology*, **285**: 237–47.

Gandin, A. & Debrenne, F., 2010. Distribution of archaeocyath-calcimicrobial bioconstructions on the Early Cambrian shelves. *Palaeoworld*, **19**: 222–41.

Garcia-Alix, A., Minwer-Barakat, R., Suarez, E.M. et al., 2008. Late Miocene-Early Pliocene climatic evolution of the Grenada basin (southern Spain) deduced from the paleoecology of the micromammal associations. *Palaeogeography, Palaeoclimatology, Palaeoecology*, **265**: 214–25.

Garcia-Bellido, D.C. & Rodriguez, S., 2005. Palaeobiogeographical relationships of poriferan and coral assemblages during the Late Carboniferous and the closure of the western Palaeotethys Sea-Panthalassan Ocean connection. *Palaeogeography, Palaeoclimatology, Palaeoecology*, **219**: 321–31.

Garrison, E.G., 2000. *Techniques in Archaeological Geology*. Springer.

Garrett, R.A. & Klenk, H.-P. (eds.), 2007. *Archaea: Evolution, Physiology, and Molecular Biology*. Blackwell Publishing.

Garrod, D., 1932. A new Mesolithic industry: the Natufian of Palestine. *Journal of the Royal Anthropological Institute*, **62**: 257–69.

Gaston, K.J. & Spicer, J.I., 2004. *Biodiversity: An Introduction*, 2nd Edition. Blackwell.

Gattuso, J.-P., Frankignoulle, M., Bourge, I. et al., 1998. Effect of calcium carbonate saturation of seawater on coral calcification. *Global and Planetary Change*, **18**: 37–46.

Gaucher, C., Sial, A.N., Halverson, G.P. & Frimmel, H.E. (eds.), 2010. *Neoproterozoic–Cambrian Tectonics, Global Change and Evolution*. Elsevier (Developments in Precambrian Geology Series No. 16).

Gaucher, C. & Sprechman, P., 1999. Upper Vendian skeletal fauna of the Arroyo de Soldado group, Uruguay. *Beringeria*, **23**: 55–91.

Gautier, A., 1990. *La domestication. Et l'homme crea ses animaux*. Errance.

Gay, S.A. & Cruickshank, A.R.I., 1999. Biostratigraphy of the Permian tetrapod faunas from the Ruhuhu valley, Kenya. *Journal of African Earth Sciences*, **29**(1): 195–210.

Gayer, R. & Harris, I. (eds.), 1996. *Coalbed Methane and Coal Geology*. The Geological Society, London (Special Publication No. 109).

Gazeau, F., Quiblier, C., Jansen, J.M. et al., 2007. Impact of elevated CO_2 on shellfish calcification. *Geophysical Research Letters*, **34**: L07603.

Gensel, P.G. & Edwards, D. (eds.), 2001. *Plants Invade the Land: Evolutionary and Environmental Perspectives*. Columbia University Press.

George, S.C., Dutkiewicz, A., Volk, H. et al., 2007. Eukaryote-derived steranes in Precambrian oils and rocks: fact or fiction. *Abstracts, International Meeting of Geochemists*, Torquay, Devon: O4.

Geraads, D., Kaya, T. & Mayda, S., 2005. Late Miocene large mammals from Yulafli, Thrace region, Turkey, and their biogeographic implications. *Acta Palaeontologica Polonica*, **50**(3): 523–44.

Geyer, G. & Landing, E., 2004. A unified Lower-Middle Cambrian chronostratigraphy for west Gondwana. *Acta Geologica Polonica*, **54**(2): 179–218.

Ghavidel-syooki, M. & Vecoli, M., 2008. Palynostratigraphy of Middle Cambrian to lowermost Ordovician stratal sequences in the High Zagros Mountains, southern Iran: regional stratigraphic implications and palaeobiogeographical significance. *Review of Palaeobotany and Palynology*, **150**: 97–114.

Gheerbrant, E. & Rage, J.-C., 2006. Paleobiogeography of Africa: how distinct from Gondwana and Laurasia? *Palaeogeography, Palaeoclimatology, Palaeoecology*, **241**: 224–46.

Ghosh, S.C., 1984. Estheriid biozones and the Permo-Triassic boundary in peninsular Gondwana basins. *Proceedings, Xth Indian Colloquium on Micropalaeontology and Stratigraphy*: 159–70.

Gibbons, A., 2009. A new kind of ancestor: *Ardipithecus* unveiled. *Science*, **326** (5949): 36–40.

Gibson, R.N. & Atkinson, R.J.A. (eds.), 2003. *Oceanography and Marine Biology: An Annual Review*. Taylor & Francis.

Gierlowski-Kordesch, E.H. & Kelts, K.R. (eds.), 2000. *Lake Basins through Space and Time*. American Association of Petroleum Geologists (Studies in Geology No. 46).

Gilbert, W.H. & Asfaw, B. (eds.), 2008. *Homo erectus: Pleistocene evidence from the Middle Awash, Ethiopia*. University of California Press.

Gill, F.L., Harding, I.C., Little, C.T.S. & Todd, J.A., 2005. Palaeogene and Neogene 'cold seep' communities in Barbados, Trinidad and Venezuela: an overview. *Palaeogeography, Palaeoclimatology, Palaeoecology*, **227**: 191–209.

Gischler, E., Sandy, M.R. & Peckmann, J., 2003. *Ibergirhyncia contraria* (F.A. Roemer, 1850), an Early Carboniferous seep-related brachiopod from the Harz Mountains, Germany – a possible ancestor to *Dzieduszyckia*? *Journal of Paleontology*, **77**: 293–303.

Gluyas, J.G. & Hichens, H.M. (eds.), 2003. *United Kingdom Oil and Gas Fields. Commemorative Millennium Volume*. The Geological Society, London (Memoir No. 20).

Gluyas, J. & Swarbrick, R, 2004. *Petroleum Geoscience*. Blackwell.

Godwin, H., 1940. Pollen analysis and forest history of England and Wales. *New Phytologist*, **39**: 370–400.

Goedert, J.L. & Squires, R.L., 1990. Eocene deep-sea communities in localized limestones formed by subduction-related methane seeps, southwestern Washington. *Geology*, **18**: 1182–5.

Goedert, J.L., Thiel, V., Schmale, O. et al., 2003. The Late Eocene 'Whiskey Creek' methane-seep deposit (western Washington state). Part I: Geology, palaeontology and molecular geobiology. *Facies*, **48**: 223–40.

Goff, J., Shamshiri, M., Jahani, S. et al., 2004. Discovery and geology of a giant fossil Jurassic oil field in the Zagros mountains, south-west Iran. *GeoArabia*, **9**(1): 60–61.

Goldberg, P. & MacPhail, R.I., 2006. *Practical and Theoretical Geoarchaeology*. Blackwell.

Goldring, R., 1999. *Field Palaeontology*, 2nd edition. Longman.

Gombos, A.M., Jr., Powell, W.G. & Norton, I.O., 1995. The tectonic evolution of western India and its impact on hydrocarbon occurrences: an overview. *Sedimentary Geology*, **96**: 119–29.

Gomes, P.O., Kilsdonk, B., Minken, J. et al., 2009. The outer high of the Santos basin, southern Sao Paulo plateau, Brazil: pre-salt exploration outbreak, paleogeographic setting, and evolution of the syn-rift structures. *Search and Discovery* Article No.10193. http://www.searchanddiscovery.com

Gonzalez-Guzman, E., 1985. Commentarios bioestratigraficos en relacion al Terciario en areas adyacentes a los Andes Colombo-Venezolano. *II Simposio Bolivariano Exploracion Petrolera en las Cicas Subandinas*, Bogota, Colombia.

Gonzalez de Juana, C., Iturralde, J.M. & Picard, X., 1980. *Geologia de Venezuela y de sus cuencas petroliferas*. Ediciones Foninves.

Goodall, J.G.S., Coles, G.P. & Whitaker, M.E., 1992. An integrated palynological, palynofacies and micropalaeontological study of the pre-salt formations of the south Gabon sub-basin and the Congo basin. *Geologie Africaine: Colloque Géologique Libreville, recueil des Communications*: 365–99.

Gordon, J.B., Pemberton, S.G., Gingras, M.K. & Konhauser, K.O., 2010. Biogenically enhanced permeability: a petrographic analysis of *Macaronichnus segregatus* in the Lower Cretaceous Bluesky formation, Alberta, Canada. *American Association of Petroleum Geologists Bulletin*, **94**(11): 1779–95.

Goren-Inbar, N. & Speth, J.D. (eds.), 2004. *Human Paleoecology in the Levantine Corridor*. Oxbow Books.

Goswami, B.G., Singh, H., Bhatnagar, A.K., Sinha, A.K. & Singh, R.R., 2007. Petroleum systems of the Mumbai offshore basin, India. *Search and Discovery* Article No. 10154. http://www.searchanddiscovery.com

Goswami, S., 2008. Marine influence and incursion in the Gondwana basins of Orissa, India: a review. *Palaeoworld*, **17**: 21–32.

Gourley, T.L. & Gallagher, S.J., 2005. Foraminiferal biofacies of the Miocene warm to cool climatic transition in the Port Phillip basin, southeastern Australia. *Journal of Foraminiferal Research*, **34**(4): 294–307.

Govindan, A., 2004. Miocene deep water agglutinated foraminifera from offshore Krishna–Godavari basin, India. *Micropaleontology*, **50**(3): 213–52.

Gradstein, F.M., Gibling, M.R., Jansa, L.F. et al., 1989. *Mesozoic Stratigraphy of Thakkhola, Central Nepal*. Centre for Marine Geology, Dalhousie University (Special Report No. 1).

Gradstein, F.M., Kaminski, M.A. & Berggren, W.A., 1988. Cenozoic foraminiferal biostratigraphy of the central North Sea. *Abhandlungen der Geologischen Bundesanstalt*, **41**: 97–108.

Gradstein, F., Ogg, J. & Smith, A., 2004. *A Geologic Time Scale 2004*. Cambridge University Press.

Graham, A., 1999. *Late Cretaceous and Cenozoic History of North American Vegetation*. Oxford University Press.

Graham, A., 2009. The Andes: a geological overview from a biological perspective. *Annals of the Missouri Botanical Garden*, **96**(3): 371–85.

Graham, J.B., Dudley, R., Aguilar, N.M. & Gans, C., 1995. Implications of the Late Palaeozoic oxygen pulse for physiology and evolution. *Nature*, **375**: 117–20.

Grant, S.F., Stewart, I.J. & Jones, R.W., 2000. Lower Congo basin chronostratigraphy. *GeoLuanda 2000 International Conference (14th African Colloquium on Micropalaeontology/Fourth Colloquium on the Stratigraphy and Palaeogeography of the South Atlantic)*, Luanda, Angola, *Abstracts*: 76.

Gray, J., Massa, D. & Boucot, A.J., 1982. Caradocian land plant microfossils from Libya. *Geology*, **10**: 197–201.

Grauvogel-Stamm, L. & Ash, S.R., 2005. Recovery of the Triassic land flora from the end-Permian life crisis. *Comptes Rendus Palevol*, **4**: 525–40.

Greb, S.F. & diMichele, W.A. (eds.), 2006. *Wetlands through Time*. Geological Society of America (Special Publication No. 399).

Greb, S.F., Eble, C.F., Hower, J.C. & Andrews, W.M., 2002. Multiple bench architecture and interpretations of original mire phases in Middle Pennsylvanian coal seams. *International Journal of Coal Geology*, **49**: 147–75.

Green, J.L., Semprebon, G.M. & Solounias, N., 2005. Reconstructing the palaeodiet of Florida *Mammut americanum* via low-magnification stereomicroscopy. *Palaeogeography, Palaeoclimatology, Palaeoecology*, **223**: 34–48.

Gregory-Wodzicki, K.M., 2000. Uplift history of the central and northern Andes: a review. *Geological Society of America Bulletin*, **112**(7): 1091–105.

Grekoff, N., 1953. Contribution a l'etude des ostracodes du Mesozoique Moyen (Bathonien-Valanginien) du basin de Mahunga, Madagascar. *Revue de l'Institut Français de Pétrole*, **XVIII**(12): 1709–62.

Grekoff, N. & Krommelbein, K., 1967. Etude compare des ostracodes Mesozoiques continentaux des basins Atlantiques: serie de Cocobeach, Gabon et series de Bahia, Brasil. *Revue de l'Institut Français de Pétrole*, **22**(9): 1307–53.

Griffiths, H.I., 2001. Ostracod evolution and extinction – its biostratigraphic value in the European Quaternary. *Quaternary Science Reviews*, **20**: 1743–51.

Grimes, S. T., Collinson, M. E., Hooker, J. J. & Mattey, D. P., 2008. Is small beautiful? A review of the advantages and limitations of using small mammal teeth and the direct laser fluorination analysis technique in the isotope reconstruction of past climatic change. *Palaeogeography, Palaeoclimatology, Palaeoecology,* **266**: 39–50.

Grine, F. E., Fleagle, J. G. & Leakey, R. E. (eds.), 2009. *The First Humans: Origin and Early Evolution of the Genus* Homo. Springer (Vertebrate Paleobiology and and Paleoanthropology Series).

Grosdidier, E., 1967. Quelques ostracodes nouveaux de la serie ante-salifere ('Wealdienne') des basins cotiers du Gabon et du Congo. *Revue de Micropaléontologie,* **10**: 107–18.

Grosdidier, E., Braccini, E., Dupont, G. & Moron, J.-M., 1996. Biozonation du Cretace Inferieur non marin des basins du Gabon et du Congo. *Géologie de l'Afrique et de l'Atlantique sud (Actes Colloques Angers, 1994):* 67–82.

Groves, J. R. & Altiner, D., 2005. Survival and recovery of calcareous foraminifera pursuant to the end-Permian mass extinction. *Comptes Rendus Palevol,* **4**: 419–32.

Groves, J. R. & Brenckle, P. I., 1997. Graphic correlation in frontier petroleum provinces: application to Upper Paleozoic sections in the Tarim basin, western China. *American Association of Petroleum Geologists Bulletin,* **81**(8): 1259–66.

Groves, J. R. & Lee, A., 2008. Accelerated rates of foraminiferal origination and extinction during the Late Paleozoic Ice Age. *Journal of Foraminiferal Research,* **38**(1): 78–84.

Grun, W., Lauer, G., Niedermayr, G. & Schnabel, W., 1964. Die Kreide-Tertiar Grenze im Wienerwald-flysch bei Hochstrass (Niederosterreich). *Verhandlungen der Geologischen Bundesanstalt,* **1964**(2): 226–83.

Guensburg, T. E. & Sprinkle, J., 2009. Solving the mystery of crinoids ancestry: new fossil evidence of arm origin and development. *Journal of Paleontology,* **83**(3): 350–64.

Guidoboni, E. & Ebel, J. E., 2009. *Earthquakes and Tsunamis in the Past: A Guide to Techniques in Historical Seismology.* Cambridge University Press.

Gunding, T., 2005. *First in Line: Tracing our Ape Ancestry.* Yale University Press.

Gunnell, G. F., Gingerich, P. D., Ul-Haq, M. *et al.*, 2008. New primates (Mammalia) from the Early and Middle Eocene of Pakistan and their paleobiogeographic implications. *Contributions from the Museum of Paleontology, The University of Michigan,* **32**(1): 1–14.

Gupta, R. P., Husain, R., Maurya, S. N. & Lal, N. K., 2001. Gondwana sediments in Godavari offshore – implications for tectonostratigraphic evolution and hydrocarbon prospectivity of east coast of India. *Gondwana Research,* **4**(4): 624–5.

Gupta, S. K., 2006. Basin architecture and petroleum systems of Krishna–Godavari basin, east coast of India. *The Leading Edge, July* 2006: 830–7.

Gutak, J. M., Tolokonnikova, Z. A. & Ruban, D. A., 2008. Bryozoan diversity in southern Siberia at the Devonian–Carboniferous transition: new data confirm a resistivity to two mass extinctions. *Palaeogeography, Palaeoclimatology, Palaeoecology,* **264**: 93–9.

Hacquebard, P. S. & Barss, M. S., 1966. *Views on Remaining Mining Reserves in the Pictou Coalfield, Nova Scotia.* Geological Survey of Canada (internal report).

Hagdorn, H., Xiaofeng Wang & Chuanshang Wang, 2007. Palaeoecology of the pseudoplanktonic Triassic crinoid *Traumatocrinus* from southwest China. *Palaeogeography, Palaeoclimatology, Palaeoecology,* **247**: 181–96.

Haig, D. W., 2010. Untangling an orogeny: foraminiferal calibration of Neogene phases of Timor collision. *Abstracts, Forams 2010 International Symposium on Foraminifera,* Bonn: 99.

Haile-Selassie, Y. & WoldeGabriel, G. (eds.), 2009. Ardipithecus kadabba: *Late Miocene Evidence from the Middle Awash, Ethiopia.* University of California Press.

Halbouty, M. T. (ed.), 1980. *Giant Oil and Gas Fields of the Decade 1968–1978.* American Association of Petroleum Geologists (Memoir No. 30).

Hall, R., 2006. The geological, oceanographic and climatic controls on the Wallace Line. *Programme with Abstracts, Palaeogeography and Palaeobiogeography ('Biodiversity in Space and Time') Conference,* Cambridge: 13–4.

Hall, B. K. (ed.), 2007. *Fins into Limbs.* Chicago University Press.

Hallam, A. (ed.), 1967. Depth indicators in marine sedimentary environments. *Marine Geology Special Issue,* **5**(5/6).

1973. *Atlas of Palaeobiogeography.* Elsevier.

Hallam, A., 1994. *An Outline of Phanerozoic Biogeography.* Oxford University Press.

2004. *Catastrophes and Lesser Calamities: The Causes of Mass Extinctions.* Oxford University Press.

Hallam, A. & Wignall, P., 1997. *Mass Extinctions and their Aftermath.* Oxford University Press.

Hallett, D., 2002. *Petroleum Geology of Libya.* Elsevier.

Hallock, P. & Glenn, E. C., 1986. Larger foraminifera: a tool for paleoenvironmental analysis of Cenozoic carbonate depositional facies. *Palaios,* **1**(1): 55–64.

Hallock, P., Lidz, B. H., Cockey-Burkhard, E. M. & Donnelly, K., 2003. Foraminifera as bioindicators in coral reef assessment and monitoring: the FORAM Index. *Environmental Monitoring and Assessment,* **81**(1–3): 221–38.

Hammer, O. & Harper, D., 2006. *Paleontological Data Analysis.* Blackwell.

Hanai, T., Ikeya, N. & Ishikazi, K. (eds.), 1988. *Evolutionary biology of Ostracoda (Proceedings of the 9th International Symposium on Ostracoda, Shizuoka, Japan, 1985).* Elsevier.

Hanchiao Jiang & Zhongli Ding, 2008. A 20 Ma pollen record of East Asian summer monsoon evolution from Guyuan, Ningxia, China. *Palaeogeography, Palaeoclimatology, Palaeoecology,* **265**: 30–8.

Hancox, P. J. & Rubidge, B. S., 2001. Breakthroughs in the biodiversity, biogeography, biostratigraphy and basin analysis of the Beaufort group. *African Earth Sciences,* **33**: 563–77.

Hanlon, R.T. & Messenger, J.B., 1996. *Cephalopod Behaviour*. Cambridge University Press.

Haq, B.U., Hardenbol, J. & Vail, P.R., 1987. Chronology of fluctuating sea levels since the Triassic. *Science*, **235**: 1153–65.

Haq, B.U. & Boersma, A. (eds.), 1978. *Introduction to Marine Micropaleontology*. Elsevier.

Haque, A.F.M.H., 1956. The foraminifera of the Ranikot and the Laki of the Nammal gorge, Salt range. *Memoirs of the Geological Survey of Pakistan*, **1**: 1–300.

Hardy, M.J. & Wrenn, J.H., 2009. Palynomorph distribution in modern tropical delta and shelf sediments, Mahakam delta, Borneo, Indonesia. *Palynology*, **33**(2): 19–42.

Harland, W.B., Armstrong, R.L., Cox, A.V. *et al.*, 1990, *A Geologic Time Scale 1989*. Cambridge University Press.

Harland, W.B., Cox, A.V., Llewellyn, P.G. *et al.*, 1982. *A Geologic Time Scale*. Cambridge University Press.

Harper, D.A.T., 2006. The Ordovician biodiversification: setting an agenda for marine life. *Palaeogeography, Palaeoclimatology, Palaeoecology*, **232**: 148–66.

Harper, E.M., 2006. Dissecting post-Palaeozoic arms races. *Palaeogeography, Palaeoclimatology, Palaeoecology*, **232**: 322–43.

Harries, P.J. (ed.), 2008. *High-Resolution Approaches in Stratigraphic Paleontology*. Springer.

Harris, F. (ed.), 2004. *Global Environmental Issues*. John Wiley & Sons.

Harris, N., 2006. The elevation history of the Tibetan Plateau and its implications for the Asian monsoon. *Palaeogeography, Palaeoclimatology, Palaeoecology*, **241**: 4–15.

Harris, N.B., 2000. Evolution of the Congo rift basin, west Africa: an inorganic geochemical record in lacustrine shales. *Basin Research*, **12**: 425–45.

(ed.), 2005. *The Deposition of Organic-Carbon-Rich Sediments: Models, Mechanisms and Consequences*. Society of Economic Paleontologists and Mineralogists (Special Publication No. 82).

Harris, N.B., Freeman, K.H., Pancost, R.D. *et al.*, 2004. The character and origin of lacustrine source rocks in the Lower Cretacsous synrift section, Congo basin, west Africa. *American Association of Petroleum Geologists Bulletin*, **88**(8): 1163–84.

Harrison, R.M. (ed.), 2001. *Pollution: Causes, Effects and Control*, 4th edition. Royal Society of Chemistry.

Hart, D. & Sussman, R.W., 2005. *Man the Hunted: Primates, Predators and Human Evolution*. Westview Press.

Hart, M.B. (ed.), 1987. *Micropalaeontology of Carbonate Environments*. Ellis Horwood Ltd.

1996. *Biotic Recovery from Mass Extinction Events*. The Geological Society, London (Special Publication No. 102).

Hart, M.B., 2000. Foraminifera, sequence stratigraphy and regional correlation: an example from the uppermost Albian of southern England. *Revue de Micropaléontologie*, **43**(1–2): 27–45.

Hart, M.B., Kaminski, M.A. & Smart, C.W. (eds.), 2000. *Proceedings of the 5th International Workshop on Agglutinated Foraminifera*. The Grzybowski Foundation (Special Publication No. 7).

Hart-Davis, D., 2002. *Fauna Britannica*. Weidenfeld & Nicholson.

Harvati, K. & Harrison, T. (eds.), 2007. *Neanderthals Revisited*. Springer.

Haslett, S.K., 2002. *Quaternary Environmental Micropalaeontology*. Arnold.

2004. Late Neogene–Quaternary radiolarian biostratigraphy: a brief review. *Journal of Micropalaeontology*, **23**(1): 39–47.

Hassan, F.A., 1988. The predynastic of Egypt. *Journal of World Prehistory*, **2**: 136–85.

Haynes, J.R., 1981. *Foraminifera*. Macmillan.

Hayward, B.W., 1990. Use of foraminiferal data in analysis of Taranaki basin, New Zealand. *Journal of Foraminiferal Research*, **20**(1): 71–83.

Hayward, B.W., Grenfell, H.R., Sabaa, A.T. *et al.*, 2006. Micropaleontological evidence of large earthquakes in the past 7200 years in southern Hawke's Bay, New Zealand. *Quaternary Science Reviews*, **25**: 1186–207.

Haywood, A.M. & Valdes, P.J., 2006. Vegetation cover in a warmer world simulated using a dynamic global vegetation model for the mid-Pliocene. *Palaeogeography, Palaeoclimatology, Palaeoecology*, **237**(2–4): 412–27.

Head, M.J. & Gibbard, P.L. (eds.), 2005. *Early–Middle Pleistocene Transitions: The Land–Ocean Evidence*. The Geological Society, London (Special Publication No. 247).

Hedberg, H.D., 1937. Foraminifera of the middle Tertiary Carapita formation of northeastern Venezuela. *Journal of Paleontology*, **11**(8): 661–97.

Hedges, S.B. & Kumar, S. (eds.), 2009. *The Timetree of Life*. Oxford University Press.

Heimhofer, U. & Hochuli, P.-A., 2010. Early Cretaceous angiosperm pollen from a low-latitude succession (Araripe basin, N.E. Brazil). *Review of Palaeobotany and Palynology*, **161**: 105–26.

Helenes, J., de Guerra, C. & Vasquel, J., 1998. Palynology and chronostratigraphy of the Upper Cretaceous in the subsurface of the Barinas region, western Venezuela. *American Association of Petroleum Geologists Bulletin*, **82**(7): 1308–28.

Helenes, J. & Somoza, D., 1999. Palynology and sequence stratigraphy of the Cretaceus of eastern Venezuela. *Cretaceous Research*, **20**: 447–63.

Hembree, D.L., Hasiotis, S.T. & Martin, L.D., 2006. *Torridorefugium eskridgensis* (new ichnogenus and ichnospecies): amphibian aestivation burrows from the Lower Permian Speiser shale of Kansas. *Journal of Paleontology*, **79**(3): 583–93.

Hemleben, C., Kaminski, M.A., Kuhnt, W. & Scott, D.B. (eds.), 1990. *Paleoecology, Biostratigraphy, Paleoceanography and Taxonomy of Agglutinated Foraminifera*. Kluwer.

Henriksen, N., 2008. *Geological History of Greenland*. GEUS.

Hershkovitz, M.A., Arroyo, M.T.K., Bell, C. & Hinojosa, L.F., 2006. Phylogeny of *Chaetanthera* (Asteraceae; Mutisieae) reveals both ancient and recent origins of high altitude lineages. *Molecular Phylogenetics and Evolution*, **41**: 594–605.

Herz, N. & Garrison, E.G., 1998. *Geological Methods for Archaeology*. Oxford University Press.

Hess, S., Alve, E. & Trannum, H.C., 2010. Effects of water-based drill cuttings vs physical burial on benthic foraminifera and macrofauna: a mesocosm experiment. *Abstracts, Forams 2010 International Symposium on Foraminifera*, Bonn: 106.

Hesselbo, S.P., Grocke, D.R., Jenkyns, H.C. *et al.*, 2000. Massive disocciation of gas hydrate during a Jurassic oceanic anoxic event. *Nature*, **406**: 392–5.

Hesselbo, S.P., Robinson, S.A., Surlyk, F. & Piasecki, S., 2002. Terrestrial and marine extinction at the Triassic–Jurassic boundary synchronized with major carbon-cycle perturbation: a link to initiation of massive volcanism. *Geology*, **30**: 251–4.

Hetherington, R. & Reid, R.G.B., 2010. *The Climate Connection: Climate Change and Modern Human Evolution*. Cambridge University Press.

Heydari, E., Arzani, N. & Hassanzadeh, J., 2008. Mantle plume: the invisible serial killer – application to the Permian-Triassic boundary mass extinction. *Palaeogeography, Palaeoclimatology, Palaeoecology*, **264**: 147–62.

Higgins, G.E., 1996. *A History of Trinidad Oil*. Trinidad Express Newspapers Ltd.

Highton, P.J.C., Racey, A., Wakefield, M.I. *et al.*, 1997. Quantitative biostratigraphy: an example from the Neogene of the Gulf of Thailand. *Proceedings, International Conference on Stratigraphy and Tectonic Evolution of Southeast Asia and the South Pacific*, Bangkok, Thailand, *1997*: 563–85.

Hillaire-Marcel, C. & de Vernal, A. (eds.), 2007. *Proxies in Late Cenozoic Paleoceanography*. Elsevier (Developments in Marine Geology No. 1).

Hillson, S., 2005. *Teeth*, 2nd edition. Cambridge University Press.

Himmler, T., Freiwald, A., Stollhofen, H. & Peckmann, J., 2008. Late Carboniferous hydrocarbon-seep carbonates from the glaciomarine Dwyka group, southern Namibia. *Palaeogeography, Palaeoclimatology, Palaeoecology*, **257**: 185–97.

Hochuli, P.A. & Feist-Burkhardt, S., 2004. A Boreal early cradle of angiosperms? Angiosperm-like pollen from the Middle Triassic of the Barents Sea (Norway). *Journal of Micropalaeontology*, **23**: 97–104.

Hoelzel, A.R. (ed.), 2002. *Marine Mammal Ecology: An Evolutionary Approach*. Blackwell.

Hofker, J., 1976. *Further Studies on Caribbean Foraminifera*. Natuurwet. Studiekring Suriname et Ned. Antillen, Plompetorengracht, 9–11 (Studies on the Fauna of Curacao and Other Caribbean Islands No. 162).

Hoffman, P.F. & Li, Z.-X., 2009. A palaeogeographic context for Neoproterozoic glaciations. *Palaeogeography, Palaeoclimatology, Palaeoecology*, **277**: 158–72.

Holl, A.F.C., 2004. *Holocene Saharans: An Anthropological Perspective*. Continuum.

Holland, C.H. (ed.), 1981. *Lower Palaeozoic of the Middle East, Eastern and Southern Africa and Antarctica*. Volume 3 of *Lower Palaeozoic Rocks of the World*. John Wiley.

Holmes, J.A., Atkinson, T., Darbyshire, F. *et al.*, 2010. Middle Pleistocene climate and hydrological environment at the Boxgrove hominin site (West Sussex, UK) from ostracod records. *Quaternary Science Reviews*, **29**(13–14): 1515–27.

Holmes, J.A. & Chivas, A.R. (eds.), 2002. *The Ostracoda: Applications in Quaternary Research*. American Geophysical Union.

Holmes, K.M., 2007. Using Pliocene palaeoclimatic data to postulate dispersal pathways of early hominins. *Palaeogeography, Palaeoclimatology, Palaeoecology*, **247**: 96–108.

Holz, M., Kalkreuth, W. & Banerjee, I., 2002. Sequence stratigraphy of paralic coal-bearing srata: an overview. *International Journal of Coal Geology*, **48**: 147–79.

Holzmann, M. & Pawlowski, J., 2002. Freshwater foraminiferans from Lake Geneva: past and present. *Journal of Foraminiferal Research*, **32**: 344–50.

Hongfu Yin, Quinglai Feng, Xulong Lai *et al.*, 2007. The protracted Permo-Triassic crisis and multi-episode extinction around the Permian-Triassic boundary. *Global and Planetary Change*, **55**: 1–20.

Hongo, C. & Kayanne, H., 2010. Holocene sea-level record from corals: reliability of paleodepth indicators at Ishigaki island, Ryukyu islands, Japan. *Palaeogeography, Palaeoclimatology, Palaeoecology*, **287**: 143–51.

Hooghiemstra, H., Wijninga, V.M. & Cleef, A.M., 2006. The palaeobotanical record of Colombia: implications for biogeography and biodiversity. *Annals of the Missouri Botanical Garden*, **93**: 297–324.

Hooker, J.H. & Millbank, C., 2001. A Cernaysian mammal from the Upnor formation (Late Palaeocene, Herne Bay, UK) and its implications for correlation. *Proceedings of the Geologists' Association*, **112**: 331–8.

Hoorn, C., Ohja, T. & Quade, J., 2000. Palynological evidence for vegetation development and climate change in the sub-Himalayan zone (Neogene, central Nepal). *Palaeogeography, Palaeoclimatology, Palaeoecology*, **163**: 133–61.

Hoorn, C. & Wesselingh, F., 2010. *Amazonia: Landscape and Species Evolution*. Wiley-Blackwell.

Hoorn, C., Gurrero, J., Sarmiento, G.A. & Lorente, M.A., 1995. Andean tectonics as a cause for changing drainage patterns in Miocene northern South America. *Geology*, **23**(3): 237–40.

Hopley, P.J., Latham, A.G. & Marshall, J.D., 2006. Palaeoenvironments and palaeodiets of mid-Pliocene micromammals from Makapansgat limeworks, South Africa: a stable isotope and dental micro-wear approach. *Palaeogeography, Palaeoclimatology, Palaeoecology*, **233**: 235–51.

Horne, D.J., 2007. A mutual temperature range nmethod for Quaternary palaeoclimatic analysis using European non-marine Ostracoda. *Quaternary Science Reviews*, **26**(9–10): 1398–1415.

Horne, D., 2010. Best of three? Ostracod, chironomid and beetle palaeothermometry of British Pleistocene archaeological sites. *Abstracts, The Micropalaeontological Society Annual General Meeting*.

Horne, D., Brooks, S., Cooper, R. & Whittaker, J., 2010. Hoxnian palaeotemperature estimates. *Abstracts, Ancient Human Occupation of Britain Conference*, British Museum, April 8th–10th 2010: 15.

Horsfield, R. (2010). The British Lower Palaeolithic of the early Middle Pleistocene. *Quaternary Science Reviews*.

Horton, B.P. & Edwards, R.J., 2006. *Quantifying Holocene Sea-Level Change Using Intertidal Foraminifera: Lessons from the British Isles*. Cushman Foundation for Foraminiferal Research (Special Publication No. 40).

Horton, B.P., Yongqiang Zong, Hillier, C. & Engelhart, S., 2007. Diatoms from Indonesian mangroves and their suitability as sea-level indicators for tropical environments. *Marine Micropaleontology*, **63**: 155–68.

Houghton, J., 2004. *Global Warming: The Complete Briefing*, 3rd edition. Cambridge University Press.

Hounslow, M.W. & McIntosh, G., 2003. Magnetostratigraphy of the Sherwood Sandstone group (Lower and Middle Triassic), south Devon, UK: detailed correlation of the marine and non-marine Anisian. *Palaeogeography, Palaeoclimatology, Palaeoecology*, **193**: 325–48.

Hovland, M. & Thomsen, E., 1997. Cold-water corals – are they hydrocarbon seep related? *Marine Geology*, **137**: 159–64.

Hsu, K.J., 1999. *Caribbean Basins*. Elsevier (Sedimentary Basins of the World Series, No. 4).

Hublin, J.-J. & Richards, M. (eds.), 2009. *The Evolution of Hominin Diets: Integrating Approaches to the Study of Palaeolithic Sustenance*. Springer (Vertebrate Paleobiology and Paleoanthropology Series).

Hughes, G.W., 2004. Middle to Upper Jurassic Saudi Arabian carbonate petroleum reservoirs: biostratigraphy, micropalaeontology and palaeoenvironments. *GeoArabia*, **9**(3): 7–114.

2006. Calcareous algae of Saudi Arabian Permian to Cretaceous carbonates. *Revista Espanola de Micropaleontologia*, **37**(1): 131–40.

2009. Biofacies and palaeoenvironments of the Upper Jurassic Shaqra group, Saudi Arabia. *Volumina Jurassica*, **VI**: 33–45.

Hughes, G.W., Al-Khaled, M. & Varol, O., 2009. Oxfordian biofacies and palaeoenvironments of Saudi Arabia. *Volumina Jurassica*, **VI**: 47–60.

Hughes, G.W., Enezy, S.S. & Rashid, S., 2010a. Biosteering the Upper Permian Khuff C reservoirs in Saudi Arabia. *Abstracts, GEO2010 Middle East Geoscience Exhibition*, Manama, Bahrain.

2010b. Using foraminifera to biosteer the Upper Permian Khuff reservoirs in Saudi Arabia. *Abstracts, Forams 2010 International Symposium on Foraminifera*, Bonn: 112.

Hughes, G.W. & Naji, N., 2009. Sedimentological and micropalaeontological evidence to elucidate post-evaporitic carbonate palaeoenvironments of the Saudi Arabian latest Jurassic. *Volumina Jurassica*, **VI**: 61–73.

Hunter, J. & Ralston, I. (eds.), 2009. *The Archaeology of Britain*, 2nd edition. Routledge.

Hunter, V.F., 1976. Benthonic microfaunal shelfal assemblages and Neogene depositional pattern from northern Venezuela. *Maritime Sediments Special Publication*, **1** ('*Benthonics 75*'): 459–66.

Hunter, V.F., 1978. Foraminiferal correlation of Tertiary mollusc horizons of the southern Caribbean area. *Geologie en Mijnbouw*, **57**: 193–203.

Husain, R., Gupta, R.P. & Lal, N.K., 2000. Tectono-stratigraphic evolution and petroleum systems of Krishna–Godavari basin, India. *Proceedings, 5th International Conference & Exhibition, Petroleum Geochemistry & Exploration in the Afro-Asian Region*, New Delhi: 443–58.

Hustedt, F., 1953. Die Systematik der Diatomeen in ihren Beziehungen zur Geologie und Okologie nebst einer Revision des Halobien-systems. *Sveriges Botaniska Tidskrift*, **47**: 509–19.

1957. Die Diatomeenflora des Fluss-system der Weser im Gebiet der Hansestadt Bremen. *Abhandlungen Naturwissenschaftlicher Verein zu Bremen*, **34**: 181–440.

Huynh, T.T. & Poulsen, C.J., 2005. Rising atmospheric CO_2 as a possible trigger for the end-Triassic mass extinction. *Palaeogeography, Palaeoclimatology, Palaeoecology*, **217**: 223–42.

Iglesias-Rodriguez, M.D., Halloran, P.R., Rickaby, R.E. *et al.*, 2008. Phytoplankton calcification in a high CO_2 world. *Science*, **320**: 336–40.

Imam, B., 2005. *Energy Resources of Bangladesh*. University Grants Commission of Bangladesh.

Imam, M.B. & Hussain, M., 2002. A review of hydrocarbon habitats in Bangladesh. *Journal of Petroleum Geology*, **25**(1): 31–52.

International Committee for Coal and Organic Petrology (ICCP), 1998. The new vitrinite classification (ICCP System 1994). *Fuel*, **77**(5): 349–58.

Iordanidis, A. & Georgakopoulos, A., 2003. Pliocene lignites from Apofysis mine, Amynteo basin, northwestern Greece: petrographical characteristics and depositional environment. *International Journal of Coal Geology*, **54**: 57–68.

Iqbal, N., Broutin, J., Izart, A. *et al.*, 1998. Quelques donnees stratigraphiques sur le Permien Inferieur du Salt Range (Pakistan). *Geodiversitas*, **20**(4): 723–30.

Isokazi, Y., 2009a. Illawarra reversal: the fingerprint of a superplume that triggered Pangean breakup and the end-Guadalpuian (Permian) mass extinction. *Gondwana Research*, **15**: 421–32.

2009b. Integrated 'plume winter' scenario for the double-phased extinction during the Paleozoic-Mesozoic transition: the G-LB and P-TB events from a Panthalassan perspective. *Journal of Asian Earth Sciences*, **36**: 459–80.

Isokazi, Y. & Aljinovic, D., 2009. End-Guadalupian extinction of the Permian gigantic bivalve Alatoconchidae: end of gigantism in tropical seas by cooling. *Palaeogeography, Palaeoclimatology, Palaeoecology*, **284**: 11–21.

Ivanov, D.A., Ashraf, A.R. & Mosbrugger, V., 2007. Late Oligocene and Miocene climate and vegetation in the Eastern

Paretethys area (northeast Bulgaria), based on pollen data. *Palaeogeography, Palaeoclimatology, Palaeoecology*, **255**: 342–60.

Ivanov, D. A., Ashraf, A. R., Mosbrugger, V. & Palamarev, E., 2002. Palynological evidence for Miocene climate change in ther Forecarpathian basin (central Paretethys, NW Bulgaria). *Palaeogeography, Palaeoclimatology, Palaeoecology*, **178**: 19–37.

Ivany, L. C., Portell, R. W. & Jones, D. S., 1990. Animal-plant relationships of an Eocene seagrass community from Florida. *Palaois*, **5**: 244–58.

Iyeka, N., Kayima, T. & Cronin, T. M. (eds.), 2001. Earth environments and dynamics of Ostracoda (collected papers from the symposium 'Towards the New Ostracodologist in the 21st Century', Shizuoka, Japan). *Palaeogeography, Palaeoclimatology, Palaeoecology*, **225**(1–4).

Izart, A., Vaslet, D., Briand, C. *et al.*, 1998. Stratigraphic correlations between the continental and marine Tethyan and peri-Tethyan basins during the Late Carboniferous and the Early Permian. *Geodiversitas*, **20**(4): 521–93.

Jackson, J. B. C. & Erwin, D. H., 2006. What can we learn about ecology and evolution from the fossil record? *Trends in Ecology and Evolution*, **21**(6): 322–8.

Jacobs, B., Kingston, J. B. & Jacobs, L. L., 1999. The origin of grass-dominated ecosystems. *Annals of the Missouri Botanical Gardens*, **86**(2): 590–643.

Jacobs, B. F., 2004. Palaeobotanical studies from tropical Africa: relevance to the evolution of forest, woodland and savannah biomes. *Philosophical Transactions of the Royal Society B*, **359**: 1573–83.

Jacobsen, J. D., Twitchett, R. J. & Price, G. D., 2009. Temperature change and the Late Permian mass extinction event. *Abstracts, Palaeontological Association Annual Meeting*, Birmingham.

James, K. H., Lorente, M. A. & Pindell, J. L. (eds.), 2009. *The Origin and Evolution of the Caribbean Plate*. The Geological Society, London (Special Publication No. 328).

Jan, I. U., Stephenson, M. H. & Khan, F. R., 2009. Palynostratigraphic correlation of the Sardhai formation (Permian) of Pakistan. *Review of Palaeobotany and Palynology*, **158**: 72–2.

Janevsky, G. A. & Baumiller, T. K., 2010. Could a stalked crinoid swim? A biomechanical model and characteristics of swimming crinoids. *Palaios*, **25**: 588–96.

Jansonius, J. & MacGregor, D. C. (eds.), 1996. *Palynology: Principles and Applications*. American Association of Stratigraphic Palynologists Foundation.

Janvier, P., 2003. Vertebrate characters and the Cambrian vertebrates. *Comptes Rendus Palevol*, **2**: 523–31.

2009. Les premiers vertebres et les premieres etapes de l'evolution du crane. *Comptes Rendus Palevol*, **8**: 209–19.

Janz, H. & Vennemann, T. W., 2005. Isotopic composition (O, C, Sr and Nd) and trace element ratios (Sr/Ca, Mg/Ca) of Miocene marine and brackish ostracods from North Alpine Foreland deposits (Germany and Austria) as indicators for palaeoclimate. *Palaeogeography, Palaeoclimatology, Palaeoecology*, **225**: 216–47.

Jaramillo, C. A., Rueda, M. J. & Mora, G., 2006. Cenozoic plant diversity in the Neotropics. *Science*, **311**: 1893–6.

Jardine, S., Kieser, G. & Reyre, Y., 1974. L'individualisation progressive du continent Africain vue a travers les donnees palynologiques de l'ere secondaire. *Bulletin des Sciences Géologiques, Strasbourg*, **27**(1/2): 69–85.

Jasper, K., Hartkopf-Froder, C., Flajs, G. & Littke, R., 2010. Evolution of Pennsylvanian (Late Carboniferous) peat swamps of the Ruhr basin, Germany: comparison of palynological, coal petrological and organic geochemical data. *International Journal of Coal Geology*, **83**: 346–65.

Jassim, S. Z. & Goff, J. C., 2006. *Geology of Iraq*. Moravian Museum and Dolin.

Jastrow, R. & Rampino, M., 2008. *Origins of Life in the Universe*. Cambridge University Press.

Javaux, E. J. & Marshal, C. P., 2006. A new approach in deciphering early protist paleobiology and evolution: combined microscopy and microchemistry of single Proterozoic acritarchs. *Review of Palaeobotany and Palynology*, **139**: 1–15.

Jeffrey, L. S., 2005. Characterization of the coal resources of South Africa – a 2002 review. *The Journal of the South African Institute of Mining and Metallurgy*, February 2005: 95–102.

Jell, P. A. (ed.), 1987. *Studies in Australasian Mesozoic Palynology*. Association of Australasian Palaeontologists.

Jenkins, D. G. (ed.), 1993. *Applied Micropalaeontology*. Kluwer.

Jenkins, D. G. & Murray, J. W. (eds.), 1989. *Stratigraphical Atlas of Fossil Foraminifera*, 2nd edition. Ellis Horwood.

Jermannaud, P., Rouby, D., Robin, C. *et al.*, 2010. Plio-Pleistocene sequence stratigraphic architecture of the eastern Niger Delta: a record of eustasy and aridification of Africa. *Marine and Petroleum Geology*, **27**: 810–21.

Jimenez-Moreno, G., 2006. Progressive substitution of a subtropical forest for a temperate one during the Middle Miocene climate cooling in Central Europe according to palynological data from cores Tengelic-2 and Hidas-53 (Pannonian Basin, Hungary). *Palaeogeography, Palaeoclimatology, Palaeoecology*, **142**: 1–14.

Jimenez-Moreno, G., de Leeuw, A., Mandic, O. *et al.*, 2009. Integrated stratigraphy of the Early Miocene lacustrine deposits of Pag Island (SW Croatia): palaeovegetation and environmental changes in the Dinaride lake system. *Palaeogeography, Palaeoclimatology, Palaeoecology*, **280**: 193–206.

Jimenez-Moreno, G., Rodriguez-Tovar, F. J., Pardo-Iguzquiza, E. *et al.*, 2005. High-resolution palynological analysis in late Early-Middle Miocene core from the Pannonian Basin, Hungary: climatic changes, astronomical forcing and eustatic fluctuations in the Central Parathys. *Palaeogeography, Palaeoclimatology, Palaeoecology*, **216**: 73–97.

Jimenez-Moreno, G. & Suc, J.-P., 2007. Middle Miocene latitudinal climatic gradient in western Europe: evidence from pollen records. *Palaeogeography, Palaeoclimatology, Palaeoecology*, **253**: 224–41.

Jimin Sun & Zhenqing Zhang, 2008. Palynological evidence for the Mid-Miocene Climatic Optimum recorded in Cenozoic sediments of the Tian Shan Range, northwestern China. *Global and Planetary Change*, **64**: 53–68.

Jin-Feng Li, Ferguson, D. K., Jian Yang *et al.*, 2009. Early Miocene vegetation and climate in Weichang district, north China. *Palaeogeography, Palaeoclimatology, Palaeoecology*, **280**: 47–63.

Jing-Xian Xu, Ferguson, D. K., Cheng-Sen Li & Yu-Fei Wang, 2008. Late Miocene vegetation and climate of the Luhe region in Yunnan, southwest China. *Review of Palaeobotany and Palynology*, **148**: 36–59.

Johnson, D., 2004. *The Geology of Australia*. Cambridge University Press.

2006. *Australia's Mammal Extinctions*. Cambridge University Press.

Jolley, D. W. & Bell, B. R. (eds.), 2002. *The North Atlantic Igneous Province: Tectonic, Volcanic and Magmatic Processes*. The Geological Society, London (Specal Publication No. 197).

Jolley, S. J., Fisher, Q. J., Ainsworth, R. B. *et al.*, 2010. *Reservoir Compartmentalization*. The Geological Society, London (Special Publication No. 347).

Jolly, A., 1999. *Lucy's Legacy: Sex and Intelligence in Human Evolution*. Harvard University Press.

Joly, C., Barille, L., Barreau, M. *et al.*, 2007. Grain and annulus diameter as criteria for distinguishing pollen grains of cereal from wild grasses. *Review of Palaeobotany and Palynology*, **146**: 221–33.

Jones, R. W., 1994. *The Challenger Foraminifera*. Oxford University Press.

1996. *Micropalaeontology in Petroleum Exploration*. Oxford University Press.

1997. Aspects of the Cenozoic stratigraphy of the northern Sulaiman Ranges, Pakistan. *Journal of Micropalaeontology*, **16**: 51–8.

2000. Proterozoic to Palaeozoic sequence stratigraphy of south-west Iran. *Abstracts, MEGSTRAT1 Workshop, Dubai*: 4–6.

2001. Biostratigraphic characterisation of submarine fan sub-environments, deep-water offshore Angola. *Abstracts, 6th International Workshop on Agglutinated Foraminifera*, Prague.

2003a. Micropalaeontological characterisation of submarine fan/channel sub-environments, deep-water Angola. *Abstracts, William Smith Conference ('Wrestling with Mud')*. The Geological Society, London.

2003b. Micropalaeontological characterisation of mudrock seal capacity. *Abstracts, William Smith Conference ('Wrestling with Mud')*. The Geological Society, London.

2006. *Applied Palaeontology*. Cambridge University Press.

2009. Stratigraphy, palaeoenvironmental interpretation and uplift history of Barbados based on foraminiferal and other palaeontological evidence. *Journal of Micropalaeontology*, **28**: 37–44.

Jones, R. W. & Charnock, M. A., 1985. 'Morphogroups' of agglutinating foraminifera, their life positions and feeding habits and potential applicability in (paleo)ecological studies. *Revue de Paléobiologie*, **4**(2): 311–20.

Jones, R. W. & Simmons, M. D. (eds.), 1999. *Biostratigraphy in Production and Development Geology*. The Geological Society, London (Special Publication No. 152).

Jones, R. W., Simmons, M. D. & Whittaker, J. E., 2006. On the stratigraphic and palaeobiogeographic significance of *Borelis melo* (Fichtel & Moll, 1798) and *B. melo curdica* (Reichel, 1937) (Foraminifera, Miliolida, Alveolinidae). *Journal of Micropalaeontology*, **25**(2): 175–85.

Jorissen, F., 2010. Foraminifera and pollution monitoring monitoring – room for improvement. *Abstracts, Forams 2010 International Symposium on Foraminifera, Bonn*: 116–7.

Jorissen, F. J., Bicchi, E., Duchemin, G. *et al.*, 2009. Impact of oil-based drill mud disposal on benthic foraminiferal assemblages on the continental margin of Angola. *Deep-Sea Research II*, **56**: 2270–91.

Jorissen, F. J., de Stigter, H. C. & Widmark, J. G. V., 1995. A conceptual model explaining benthic foraminiferal microhabitats. *Marine Micropaleontology*, **26**: 3–15.

Jorry, S., Davaud, E. & Caline, B., 2003a. Controls on the distribution of nummulite facies: a case study from the late Ypresian El Garia formation (Kesra plateau, central Tunisia). *Journal of Petroleum Geology*, **26**(3): 283–306.

2003b. Depositional facies and sequence stratigraphy of reservoir nummulite bodies in central Tunisia (El Garia formation, upper Ypresian): results of a field analogue study from the Kesra plateau). *Abstracts, AAPG International Conference, Barcelona, Spain*.

Jun-Yuan Chen, Bottjer, D. J., Davidson, E. H. *et al.*, 2009. Phase contrast synchrotron X-ray microtomography of Ediacaran (Doushantuo) metazoan microfossils: phylogenetic diversity and evolutionary implications. *Precambrian Research*, **173**: 191–200.

Jung-Hyun Kim, Schouten, S., Hopmans, E. C. *et al.*, 2008. Global sediment core-top calibration of the TEX86 paleothermometer in the ocean. *Geochimica et Cosmochimica Acta*, **72**: 1154–73.

Jun Li, Servais, T., Kui Yan & Huaicheng Zhu, 2004. A nearshore-offshore trend in acritarch distribution from the Early-Middle Ordovician of the Yangtze platform, south China. *Review of Palaeobotany and Palynology*, **130**:141–61.

Kaakinen, A., Sonninen, E. & Lunkka, J. P., 2006. Stable isotope record in paleosol carbonates from the Chinese loess plateau: implications for late Neogene paleoclimate and paleovegetation. *Palaeogeography, Palaeoclimatology, Palaeoecology*, **237**(2–4): 359–69.

Kaiho, K., 1994. Benthic foraminiferal dissolved oxygen index and dissolved oxygen levels in the modern ocean. *Geology*, **22**: 719–22.

Kaim, A., Bitner, M. A., Jenkins, R. G. & Hikida, Y., 2010. A monospecific assemblage of terebratulide brachiopods in the Upper Cretaceous seep deposits of Omagari, Hokkaido, Japan. *Acta Palaeontologica Polonica*, **55**(1): 73–84.

Kaim, A., Jenkins, R.G. & Hikida, Y., 2009. Gastropods from Late Cretaceous Omagari and Yasukawa hydrocarbon seep deposits in the Nakagawa area, Hokkaido, Japan. *Acta Palaeontologica Polonica*, **54**(3): 463–90.

Kaiser, S.I., Steuber, T., Becker, R.T. & Joachimski, M.M., 2006. Geochemical evidence for major environmental change at the Devonian-Carboniferous boundary in the Carnic Alps and the Rhenish Massif. *Palaeogeography, Palaeoclimatology, Palaeoecology*, **240**: 146–60.

Kaiser, T.M. & Brinkmann, G., 2006. Measuring dental wear equilibriums – the use of industrial surface texture parameters to infer the diet of fossil mammals. *Palaeogeography, Palaeoclimatology, Palaeoecology*, **239**(3–4): 221–40.

Kakuwa, Y. & Matsumoto, R., 2006. Cerium negative anomaly just before the Permian and Triassic boundary event – the upward expansion of anoxia in the water column. *Palaeogeography, Palaeoclimatology, Palaeoecology*, **229**: 335–44.

Kalkreuth, W.D., Marchioni, D.L., Calder, J.H. et al., 1991. The relationship between coal petrography and depositional environments from selected coal basins in Canada. *International Journal of Coal Geology*, **19**: 21–76.

Kamikuri, S.-I., Motoyama, I., Nishi, H. & Iwai, M., 2009. Evolution of eastern Pacific warm pool and upwelling processes since the Middle Miocene based on analysis of radiolarian assemblages: response to Indonesian and Central American seaways. *Palaeogeography, Palaeoclimatology, Palaeoecology*, **280**: 469–79.

Kaminski, M.A. & Coccioni, R., 2008. *Proceedings of the 7th International Workshop on Agglutinated Foraminifera*. The Grzybowski Foundation (Special Publication No. 13).

Kaplan, J.O., Krumhardt, K.M. & Zimmermann, N., 2009. The prehistoric and preindustrial deforestation of Europe. *Quaternary Science Reviews*, **28**: 3016–34.

Kapoor, P.N. & Nanjundaswamy, S., 2007. Paleoenvironment and source rock palynological studies for hydrocarbon prospects in Mumbai offshore basin. *Association of Petroleum Geologists Bulletin*, **1**: 123–8.

Karabiyikoglu, M., Tuzcu, S., Ciner, A. et al., 2005. Facies and environmental setting of the Miocene coral reefs in the late orogenic fill of the Antalya basin, western Taurides, Turkey: implications for tectonic control and sea-level changes. *Sedimentary Geology*, **173**: 345–71.

Kardong, K.V., 2006. *Vertebrates: Comparative Anatomy, Function, Evolution*, 4th edition. McGraw-Hill.

Kassi, A.M., Kelling, G., Kasi, A.K. et al., 2009. Contrasting Late Cretaceous-Palaeocene lithostratigraphic successions across the Bibai Thrust, western Sulaiman fold-thrust belt, Pakistan: their significance in deciphering the early collisional history of the NW Indian plate margin. *Journal of Asian Earth Sciences*, **35**: 435–44.

Katz, B.J. (ed.), 1994. *Petroleum Source Rocks*. Springer Verlag.

Katz, B.J. & Pratt, L.M. (eds.), 1993. *Source Rocks in a Sequence Stratigraphic Framework*. American Association of Petroleum Geologists (Studies in Geology No. 73).

Kaufmann, B., 2006. Calibrating the Devonian time scale: a synthesis of U-Pb ID-TIMS ages and conodont stratigraphy. *Earth-Science Reviews*, **76**: 175–90.

Kaufmann, E.G. & Johnson, C.C., 1988. The morphological and ecological evolution of middle and upper Cretaceous reef-building rudistids. *Palaios*, **3**: 194–216.

Kay, R.F., Hadden, R.H., Cifelli, R.L. & Flynn, J.J. (eds.), 1997. *Vertebrate Paleontology in the Tropics: The Miocene Fauna of La Venta, Colombia*. Smithsonian Press.

Ke Xia, Tao Su, Yu-Sheng Liu et al., 2009. Quantitative climate reconstruction of the Late Miocene Xiaolongtan megaflora from Yunnan, southwest China. *Palaeogeography, Palaeoclimatology, Palaeoecology*, **276**: 80–6.

Keller, G., Zenker, C.E. & Stone, S.M., 1989. Late Neogene history of the Pacific-Caribbean gateway. *Journal of South American Earth Sciences*, **2**(1): 73–108.

Keller, G., Khosla, S.C., Sharma, R. et al., 2009. Early Danian planktic foraminifera from Cretaceous–Tertiary Intertrappean beds at Jhilmili, Chhindwara district, Madhya Pradesh, India. *Journal of Foraminiferal Research*, **39**(1): 40–55.

Kelley, P.H., Kowalewski, M. & Hansen, T.A. (eds.), 2003. *Predator–Prey Interactions in the Fossil Record*. Kluwer Academic/Plenum Publishers.

Kennedy, B.A. (eds.), 1990. *Surface Mining*, 2nd edition. Society for Mining, Metallurgy and Exploration.

Kennedy, W.J., Phansalkar, V.G. & Walaszczyk, I., 2003. *Prionocyclus germari* (Reuss, 1845), a Late Turonian marker fossil from the Bagh beds of central India. *Cretaceous Research*, **24**: 433–8.

Kerney, M., 1999. *Atlas of the Land and Freshwater Molluscs of Britain and Ireland*. Harley Books.

Khan, M.A., Ahmed, R., Raza, H.A. & Kemal, A., 1986. Geology of petroleum in Kohat-Potwar depression, Pakistan. *American Association of Petroleum Geologists Bulletin*, **70**(4): 396–414.

Khowaja-Ateequzzaman, Garg, R. & Mehrotra, N.C., 2006. *A Catalogue of Dinoflagellate Cysts from India*. Birbal Sahni Institute of Palaeobotany.

Kidder, D.L. & Worsley, T.R., 2004. Causes and consequences of extreme Permo-Triassic warming to globally equable Climate and relation to the Permo-Triassic extinction and recovery. *Palaeogeography, Palaeoclimatology, Palaeoecology*, **203**(3–4): 207–37.

2010. Phanerozoic Large Igneous Provinces (LIPs), HEATT (Haline Euxinic Acidic Thermal Transgression) episodes, and mass extinctions. *Palaeogeography, Palaeoclimatology, Palaeoecology*, **295**: 163–96.

Kiel, S., 2008. Fossil vidence for micro- and macrofaunal utilization of large nekton falls: examples from early Cenozoic deep-water sediments in Washington State, USA. *Palaeogeography, Palaeoclimatology, Palaeoecology*, **267**: 161–74.

2010. On the potential generality of depth-related ecologic structure in cold-seep communities: evidence from

Cenozoic and Mesozoic examples. *Palaeogeography, Palaeoclimatology, Palaeoecology*, **295**: 245–57.

Kiel, S., Amano, K. & Jenkins, R. G., 2008. Bivalves from Cretaceous cold-seep deposits on Hokkaido, Japan. *Acta Palaeontologica Polonica*, **53**(3): 525–37.

Kiel, S. & Campbell, K. A., 2005. *Lithomphalus enderlini* gen. et sp. nov. from cold-seep carbonates in California – a Cretaceous neomphalid gastropod? *Palaeogeography, Palaeoclimatology, Palaeoecology*, **227**: 232–41.

Kiel, S., Campbell, K. A., Elder, W. P. & Little, C. T. S., 2008. Jurassic and Cretaceous gastropods from hydrocarbon seeps in forearc basins and accretionary prism settings, California. *Acta Palaeontologica Polonica*, **53**(4): 679–703.

Kiel, S. & Dando, P. R., 2009. Chaetopterid tubes from vent and seep sites: implications for fossil record and evolutionary history of vent and seep annelids. *Acta Palaeontologica Polonica*, **54**(3): 443–8.

Kiel, S. & Goedert, J. L., 2006. Deep-sea food bonanzas: early Cenozoic whale-fall communities resemble wood-fall rather than seep communities. *Proceedings of the Royal Society, B*, **273**: 2625–31.

Kiel, S. & Little, C. T. S., 2006. Cold-seep molluscs are older than the general marine mollusc fauna. *Science*, **313**: 1429–31.

Kiessling, W., 2006. Geographic range and extinction risk: new lessons from ancient marine benthic organisms. *Programme with Abstracts, Palaeogeography and Palaeobiogeography ('Biodiversity in Space and Time') Conference*, Cambridge: 9.

Kimbel, W. H., Rak, Y. & Johanson, D. C., 2004. *The Skull of Australopithecus afarensis*. Oxford University Press.

Kingdon, J., 2003. *Lowly Origin: Where, When and Why our Ancestors First Stood Up*. Princeton University Press.

Kingston, J. D. & Harrison, T., 2007. Isotopic dietary reconstructions of Pliocene herbivores at Laetoli: implications for early hominin paleoecology. *Palaeogeography, Palaeoclimatology, Palaeoecology*, **243**: 272–306.

Kirby, M. X. & MacFadden, B., 2005. Was southern Central America an archipelago or a peninsula in the Middle Miocene? A test using land-mammal body size. *Palaeogeography, Palaeoclimatology, Palaeoecology*, **228**: 193–202.

Kirchman, D. L. (ed.), 2008. *Microbial Ecology of the Oceans*, 2nd edition. John Wiley.

Kirk, N., 1969. Some thoughts on the ecology, mode of life, and evolution of the Graptolithina. *Proceedings of the Geological Society, London*, **1659**: 273–92.

1972. More thoughts on the automobility of the graptolites. *Quarterly Journal of the Geological Society, London*, **128**: 127–33.

Kjennerud, T. & Gillmore, G. K., 2003. Integrated Palaeogene palaeobathymerty of the northern North Sea. *Petroleum Geoscience*, **9**: 125–32.

Klemme, H. D. & Ulmishek, G. F., 1991. Effective petroleum source rocks of the world: stratigraphhic distribution and controlling depositional factors. *American Association of Petroleum Geologists Bulletin*, **75**(12): 1809–51.

Klitzsch, E., Lejal-Nicol, A. & Massa, D., 1973. Le Siluro-Devonien a psilophytes et lycophytes du basin de Mourzouk (Libye). *Compte Rendu des Seances de l'Academie des Sciences, Paris, Serie D*, **277**: 2465–7.

Knaust, D., 2009. Characterisation of a Campanian deep-sea an in the Norwegian Sea by means of ichnofabrics. *Marine and Petroleum Geology*, **26**: 1199–211.

Knoll, A.H., Bambach, R. K., Canfield, D. E. & Grotzinger, J. P., 1996. Comparative earth history and Late Permian mass extinction. *Science*, **273**: 452–7.

Knowles, T., Taylor, P. D., Wlliams, M. *et al.*, 2009. Pliocene seasonality across the North Atlantic inferred from cheilostome bryozoans. *Palaeogeography, Palaeoclimatology, Palaeoecology*, **277**: 226–35.

Kodrans-Nsiah, M., de Lange, G. J. & Zonneveld, K. A. F., 2008. A natural exposure experiment on short-term species-selective aerobic degradation of dinoflagellate cysts. *Review of Palaeobotany and Palynology*, **152**: 32–9.

Kogbe, C.A. & Mehes, K., 1986. Micropalaeontology and biostratigraphy of the coastal basins of west Africa. *Journal of African Earth Sciences*, **5**(1): 1–100.

Kohn, M., 2006. Made in Savannahstan. *New Scientist*, 1 July, 2006: 34–9.

Koho, K. A., 2008a. Benthic foraminifera: ecological indicators of past and present oceanic environments – a glance at the modern assemblages from the Portuguese submarine canyons. *Geologi*, **60**: 161–6.

2008b. *The Dynamic Balance between Food Abundance and Habitat Instability: Benthic Foraminifera of Portuguese Margin Canyons*. Geologica Ultraiectina (Mededelingen van de Faculteit Geowetenschappen Universiteit Utrecht) No. 286.

Koho, K. A., Kouwenhouven, T. J., de Stigter, H. C. & van der Zwaan, G. J., 2007. Benthic foraminifera in the Nazare canyon, Portuguese continental shelf: sedimentary environments and disturbance. *Marine Micropaleontology*, **66**: 27–51.

Kolcon, L. & Sachsenhofer, R., 1999. Petrology, palynology and depositional environment of the Early Miocene Oberdorf lignite seam (Styrian basin, Austria). *International Journal of Coal Geology*, **41**: 275–308.

Konert, G., Afifi, A. M., Al-Hajri, S. A. & Droste, H. J., 2001. Paleozoic stratigraphy and hydrocarbon habitat of the Arabian plate. *GeoArabia*, **6**(3): 407–42.

Konhauser, K., 2007. *Introduction to Geomicrobiology*. Blackwell.

Koshal, V. N. & Uniyal, S. N., 1984. Palaeocene–Early Eocene palynofossils in the subsurface of north Cambay basin, Gujrat (western India). *Proceedings of the Xth Indian Colloquium on Micropalaeontology & Stratigraphy*: 233–44.

Kothe, A., 1988. Biostratigraphy of the Surghar range, Salt range, Sulaiman range and the Kohat area, Pakistan, according to Jurassic through Paleogene calcareous nannofossils and Paleogene dinoflagellates. *Geologisches Jahrbuch*, **B71**: 3–87.

Koutsoukos, E. A. M., 1985. Distribucao paleobatimetrica de foraminiferos bentonicos do Cenozoico margem continental Atlantica. *Trabahos VIII Congreso Brasiliero de Paleontologia*, Brasilia: 355–70.

Koutsoukos, E. A. M. (ed.), 2005. *Applied Stratigraphy*. Springer.

Koutsoukos, E. A. M., Leary, P. N. & Hart, M. B., 1990. Latest Cenomanian-earliest Turonian low-oxygen toleant benthonic foraminifera: a case-study from the Sergipe basin (NE Brazil) and the western Anglo-Paris basin (southern England). *Palaeogeography, Palaeoclimatology, Palaeoecology*, **77**(2): 145–79.

Koutsoukos, E. A. M., Mello, M. R., de Azambuja Filho, N. C., Hart, M. B. & Maxwell, J. R., 1991. The upper Aptian-Albian succession of the Sergipe basin, Brazil: an integrated paleoenvironmental assessment. *American Association of Petroleum Geologists Bulletin*, **75**(3): 479–98.

Koutsoukos, E. A. M. & Merrick, K. A., 1986. Foraminiferal paleoenvironments from the Barremian to Maestrichtian of Trinidad, West Indies. *Transactions of the 1st Geological Conference of the Geological Society of Trinidad & Tobago*, **1985**: 85–101.

Kovacs-Endrody, E., 1991. *On the Late Permian Age of Ecca Glossopteris floras in the Transvaal Province, with a Key to and Description of Twenty-five Glossopteris Species*. Geological Survey of South Africa (Memoir No. 77).

Kovar-Eder, J., Kvacek, Z., Martinetto, E. & Roiron, P., 2006. Late Miocene to Early Pliocene vegetation of southern Europe (7–4 Ma) as reflected in the megafossil plant record. *Palaeogeography, Palaeoclimatology, Palaeoecology*, **238**(1–4): 321–39.

Krassilov, V. A., 1997. *Angiosperm Origins: Morphological and Ecological Aspects*. Pensoft.

2003. *Terrestrial Paleoecology and Global Change*. Sofia; Pensoft.

Krommelbein, K., 1962. Zur taxinomie und Biochronologie stratigraphisch wichtiger Ostracoden-Arten aus der Oberjurassisch?-Unterkretazischen Bahia Serie (Wealden-Fazies) N.E. Brasilien. *Senckenbergiana Lethaea*, **43**(6): 437–528.

1963. Ilhasina n.g. und *Salvadoriella* n.g., zwei neue Ostracoden-Gattung aus der Bahia Serie (nicht-marinen Oberjura?-Unterkreide, N.E. Brasilien. *Zoologischer Anzeiger*, **171**(9/10): 376–90.

1964a. Neue Arten der Ostracoden-Gattung *Paracypridea* Swain aus der Bahia-Serie des Reconcavo Baluiano (Oberjura?-Unterkreide, Wealden-Fazies, N.E. Brasilien. *Senckenbergiana Lethaea*, **45**: 29–41.

1964b. Uber einege neue Arten der Ostracoden-Gattung *Reconcavoana* Krommelbein, 1962 aus der N.E. Brasilienischen Bahia-Serie (nicht-mariner Oberjura?-Unterkreide). *Boletim Paranense de Geografia, Universidade de Parana, Curitaba*, **1964**(10–15): 139–60.

1965a. Neue, fur Vergleiche mit West-Afrika wichtige, Ostracoden-Arten der Brasilienischen Bahia-Serie (Ober-Jura?)/Unter-Kreide in Wealden Fazies. *Senckenbergiana Lethaea*, **46**: 177–213.

1965b. Ostracoden aus der nicht-marinen Unter-Kreide ('Westafrikanischer Wealden') des Congo-Kustenbeckens. *Meyniana*, **15**: 59–74.

1968. The first non-marine Lower Cretaceous ostracods from Ghana, west Africa. *Palaeontology*, **11**(2): 259–63.

Krommelbein, K. & Weber, R., 1971. Ostracoden des 'Nordost-Brasilienischen Wealden'. *Beihefte zum Geologisches Jahrbuch*, **115**: 1–93.

Krzyzaniak, L. & Kobusiewicz, M. (eds.), 1989. *Late Prehistory of the Nile Basin and the Sahara*. Poznan.

Kumar, A. & Saxena, R.K., 1996. Dinoflagellate cysts and calcareous nannoplankton from the Kaikalur Claystone lithounit (Chintalapalli Shale formation) of Kaikalur well-A, Krishna–Godavari basin, India. *Geoscience Journal*, **17**(2): 95–111.

Kumar, K., Rana, K.S. & Singh, H., 2007. Fishes of the Khuiala formation (Early Eocene) of the Jaisalmer basin, western Rajasthan, India. *Current Science*, **93**(4): 553–9.

Kumar, 1984. Middle Eocene–Early Oligocene biofacies and palaeoecology of the northern part of Cambay basin. *Proceedings of the Xth Indian Colloquium on Micropalaeontology & Stratigraphy*: 289–98.

Kump, L. R., Pavlov, A. & Arthur, M. A., 2005. Massive release of hydrogen sulphide to the surface ocean and atmosphere during intervals of oceanic anoxia. *Geology*, **33**(5): 397–400.

Kuper, R. & Kropelin, S., 2006. Climate-controlled Holocene occupation of the Sahara: motor of Africa's evolution., *Science*, **313**: 803–7.

Kureshy, A. A., 1977a. The Cretaceous larger foraminiferal biostratigraphy of Pakistan. *Journal of the Geological Society of India*, **18**(12): 662–7.

1977b. The Cretaceous planktonic foraminiferal biostratigraphy of Pakistan. *Palaeontological Society of Japan Special Papers*, **21**: 223–31.

1978a. The biostratigraphic correlation of sedimentary basins of West Pakistan. *Annales des Mines et de la Géologie*, **28**(2): 327–36.

1978b. Tertiary larger foraminiferal zones of Pakistan. *Revista Espanola de Micropaleontologia*, **10**(3): 467–83.

Kuroyanagi, A., Kawahita, H., Suzuki, A. *et al.*, 2009. Impacts of ocean acidification on large benthic foraminifers: results from laboratory experiments. *Marine Micropaleontology*, **73**: 190–5.

Labandeira, C. C., 2005. Invasion of the continents: cyanobacterial crusts to tree-inhabiting arthropods. *Trends in Ecology and Evolution*, **20**(5): 253–62.

Lakshminarayana, G., 2002. Evolution in basin-fill style during the Mesozoic Gondwana continental break-up in the Godavari triple junction, SE India. *Gondwana Research*, **5**(1): 227–44.

Lal, N.K., Siawal, A. & Kaul, A.K., 2009. Evolution of east coast of India – a plate tectonic reconstruction. *Journal of the Geological Society of India*, **73**: 249–60.

Lamb, J.L., 1964. The geology and paleontology of the Rio Aragua surface section, Serrania del Interior, State of Monagas, Venezuela. *Boletin Informativo Asociacion Venezolano de Geologia, Mineria y Petroleo, Caracas*, **7**(4): 111–23.

Lamy, A., 1986. Plio-Pleistocene palynology and visual kerogen studies, Trinidad, W.I., with emphasis on the Columbus basin. *Transactions of the First Geological Conference of the Geological Society of Trinidad & Tobago, 1985*: 114–27.

Lana, C.C. & de Souza Carvalho, I., 2002. Cretaceous conchostracans from Potiguar basin (northeast Brazil): relationships with west African conchostracan faunas and palaeoecological inferences. *Cretaceous Research*, **23**(3): 357–62.

Landing, E. & Johnson, M.E. (eds.), 2003. *Silurian Lands and Seas: Paleogeography outside of Laurentia*. New York State Museum (Bulletin No. 493).

Landing, E., English, A. & Keppie, J.D., 2010. Cambrian origin of all skeletalized metazoan phyla – discovery of Earth's oldest bryozoans (Upper Cambrian, southern Mexico). *Geology*, **38**(6): 547–50.

Landing, E., Peng, S., Babcock, L.E. *et al.*, 2007. Global standard names for the lowermost Cambrian series and stage. *Episodes*, **30**(4): 287–9.

Larghi, C., 2005. *Dickinsartella* fauna from the Saiwan formation (Oman): a bivalve fauna testifying to the late Sakmarian (Early Permian) climaticv amelioration along the northeastern Gondwanan fringe. *Rivista Italiana di Paleontologia e Stratigrafia*, **111**(3): 353–75.

Launder, J.B. & Thompson, J.M.T. (eds.), 2009. *Geo-engineering Climate Change*. Cambridge University Press.

Lawton, J.H. & May, R.M. (eds.), 1995. *Extinction Rates*. Oxford University Press.

Lazarus, D., Bittniok, B., Diester-Haass, L. *et al.*, 2006. Comparison of radiolarian and sedimentologic paleoproductivity proxies in the latest Miocene-Recent Benguela upwelling system. *Marine Micropaleontology*, **60**: 269–94.

Lees, J.A., Bown, P.R. & Young, J.R., 2006. Photic zone palaeoenvironments of the Kimmeridge Clay Formation (Upper Jurassic, UK) suggested by calcareous nannoplankton palaeoecology. *Palaeogeography, Palaeoclimatology, Palaeoecology*, **235**(1–3): 110–34.

Legendre, S., Montuire, S., Maridet, O. & Escarguel, G., 2005. Rodents and climate: a new model for estimating past temperatures. *Earth and Planetary Science Letters*, **235**: 408–20.

le Herisse, A., Al-Ruwaili, M., Miller, M. & Vecoli, M., 2007. Environmental changes reflected by palynomorphs in the early Middle Ordovician Hanadir member of the Qasim formation, Saudi Arabia. *Revue de Micropaléontologie*, **50**: 3–16.

Lehman, S.M. & Fleagle, J.G., 2006. *The Biogeography of Primate Evolution*. Springer.

Lehmann, C., Goff, J. & Jones, R.W., 2004. Jurassic carbonates and evaporites of the Middle East: a new look at an old play. *American Association of Petroleum Geologists International Conference and Exhibition, Cancun, Mexico*, Paper 89625.

2006a. Facies distribution and dolomitization along the shelf margin of the Jurassic Gotnia Basin. *Abstracts, American Association of Petroleum Geologists Annual Convention, Houston, 2006*.

2006b. A detailed Jurassic sequence stratigraphic framework for the formation of dolomites along the Gotnia shelf margin. *Abstracts, 7th Middle East Geosciences Conference and Exhibition, Namama, Bahrain, 2006*.

Leonard, R., 1983. Geology and hydrocarbon accumulations, Columbus basin, offshore Trinidad. *American Association of Petroleum Geologists Bulletin*, **67**(7): 1081–93.

Leroy, S.A.G., Boyraz, S. & Gurbuz, A., 2009. High-resolution palynological analysis in Lake Sapanca as a tool to detect recent earthquakes on the North Anatolian Fault. *Quaternary Science Reviews*, **28**: 2626–32.

Leroy, S.A.G., Marco, S., Bookman, R. & Miller, C.S., 2010. Impact of earthquakes on agriculture during the Roman–Byzantine period from pollen records of the Dead Sea laminated sediment. *Quaternary Research*, **73**: 191–200.

Lesslar, P., 1987. Computer-assisted interpretation of depositional palaeoenvironments based on foraminifera. *Geological Society of Malaysia Bulletin*, **21**: 103–19.

Leturmy, P. & Robin, C. (eds.), 2010. *Tectonic and Stratigraphic Evolution of Zagros and Makran during the Mesozoic–Cenozoic*. The Geological Society, London (Special Publication No. 330).

Leveque, C., Juniper, S.K. & Limen, H., 2006. Spatial organization of food webs along habitat gradients at deep-sea hydrothermal vents on Axial Volcano, northeast Pacific. *Deep-Sea Research I*, **53**: 726–39.

Lieberman, B.S. & Kaesler, R., 2010. *Prehistoric Life – Evolution and the Fossil Record*. Wiley-Blackwell.

Liem, R.F., Bemes, W.E., Walker, W.K. Jr & Grande, L., 2001. *Functional Anatomy of the Vertebrates*, 3rd edition. Blackwell.

Lietard, C. & Pierre, C., 2009. Isotopic signatures (del^{18}O and del^{13}C) of bivalve shells from cold seeps and hydrothermal vents. *Geobios*, **42**: 209–19.

Linstadter, J. & Kropelin, S., 2004. Wadi Bakht revisited: Holocene climate change and prehistoric occupation in the Guilf Kebir region of the eastern Sahara, SW Egypt. *Geoarchaeology*, **19**(8): 753–78.

Lipps, J. (ed.), 1993. *Fossil Prokaryotes and Protists*. Blackwell.

Little, C.T.S., 2007. Fossil perspectives on chemosynthetic ecosystems. *Abstracts, 'Dark Energy and the History of Chemosynthetic Life in the Deep Sea' Conference*. The Geological Society, London.

Little, C.T.S. & Vrijenhook, R.C., 2003. Are hydrothermal vent animals living fossils? *Trends in Ecology and Evolution*, **18**: 582–8.

Liu, A.G., McIlroy, D. & Brasier, M.D., 2009. Earliest evidence for metazoan-style locomotion in the fossil record: trace fossils from the 565 Ma Mistaken Point formation, Newfoundland. *Abstracts, Palaeontological Association Annual Meeting, Birmingham*.

Lobegeier, M.K. & Sen Gupta, B.K., 2008. Foraminifera of hydrocarbon seeps, Gulf of Mexico. *Journal of Foraminiferal Research*, **38**(2): 93–116.

Lockett, J. & Andrews, R., 2010. Rediscovering the geology of the south Wales coalfield for CBM exploration. *Abstracts, International Conference, The Geology of Unconventional Gas Plays*. The Geological Society, London: 70.

Loeblich, A. R., Jr. & Tappan, H., 1987. *Foraminiferal Genera and their Classification*. Van Nostrand Reinhold.

Long, J., 2006. *Swimming in Stone: The Amazing Gogo Fossils of the Kimberley*. Fremantle Arts Centre Press.

Long, J. & Schouten, P., 2008. *Feathered Dinosaurs: The Origin of Birds*. Oxford University Press.

Long, J. A., Young, G. C., Holland, T. *et al.*, 2006. An exceptional Devonian fish from Australia sheds light on tetrapod origins. *Nature*, **444**: 199–202.

Looy, C. V., Collinson, M. E., Van Konijnenburg, V. C. *et al.*, 2005. The ultrastructure and botanical affinity of end-Permian spore tetrads. *International Journal of Plant Science*, **166**: 875–7.

Lopez-Saez, J.-A. & Dominguez-Rodrigo, M., 2009. Palynology of OGS-6a and OGS-7, two new 2.6 Ma archaeological sites from Gona, Afar, Ethiopia: insights of aspects of Late Pliocene habitats and the beginnings of stone-tool use. *Geobios*, **42**: 503–11.

Lorente, M. A., 1986. *Palynology and Palynofacies of the Upper Tertiary in Venezuela*. J. Cramer.

Louchart, A., Wesselman, H., Blumenschine, R. J. *et al.*, 2009. Taphonomic, avian and small vertebrate indicators of *Ardipithecus ramidus* habitat. *Science*, **326**: 66e1–e4.

Loucks, R. G. & Sarg, J. F., 1993. *Carbonate Sequence Stratigraphy*. American Association of Pertroleum Geologists (Memoir No. 57).

Loughlin, N. J. D. & Hillier, R. D., 2010. Early Cambrian *Teichichnus*-dominated ichnofabrics and palaeoenvironmental analysis of the Caerfai group, southwest Wales, UK. *Palaeogeography, Palaeoclimatology, Palaeoecology*, **297**: 239–51.

Love, G. D., Grosjean, E., Stalvies, C. *et al.*, 2009. Fossil steroids record the appearance of demosponges during the Cryogenian period. *Nature*, **457**: 718–21.

Lovejoy, C. O., 2009. Re-examining human origins in light of *Ardipithecus ramidus*. *Science*, **326**: 74, 74e1–e8.

Lovejoy, C. O., Latimer, B., Suwa, G. *et al.*, 2009a. Combining prehension and propulsion: the foot of *Ardipithecus ramidus*. *Science*, **326**: 72, 72e1–e8.

Lovejoy, C. O., Simpson, S. W., White, T. D. *et al.*, 2009b. Careful climbing in the Miocene: the forelimbs of *Ardipithecus ramidus* and humans are primitive. *Science*, **326**: 70, 70e1–e8.

Lovejoy, C. O., Suwa, G., Simpson, S. W. *et al.*, 2009c. The great divides: *Ardipithecus ramidus* reveals the postcrania of our last common ancestors with African apes. *Science*, **326**: 73, 100–106.

Lovejoy, C. O., Suwa, G., Spurlock, L. *et al.*, 2009d. The pelvis and femur of *Ardipithecus ramidus*: the emergence of upright walking. *Science*, **326**: 71, 71e1–e6.

Lovell, B., 2006. Climate change: conflict of observational science, theory and politics: discussion. *American Association of Petroleum Geologists Bulletin*, **90**(3): 405–7.

2009. *Challenged by Carbon: The Petroleum Industry and Climate Change*. Cambridge University Press.

Lowe, S., Chitolie, J., Pearce, J. M. & Welsh, A., 1988. Applications of well-site palynology to hydrocarbon exploration. *Proceedings of the Indonesian Petroleum Association 17th Annual Convention, Jakarta*: 301–3.

Lucas, S. G., 2009. Global Jurassic tetrapod biochronology. *Volumina Jurassica*, **VI**: 99–108.

Lucas, S. G. (ed.), 2010. *The Triassic Timescale*. The Geological Society, London (Special Publication No. 265).

Lucas, S. G., Cassinis, G. & Schneider, J. W. (eds.), 2006. *Nonmarine Permian Biostratigraphy and Biochronology*. The Geological Society, London (Special Publication No. 265).

Lucas, S. G. & Morales, M. (eds.), 1993. *The Non-marine Triassic*. New Mexico Museum of Natural History & Science.

Luer, V., Cortese, G., Neil, H. L. *et al.*, 2009. Radiolarian-based sea surface temperatures and oceanographic canges during the Late Pleistocene-Holocene in the subantarctic southwest Pacific. *Marine Micropaleontology*, **70**: 151–65.

Luisi, P. L., 2006. *The Emergence of Life*. Cambridge University Press.

Lunt, D., Valdes, P. J., Dunkley Jones, T. *et al.*, 2010. CO_2-driven ocean circulation changes as anamplifier of Palaeocene–Eocene Thermalaximum hydrate destablization. *Geology*, **38**(10): 875–8.

Luoto, T. P., 2009. A Finnish chironomid- and chaoborid- based inference model for reconstructing past lake levels. *Quaternary Science Reviews*, **28**: 1481–9.

Lyman, R. L., 1994. *Vertebrate Taphonomy*. Cambridge University Press.

2008. *Quantitative Paleozoology*. Cambridge University Press.

Macaulay, C. I., Beckett, D., Braithwaite, K. *et al.*, 2001. Constraints on diagenesis and reservoir quality in the fractured Hasdrubal field, offshore Tunisia. *Journal of Petroleum Geology*, **24**(1): 55–78.

Macellari, C. E. & de Vries, T. J., 1987. Late Cretaceous upwelling and anoxic sedimentation in northwestern South America. *Palaeogeography, Palaeoclimatology, Palaeoecology*, **59**: 279–92.

MacGregor, D. S., Moody, R. T. J. & Clark-Lowes, D. D. (eds.), 1998. *Petroleum Geology of North Africa*. The Geological Society, London (Special Publication No. 132).

MacLennan, J. & Jones, S. M., 2006. Regional uplift, gas hydrate dissociation and the origins of the Paleocene–Eocene thermal maximum. *Earth and Planetary Science Letters*, **245**: 65–80.

MacQueen, R. W. & Leckie, D. A. (eds.), 1992. *Foreland Basins and Fold Belts*. American Association of Petroleum Geologists (Memoir No. 55).

MacRae, C., 1988. *Palynostratigraphic Correlation between the Lower Karoo Sequence of the Waterberg and Pafuni Coal-Bearing Basins and the Hammanskraal Plant Macrofossil Locality, Republic of South Africa*. Geological Survey of South Africa (Memoir No. 75).

1999. *Life Etched in Stone: Fossils of South Africa.* The Geological Society of South Africa.

Madella, M. & Zurro, D., 2007. *Plants, People and Places: Recent Studies in Phytolith Analysis.* Oxbow Books.

Mae, A., Yamanaka, T. & Shimoyama, S., 2007. Stable isotope evidence for identification of chemosynthesis-based fossil bivalves associated with 'cold seep' ages. *Palaeogeography, Palaeoclimatology, Palaeoecology,* 245: 411–20.

Magnavita, L. P. & da Silva, H. T. F., 1995. Rift-border system: the interplay between tectonics and sedimentation in the Reconcavo Basin, northeastern Brazil. *American Association of Petroleum Geologists Bulletin,* 79(11): 1590–1607.

Magoon, L. B. & Dow, W. G. (eds.), 1994. *The Petroleum System – From Source to Trap.* American Association of Petroleum Geologists (Memoir No. 60).

Mahanti, G., Patra, S., Banerjee, A. *et al.*, 2006. Outcrop study in south Cambay basin, India. *The Leading Edge,* July 2006: 869–72.

Mahmudi Gharaie, M. H., Matsumoto, R., Kakuwa, Y. & Milroy, P. G., 2004. Late Devonian facies variety in Iran: volcanism as a possible trigger of the environmental perturbation near the Frasnian-Famennian boundary. *Geological Quarterly,* 48(4): 323–32.

Mailliot, S., Mattioli, E., Guex, J. & Pittet, B., 2006. The Early Toarcian anoxia, a synchronous event in the Western Tethys? An approach by quantitative biochronology (Unitary Associations), applied on calcareous nannofossils. *Palaeogeography, Palaeoclimatology, Palaeoecology,* 240(3/4): 562–86.

Majima, R., Nobuhara, T. & Kitazaki, T., 2005. Review of fossil chemosynthetic assemblages in Japan. *Palaeogeography, Palaeoclimatology, Palaeoecology,* 227: 86–123.

Malkoc, M. & Mutterlose, J., 2010. The Early Barremian warm pulse and the Late Barremian cooling: a high-resolution geochemical record of the Boreal realm. *Palaios,* 25: 14–23.

Malone, C., 2001. *Neolithic Britain and Ireland.* Tempus Publishing Ltd.

Maltby, M. (ed.), 2006. *Integrating Zooarchaeology.* Oxbow Books.

Mamo, B., Strotz, L. & Dominey-Howes, D., 2009. Tsunami sediments and their foraminiferal assemblages. *Earth-Science Reviews,* 96: 263–78.

Mander, L., Kurschner, W. M. & McElwain, J. C., 2010. Response of terrestrial vegetation to Triassic–Jurassic climate change, east Greenland. *Abstracts, International Palaeontological Congress, London:* 261.

Mann, K. O. & Lane, H. R., 1995. *Graphic Correlation.* Society of Economic Paleontologists and Mineralogists/Society for Sedimentary Geology (Special Publication No. 53).

Mann, P. & Escalona, A.(eds.) (2010). Tectonics, basinal framework and petroleum systems of eastern Venezuela, the Leeward Islands, Trinidad and Tobago, and offshore areas. *Marine and Petroleum Geology (Special Issue),* 28(1): 1–278.

Mann, U. & Stein, R., 1997. Organic facies variations, source rock potential and sea level changes in Cretaceous black shales of the Quebrada Ocal, upper Magdalena valley, Colombia. *American Association of Petroleum Geologists Bulletin,* 81(4): 556–76.

Markov, A. V. & Korotayev, A. V., 2007. Phanerozoic marine diversity follows a hyperbolic trend. *Palaeoworld,* 16: 311–8.

Markwick, P. J., 1998. Fossil crocodilians as indicators of Late Cretaceous and Cenozoic climates: implications for using palaeontological data in reconstructing palaeoclimate. *Palaeogeography, Palaeoclimatology, Palaeoecology,* 137: 205–71.

Marquez, L., 2005. Foraminiferal fauna recovered after the Late Permian extinctions in Iberia and the westernmost Tethys area. *Palaeogeography, Palaeoclimatology, Palaeoecology,* 229: 137–57.

Marsh, P. E. & Cohen, A. D., 2008. Identifying high-level salt marshes using a palynomorphic fingerprint with potential implications for tracking sea level change. *Review of Palaeobotany and Palynology,* 148: 60–9.

Marshall, P. R., 2010. Geosteering with minimal LWD capability: biosteering during underbalanced, coiled tubing drilling operations for gas in the Middle East and the development of the Stratsteer concept. *Abstracts, EAGE Geosteering & Well Placement Workshop, Dubai.*

Marshall, P. R., Burchette, T. & Ali, K. S., 2007. Integrated subsurface geology and biosteering: a case study from the Sajaa field, Sharjah, UAE. *Abstracts, Integrated Petroleum Technology Conference, Dubai.*

Martill, D. M., Bechly, G. & Loveridge, R. F. (eds.), 2007. *The Crato Fossil Beds of Brazil.* Cambridge University Press.

Martin, P. S., 2005. *Twilight of the Mammoths: Ice Age Extinction and the Rewilding of America.* University of California Press.

Martin, R. A., Nesbitt, E. A. & Campbell, K. A., 2007. Carbon stable isotopic composition of benthic foraminifera from Pliocene cold methane seeps, Cascadia accretionary margin. *Palaeogeography, Palaeoclimatology, Palaeoecology,* 246: 260–77.

2010. The effects of anaerobic methane oxidation on benthic foraminiferal assemblages and stable isotopes on Hikurangi margin of eastern New Zealand. *Marine Geology,* 272: 270–84.

Martin, R. E. (ed.), 2000. *Environmental Micropaleontology.* Kluwer Academic/Plenum Press (Topics in Geobiology Series No. 15).

Martin-Closas, C., Serra-Kiel, J., Busquets, P. & Ramos-Guerrero, E., 1999. New correlation between charophyte and larger foraminifer biozones (Middle Eocene, southeastern Pyrenees). *Geobios,* 31(1): 5–18.

Martinez, A. R., 1989. *Venezuelan Oil – Development and Chronology.* Elsevier Applied Science.

Martinez, J. I., 1995. Microfossiles del grupo Guadalupe y la formacion Guaduas (Campaniano-Maastrichtiano) en la seccion de Tausa, Cundinamarea, Colombia. *Cienca, Tecnologia y Futuro,* 1(1): 65–81.

2003. The paleoecology of Late Cretaceous upwelling events from the upper Magdalena basin, Colombia. *Palaios,* 18: 305–20.

Martinez, M. A., Pramparo, M. B., Quattrocchio, M. E. & Zavala, C. A., 2008. Depositional environments and hydrocarbon potential of the Middle Jurassic Las Molles formation, Neuquen basin, Argentina: palynofacies and organic geochemical data. *Revista Geologica de Chile*, **35**(2): 279–305.

Martinsen, O. J. & Dreyer, T. (eds.), 2001. *Sedimentary Environments Offshore Norway – Palaeozoic to Recent*. Elsevier (under the auspices of the Norwegian Petroleum Society).

Maruyama, S. & Santosh, M., 2008. Models of Snowball Earth and Cambrian explosion: a synthesis. *Gondwana Research*, **14**: 22–32.

Marzoli, A., Bertrand, H., Knight, K. *et al.*, 2004. Synchrony of the Central Atlantic Magmatic Province and the Triassic–Jurassic boundary climatic and biotic crisis. *Geology*, **32**: 973–6.

Mascarelli, A. L., 2009. Quaternary geologists win timescale vote. *Nature*, **459**: 624.

Masse, J.-P. & Fenerci-Masse, M., 2008. Time contrasting palaeobiogeographies among Hauterivian–lower Aptian rudist bivalves from the Mediterranean Tethys, their climatic control and palaeoecological implications. *Palaeogeography, Palaeoclimatology, Palaeoecology*, **269**: 54–65.

Masters, G. M. & Ela, W. P. (eds.), 2008. *Inroduction to Environmental Engineering and Science*, 3rd edition. Pearson Education Inc.

Masure, E. & Vrielynck, B., 2009. Late Albian dinoflagellate cyst paleobiogeoraphy as indicator of asymmetric sea surface temperature gradient on both hemispheres with southern high latitudes warmer then northern ones. *Marine Micropaleontology*, **70**: 120–33.

Matthews, R. K. & Frohlich, C., 2002. Maximum flooding surfaces and sequence boundaries: comparsons between observations and orbital forcing in the Cretaceous and Jurassic (65–190 Ma). *GeoArabia*, **7**(3): 503–38.

Mattner, J. & Al-Husseini, M., 2002. Essay: applied cyclostratigraphy for the Middle East E & P industry. *GeoArabia*, **7**(4): 734–44.

Mayhew, P. J., Jenkins, G. B. & Benton, T. G., 2007. A long-term association between global temperature and biodiversity, origination and extinction in the fossil record. *Proceedings of the Royal Society, B*, DOI: 10.1098/rspb.**2007**.1302.

Mays, S., 1998. *The Archaeology of Human Bones*. Routledge.

McCabe, P. J. & Parrish, J. T. (eds.), 1992. *Controls on the Distribution and Quality of Cretaceous Coals*. Geological Society of America (Special Publication No. 267).

McCall, G. J. H., 2006. The Vendian (Ediacaran) in the geological record: enigmas in geology's prelude to the Cambrian explosion. *Earth-Science Reviews*, **77**: 1–129.

McCall, J., 2010. Lake Bogoria, Kenya: hot and warm springs, geysers and Holocene stromatolites. *Earth-Science Reviews*, **103**: 71–9.

McCall, J., Rosen, B. & Darrell, J., 1994. Carbonate deposition in accretionary prism settings: Early Miocene coral limestones and corals of the Makran mountain range in southern Iran. *Facies*, **31**: 141–78.

McCarthy, T. S. & Pretorius, K., 2009. Coal mining on the Highveld and its implications for future water quality in the Vaal river system. *Abstracts of the International Mine Water Conference Proceedings, Pretoria*: 56–65.

McCourt, R. M., Delwiche, C. F. & Karol, K. G., 2004. Charophyte algae and land plant origins. *Trends in Ecology and Evolution*, **19**(12): 661–6.

McElwain, J. C., Beerling, D. J. & Woodward, F. I., 1999. Fossil plants and global warming at the Triassic-Jurassic boundary. *Science*, **285**: 1386–91.

McElwain, J. C., Wagner, P. J. & Hesselbo, S. P., 2009. Fossil plant relative abundances indicate sudden loss of Late Triassic biodiversity in east Greenland. *Science*, **324**: 1554–6.

McFadden, K. A., Shuhai Xiao, Chuanming Zhou & Kowalewski, M., 2009. Quantitative evaluation of the biostratigraphic distribution of acanthomorphic acritarchs in the Ediacaran Doushantuo formation in the Yangtze Gorges area, south China. *Precambrian Research*, **173**: 170–90.

McGann, M., Vrijenhoek, R. C., Johnson, S. *et al.*, 2010. Foraminiferal response to whale falls in the northeastern Pacific Ocean. *Abstracts, Forams 2010 International Symposium on Foraminifera, Bonn*: 140.

McGowan, C., 2001. *The Dragon Seekers: The Discovery of Dinosaurs during the Prelude to Darwin*. Perseus Books.

McGowran, B., 2005. *Biostratigraphy: Principles and Practice*. Cambridge University Press.

Mchedlishvili, P. A., 1951. The palaeogeography of the Caucasus in Karagan time in association with palaeobotanic data. *Doklady Akademii Nauk Soyuza Sovetskikh Sotsialisticheskikh Respublik*, **78**(5): 921–3 (in Russian).

McKee, B. A., Aller, R. C., Allison, M. A. *et al.*, 2004. Transport and transformation of dissolved and particulate materials on continental margins influenced by major rivers: benthic boundary layer and seabed processes. *Continental Shelf Research*, **24**: 899–926.

McKenna, M. C., 1983. Holarctic landmass rearrangement, cosmic events, and Cenozoic terrestrial organisms. *Annals Missouri Botanical Gardens*, **70**: 459–89.

McKerrow, W. S. (ed.), 1978. *The Ecology of Fossils*. Duckworth.

McKerrow, W. S. & Scotese, C. R. (eds.), 1990. *Palaeozoic Palaeogeography and Biogeography*. The Geological Society, London (Memoir No. 12).

McLaughlin, P. I., Brett, C. E., McLaughlin, S. L. T. & Cornell, S. R., 2004. High-resolution sequence stratigraphy of a mixed carbonate-siliciclastic, cratonic ramp (Upper Ordovician; Kentucky-Ohio, USA): insights into the relative influence of eustasy and tectonics through analysis of facies gradients. *Palaeogeography, Palaeoclimatology, Palaeoecology*, **210**(2–4): 267–94.

McLoughlin, S. & Pott, C., 2009. The Jurassic flora of Western Australia. *GFF*, **131**(1): 113–36.

McNamara, K. J. (ed.), 1990. *Evolutionary Trends*. Belhaven.

McNeil, D. H., Issler, D. R. & Snowdon, L. R., 1996. Colour alteration, thermal maturity, and burial diagenesis in fossil foraminifers. *Geological Survey of Canada Bulletin*, **499**: 1–34.

Mead, J. G., 1975. A fossil beaked whale (Cetacea: Ziphiidae) from the Miocene of Kenya. *Journal of Paleontology*, **49**: 745–51.

MEEI (Ministry of Energy and Energy Industries) 2009. *Trinidad and Tobago – Celebrating a Century of Commercial Oil Production*. The Ministry of Energy and Energy Industries.

Megirian, D., Prideaux, G. J., Murray, P. F. & Smit, N., 2010. An Australian Land Mammal Age biochronological scheme. *Paleobiology*, **36**(4): 658–71.

Mehrotra, N. C. & Aswal, H. S., 2003. *Late Jurassic–Cretaceous dinoflagellate cysts*. Volume I of *Atlas of dinoflagellate cysts from Mesozoic–Tertiary sediments of Krishna–Godovari basin*. K.D. Malaviya Institute of Petroleum Exploration, Oil and Natural Gas Corporation Limited (Paleontographica Indica, 7).

Mehrotra, N. C. & Singh, K., 2003. *Tertiary Dinoflagellate Cysts*. Volume II of *Atlas of Dinoflagellate Cysts from Mesozoic–Tertiary Sediments of Krishna–Godovari basin*. K.D. Malaviya Institute of Petroleum Exploration, Oil and Natural Gas Corporation Limited (Paleontographica Indica, 8).

Mehrotra, N. C., Venkatachala, B. S. & Kapoor, P. N., 2005. *Palynology in Hydrocarbon Exploration (the Indian Scenario). Part II: Spatial and Temporal Distribution of Significant Spores, Pollen and Dinoflagellate Cysts in the Mesozoic–Cenozoic Sediments of Petroliferous Basins*. Geological Society of India (Memoir No. 61).

2010. Palynology in hydrocarbon exploration: high impact palynological studies in western offshore and Krishna–Godavari basins. *Journal of the Geological Society of India*, **75**: 364–79.

Mehrotra, N. C., Venkatachala, B. S., Swamy, S. N. & Kapoor, P. N., 2002. *Palynology in Hydrocarbon Exploration (the Indian Scenario). Part I: Category I Basins*. Geological Society of India (Memoir No. 48).

Mei-Fu Zhou, Malpas, J., Xie-Yan Song et al., 2002. A temporal link between the Emeishan Large Igneous Province (SW China) and the end-Guadalupian mass extinction. *Earth and Planetary Science Letters*, **196**(3–4): 113–22.

Meisterfeld, R., Holzmann, M. & Pawlowski, J., 2001. Morphological and molecular characterization of a new terrestrial allogromiid species: *Edaphoallogromia australica gen. et sp. nov.* (Foraminifera) from northern Queensland (Australia). *Protist*, **152**: 185–92.

Mejri, F., Burollet, P. F. & Ben Ferjani, A., 2006. *Petroleum Geology of Tunisia: A Renewed Synthesis*. ETAP (Memoir No. 22).

Melchin, M. J. & DeMont, M. E., 1995. Possible propulsion modes in Graptoloidea: a new model for graptoloid locomotion. *Paleobiology*, **21**(1): 110–20.

Mellars, P., 1995. *The Neanderthal Legacy: An Archaeological Perspective of Western Europe*. Princeton University Press.

Mello, M. R. & Katz, B. J. (eds.), 2000. *Petroleum Systems of South Atlantic Margins*. American Association of Petroleum Geologists (Memoir No. 73).

Melluso, L., Sheth, H. C., Mahoney, J. J. et al., 2009. Correlations between silicic volcanic rocks of the St Mary's Islands (southwestern India) and eastern Madagascar: implications for Late Cretaceous India–Madagascar reconstructions. *Journal of the Geological Society of London*, **166**: 283–94.

Melott, A. L. & Thomas, B. C., 2009. Late Ordovician geographic patterns of extinction compared with simulations of astrophysical ionizing radiation damage. *Paleobiology*, **35** (3): 311–20.

Meltzer, D. J., 2009. *First Peoples in a New World: Colonizing Ice Age America*. University of Californa Press.

Menanteau, L., Vanney, J. R. & Zazo, C., 1983. *Belo II: Belo et son environnement (Detroit de Gibraltar). Etude physique d'une site antique*. Broccard (Serie Archeologie No. 4).

Merceron, G., de Bonis, L., Viriot, L. & Blondel, C., 2005. Dental micro-wear of fossil bovids from northern Greece: paleoenvironmental conditions in the eastern Mediterranean during the Messinian. *Palaeogeography, Palaeoclimatology, Palaeoecology*, **217**: 173–85.

Merceron, G., Zazzo, A., Spassov, N. et al., 2006. Bovid paleoecology and paleoenvironments from the Late Miocene of Bulgaria: evidence from dental micro-wear and stable isotopes. *Palaeogeography, Palaeoclimatology, Palaeoecology*, **241**: 637–54.

Mesolella, K. J., 1967. Zonation of uplifted Pleistocene coral reefs on Barbados, West Indies. *Science*, **156**: 638–40.

Mesolella, K. J., Sealy, H. A. & Matthews, R. K., 1970. Facies geometries within Pleistocene reefs of Barbados, West Indies. *American Association of Petroleum Geologists Bulletin*, **54**: 1890–1917.

Messager, E., Lordkipanidze, D., Delhon, C. & Ferring, C. R., 2010. Palaeoecological implications of the Lower Pleistocene phytolith record from the Dmanisi site (Georgia). *Palaeogeography, Palaeoclimatology, Palaeoecology*, **288**: 1–13.

Metais, G., Antoine, P.-O., Baqri, S. R. H. et al., 2009. Lithofacies, depositional environments, regional biostratigraphy and age of the Chitarwata formation in the Bugti Hills, Balochistan, Pakistan. *Journal of Asian Earth Sciences*, **34**: 154–67.

Mette, W., 2004. Middle to Upper Jurassic sedimentary sequences and marine biota of the early Indian Ocean (southwest Madagascar): some biostratigraphic, palaeoecologic and palaeobiogeographic conclusions. *Journal of African Earth Sciences*, **38**: 331–42.

Meyerhoff, A. A., Boucot, A. J., Meyerhoff Hull, D. & Dickins, J. M., 1996. *Phanerozoic Faunal & Floral Realms of the Earth: The Intercalary Relations of the Malvinokaffric and Gondwana Faunal Realms with the Tethyan Faunal Realm*. Geological Society of America (Memoir No. 189).

Miall, A. D., 2010. *The Geology of Stratigraphic Sequences*. Springer

Midant-Reynes, B., 2000. *The Prehistory of Egypt*. Blackwell (English translation of French original).

Mienis, F., van der Land, C., de Stigter, H. C. et al., 2009. Sediment accumulation on a cold-water carbonate mound at the southwest Rockall Trough margin. *Marine Geology*, **265**: 40–50.

Miller, A.I., 1997. Dissecting global diversity patterns: examples from the Ordovician radiation. *Annual Reviews of Ecological Systems*, **28**: 85–104.

Miller, B.G., 2005. *Coal Energy Systems*. Elsevier.

Miller, C.G., Richter, M. & do Carmo, D.A., 2002. Fish and ostracod remains from the Santos basin (Cretaceous to Recent), Brazil. *Geological Journal*, **37**: 297–316.

Miller, M. & Melvin, J., 2005. Significant new biostratigraphic horizons in the Qusaiba member of the Silurian Qalibah formation of central Saudi Arabia, and their sedimentologic expression in a sequence stratigraphic context. *GeoArabia*, **10**(1): 49–92.

Millo, C., Sarnthein, M., Erlenkauser, H. *et al.*, 2006. Methane-induced early diagenesis of foraminiferal tests in the southwestern Greenland Sea. *Marine Micropaleontology*, **58**: 1–12.

Milsom, C. & Rigby, S., 2004. *Fossils at a Glance*. Blackwell.

Minter, N.J., Krainer, K., Lucas, S.G. *et al.*, 2007. Palaeoecology of an Early Permian playa lake trace fossil assemblage from Castle Peak, Texas, USA. *Palaeogeography, Palaeoclimatology, Palaeoecology*, **246**: 390–423.

Mintz, J.S., Driese, S.G. & White, J.D., 2010. Environmental and ecological variability of Middle Devonian (Givetian) forests in Appalachian Basin paleosols, New York, United States. *Palaios*, **25**: 85–96.

Minwer-Barakat, R., Garcia-Alix, A., Agusti, J. *et al.*, 2009. The micromammal fauna from Negratin-1 (Guadix basin, southern Spain): new evidence of African-Iberian mammal exchanges during the Late Miocene. *Journal of Paleontology*, **83**(6): 854–79.

Miranda, A.M., Rodrigues, C.F. & Lemos de Sousa, M.J., 2009. Process for carbon dioxide sequestration/storage by injection in coal seams. *3rd Symposium on Gondwana Coals, Pontifical Catholic University of Rio Grande do Sul (PUCRS), Porto Alegre, Brazil*.

Mischke, S. & Holmes, J.A., 2008. Applications of lacustrine and marginal marine Ostracoda to palaeoenvironmental reconstruction. *Palaeogeography, Palaeoclimatology, Palaeoecology*, **264**: 21–2.

Mishra, P.K., 1996. Study of Miogypsinidae and associated planktonics from Cauvery, Krishna–Godavari and Andaman basins of India. *Geoscience Journal*, **47**(2): 123–251.

Misumi, K. & Yamanaka, Y., 2008. Ocean anoxic events in the mid-Cretaceous simulated by a 3-D biogeochemical general circulation model. *Cretaceous Research*, **29**: 893–900.

Mitchell, C.E., Melchin, M.J., Cameron, C.B. & Maletz, J., 2010. Phylogeny of the tube-building Hemichordata reveals that *Rhabdopleura* is an extant graptolite. *Abstracts, International Palaeontological Congress, London*: 283.

Mitchell-Jones, A.J., Amuri, G., Bogdanowicz, W. *et al.* (eds.), 1999. *The Atlas of European Mammals*. Poyser Natural History.

Moczydlowka, M., 2008. The Ediacaran microbiota and the survival of Snowball Earth conditions. *Precambrian Research*, **167**: 1–15.

Modie, B.N. & le Herisse, A., 2009. Late Palaeozoic palynomorph assemblages from the Karoo supergroup and their potential for biostratigraphic correlation, Kalahari Karoo basin, Botswana. *Bulletin of Geosciences*, **84**(2): 337–58.

Moe, A.P. & Smith, D.M., 2005. Using pre-Quaternary Diptera as indicators of paleoclimate. *Palaeogeography, Palaeoclimatology, Palaeoecology*, **221**: 203–14.

Mohinuddin, S.K., Satyanarayana, K. & Rao, G.N., 1991. Cretaceous sedimentation in the sub-surface of Krishna–Godavari basin. *Journal of the Geological Society of India*, **41**: 533–9.

Moine, O. & Rousseau, D.-D., 2002. Mollusques terrestres et temperatures: une nouvelle fonction de transfert quantitative. *Comptes Rendus Palevol*, **1**: 145–151.

Mojtahid, M., Jorissen, F., Durrieu, J. *et al.*, 2006. Benthic foraminifera as bio-indicators of drill cutting disposal in tropical east Atlantic outer shelf environments. *Marine Micropaleontology*, **61**: 58–75.

Mojtahid, M., Jorissen, F. & Pearson, T.H., 2008. Comparison of benthic foraminiferal and macrofaunal responses to organic pollution in the Firth of Clyde (Scotland). *Marine Pollution Bulletin*, **56**: 42–70.

Moldowan, J.M., Dahl, J., Huizinga, B.J. *et al.*, 1994. The molecular fossil record of oleanane and its relation to angiosperms. *Science*, **265**: 768–71.

Moldowan, J.M., Dahl, J., Jacobson, S.R. *et al.*, 1996. Chemostratigraphic reconstruction of biofacies: molecular evidence linking cyst-forming dinoflagellates with pre-Triassic ancestors. *Geology*, **24**(2): 159–62.

Molina, E., Alegret, L., Arenillas, I. *et al.*, 2006. The Global Stratotype Section and Point for the base of the Danian stage (Paleocene, Paleogene, 'Tertiary', Cenozoic) at El Kefm Tunisia – original definition and revision. *Episodes*, **29**(4): 263–273.

Molineux, A., Scott, R.W., Ketcham, R.A. & Maisano, J.A., 2007. Rudist taxonomy using X-ray computed tomography. *Palaeontologia Electronica*, **10**(3).

Molnar, P., 2005. Mio-Pliocene growth of the Tibetan Plateau and evolution of east Asian climate. *Palaeontologia Electronica*, **8**(1): 23pp.

Molyneux, S., 2009. Acritarch (marine microplankton) diversity in an Early Ordovician deep-water setting (the Skiddaw group, northern England): implications for the relationship between sea-level change and phytoplankton diversity. *Palaeogeography, Palaeoclimatology, Palaeoecology*, **275**: 59–76.

Molyneux, S., Osterloff, P., Penney, R. & Spaak, P., 2006. Biostratigraphy of the Lower Palaeozoic Haima supergroup, Oman; its application in sequence stratigraphy and hydrocarbon exploration. *GeoArabia*, **11**(2): 17–48.

Montaggioni, L.F., 2005. History of Indo-Pacific coral reef systems since the last glaciation: development patterns and controlling factors. *Earth-Science Reviews*, **71**: 1–75.

Montenegro, A., Hetherington, R., Eby, M. & Weaver, A.J., 2006. Modelling pre-historic transoceanic crossings into the Americas. *Quaternary Science Reviews*, **25**: 1323–38.

Montoya, E., Rull, V. & van Geel, B., 2010. Non-pollen palyno-morphs from surface sediments along an altitudinal transect of the Venezuelan Andes. *Palaeogeography, Palaeoclimatology, Palaeoecology*, **297**: 169–83.

Montuire, S., Maridet, O. & Legendre, S., 2006. Late Miocene–Early Pliocene temperature estimates in Europe using rodents. *Palaeogeography, Palaeoclimatology, Palaeoecology*, **238**: 247–62.

Moreira, J.L.P., Madeira, C.V., Gil, J.A. & Machado, M.A.P., 2007. Bacia do Santos. *Boletim Geociencias Petrobras, Rio de Janeiro*, **15**(2): 531–49.

Moran-Zenteno, D., 1994. *The Geology of the Mexican Republic*. The American Association of Petroleum Geologists (Studies in Geology No. 39).

Moreno-Vasquez, J., 1995. Neogene biofacies in eastern Venezuela and their calibration with seismic data. *Marine Micropaleontology*, **26**: 287–302.

Morley, R.J., 1991. Tertiary stratigraphic palynology in southeast Asia: current status and new directions. *Geological Society of Malaysia*, **28**: 1–36.

Morris, P. & Therivel, R. (eds.), 2001. *Methods of Environmental Impact Assessment*, 2nd edition. Spon Press.

Mortensen, P.B., Buhl-Mortensen, L., Gebruk, A.V. & Krylova, E.M., 2008. Occurrence of deep-water corals on the Mid-Atlantic Ridge based on MAR-ECO data. *Deep-Sea Research II*, **55**: 142–52.

Morton, A.C. (ed.), 1992. *Geology of the Brent Group*. The Geological Society, London (Special Publication No. 61).

Morwood, M. & Oosterzee, P.V., 2007. *The Discovery of the Hobbit. The Scientific Breakthrough that Changed the Face of Human History*. Random House Australia.

Morwood, M., Sutikna, T. & Roberts, R., 2005. The people time forgot. *National Geographic Magazine, April* **2005**: 2–15.

Moss, G.D., Cathro, D.L. & Austin, J.A. Jr, 2004. Sequence biostratigraphy of prograding clinoforms, northern Carnarvon basin, western Australia: a proxy for variations in Oligocene to Pliocene global sea level? *Palaios*, **19**(3): 206–26.

Motoyama, I. & Maruyama, T., 1998. Neogene diatom and radiolarian biochronology for the middle to high latitudes of the northwest Pacific region: calibration to the Cande and Kent's geomagnetic polarity time scales (CK92 and CK95). *Journal of the Geological Society of Japan*, **104**: 171–83.

Moura, J.A., 1972. Algumas species e subspecies novas de ostracodes da Bacia Reconcavo/Tucano. *Boletim Tecnico de Petrobras*, **15**: 245–263.

Mukherjee, A., Fryar, A.E. & Thomas, W.A., 2009. Geologic, geomorphic and hydrologic framework and evolution of the Bengal basin, India and Bangladesh. *Journal of Asian Earth Sciences*, **34**: 227–44.

Muller, J., di Giacomo, E. de & van Erve, A.W., 1987. A palynological zonation for the Cretaceous, Tertiary and Quaternary of northern South America. *American Association of Stratigraphic Palynologists Contributions Series*, **19**: 7–76.

Munnecke, A. & Servais, T., 2007. What caused the Ordovician biodiversification? *Palaeogeography, Palaeoclimatology, Palaeoecology*, **245**: 1–4.

Murchison, D.G. & Westall, T.S. (eds.), 1968. *Coal and Coal-Bearing Strata*. Oliver & Boyd.

Murray, A.M., 2000. The Palaeozoic, Mesozoic and Cenozoic fossil fishes of Africa. *Fish and Fisheries*, **1**: 111–45.

Mussard, J.-M., Gerard, J., Ducazeuax, J. & Begouen, V., 1994. Quantitative palynology: a tool for the recognition of genetic depositional sequences: application to the Brent group. *Bulletin du Centre de Recherches Exploration-Production Elf-Aquitaine*, **18**(2): 463–74.

Mutterlose, J. & Wiedenroth, K., 2008. Early Cretaceous (Valanginian-Hauterivian) belemnites from western Morocco: stratigraphy and palaeoecology. *Cretaceous Research*, **29**: 814–29.

Nagendra, R., Kamalak Kannan, B.V., Sen, G. *et al.*, 2010. Sequence surfaces and paleobathymetric trends in Albian to Maasrichtian sediments of Ariyalur area, Cauvery basin, India. *Marine and Petroleum Geology*.

Nagy, J., 1992. Environmental significance of foraminiferal morphogroups in Jurassic North Sea deltas. *Palaeogeography, Palaeoclimatology, Palaeoecology*, **95**: 111–34.

Nagy, J., Hess, S. & Alve, E., 2010. Environmental significance of foraminiferal assemblages dominated by small-sized *Ammodiscus* and *Trochammina* in Triassic and Jurassic delta-influenced deposits. *Earth-Science Reviews*, **9**: 31–49.

Nairn, A.E.M. & Stehli, F.G. (eds.), 1973. *The South Atlantic*. Volume 1 of *The Ocean Basins and Margins*. Plenum Press.

Nanda, A.C. & Mathur, A.K. (co-conveners), 2000. *Neogene Sequences of India*. Geological Society of India.

Natural History Museum, Basel, 1996. *Treatise on the Geology of Trinidad – Detailed Geological Maps and Sections*. Natural History Museum, Basel.

Neale, J.W. & Brasier, M.D., 1981. *Microfossils from Recent and Fossil Shelf Seas*. Ellis Horwood.

Negi, A.S., Sahu, S.K., Thomas, P.D. *et al.*, 2006. Fusing geologic knowledge and seismic in searching for subtle hydrocarbon traps in India's Cambay basin. *The Leading Edge*, July 2006: 872–80.

Nelson, C.H. & Damuth, J.E., 2003. Myths of turbidite system control: insights provided by modern turbidite studies. *Abstracts, International Conference on Deep-Water Processes in Modern and Ancient Environments, Barcelona*: 32.

Nelson, S.V., 2005. Paleoseasonality inferred from equid teeth and intra-tooth isotopic variability. *Palaeogeography, Palaeoclimatology, Palaeoecology*, **222**: 122–44.

Neuweiler, F., Turner, E.C. & Buridge, D.J., 2009. Early Neoproterozoic origin of the metazoan clade recorded in carbonate rock texture. *Geology*, **37**(5): 475–8.

Nevle, R.J. & Bird, D.K., 2008. Effects of syn-pandemic fire reduction and reforestation in the tropical Americas on atmospheric CO_2 during the European conquest. *Palaeogeography, Palaeoclimatology, Palaeoecology*, **264**: 25–38.

Nichols, D.J. & Johnson, K.R., 2008. *Plants and the K–T Boundary*. Cambridge University Press.

Nicholls, K., 2009. Trace fossil assemblage associated with the end-Ordovician 'extinction' event, Llangrannog, Dyfed. *Abstracts, Palaeontological Association Annual Meeting, Birmingham.*

Niedzwiedzki, G., Szrek, P., Narkiewicz, K. *et al.*, 2010. Tetrapod trackways from the early Middle Devonian period of Poland. *Nature*, **463**: 43–8.

Nigam, R., 2005. Addressing environmental issues through foraminifera – case studies from the Arabian Sea. *Journal of the Palaeontological Society of India*, **50**(2): 25–36.

Nigrini, C. & Caulet, J.P., 1992. Late Neogene radiolarian assemblages characteristic of Indo-Pacific areas of upwelling. *Micropaleontology*, **38**(2): 139–64.

Niklas, K.J., Tiffney, B.H. & Knoll, A.H., 1983. Patterns of vascular land plant diversification. *Nature*, **303**: 614–616.

Nomade, S., Knight, K.B., Beutel, E. *et al.*, 2007. Chronology of the Central Atlantic Magmatic Province: implications for the Central Atlantic rifting processes and the Triassic–Jurassic biotic crisis. *Palaeogeography, Palaeoclimatology, Palaeoecology*, **244**: 326–44.

Norman, M., 2003. *Cephalopods: A World Guide*, 2nd edition. Conch Books.

Nudds, J.R. & Selden, P.A., 2008. *Fossil Ecosystems of North America: A Guide to the Sites and their Extraordinary Fossils.* Manson Publishing.

Nunn, C.V. & Price, D.G., 2010. Late Jurassic (Kimmeridgian–Tithnian) stable isotopes (del^{18}O, del^{13}C) and Mg/Ca ratios: new palaeoclimatic data from Helmsdale, northeast Scotland. *Palaeogeography, Palaeoclimatology, Palaeoecology*, **292**: 325–35.

Nunn, C.V., Price, D.G., Grocke, D.R. *et al.*, 2010. The Valanginian positive carbon isotope event in Arctic Russia: evidence from terrestrial and marine isotope records and implications for global carbon cycles. *Cretaceous Research*, **31**: 577–92.

O'Connor, T. & Evans, J.G., 2005. *Environmental Archaeology: Principles and Methods.* Sutton.

Odada, E.O. & Olago, D.O. (eds.), 2002. *The East African Great Lakes: Limnology, Palaeolimnology and Biodiversity.* Kluwer Academic Publishers.

Oeggl, K., Jofler, W., Schmidl, A. *et al.*, 2007. The reconstruction of the last itinerary of 'Otzi', the Neolithic Iceman, by pollen analyses from sequentially sampled gut extracts. *Quaternary Science Reviews*, **26**: 853–61.

Oertli, H. (ed.), 1984. *Benthos'83 (Proceedings of the 2nd International Symoposium on Benthic Foraminifera, Pau, 1983).* Elf Aquitaine, Esso REP, and Total CFP.

Ogden, C.G. & Hedley, R.H., 1980. *An Atlas of Freshwater Testate Amoebae.* British Museum (Natural History)/Oxford University Press.

Ogg, J.G., Ogg, G. & Gradstein, F.M., 2008. *The Concise Geologic Time Scale.* Cambridge University Press.

Ojeda, H.A.O., 1982. Structural framework, stratigraphy and evolution of Brazilian margin basins. *American Association of Petroleum Geologists Bulletin*, **66**(6): 732–49.

Olivarez, C.A., Anton, M.G., Manzaneque, F.G. & Guaristi, C.M., 2005. Palaeoenvironmental interpretation of the Neogene locality Caranjeca (Reocin, Cantabria, N. Spain) from comparative studies of wood, charcoal and pollen. *Review of Palaeobotany and Palynology*, **132**: 133–57.

Oliveri, S.R. & Bohacs, K. (eds.), 2005. *Field Safety in Uncontrolled Environments: A Process-Based Guidebook.* American Association of Petroleum Geologists.

Ollivier, B. & Magot, M. (eds.), 2005. *Petroleum Microbiology.* American Society for Microbiology Press.

Olson, H.C. & Leckie, R.M. (eds.), 2003. *Micropaleontologic Proxies for Sea-Level Change and Stratigraphic Discontinuities.* Society of Economic Paleontologists and Mineralogists (Special Publication No. 75).

Olsson, P.K. & Nyong, E.E., 1984. A paleoslope model for Campanian-lower Maestrichtian foraminifera of New Jersey and Delaware. *Journal of Foraminiferal Research*, **14**(1): 50–68.

Olu-Le Roy, K., von Cosel, R., Hourdez, S. *et al.*, 2007. Amphi-Atlantic cold-seep *Bathymodiolus* species complexes across the equatorial belt. *Deep-Sea Research, I*, **54**: 1890–911.

Orr, J.C., Fabry, V.J., Aumont, O. *et al.*, 2005. Anthropogenic ocean acidification over the twenty-first century and its impact on calcifying organisms. *Nature*, **437**: 681–6.

Ortner, D.J. (ed.), 2003. *Identification of Pathological Conditions in Human Skeletal Remains.* Elsevier.

Osborne, C.P. & Beerling, D.J., 2006. Nature's green revolution: the remarkable evolutionary rise of C4 plants. *Philosophical Transactions of the Royal Society, B*, **361**: 173–94.

OSM, 2001. *Technical Measures for the Investigation and Mitigation of Fugitive Methane Hazards in Areas of Coal Mining.* Office of Surface Mining, Reclamation and Enforcement, US Department of the Interior.

Over, D.J., Morrow, J.R. & Wignall, P.B. (eds.), 2005. *Understanding Late Devonian and Permian–Triassic Biotic and Climatic Events.* Elsevier (Developments in Palaeontology & Stratigraphy, 20).

Owens, B., McLean, D. & Bodman, D., 2004. A revised palynozonation of British Namurian deposits and comparisons with eastern Europe. *Micropaleontology*, **50**(1): 89–103.

Packard, A., 1972. Cephalopods and fish: the limits of convergence. *Biological Reviews*, **47**: 247–301.

Paik, I.S., 2005. The oldest record of microbial-caddisfly bioherms from the Early Cretaceous Jinju formation, Korea: occurrence and palaeoenvironmental implications. *Palaeogeography, Palaeoclimatology, Palaeoecology*, **218**: 301–15.

Palacios, T., Jensen, S. & Mus, M.M., 2009. Morphological evidences of the presences of dinoflagellates in Cambrian oceans. *Abstracts, Palaeontological Association Annual Meeting, Birmingham*, 68–9.

Palmer, D. & Rickards, B. (eds.), 1991. *Graptolites.* Boydell Press.

Pande, D.K., Singh, R.R. & Chandra, K., 2008. Source rocks in deep water depositional systems of east and west coasts of India. *Search and Discovery* Article 10169. http://www.searchanddiscovery.com

Pandey, J., Azmi, R. J., Bhandari, A. & Dave, A. (eds.), 1996. *Contributions from the XVth Indian Colloquium on Micropalaeontology and Stratigraphy, Dehra Dun.* Dehra Dun.

Pandey, J. & Choubey, M. S., 1984. Middle Late Eocene palaeoecology and sediments distribution pattern in the south Cambay basin. *Proceedings of the Xth Indian Colloquium on Micropalaeontology & Stratigraphy*: 339–50.

Pandey, S. C., Pande, A. & Sharma, B. K., 2000. Source organics and depositional environment in Ratnagiri and Mukta blocks, Bombay offshore basin, India. *Proceedings of the 5th International Conference & Exhibition, Petroleum Geochemistry & Exploration in the Afro-Asian Region, New Delhi*: 325–9.

Panieri, G., 2005. Benthic foraminifera associated with a hydrocarbon seep in the Rockall Trough (NE Atlantic). *Geobios*, **38**: 247–55.

2006. Foraminiferal response to an active methane seep environment: a case study from the Adriatic Sea. *Marine Micropaleontology*, **61**: 116–30.

Panieri, G. & Camerlanghi, A., 2010. Benthic foraminifera as proxy of methane emissions in the marine environment. *Abstracts, Forams 2010 International Symposium on Foraminifera*, Bonn: 154.

Panieri, G., Camerlenghi, A., Conti, S. *et al.*, 2009. Methane seepages recorded in benthic foraminifera from Miocene seep carbonates, northern Apennines, Italy. *Palaeogeography, Palaeoclimatology, Palaeoecology*, **284**: 271–82.

Panieri, G. & Sen Gupta, B. K., 2008. Benthic foraminifera from the Blake Ridge hydrate mound, western North Atlantic Ocean. *Marine Micropaleontology*, **66**: 91–102.

Papazzoni, C. A., 2010. Taphonomic index and transportation in nummulite banks and in nummulitic limestones. *Abstracts, Forams 2010 International Symposium on Foraminifera*, Bonn: 154.

Parente, M., Frijia, D., di Lucia, M. *et al.*, 2008. Stepwise extinction of larger foraminifers at the Cenomanian-Turonian boundary: a shallow-water perspective on nutrient fluctuations during Oceanic Anoxic Event 3 (Bonarelli Event). *Geology*, **36**(9): 715–8.

Parfitt, S. A., Ashton, N. M., Lewis, S. G. *et al.*, 2010. Early Pleistocene human occupation at the edge of the boreal zone in northwest Europe. *Nature*, **466**: 229–33.

Park, L. E. & Gierlowski-Kordesch, E. H., 2007. Paleozoic lake faunas: establishing aquatic life on land. *Palaeogeography, Palaeoclimatology, Palaeoecology*, **250**: 160–79.

Parker, J. R. (ed.), 1993. *Petroleum Geology of North-West Europe: Proceedings of the 4th Conference.* The Geological Society, London.

Parras, A. & Casadio, S., 2005. Taphonomy and sequence stratigraphic significance of oyster-dominated concentrations from the San Julian formation, Oligocene of Patagonia, Argentina. *Palaeogeography, Palaeoclimatology, Palaeoecology*, **217**: 47–66.

Parrish, J. T., Ziegler, A. M. & Scotese, C. R., 1982. Rainfall patterns and the distribution of coals and evaporites in the Mesozoic and Cenozoic. *Palaeogeography, Palaeoclimatology, Palaeoecology*, **40**: 67–101.

Pashin, J. C. & Gastaldo, R. A. (eds.), 2004. *Sequence Stratigraphy, Paleoclimate and Tectonics of Coal-Bearing Strata.* American Association of Petroleum Geologists (Studies in Geology No. 51).

Patterson, C., 1999. *Evolution*, 2nd edition. The Natural History Museum, London.

Patterson, R. T., McKillop, W. B., Kroker, S. *et al.*, 1997. Evidence for rapid avian-mediated foraminiferal colonization of Lake Winnipegosis, Manitoba, during the Holocene hypsithermal. *Journal of Paleolimnology*, **18**(2): 131–43.

Pawlowski, J., Lecroq, B. & Lejzerowicz, F., 2010. High diversity of early foraminifera unveiled by environmental DNA surveys. *Abstracts, Forams 2010 International Symposium on Foraminifera*, Bonn: 156.

Pearson, T. H. & Rosenberg, R., 1978. Macrobenthic succession in relation to organic enrichment and pollution of the marine environment. *Oceanography and Marine Biology Annual Review*, **16**: 229–311.

Peatfield, D., 2003. Coal and coal preparation in South Africa – a 2002 review. *The Journal of the South African Institute of Mining and Metallurgy*, July/August 2003: 355–72.

Peckmann, J. & Goedert, J. L., 2005. Geobiology of ancient and modern methane seeps. *Palaeogeography, Palaeoclimatology, Palaeoecology*, **227**: 1–5.

Peckmann, J., Little, C. T. S., Gill, F. & Reitner, J., 2005. Worm tube fossils from the Hollard Mound hydrocarbon-seep deposit, Middle Devonian, Morocco: Palaeozoic seep-related vestimentiferans. *Palaeogeography, Palaeoclimatology, Palaeoecology*, **227**: 242–57.

Penkman, K., Demarchi, B., Preece, R. *et al.*, 2010. Dating the early Palaeolithic: the new aminostratigraphy. *Abstracts, Ancient Human Occupation of Britain Conference, British Museum, 8–10 April 2010*: 20–1.

Penney, R. A., Al Barram, I. & Stephenson, M. H., 2008. A high resolution palynozonation for the Al Khlata formation (Pennsylvanian to Lower Permian), south Oman. *Palynology*, **32**: 213–29.

Peppers, R. A., 1996. *Palynological Correlation of Major Pennsylvanian (Middle and Upper Carboniferous) Chronostratigraphic Boundaries in the Illinois and Other Coal Basins.* Geological Society of America (Memoir No. 188).

Perch-Nielsen, K., 1973. Fossil coccoliths as indicators of the origin of Late Cretaceous chalk used in medieval Norwegian art. *Universitetets Oldsaksamlings Arbok*, **1970–71**: 161–9.

Peres, W. E., 1993. Shelf-fed turbidite system model and its application to the Oligocene deposits of the Camps basin, Brazil. *American Association of Petroleum Geologists Bulletin*, **77**(1): 81–101.

Perez-Huerta, A. & Sheldon, N., 2006. Pennsylvanian sea level cycles, nutrient availability and brachiopod paleoecology. *Palaeogeography, Palaeoclimatology, Palaeoecology*, **230**: 264–79.

Peryt, T. M., 2006. The beginning, development and termination of the Middle Miocene Salinity Crisis in central Paratethys. *Sedimentary Geology*, **188-189**: 379-96.

Peters, K. E., Clark, M. E., Das Gupta, U. *et al.*, 1995. Recognition of an Infracambrian source rock based on biomarkers in the Baghewala-1 oil, India. *American Association of Petroleum Geologists Bulletin*, **79**(10): 1481-94.

Peters-Kottig, W., Strauss, H. & Kerp, H., 2006. The land plant del^{13}C record and land plant evolution in the Palaeozoic. *Palaeogeography, Palaeoclimatology, Palaeoecology*, **240**: 237-52.

Petri, S. & Campanha, V. A., 1981. Brazilian continental Cretaceous. *Earth Science Reviews*, **17**: 69-85.

Peybernes, B., Kamoun, F., Ben Youssef, M. *et al.*, 1993. Sequence stratigraphy and micropaleontology of the Triassic series from the southern part of Tunisia. *Journal of African Earth Sciences*, **17**(3): 293-305.

Pfeiffer, M., Dullo, W.-C. & Eisenhauer, A., 2004. Variability of the Intertropical Convergence Zone recorded in coral isotopic records from the central Indian Ocean (Chagos Archipelago). *Quaternary Research*, **61**: 245-55.

Philip, J. M., 2003. Paleoecological and paleobiogeographic significance of the rudist *Macgillavryia chubbii* sp. n. in the Campanian of Oman. *GeoArabia*, **8**(1): 129-46.

Philippe, M., Cvuny, G., Bamford, M. *et al.*, 2003. The palaeoxylological record of *Metapodocarpoxylon libanoticum* (Edwards) Duperon-Laudoueneix et Pons and the Gondwana Late Jurassic-Early Cretaceous continental biogeography. *Journal of Biogeography*, **30**: 389-400.

Philippot, P., van Kranendonk, M., van Zuilen, M. *et al.*, 2009. Early traces of life investigations in drilling Archaean hydrothermal and sedimentary rocks of the Pilbara craton, Western Australia and Barberton greenstone belt, South Africa. *Comptes Rendus Palevol*, **8**: 649-63.

Pickering, K. T. & Owen, L. A., 1994. *An Introduction to Global Environmental Issues*. Routledge.

Pickford, M., 1995. Fossil land snails of east Africa and their palaeoecological significance. *Journal of African Earth Sciences*, **20**(3-4): 167-226.

2004. Palaeoeonvironments of Early Miocene hominid-bearing deposits at Napak, Uganda, based on terrestrial molluscs. *Annales de Paleontologie*, **90**: 1-12.

Pickford, M., Senut, B., Gommery, D. & Treil, J., 2002. Bipedalism in *Orrorin tugenensis* revealed by its femora. *Comptes Rendus Palevol*, **1**: 191-203.

Pickford, M., Wanas, H. & Soliman, H., 2006. Indications for a humid climate in the Western Desert of Egypt 11-10 Myr ago: evidence from Galagidae (Primates, Mammalia). *Comptes Rendus Palevol*, **5**: 935-43.

Pierce, P., 2006. *Jurassic Mary: Mary Anning and the Primeval Monsters*. Sutton Publishing.

Piller, W. E., Reuter, M., Harzhauser, M. *et al.*, 2010. The Quilon limestone (Kerala basin, India) - an archive for Miocene Indo-Pacific seagrass beds. *Geophysical Research Abstracts*, **12**.

Pina-Ochoa, E., Hogslund, S., Geslin, E. *et al.*, 2010. Widespread occurrence of nitrate storage and denitrification among foraminifera. *Abstracts, Forams 2010 International Symposium on Foraminifera*, Bonn: **160**.

Pindell, J. L. & Drake, C. (eds.), 1998. *Paleogeographic Evolution and Non-Glacial Eustasy, Northern South America*. Society of Economic Paleontologists and Mineralogists (Special Publication No. 58).

Pinheiro, H. J. & Hancox, P. J., 2009. Coal in South Africa. *3rd Symposium on Gondwana Coals, Pontifical Catholic University of Rio Grande do Sul (PUCRS)*, Porto Alegre, Brazil.

Pint, A. & Frenzel, P., 2010. Occurrence of Holocene *Haplophragmoides canariensis* (d'Orbigny, 1839) (Haplophragmiina, Foraminifera) in central Germany and its significance for the identification of athalassic sediments. *Abstracts, Forams 2010 International Symposium on Foraminifera*, Bonn: 161.

Piper, K. G., Fitzgerald, E. M. G. & Rich, T. H., 2006. Mesozoic to early Quaternary mammal faunas of Victoria, south-east Australia. *Palaeontology*, **49**(6): 1237-62.

Piperno, D. R., 2006. *Phytoliths*. Altamira Press.

2009. Identifying crop plants with phytoliths (and starch grains) in central and South America: a review and an update of the evidence. *Quaternary International*, **193**: 146-59.

Pique, A., 2001. *Geology of Northwest Africa*. Gebruder Borntraeger.

Pires, E. F., Sommer, M. G. & dos Santos Scherer, C. M., 2005. Late Triassic climate in southernmost Parana basin (Brazil): evidence from dendrochronological data. *Journal of South American Earth Sciences*, **18**: 213-21.

Pisera, A. & Saez, A., 2003. Paleoenvironmental significance of a new species of freshwater sponge from the Late Miocene Quillaga formation (N. Chile). *Journal of South American Earth Sciences*, **15**: 847-52.

Plumstead, E. P., 1969. *Three Thousand Million Years of Plant Life in Africa*. Geological Survey of South Africa (Alex L. Du Toit Memorial Lectures No. 11).

Poag, C. W., 1985. Benthic foraminifera as indicators of potential petroleum sources. *Proceedings of the 4th Annual Research Conference of the Gulf Coast Section of the Society of Economic Paleontologists and Mineralogists Foundation*: 275-84.

Poinar, G. Jr, Lambert, J. B. & Yuyang Wu, 2004. NMR analysis of amber in the Zubair formation, Khafji oilfield (Saudi Arabia-Kuwait): coal as an oil source rock? *Journal of Petroleum Geology*, **27**(2): 207-9.

Pollard, A. M. (ed.), 1999. *Geoarchaeology*. The Geological Society, London (Special Publication No. 165).

Popov, S. J., Shcherba, I. G., Nevesskaya, L. A. *et al.*, 2006. Late Miocene to Pliocene palaeogeography of the Paratethys and its relation to the Mediterranean. *Palaeogeography, Palaeoclimatology, Palaeoecology*, **238**(1-4): 5-14.

Pospelova, V., de Vernal, A. & Pedersen, T. F., 2008. Distribution of dinoflagellate cysts in surface sediments from the northeastern Pacific Ocean in relation to sea-surface temperature, salinity, productivity and coastal upwelling. *Marine Micropaleontology*, **68**: 21–48.

Potter, J., McIlreath, I. & Natras, T., 2009. Lithofacies, macerals and coal facies studies of Lower Cretaceous Medicine River coals in south-central Alberta: applications in CBM exploration, depositional environments and tectonic history studies. *CSPG CSEG CWLS Convention*: 777–80.

Poumot, C., 1989. Palynological evidence for eustatic events in the tropical Neogene. *Bulletin du Centre de Recherches Exploration-Production Elf-Aquitaine*, **13**(2): 437–53.

Pound, M.J., Haywood, A.M., Riding, J. *et al.*, 2009. Land cover in a warmer world: a global Late Miocene vegetation reconstruction. *Abstracts, Palaeontological Association Annual Meeting*, Birmingham, 69–70.

Powell, A.J. (ed.), 1992. *A Stratigraphical Atlas of Dinoflagellate Cysts*. Chapman & Hall.

Powell, A.J. & Riding, J. (eds.), 2005. *Recent Developments in Applied Biostratigraphy*. The Micropalaeontological Society (Special Publication).

Powers, L.A., Werne, J.P., Johnson, T.C. *et al.*, 2004. Crenarchaeotal membrane lipids in lake sediments: a new paleotemperature proxy for continental paleoclimate reconstruction. *Geology*, **32**: 613–6.

Prakash, N., 2008. Biodiversity and palaeoclimatic interpretation of Early Cretaceous flora of Dubrajpur formation, Satpura basin, India. *Paleoworld*, **17**: 253–63.

Prasad, B., 1997. Palynology of the subsurface Triassic sediments of Krishna–Godavari basin, India. *Palaeontographica B*, **242**(4–6): 91–125.

Prasad, B., Jain, A.K. & Mathur, Y.K., 1995. A standard palynozonation scheme for the Cretaceous and pre-Cretaceous subsurface sediments of Krishna–Godavari basin, India. *Geoscience Journal*, **XVI**(2): 155–233.

Prasad, B. & Pundeer, B.S., 2002. Palynological events and zones in Cretaceous-Tertiary boundary sediments of Krishna–Godavari and Cauvery basins, India. *Palaeontographica Abteilung B*, **262**(1–4): 39–70.

Prasad, B., Pundeer, B.S. & Swamy, S.N., 2000. Age, depositional environment, source potential and distribution of pay sands in Mandapeta subbasin, with remarks on the biostratigraphic classification of Gondwana sediments in Krishna–Godavari basin. *Indian Journal of Petroleum Geology*, **9**(1): 1–19.

Prasad, B. & Pundir, B.S., 1999. Biostratigraphy of the exposed Gondwana and Cretaceous rocks of Krishna–Godavari basin, India. *Journal of the Palaeontological Society of India*, **44**: 91–117.

Prasad, G.V.R. & Bajpai, S., 2008. Agamid lizards from the Early Eocene of western India: oldest Cenozoic lizards from south Asia. *Palaeontologia Electronica*, **11**(1): 14pp.

Prevec, R., Labandeira, C.C., Neveling, J. *et al.*, 2009. Portrait of a Gondwanan ecosystem: a new Late Permian locality from Kwazulu-Natal, South Africa. *Review of Palaeobotany and Palynology*, **156**: 454–93.

Price, G.D., 2010. Carbon isotope stratigraphy and temperature change during the Early-Middle Jurassic (Toarcian-Aalenian), Raasay, Scotland, UK. *Palaeogeography, Palaeoclimatology, Palaeoecology*, **285**: 255–63.

Price, G.D. & Mutterlose, K., 2004. Isotopic signals from Late Jurassic-Early Cretaceous (Volgian–Valanginian) sub-Arctic belemnites, Yatria river, western Siberia. *Journal of the Geological Society, London*, **161**(6): 959–68.

Price, G.D., Ruffell, A.H., Jones, C.E. *et al.*, 2000. Isotopic evidence for temperature variation during the early Cretaceous (late Ryazanian-mid Hauterivian). *Journal of the Geological Society, London*, **157**: 335–43.

Pritchard, P.H., Mueller, J.G., Rogers, J.C. *et al.*, 1992. Oil spill bioremediation: experiences, lessons and results from the Exon Valdez oil spill in Alaska. *Bioremediation*, **3**: 315–35.

Prothero, D.R., 2006. *After the Dinosaurs: The Age Of Mammals*. Indiana University Press.

Prothero, D.R., Ivany, L.C. & Nesbitt, E.A. (eds.), 2003. *From Greenhouse to Icehouse: The Marine Eocene–Oligocene Transition*. Columbia University Press.

Pruss, S.B. & Bottjer, D.J., 2005. The reorganization of reef communities following the end-Permian mass extinction. *Comptes Rendus Palevol*, **4**: 485–500.

Pruss, S.B., Bottjer, D.J., Corsetti, F.A. & Baud, A., 2006. A global marine sedimentary response to the end-Permian mass extinction: examples from southern Turkey and the western United States. *Earth-Science Reviews*, **78**: 193–206.

Punyasena, S.W., 2008. Estimating Neotropical palaeotemperature and palaeoprecipitation using plant family climatic optima. *Palaeogeography, Palaeoclimatology, Palaeoecology*, **265**: 226–37.

Purser, B.H. & Bosence, D.W.J. (eds.), 1998. *Sedimentation and Tectonics in Rift Basins, Red Sea–Gulf of Aden*. Chapman & Hall.

Pye, K. & Croft, D.J. (eds.), 2004. *Forensic Geoscience: Principles, Techniques and Applications*. The Geological Society, London (Special Publication No. 232).

Quadri, V.U.N. & Quadri, S.M.J.G., 1998. Pakistan has unventured regions, untested plays. *Oil & Gas Journal*, **5** January: 62–5.

Quadri, V.U.N. & Shuaib, S.M., 1986. Hydrocarbon prospects of southern Indus basin, Pakistan. *American Association of Petroleum Geologists Bulletin*, **70**(6): 730–47.

Quinn, P.S. & Day, P.M., 2007. Ceramic micropalaeontology: the analysis of microfossils in ancient ceramics. *Journal of Micropalaeontology*, **26**(2): 159–68.

Quinterno, P.J. & Gardner, J.V., 1987. Benthic foraminifers on the continental shelf and upper slope, Russian River area, northern California. *Journal of Foraminiferal Research*, **17**(2): 132–52.

Racey, A., 2001. A review of Eocene nummulite accumulations: structure, formation and reservoir potential. *Journal of Petroleum Geology*, **24**(1): 79–100.

Racey, A., Bailey, H.W., Beckett, D. *et al.*, 2001. The petroleum geology of the Early Eocene El Garia formation in the

Hasdrubal field, offshore Tunisia. *Journal of Petroleum Geology*, **24**(1): 29–53.

Rage, J.-C., Folie, A., Rana, R.S. et al., 2008. A diverse snake fauna from the Early Eocene of Vastan lignite mine, Gujurat, India. *Acta Palaeontologica Polonica*, **53**(3): 391–403.

Rahmani, R.A. & Flores, R.M. (eds.), 1984. *Sedimentology of Coal and Coal-Bearing Sequences*. International Association of Sedimentologists (Special Publication No. 7).

Raju, A.T.R., 1968. Geological evolution of Assam and Cambay Tertiary basins of India. *American Association of Petroleum Geologists Bulletin*, **52**(12): 2422–37.

Raju, D.S.N., 1974. Study of Indian Miogypsinidae. *Utrecht Micropaleontological Bulletins*, **9**: 1–149.

2008. High-resolution bio-, chrono- and bio-sequence stratigrapy and sea level changes:global vs. Indian record. *ONGC Bulletin*, **43**(1): 10–43.

Raju, D.S.N., Peters, J., Shanker, R. & Kumar, G., 2005. *An Overview of Litho-, Bio-, Chrono-sequence Stratigraphy and Sea-Level Changes of Indian Sedimentary Basins*. Association of Petroleum Geologists (Special Publication No. 1).

Raju, D.S.N., Bhandari, A. & Ramesh, P., 1999. Relative sea-level fluctuations during Cretaceous and Cenozoic in India. *Bulletin of the Oil and Natural Gas Corporation Limited*, **36**(1): 185–202.

Ramakrishnan, M. & Vaidyanadhan, R., 2008. *Geology of India*, Volume 1. Geological Society of India.

Ramos, M.I.F., de Rosetti, D. & Paz, J.D.S., 2006. Caracterizacao e significado paleoambiental da fauna de ostracodes de formacao Codo (Neoaptiano), leste da bacia de Grujau, MA, Brazil. *Revista Brasíliera de Paleontologia*, **9**(3): 339–48.

Ramsbottom, W.H.C., Calver, M.A., Eagar, R.M.C. et al., 1978. *A Correlation of Silesian Rocks in the British Isles*. The Geological Society, London (Special Report No. 10).

Rana, R.S., Kumar, K, Escarguel, G. et al., 2008. An ailuavine rodent from the Lower Eocene Cambay formation at Vastan, western India, and its palaeobiogeographic implications. *Acta Palaeontologica Polonica*, **53**(1): 1–14.

Rana, R.S., Kumar, K. & Singh, H., 2004. Vertebrate fauna from the subsurface Cambay Shale (Lower Eocene), Vastan lignite mine, Gujurat, India. *Current Science*, **87**(12): 1726–33.

Rao, G.N., 2001. Sedimentation, stratigraphy and petroleum potential of Krishna–Godavari basin, east coast of India. *American Association of Petroleum Geologists Bulletin*, **85**(9): 1623–43.

Rao, G.N., Manmohan, Nirupama & Prasad, P., 1996. A study of Paleocene depositional environments in Krishna–Godavari basin – an integrated approach. *Indian Journal of Petroleum Geology*, **5**(2): 1–20.

Rasanen, M., Linna, A.M., Santos, J.C.R. & Negri, F.R., 1995. Late Miocene tidal deposits in the Amazonas foreland basin. *Science*, **269**: 386–9.

Rathore, S.S., 2007. *Exploring Exploration*. Association of Petroleum Geologists (Bulletin No. 1).

Raup, D.M. & Sepkoski, J.J. Jr, 1982. Mass extinctions in the marine fossil record. *Science*, **215**: 1501–3.

1984. Periodicity of extinctions in the geological past. *Proceedings of the National Academy of Sciences*, **81**: 801–5.

Raymond, A., Gensel, P. & Stein, W.E., 2006. Phytogeography of Late Silurian macrofloras. *Review of Palaeobotany and Palynology*, **142**: 165–92.

Raza, H.A., Ali, S.M. & Ahmed, R., 1990. Petroleum geology of Kirthar sub-basin and part of Kutch basin. *Pakistan Journal of Hydrocarbon Research*, **2**(1): 29–73.

Raza Khan, M.S., Sharma, A.K., Sahota, S.K. & Mathur, M., 2000. Generation and hydrocarbon entrapment within Gondwanan sediments of the Mandapeta area, Krishna–Godavari basin, India. *Organic Geochemistry*, **31**: 1495–1507.

Reddy, A.N., Satyanarayana, K., Bhaktavatsala, K.V. et al., 2005. Sequence stratigraphy and depositional processes of Miocene sediments in KD structure, deep waters of Krishna–Godavari basin, India. *Journal of the Geological Society of India*, **66**: 42–58.

Regali, M. da Silva Pares, Uesugui, N. & Santos, A.da Silva, 1974a. Palinologia dos sedimentos Meso-Cenozoicos do Brasil (I). *Boletim Tecnico de Petrobras*, **17**(3): 177–91.

1974b. Palinologia dos sedimentos Meso-Cenozoicos do Brasil (I). *Boletim Tecnico de Petrobras*, **17**(4): 263–301.

Reggiani, L., Mattioli, E., Pittet, B. et al., 2010. Pliensbachian (Early Jurassic) calcareous nannofossils in the Peniche section (Lusitanian basin, Portugal): a clue for palaeoenvironmental reconstructions. *Marine Micropaleontology*, **75**: 1–16.

Rehanek, J. & Cecca, F., 1993. Calcareous dinoflagellate cysts, biostratigraphy in upper Kimmeridgian-lower Tithonian pelagic limestones of Marches Apennines (central Italy). *Revue de Micropaléontologie*, **36**(2): 146–63.

Reicherter, K., Michetti, A. & Silva, P.G. (eds.), 2009. *Palaeoseismology: Historical and Prehistoric Records of Earthquake Ground Effects for Seismic Hazard Assessment*. The Geological Society, London (Special Publication No. 16).

Reichow, M.K., Pringle, M.S., Al'Mukamedov, A.J. et al., 2009. The timing and extent of the eruption of the Siberian Traps large igneous province: implications for the end-Permian environmental crisis. *Earth and Planetary Science Letters*, **277**: 9–20.

Reid, C., 1913. *Submerged Forests*. Cambridge Series of Manuals of Literature and Science.

Reiss, Z. & Hottinger, L., 1984. *The Gulf of Aqaba – Ecological Micropaleontology*. Springer Verlag.

Renaud, S., Michaux, J., Schmidt, D.N. et al., 2005. Morphological evolution, ecological diversification and climate change in rodents. *Proceedings of the Royal Society B*, **272**: 609–17.

Renema, W. (ed.), 2007. *Biogeography, Time and Place: Distributions, Barriers and Islands*. Springer.

Renema, W., Bellwood, D.R., Braga, J.C. et al., 2008. Hopping hotspots: global shifts in marine biodiversity. *Science*, **321**: 654–7.

Renfrew, C. & Bahn, P., 2000. *Archaeology*, 3rd edition. Thames & Hudson.

Renfrew, C. & Boyle, K. (eds.), 2000. *Archaeogenetics: DNA and the Population History of Europe.* McDonald Institute for Archaeological Research.

Renz, H. H., 1948. Stratigraphy and fauna of the Agua Salada group, State of Falcon, Venezuela. *Memoirs, Geological Society of America,* **32.**

1962. Stratigraphy and paleontology of the type section of the Santa Anita group and the overlying Merecure group, Rio Querecual, northeastern Venezuela. *Boletin Informativo Asociacion Venezolano de Geologia, Mineria y Petroleo, Caracas,* **5:** 89–108.

Renz, O., 1982. *The Cretaceous Ammonites of Venezuela.* Birkhauser Verlag.

Requejo, A. G., Wielchowsky, C. C., Klosterman, M. J. & Sassen, R., 1994. Geochemical characterisation of lithofacies and organic facies in Cretaceous organic-rich rocks from Trinidad, east Venezuela basin. *Organic Geochemistry,* **22**(3–5): 441–59.

Retallack, G., 1991. *Miocene Paleosols and Ape Habitats from Pakistan and Kenya.* Oxford University Press.

1995. Paleosols of the Siwalik group as a 15 Myr record of south Asian palaeoclimate. *Memoirs, Geological Society of India,* **32:** 36–51.

Retallack, G. J., Veevers, J. J. & Morante, R., 1996. Global coal gap between Permian–Triassic extinction and Middle Triassic recovery of peat-forming plants. *Geological Society of America Bulletin,* **108**(2): 195–207.

Reumer, J. W. F. & Wessels, W. (eds.), 2003. *Distribution and Migration of Tertiary Mammals in Eurasia.* Deinsea.

Rexfort, A. & Mutterlose, J., 2009. The role of biogeography and ecology on the isotope signature of cuttlefishes (Cephalopoda, Sepiidae) and the impact on belemnite studies. *Palaeogeography, Palaeoclimatology, Palaeoecology,* **284:** 153–63.

Reynolds, C., 2006. *Ecology of Phytoplankton.* Cambridge University Press.

Rickards, R. B., 1975. Palaeoecology of the Graptolithina, an extinct class of the phylum Hemichordata. *Biological Reviews,* **50:** 397–436.

Ricketts, E. R., Kennett, J. P., Hill, T. M. & Barry, J. P., 2009. Effects of carbon dioxide sequestration on California margin deep-sea foraminiferal assemblages. *Marine Micropaleontology,* **72:** 165–75.

Riding, J. B. & Kyffin-Hughes, J. E., 2006. Further testing of a non-acid palynological preparation technique. *Palynology,* **30:** 69–87.

Riding, J. B., Mantle, D. J. & Backhouse, J., 2010. A review of the chronostratigraphical ages of Middle Triassic to Late Jurassic dinoflagellate cyst biozones of the north-west shelf of Australia. *Review of Palaeobotany and Palynology,* **162:** 543–75.

Riding, R., 2005. Phanerozoic reefal microbial carbonate abundance: comparisons with metazoan diversity, mass extinction events, and seawater saturation state. *Revista Espanola de Micropaleontologia,* **37**(1): 23–39.

Riebesell, U., Zondervan, I., Rost, B. *et al.,* 2000. Reduced calcification of marine plankton in response to increased atmospheric CO_2. *Nature,* **407:** 364–7.

Riegel, W., 2008. The Late Palaeozoic phytoplankton blackout – artefact or evidence of global change? *Review of Palaeobotany and Palynology,* **148:** 73–80.

Rigby, S., 1991. Feeding strategies in graptoloids. *Palaeontology,* **34:** 797–814.

Rigby, S. & Rickards, R. B., 1989. New evidence for the life habit of graptoloids from physical modelling. *Palaeobiology,* **15:** 402–13.

Risgaard-Petersen, N., Langezaal, A. M., Ingvardsen, S. *et al.,* 2007. Evidence for complete denitrification in a benthic foraminifer. *Nature,* **443:** 93–6.

Rivals, F., Mihlbachler, M. C., Solounias, N. *et al.,* 2010. Palaeoecology of the Mammoth Steppe fauna from the Late Pleistocene of the North Sea and Alaska: separating species preferences from geographic influence in palaeoecological dental wear analysis. *Palaeogeography, Palaeoclimatology, Palaeoecology,* **286:** 42–54.

Roberts, A., 2009. *The Incredible Human Journey. The Story of How We Colonised the Planet.* Bloomsbury.

Roberts, C. & Manchester, K., 2005. *The Archaeology of Disease,* 3rd edition. Sutton Publishing.

Roberts, J. M., Wheeler, A., Freiwald, A. & Cairns, S., 2009. *Cold-Water Corals.* Cambridge University Press.

Roberts, M. B. & Parfitt, S. A. (eds.), 1999. *Boxgrove: A Middle Pleistocene Hominid Site at Eartham Quarry, Boxgrove, West Sussex.* English Heritage (Archaeological Report 17).

Robinson, G. S. & Kohl, B., 1978. Computer assisted paleoecologic analyses and application to petroleum exploration. *Transactions – Gulf Coast Association of Geological Societies,* **XXVIII:** 433–47.

Robison, C. R., Smith, M. A. & Royle, R. A., 1999. Organic facies in Cretaceous and Jurassic hydrocarbon source rocks, southern Indus basin, Pakistan. *International Journal of Coal Geology,* **39:** 205–23.

Rodrigues, K., 1993. The Naparima Hill–Cruse/Forest/Gros Morne petroleum system of Trinidad: a quantitative evaluation of petroleum generated. *Extended Abstract, American Association of Petroleum Geologists Annual Convention,* New Orleans.

Roehl, P. O. & Choquette, P. W. (eds.), 1985. *Carbonate Petroleum Reservoirs.* Springer-Verlag.

Rogerson, M., Kouwenhaven, T. J., van der Zwaan, G. J., O'Neill, B. J. *et al.,* 2006. Benthic foraminifera of a Miocene canyon and fan. *Marine Micropaleontology,* **60:** 295–318.

Rohr, G. M., 1991. Paleogeographic maps, Maturin basin of E. Venezuela and Trinidad. *Transactions of the Second Geological Conference of the Geological Society of Trinidad & Tobago:* 88–105.

Rolland, N., Larocque, I., Francus, P. *et al.,* 2009. Evidence for a warmer period during the 12th and 13th centuries AD from chironomid assemblages in Southampton Island, Nunavut, Canada. *Quaternary Research,* **72:** 27–37.

Rojas, R., Iturralde-Vinent, M. & Skelton, P.W., 1995. Stratigraphy, composition and age of Cuban rudist-bearing deposits. *Revista Mexicana de Ciencias Geologicas*, **12**(2): 272–91.

Rong Jia-Yu, Chen Xu & Harper, D.A.T., 2002. The latest Ordovician *Hirnantia* fauna (Brachiopoda) in time and space. *Lethaia*, **35**: 231–49.

Rook, L., Gallai, G. & Torre, D., 2006. Lands and endemic mammals in the Late Miocene of Italy: constraints for paleogeographic outlines of Tyrrhenian area. *Palaeogeography, Palaeoclimatology, Palaeoecology*, **238**(1–4): 263–9.

Rose, K.D., Rana, R.S., Sahni, A. & Smith, T., 2007. A new adapoid primate from the Early Eocene of India. *Contributions from the Museum of Paleontology, The University of Michigan*, **31**(14): 379–85.

Rose, K.D., Rana, R.S., Sahni, A. *et al.*, 2009. First tillodont from India: additional evidence for an Early Eocene faunal connection between Europe and India. *Acta Palaeontologica*, **54**(2): 351–5.

Ross, D.J.K. & Bustin, R.M., 2009. The importance of shale composition and pore structure upon gas storage potential of shale gas reservoirs. *Marine and Petroleum Geology*, **26**: 916–27.

Rothschild, L.J. & Lister, A.M. (eds.), 2003. *Evolution on Planet Earth: The Impact of the Physical Environment*. Academic Press.

Roy, P., Bardhan, S., Mitra, A. & Jana, S.K., 2007. New Bathonian (Middle Jurassic) ammonite assemblages from Kutch, India. *Journal of Asian Earth Sciences*, **30**: 629–51.

Roychoudhury, S.C. & Deshpande, S.V., 1982. Regional distribution of carbonate facies, Bombay offshore region, India. *American Association of Petroleum Geologists Bulletin*, **66**(10): 1483–96.

Royer, D.L., 2006. CO$_2$-forced climate thresholds during the Phanerozoic. *Geochimica et Cosmochimica Acta*, **70**: 5665–75.

Ruban, D.A., 2009. Phanerozoic changes in the high rank suprageneric diversity structure of brachiopods: linear and non-linear effects. *Paleoworld*, **18**: 263–77.

Rubidge, B. (ed.), 1995. Biostratigraphy of the Beaufort group (Karoo sequence), South Africa. *Geological Journal of South Africa, Biostratigraphic Series*, **1**: 1–45.

Ruddiman, W.F., 2003. The Anthropogenic era began thousands of years ago. *Climate Change*, **61**: 261–93.

2005. *Plows, Plagues & Petroleum: How Humans Took Control of Climate*. Princeton University Press.

Rudling, D. (ed.), 2003. *The Archaeology of Sussex to 2000 AD*. Heritage Marketing & Publications Ltd.

Rudwick, M.J.S., 1961. The feeding mechanism of the Permian brachiopod *Prorichthofenia*. *Palaeontology*, **3**: 450–7.

1964. The inference of structure from function in fossils. *British Journal for the Philosophy of Science*, **15**: 27–40.

Ruffine, L., Donval, J.P., Charlou, J.L. *et al.*, 2010. Experimental study of gas hydrate formation and destabilisation using a novel high-pressure apparatus. *Marine and Petroleum Geology*, **27**: 1157–65.

Ruiz, C., Matthews, S., Goff, J. *et al.*, 2004. Mid-Miocene–Recent tectonostratigraphic evolution of the northern Dezful embayment, southwest Iran. *GeoArabia*, **9**(1): 121.

Ruiz, F., Abad, M., Bodergat, A.N. *et al.*, 2005. Marine and brackish-water ostracods as sentinels of anthropogenic impacts. *Earth-Science Reviews*, **72**: 89–111.

Ruiz, F., Abad, M., Caceres, L.M. *et al.*, 2010. Ostracods as tsunami tracers in Holocene sequences. *Quaternary Research*, **73**: 130–5.

Rull, V., 2002. High-impact palynology in petroleum geology: applications from Venezuela (northern South America). *American Association of Petroleum Geologists Bulletin*, **86**(2): 279–300.

Ruse, M., 2005. *The Evolution–Creation Struggle*. Harvard University Press.

Ryves, D.B., Battarbee, R.W. & Fritz, S.C., 2009. The dilemma of disappearing diatoms: incorporating diatom dissolution data into palaeoenvironmental modelling and reconsruction. *Quaternary Science Reviews*, **28**: 120–36.

Sacal, V. & Cuvillier, J., 1963. *Microfacies du Paleozoique saharien*. Compagnie Francaise des Petroles.

Sadooni, F.N., 2005. The nature and origin of Upper Cretaceous basin-margin rudist build-ups of the Mesopotamian basin, southern Iraq, with consideration of possible stratigraphic entrapment. *Cretaceous Research*, **26**: 213–24.

Sahling, H., Bohrmann, G., Spiess, V. *et al.*, 2008. Pockmarks in the northern Congo Fan area, SW Africa: complex seafloor features shaped by fluid flow. *Marine Geology*, **249**: 206–25.

Sahni, A. (ed.), 1996. *Cretaceous Stratigraphy and Palaeoenvironments*. Geological Society of India.

Said, R. (ed.), 1990. *The Geology of Egypt*. A.A. Balkema.

Saint-Martin, S., Conesa, G. & Saint-Martin, J.-P., 2003. Paleoecological significance of Messinian diatom assemblages from the Melilla–Nador basin (northeastern Rif, Morocco). *Revue de Micropaléontologie*, **46**: 161–90.

Saint-Martin, S. & Saint-Martin, J.-P., 2005. Enregistrement par les diatomees des variations paleoenvironmentales durant le Sarmatien dans l'aire paratethysienne (Roumanie). *Comptes Rendus Palevol*, **4**: 191–201.

Salem, M.J. & Belaid, M.N. (eds.), 1991. *The Geology of Libya*. Elsevier.

Salem, M.J. & Oun, K.M. (eds.), 2003. *The Geology of Northwest Libya*. Earth Science Society of Libya.

Saller, A., Lin, R. & Dunham, J., 2006. Leaves in turbidite sands: the main source of oil and gas in the deep-water Kutei basin, Indonesia. *American Association of Petroleum Geologists Bulletin*, **90**(10): 1585–1608.

Samant, B. & Bajpai, S., 2001. Fish otoliths from the subsurface Cambay Shale (Lower Eocene), Surat lignite field, Gujarat (India). *Current Science*, **81**(7): 75–9.

Samanta, B.K., 1973. Planktonic foraminifera from the Palaeocene-Eocene succession in the Rakhi Nala, Sulaiman range, Pakistan. *Bulletin of the British Museum (Natural History)*, **22**(6): 423–482.

Sampsell, B. M., 2003. *A Traveler's Guide to the Geology of Egypt*. The American University in Cairo Press.

Sams, R. H., 1995. Interpreted sequence stratigraphy of the Los Jabillos, Areo and (subsurface) Naricual formations, northern Monagas area, eastern Venezuelan basin. *Boletin de la Sociedad Venezolana de Geologia*, **20**(1–2): 30–40.

Sanchez-Villagra, M. R. & Aguilera, O. A., 2006. Neogene vertebrates from Urumaco, Falcon state, Venezuela: diversity and significance. *Journal of Systematic Palaeontology*, **4**(3): 213–20.

Sanfilippo, A. & Nigrini, C., 1998. Code numbers for Cenozoic low latitude radiolarian biostratigraphic zones and GPTS conversion tables. *Marine Micropaleontology*, **33**: 109–56.

Sansom, I., 2005. Fishing in the Ordovician: microinvertebrates and macroevolution. *Abstracts, The Micropalaeontological Society Annual General Meeting Presentations*, **2005**.

Sanyal, P., Sarkar, A., Bhattacharya, S. K. *et al.*, 2010. Intensification of monsoon, microclimate and asynchronous C4 appearance: isotopic evidence from the Indian Siwalik sediments. *Palaeogeography, Palaeoclimatology, Palaeoecology*, **296**: 165–73.

Sanz-Montero, M. E., Rodriguez-Aranda, J. P. & Garcia del Cura, M. A., 2009. Bioinduced precipitation of barite and celestite in dolomite microbialites: examples from Miocene lacustrine sequences in the Madrid and Duero basins, Spain. *Sedimentary Geology*, **222**: 138–48.

Sarkoh, N. A., Al-Hasan, R. H., Khaferer, M. & Radwan, S. S., 1995. Establishment of oil-degrading Bacteria associated with Cyanobacteria in oil-polluted soil. *Journal of Applied Bacteriology*, **78**: 194–9.

Sarmiento, E., Sawyer, G. J., Milner, R. *et al.*, 2007. *The Last Human*. Yale University Press.

Sastri, V. V., Sinha, R. N., Singh, G. & Murti, K. V. S., 1973. Stratigraphy and tectonics of sedimentary basins of east coast of peninsular India. *American Association of Petroleum Geologists Bulletin*, **57**(4): 65–78.

Saunders, J. B., 1958. Recent foraminifera of mangrove swamps and their fossil counterparts in Trinidad. *Micropaleontology*, **4**(1): 79–92.

  (ed.), 1968. *Transactions of the Fourth Caribbean Geological Conference, Port-of-Spain, Trinidad & Tobago, 1965*. Caribbean Printers.

Saunders, J. B. & Bolli, H. M., 1985. Trinidad's contributions to world biostratigraphy. *Transactions of the Fourth Latin American Geological Congress, Trinidad & Tobago, 1979*: 781–95.

Saxena, R. K. & Misra, C. M., 1995. Campanian–Maastrichtian nannoplankton biostratigraphy of the Narsapur Claystone formation, Krishna–Godavari basin, India. *Journal of the Geological Society of India*, **45**: 323–9.

Savazzi, E. (ed.), 1999. *Functional Morphology of the Invertebrate Skeleton*. John Wiley & Sons.

Sawyer, M. J. & Keegan, J. B., 1996. Use of palynofacies characterization in sand-dominated sequences, Brent group, Ninian field, UK North Sea. *Petroleum Geoscience*, **2**: 289–97.

Schellmann, G. & Radtke, U., 2004. A revised morpho- and chronostfratigraphy of the Late and Middle Pleistocene coral reef terraces on southern Barbados (West Indies). *Earth-Science Reviews*, **64**: 157–187.

Schenck, H. G., 1940. Applied paleontology. *Bulletin of the American Association of Petroleum Geologists*, **24**(1): 1752–78.

Scheuth, J. D. & Frank, T. D., 2008. Reef foraminifera as bioindicators of coral reef health: Low Isles Reef, northern Great Barrier Reef, Australia. *Journal of Foraminiferal Research*, **38**(1): 11–22.

Schlee, J. S. (ed.), 1984. *Inter-regional Unconformities and Hydrocarbon Accumation*. American Association of Petroleum Geologists (Memoir No. 36).

Schluter, T., 1997. *Geology of East Africa*. Gebruder Borntraeger. 2006. *Geological Atlas of Africa*. Springer.

Schmidt, D. N., Thierstein, H. R., Bollmann, J. & Schiebel, R., 2004. Abiotic forcing of plankton evolution in the Cenozoic. *Science*, **303**: 207–10.

Schneider, G. & Marais, C., 2004. *Passage through Time – The Fossils of Namibia*. Gamsberg MacMillan.

Schopf, J. W., 2009. The hunt for Precambrian fossils: an abbreviated genealogy of the science. *Precambrian Research*, **173**: 4–9.

Schopf, J. W., Kudrtavtsev, A. B., Czaja, A. D. & Tripathi, A. B., 2007. Evidence of archaean life: stromatolites and microfossils. *Precambrian Research*, **158**: 141–55.

Schopf, T. J. M., 1972. *Models in Paleobiology*. Freeman.

Schouten, S., Hopmans, E. C., Schefuss, E. & Sissinghe Damste, J. S., 2002. Distributional variations in marine crenarchaeotal membrane lipids: a new tool for reconstructing ancient sea temperatures. *Earth and Planetary Science Letters*, **204**: 265–74.

Schreiber, B. C., Lugli, S. & Babel, M. (eds.), 2007. *Evaporites through Space and Time*. The Geological Society, London (Special Publication No. 285).

Schreve, D. C., 2004. *The Quaternary Mammals of Southern and Eastern England*. Quaternary Research Association.

Schroder, T., 1992. A palynological zonation of the North Sea basin. *Journal of Micropalaeontology*, **11**(2): 113–26.

Schulte, P., Alegret, L. Arenillas, I. *et al.*, 2010. The Chicxulub asteroid impact and mass extinction at the Cretaceous-Paleogene boundary. *Science*, **327**(5970): 1214–8.

Schulz, H. D. & Zabel, M. (eds.), 2000. *Marine Geochemistry*. Springer-Verlag.

Scott, A. C. (ed.), 1987. *Coal and Coal-Bearing Strata: Recent Advances*. The Geological Society, London (Special Publication No. 32).

Scott, A. C, 2000. The pre-Quaternary history of fire. *Palaeogeography, Palaeoclimatology, Palaeoecology*, **164**: 281–329. 2002. Coal petrology and the origin of coal macerals: a way ahead? *International Journal of Coal Geology*, **50**: 119–34.

Scott, A. C., Anderson, J. M. & Anderson, H., 2004. Evidence of plant–insect interactions in the Upper Triassic Molteno Formation of South Africa. *Journal of the Geological Society*, **161**(3): 401–10.

Scott, A.C. & Fleet, A.J. (eds.), 1994. *Coal and Coal-Bearing Strata as Oil-Prone Source Rocks?* The Geological Society, London (Special Publication No. 77).

Scott, A.R., 2010. Historical perspective and future opportunities for worldwide coalbed methane development. *Abstracts, International Conference, The Geology of Unconventional Gas Plays.* The Geological Society, London: 67.

Scott, D.B., Medioli, F.S. & Schafer, C.T., 2001. *Monitoring in Coastal Environments Using Foraminifera and Thecamoebian Indicators.* Cambridge University Press.

Scott, D.B., Tobin, R., Williamson, M. *et al.*, 2005. Pollution monitoring in two North American estuaries: historical reconstructions using benthic foraminifera. *Journal of Foraminiferal Research*, 35(1): 65–82.

Scott, S., Anderson, B., Crosdale, P. *et al.*, 2007. Coal petrology and coal seam gas contents of the Walloon subgroup – Surat basin, Queensland, Australia. *International Journal of Coal Geology*, 70: 209–22.

Segev, A., 2002. Flood basalts, continental breakup and the dispersal of Gondwana: evidence for periodic migration of upwelling mantle flows (plumes). *EGU Stephan Mueller Special Publications Series*, 2: 171–91.

Seiglie, G.A. & Baker, M.B., 1983. Some west African Cenozoic agglutinated foraminifers with inner structures: taxonomy, age and evolution. *Micropaleontology*, 29(4): 391–403.

Seilacher, A., 1984. Constructional morphology of bivalves: evolutionary pathways in primary versus secondary soft-bottom dwellers. *Palaeontology*, 27: 207–37.

2007. *Trace Fossil Analysis.* Springer.

Seilacher, A., Buatois, L.A. & Mangano, M.G., 2005. Trace fossils in the Ediacaran-Cambrian transition: behavioral diversification, ecological turnover and environmental shift. *Palaeogeography, Palaeoclimatology, Palaeoecology*, 227: 323–56.

Seilacher, A., Luning, S., Martin, M.A. *et al.*, 2002. Ichnostratigraphic correlation of Lower Palaeozoic clastics in the Kufra basin (SE Libya). *Lethaia*, 35: 257–62.

Seiter, K., Hensen, C., Schroter, J. & Zabel, M., 2004. Organic carbon content in surface sediments – defining regional provinces. *Deep-Sea Research I*, 51: 2001–26.

Selley, R.C., 1998. *Elements of Petroleum Geology*, 2nd edition. Academic Press.

Sellwood, B.W. & Valdes, P.J., 2006/2007. Mesozoic climates: general circulation models and the rock record. *Sedimentary Geology*, 190: 269–87.

Semeniuk, T.A., 2001. Epiphytic foraminifera along a climatic gradient, Western Australia. *Journal of Foraminiferal Research*, 31(3): 191–200.

Sen-Gui Zhang, Feng-Sheng Xia, Hui-Jun Yan & Zong-Zhe Wang, 2009. Horizon of the oldest known bryozoans (Ordovician). *Paleoworld*, 18: 67–73.

Sen Gupta, B.K., Smith, L.E. & Lobegeier, M.K., 2007. Attachment of foraminifera to vestimentiferan tubeworms at 'cold seeps': refuge from seafloor hypoxia and sulphide toxicity. *Marine Micropaleontology*, 62: 1–6.

Sepkoski, J.J. Jr, 1981. A factor analytical description of the Phanerozoic marine fossil record. *Paleobiology*, 7: 36–53.

Serezhnikova, E.A. & Ivantsov, A.Y., 2007. *Fedomia mikhalii* – a new spicule-bearing organism of sponge grade from the Vendian (Ediacaran) of the White Sea, Russia. *Palaeoworld*, 16: 319–24.

Sevim, A., Begun, D.R., Gulec, E. *et al.*, 2001. A new Late Miocene hominid from Turkey. *American Journal of Physical Anthropology Supplement*, 32: 134–5.

Shapiro, R.S. & Spangler, E., 2009. Bacterial fossil record in whale-falls: petrographic evidence of microbial sulphate reduction. *Palaeogeography, Palaeoclimatology, Palaeoecology*, 274: 196–203.

Sharaf, E.F., BouDagher-Fadel, M.K., Simo, J.A. & Carroll, A.R., 2005. Biostratigraphy and strontium isotope dating of Oligocene-Miocene strata, east Java, Indonesia. *Stratigraphy*, 2(3): 1–19.

Sharland, P.R., Archer, R., Casey, D.M. *et al.*, 2001. *Arabian Plate Sequence Stratigraphy.* Gulf PetroLink (GeoArabia Special Publication No. 2).

Sharland, P.R., Casey, D.B., Davies, R.B., Simmons, M.D. & Sutcliffe, O.E., 2004. Arabian plate sequence stratigraphy – revisions to SP2. *GeoArabia*, 9(1): 199–214.

Shaw, B., Jackson, J.A., Higham, T.F.G. *et al.*, 2010. Radiometric dates of uplifted marine fauna in Greece: implications for the interpretation of recent earthquake and tectonic histories using lithophagid dates. *Earth and Planetary Science Letters*, 297: 395–404.

Shaw, T., Sinclair, P., Andah, B. & Okpoko, A. (eds.), 1993. *The Archaeology of Africa: Food, Metals and Towns.* London.

Sheehan, P.M., 2001. The Late Ordovician mass extinction. *Annual Reviews of Earth and Planetary Science*, 29: 331–64.

Shenbrot, G.I. & Krasnov, B.R., 2005. *An Atlas of the Geographic Distribution of the Arvicoline Rodents of the World (Rodentia: Muridae, Arvicolinae).* Pensoft.

Sheppard, C.R.C., Davy, S.K. & Pilling, G.M., 2009. *The Biology of Coral Reefs.* Oxford University Press.

Shome, S. & Bardhan, S., 2009. The genus *Umiaites* Spath, 1931 (Ammonoidea) from the Tithonian (Late Jurassic) of Kutch, western India. *Palaeontologia Electronica*, 12(1): 10pp.

Shu Zhong Shen, Chang-Qun Cao, Henderson, C.M. *et al.*, 2006. End-Permian mass extinction pattern in the northern peri-Gondwanan region. *Palaeoworld*, 15: 3–30.

Shuaib, S.M., 1982. Geology and hydrocarbon potential of offshore Indus basin, Pakistan. *American Association of Petroleum Geologists Bulletin*, 66(7): 940–6.

Siddiqui, N.K., 2004. Sui Main limestone: regional geology and the analysis of original pressures of a closed-system reservoir in Pakistan. *American Association of Petroleum Geologists Bulletin*, 88(7): 1007–35.

Sidell, J., Wilkinson, K., Scarfe, R. & Cameron, N., 2000. *The Holocene Evolution of the London Thames: Archaeological Excavations (1991–1998) for the London Underground Limited Jubilee*

Line Extension Project. Museum of London Archaeology Service (MoLAS Monograph 5).

Silva, P. G., Borja, F., Zazo, C. *et al.*, 2005. Archaeoseismic record at the ancient Roman city of Baelo Claudia (Cadiz, southern Spain). *Tectonophysics*, **408**: 129–46.

Simmons, M. D. (ed.), 1994. *Micropalaeontology and Hydrocarbon Exploration in the Middle East*. Chapman & Hall.

Simmons, M. D., Sharland, P. R., Casey, D. M. *et al.*, 2007. Arabian Plate sequence stratigraphy: potential implications for global chronostratigraphy. *GeoArabia*, **12**(4): 101–30.

Simms, M. J., 2007. Uniquely extensive soft-sediment deformation in the Rhaetian of the UK: evidence for earthquake or impact. *Palaeogeography, Palaeoclimatology, Palaeoecology*, **244**: 407–23.

Simpson, G. G., 1946. Tertiary land bridges. *Transactions of the New York Academy of Sciences, Series II*, **8**(8): 255–8.

Singh, K., Aswal, H. S., Swamy, S. N. & Phor, L., 2008. Sequence biostratigraphy and hydrocarbon source potential of Mesozoic sediments, Kutch basin, India. *Proceedings, GEO India Expo XXI, Noida, New Delhi*: 1–6.

Sinha, D. K. (ed.), 2007. *Micropaleontology*. Narosa Publishing House.

Sinibaldi, R. W., 2010. *What Your Fossils Can Tell You: Vertebrate Morphology, Pathology, and Cultural Modification*. University Press of Florida.

Skelton, P. (ed.), 1993. *Evolution: A Biological and Palaeontological Approach*. Addison Wesley.

Skelton, P. W. (ed.), 2003. *The Cretaceous World*. Cambridge University Press.

Slatt, R. M., 2006. *Stratigraphic Reservoir Characterization for Petroleum Geologists, Geophysicists and Engineers*. Elsevier (Handbook of Petroleum Exploration and Production Series).

Sluijs, A., Pross, J. & Brinkhuis, H., 2005. From greenhouse to icehouse; organic-walled dinoflagellate cysts as paleoenvironmental indicators in the Paleogene. *Earth-Science Reviews*, **68**: 281–315.

Smith, A. B., 1984. *Echinoid Palaeobiology*. George Allen & Unwin.

Smith, A. G. & Pickering, K. T., 2003. Oceanic gateways as a critical factor to initiate icehouse Earth. *Journal of the Geological Society, London*, **160**: 337–40.

Smith, A. H. V. & Butterworth, M. A., 1967. *Miospores in Coal Seams of the Carboniferous of Great Britain*. The Palaeontological Association (Special Papers in Palaeontology No. 1).

Smith, C. J., Collins, L. S., Jaramillo, C. & Quiroz, L. I., 2010. Marine paleoenvironments of Miocene-Pliocene formations of north-central Falcon state, Venezuela. *Journal of Foraminiferal Research*, **40**(3): 266–82.

Smith, R. & Botha, J., 2005. The recovery of terrestrial vertebrate diversity in the South African Karoo basin after the end-Permian extinction. *Comptes Rendus Palevol*, **4**: 555–68.

Smith, T., Rana, R. S., Missiaen, P. *et al.*, 2007. High bat (Chiroptera) diversity in the Early Eocene of India. *Naturwissenschaften*, **94**: 1003–9.

Smithson, T. & Wood, S., 2009. Closing Romer's gap: new tetrapods and arthropods from the basal Carboniferous

of the Scottish borders. *Abstracts, Palaeontological Association Annual Meeting, Birmingham*, 32–3.

Sobolik, K. D., 2003. *Archaeobiology*. Altamira Press.

Sola, M. A. & Worsley, D. (eds.), 2000. *Geological Exploration in the Murzuq Basin*. Elsevier.

Solounias, N., Rivals, F. & Semprebon, G., 2010. Dietary interpretation and paleoecology of herbivores from Pikermi and Samos (Late Miocene of Greece). *Paleobiology*, **36**(1): 113–36.

Soltis, D. E., Soltis, P. S., Endress, P. K. & Chase, M. W., 2005. *Phylogeny and Evolution of Angiosperms*. Sinauer Associates Inc.

Soulie-Marsche, I., Benkkadour, A., El Khiati, N. *et al.*, 2008. Charophytes, indicateurs de paleobathymetrie du Lac Tigalmamine (Moyen Atlas, Maroc). *Geobios*, **41**: 435–44 (in French with English abstract).

Spalding, M. D., 2004. *A Guide to the Coral Reefs of the Caribbean*. University of California Press.

Spicer, R. A., Valdes, P. J., Spicer, T. E. V. *et al.*, 2009. New developments in CLAMP: calibration using global gridded meteorological data. *Palaeogeography, Palaeoclimatology, Palaeoecology*, **283**: 91–8.

Spitz, K. & Trudinger, J., 2009. *Mining and the Environment: From Ore to Metal*. CRC Press (Taylor & Francis Group).

Stanley, D. J., 1960. Stratigraphy and foraminifera of lower Tertiary Vidono Shale, near Puerto La Cruz, Barcelona. *American Association of Petroleum Geologists Bulletin*, **44**: 616–27.

Stanley, S. H., 2005. *Earth System History*, 2nd edition. W.H. Freeman.

Stanley, S. M., 1970. *Relation of Shell Form to Life Habits in the Bivalvia (Mollusca)*. Geological Society of America (Memoir No. 125).

2006. Influence of seawater chemistry on biomineraliztion throughout Phanerozoic time: paleontological and experimental evidence. *Palaeogeography, Palaeoclimatology, Palaeoecology*, **233**: 214–38.

1961. Reef controlled distribution of Devonian microplankton in Alberta. *Palaeontology*, **4**: 392–424.

Steel, R. J., Felt, V. L., Johannessen, E. P. & Mathieu, C. (eds.), 1995. *Sequence Stratigraphy of the Northwest European Margin*. Norwegian Petroleum Society (Special Publication No. 5).

Stephan, J. F., Mercier de Lepinay, B., Calais, E. *et al.*, 1990. Paleogeodynamic maps of the Caribbean. *Bulletin de la Société Géologique de la France, Series 8*, **6**: 915–9.

Steiner, M., Guoxiang Li, Yi Qian *et al.*, 2007. Neoproterozoic to early Cambrian small shelly fossil assemblages and a revised biostratigraphic correlation of the Yangtze platform (China). *Palaeogeography, Palaeoclimatology, Palaeoecology*, **254**: 67–99.

Stephenson, M. H., 2006. Stratigraphic note: update of the standard Arabian Permian palynological biozonation; definition and description of OSPZ5 and 6. *GeoArabia*, **11**(3): 173–8.

2008. Spores and pollen from the middle and upper Gharif members (Permian) of Oman. *Palynology*, **32**: 157–82.

Stephenson, M.H. & McLean, D., 1999. International correlation of Early Permian palynofloras from the Karoo sediments of Morupole, Botswana. *South African Journal of Geology*, **102**(1): 3–14.

Stigall, A., 2006. Combining GIS and phylogenetics in palaeobiogeography: assessing the role of invasion events during mass extinctions. *Programme with Abstracts, Palaeogeography and Palaeobiogeography ('Biodiversity in Space and Time') Conference, Cambridge*: 8.

2010. Using GIS to assess the biogeographic impact of species invasions on native brachiopods during the Richmondian invasion in the type Cincinnatian (Late Ordovician, Cincannati region). *Palaeontologia Electronica*, **13**(1)

Strachan, T. & Read, A.P., 2004. *Human Molecular Genetics*, 3rd edition. Garland Science/Taylor & Francis.

Streel, M., Caputo, M.V., Loboziak, S. & de Melo, J.H., 2000. Late Frasnian-Famennian climates based on palynomorph analyses and the question of the Late Devonian glaciations. *Earth-Science Reviews*, **52**: 121–73.

Stricanne, L., Munnecke, A., Pross, J. & Servais, T., 2004. Acritarch distribution along an inshore-offshore transect in the Gorstian (lower Ludlow) of Gotland, Sweden. *Review of Palaeobotany and Palynology*, **130**: 195–216.

Stringer, C., 2006. *Homo britannicus: The Incredible Story of Human Life in Britain*. Allen Lane (Penguin).

Stromberg, C.A.E., Werdelin, L., Friis, E.M. & Sarac, G. 2007. The spread of grass-dominated habitats in Turkey and surrounding areas during the Cenozoic: phytolith evidence. *Palaeogeography, Palaeoclimatology, Palaeoecology*, **250**: 18–49.

Strother, P.K., Al-Hajri, S. & Traverse, A., 1996. New evidence for land plants from the lower Middle Ordovician of Saudi Arabia. *Geology*, **24**(1): 55–8.

Struckmeyer, H.I.M. & Totterdell, J.M. (co-ords.), 1990. *Australia – Evolution of a Continent*. Canberra; Australian Government Publishing Service.

Stubblefield, C.J. & Trotter, F.M., 1957. Divisions of the Coal Measures on Geological Survey maps of England and Wales. *Bulletin of the Geological Survey of Great Britain*, **13**: 1–5.

Stworziewicz, E., Szulc, J. & Pokryszko, B.M., 2009. Late Palaeozoic continental gastropods from Poland: systematic, evolutionary and paleoecological approach. *Journal of Paleontology*, **83**(6): 938–45.

Suarez-Ruiz, I. & Crelling, J.C. (eds.), 2008. *Applied Coal Petrology – The Role of Petrology in Coal Utilization*. Academic Press.

Sues, H.-D. & Fraser, N.C., 2010. *Triassic Life on Land*. Columbia University Press.

Summa, L.L., Goodman, E.D., Richardson, M. *et al.*, 2003. Hydrocarbon systems of northeastern Venezuela: plate through molecular scale analysis of the genesis and evolution of the eastern Venezuelan basin. *Marine and Petroleum Geology*, **20**(3–4): 323–49.

Surlyk, F., Milan, J. & Noe-Nygaard, N., 2008. Dinosaur tracks and possible lungfish aestivation burrows in a shallow coastal lake; lowermost Cretaceous, Bornholm, Denmark. *Palaeogeography, Palaeoclimatology, Palaeoecology*, **267**: 292–303.

Sutcliffe, O.E., Dowdeswell, J.A., Whittington, R.J. *et al.*, 2000/2001. Calibrating the Late Ordovician glaciation and mass extinction by the eccentricity cycles of Earth's orbit. *Geology*, **28**(11): 967–70.

Sutcliffe, O.E., Harper, D.A.T., Salem, A.A. *et al.*, 2000/2001. The development of an atypical *Hirnantia*-brachiopod fauna and the onset of glaciation in the late Ordovician of Gondwana. *Transactions of the Royal Society of Edinburgh*, **92**: 1–14.

2001. The development of an atypical *Hirnantia*-brachiopod fauna and the onset of glaciation in the late Ordovician of Gondwana. *Transactions of the Royal Society of Edinburgh*, **92**: 1–14.

Suwa, G., Asfaw, B., Kono, R.T. *et al.*, 2009a. The *Ardipithecus ramidus* skull and its implications for human origins. *Science*, **326** (5949): 68, 68e1–e7.

Suwa, G., Kono, R.T., Simpson, S.W. *et al.*, 2009b. Paleobiological implications of the *Ardipithecus ramidus* dentition. *Science*, **326** (5949): 69, 94–99.

Swamy, S.N. & Kapoor, P.N., 2000. Palynofacies and organic matter maturation in the Cretaceous and Tertiary sediments of Krishna–Godavari basin, India. *Proceedings, 5th International Conference & Exhibition, Petroleum Geochemistry & Exploration in the Afro-Asian Region, New Delhi*: 483–91.

Swamy, S.N. & Kapoor, P.N. (eds.), 2002. *Proceedings of the 1st Conference and Exhibition on Strategic Challenges and Paradigm Shift in Hydrocarbon Exploration with Special Reference to Frontier Basins*. Association of Petroleum Geologists.

Szaniawski, H., 2009. The earliest known venomous animals recognized among conodonts. *Acta Palaeontologica Polonica*, **54**(4): 669–76.

Tabak, J., 2009. *Coal and Oil*. Facts on File Inc. (Infobase Publishing).

Tabor, N.J. & Poulsen, C.J., 2008. Palaeoclimate across the Late Pennsylvanian–Early Permian tropical palaeolatitudes: a review of climate indicators, their distribution and relation to palaeophysiographic climatic factors. *Palaeogeography, Palaeoclimatology, Palaeoecology*, **268**: 293–310.

Tankard, A.J., Soruco, R.S. & Welsink, H.J., 1995. *Petroleum Basins of South America*. American Association of Petroleum Geologists (Memoir, No. 62).

Tao Su, Yao-Wu Xing, Yu-Sheng Liu *et al.*, 2010. Leaf margin analysis: a new equation from humid to mesis forest in China. *Palaios*, **25**: 234–8.

Tapanila, L., 2004. The earliest *Helicosalpinx* from Canada and the global expansion of commensalism in Late Ordovician sarcinulid corals (Tabulata). *Palaeogeography, Palaeoclimatology, Palaeoecology*, **215**: 99–110.

Tasker, A., Wilkinson, I.P., Williams, M. *et al.*, 2009. Using microfossils to provenance the materials for Roman mosaics: examples from Leicester and Silchester. *Abstracts, Palaeontological Association Annual Meeting, Birmingham*: 76.

Tawadros, E. E., 2001. *Geology of Egypt and Libya*. A.A. Balkema.

Taylor, E. L. & Ryberg, P. E., 2007. Tree growth at polar latitudes based on fossil tree ring analysis. *Palaeogeography, Palaeoclimatology, Palaeoecology*, **255**: 246–64.

Taylor, P., 2004. *Extinctions in the History of Life*. Cambridge University Press.

Taylor, P. D., 1995. *Field Geology of the British Jurassic*. The Geological Society.

Taylor, T. N. & Krings, M., 2010. Paleomycology: the rediscovery of the obvious. *Palaios*, **25**: 283–6.

Taylor, T. N., Taylor, E. L. & Krings, M., 2009. *Paleobotany: The Biology and Evolution of Fossil Plants*, 2nd edition. Academic Press/Elsevier.

Thomas, C., 2002. *The Archaeology of Medieval London*. Sutton Publishing Ltd.

Thomas, L., 1992. *Handbook of Practical Coal Geology*. John Wiley.

Thomason, J. J. (ed.), 1995. *Functional Morphology in Vertebrate Paleontology*. Cambridge University Press.

Thurber, A. R., Kroger, K., Neira, C. *et al.*, 2010. Stable isotope signatures and methane use by the cold seep benthos. *Marine Geology*, **272**: 260–9.

Tickell, C., 1996. *Mary Anning of Lyme Regis*. Philpot Museum.

Tiercelin, J. J., Soreghan, M., Cohen, A. S. *et al.*, 1992. Sedimentation in large rift lakes: example from the Middle Pleistocene-modern deposits of the Tanganyika trough, east African rift system. *Bulletin du Centre de Recherches Exploration-Production Elf-Aquitaine*, **16**(1): 83–111.

Tillier, A. M., 1992. Les hommes du Paleolithique moyen *et al* question de l'anciennete de l'homme moderne en Afrique. *Archeo-Nil*, **2**: 59–69.

Todd, R. & Bronnimann, P., 1957. *Recent Foraminifera and Thecamoebina from the Eastern Gulf of Paria, Trinidad*. Cushman Foundation for Foraminiferal Research (Special Publication No. 3).

Tomescu, A. M. F., Pratt, L. M., Rothwell, G. W. *et al.*, 2009. Carbon isotopes support the presence of extensive land floras pre-dating the origin of vascular plants. *Palaeogeography, Palaeoclimatology, Palaeoecology*, **283**: 46–59.

Tonkin, N. S., McIlroy, D., Meyer, R. & Moore-Turpin, A., 2010. Bioturbation influence on reservoir quality: a case study from the Cretaceous Ben Nevis formation, Jeanne d'Arc basin, offshore Newfoundland, Canada. *American Association of Petroleum Geologists Bulletin*, **94**(7): 1059–78.

Torrens, H. S., 1995. Mary Anning (1799–1847) of Lyme; 'The Greatest Fossilist the World Ever Knew'. *British Journal of the History of Science*, **28**: 257–84.

Torres, M. E., Martin, R. A., Klinkhammer, G. P. & Nesbitt, E. A., 2010. Post-depositional alteration of foraminiferal shells in cold seep settings: new insights from flow-through time-rsolved analyses of biogenic and inorganic seep carbonates. *Earth and Planetary Science Letters*, **299**: 10–22.

Torres, M. E., Mix, A. C., Kinports, K. *et al.*, 2003. Is methane venting at the seafloor recorded by del^{13}C of benthic foraminifera shells? *Paleoceanography*, **18**(3): 7-1 – 7-13.

Trauth, T. H., Larrasoana, J. C. & Mudelse, M., 2009. Trends, rhythms and events in Plio-Pleistocene African climate. *Quaternary Science Reviews*, **28**: 399–411.

Trauth, M. H., Maslin, M. A., Deino, A. & Strecker, M. R., 2005. Late Cenozoic moisture history of east Africa. *Science*, **309** (5743): 2051–3 (published online 18/9/2005, DOI: 10.1126/science.1112964).

Traverse, A., 2008. *Paleopalynology*, 2nd edition. Springer.

Tremolada, F., Erba, E., van de Schootbrugge, B. & Mattioli, E., 2006. Calcareous nannofossil changes during the late Callovian-early Oxfordian cooling phase. *Marine Micropaleontology*, **59**: 197–209.

Trewin, N. H., 2003. *The Geology of Scotland*, 4th edition. The Geological Society, London.

2008. *Fossils Alive, Or, New Walks in an Old Field*. Dunedin Academic Press.

Tribovillard, N. -P., Stephan, J. -F., Manivit, H. *et al.*, 1991. Cretaceous black shales of Venezuelan Andes: preliminary results on stratigraphy and paleoenvironmental interpretation. *Palaeogeography, Palaeoclimatology, Palaeoecology*, **81**: 313–21.

Tripathi, A., 2008. Palynochronology of Lower Cretaceous volcano-sedimentary succession of the Rajmahal formation in the Rajmahal basin, India. *Cretaceous Research*, **29**: 913–24.

Tripathi, A. & Ray, A., 2006. Palynochronology of the Dubrajpur formation (Early Triassic to Early Cretaceous) of the Rajmahal basin, India. *Palynology*, **30**: 133–49.

Trueman, A. E. & Weir, J., 1946. A monograph of British Carboniferous non-marine Lamelibranchia. *Palaeontographical Society Monographs*, **32**(1).

Truett, J. C. & Johnson, S. R. (eds.), 2000. *The Natural History of an Arctic Oil Field – Development and the Biota*. Academic Press.

Tsujita, CV. J. & Westermann, G. E. G., 1998. Ammonoid habitats and habits in the Western Interior Seaway: a case study from the Upper Cretaceous Bearpaw formation of southern Alberta, Canada. *Palaeogeography, Palaeoclimatology, Palaeoecology*, **144**: 136–60.

Turner, B. R. & Benton, M. J., 1983. Paleozoic trace fossils from the Kufra basin, Libya. *Journal of Paleontology*, **57**(3): 47–60.

Turney, C. S. M. & Brown, H., 2007. Catastrophic early Holocene sea level rise, human migration and the Neolithic transition in Europe. *Quaternary Science Reviews*, **26**: 2036–41.

Turvey, S. L., 2005. Early Ordovician (Arenig) trilobite palaeoecology and palaeobiogeography of the South China Plate. *Palaeontology*, **48**(3): 519–48.

Turvey, S. T. (ed.), 2009. *Holocene Extinctions*. Oxford University Press.

Twitchett, R. J., 2006. The palaeoclimatology, palaeoecology and palaeoenvironmental analysis of 'mass extinction' events. *Palaeogeography, Palaeoclimatology, Palaeoecology*, **232**: 190–213.

Tyson, R.V. & Pearson, T.H. (eds.), 1991. *Modern and Ancient Continental Shelf Anoxia*. The Geological Society, London (Special Publication No. 58).

Tyszka, J., Jach, R. & Bubik, M., 2010. A new vent-related foraminifera from the lower Toarcian black claystones of the Tatra Mountainas, Poland. *Acta Palaeontologica Polonica*, **55** (2): 333–42.

Tyszka, J., Oliwkiewicz-Miklasinska, M., Gedl, P. & Kaminski, M.A. (eds.), 2005. *Methods and Applications in Micropalaeontology*.

Tzedakis, P.C., 2005. Towards an understanding of the response of southern European vegetation to orbital and suborbital climate variability. *Quaternary Science Reviews*, **25**: 1585–99.

Uhl, D., Klotz, S., Traiser, C. *et al.*, 2007. Cenozoic paleotemperatures and leaf physiognomy – a European perspective. *Palaeogeography, Palaeoclimatology, Palaeoecology*, **248**: 24–31.

Underhill, J.R., 1998. *Development, Evolution and Petroleum Geology of the Wessex Basin*. The Geological Society, London (Special Publication No. 133).

Ungar, P.S., 2007. *Evolution of the Human Diet*. Oxford University Press.

Unwin, D.M., 2005. *The Pterosaurs*. Pi Press.

Upchurch, P., Nunn, C.A. & Norman, D.B., 2002. An analysis of dinosauran biogeography: evidence for the existence of vicariance and diuspersal patterns caused by geological events. *Proceedings of the Biological Society*, **269** (1491): 613–21.

Utescher, T. & Mosbrugger, V., 2007. Eocene vegetation patterns reconstructed from plant diversity – a global perspective. *Palaeogeography, Palaeoclimatology, Palaeoecology*, **247**: 243–71.

Utescher, T., Mosbruger, V. & Ashraf, A.R., 2002. Facies and paleogeography of the Tertiary of the Lower Rhine Basin – sedimentary versus climatic control. *Netherlands Journal of Geosciences/Geologie en Mijnbouw*, **81**(2): 185–91.

Vaidyanadhan, R. & Ramakrishnan, M., 2008. *Geology of India*. Volume 2. Geological Society of India.

Vail, P.R., Mitchum, R.M. Jr, Thompson, S. III *et al.*, 1977. Seismic stratigraphy and global changes of sea-level. Part 4: Global cycles of relative sea-level. *American Association of Petroleum Geologists Memoir*, **26**: 49–212.

van Andel, Tj. & Postma, H., 1954. Recent sediments of the Gulf of Paria. *Verhandelingen der Koninklijke Nederlandse Akademie van Wetenschappen, Afd. Natuurkunde, Eerste Reeks*, **XX**(5): 1–247.

van Andel, T.H. & Davies, W. (eds.), 2003. *Neanderthals and Modern Humans in the European Landscape during the Last Glaciation*. McDonald Institute Monographs.

van Buchem, F.S.P., Gerdes, K.D. & Esteban, M. (eds.), 2010. *Mesozoic and Cenozoic Carbonate Systems of the Mediterranean and the Middle East*. The Geological Society, London (Special Publication No. 329).

Van Buggenum, J.M., 1991. A probabilistic method for stratigraphic correlation applied to palynological data from the Lower Cretaceous pre-salt sequence in Gabon. *Résumés 11e Colloque Africain de Micropaleontologie, Libreville, Gabon.*

van Couvering, J.A.H., 1982. Fossil cichlid fish of Africa. *Special Papers in Palaeontology*, **29**: 1–103.

van Dam, J.A., 2006. Geographic and temporal patterns in the late Neogene (12–3 Ma) aridification of Europe: the use of small mammals as paleoprecipitation proxies. *Palaeogeography, Palaeoclimatology, Palaeoecology*, **238**: 190–218.

van den Bold, W.A., 1960. Eocene and Oligocene ostracoda of Trinidad. *Micropaleontology*, **6**(2): 145–96.

    1963. Upper Miocene and Pliocene ostracoda of Trinidad. *Micropaleontology*, **9**(4): 361–424.

    1974. Ostracode associations in the Caribbean Neogene. *Verhandlungen der Naturforchenden Gesellschaft in Basel*, **84**(1): 214–21.

van der Hammen, T. & Wymstra, T.A., 1964. A palynological study of the Tertiary and Upper Cretaceous of British Guiana. *Leidse Geologische Mededelingen*, **30**: 183–241.

van der Made, J., Morales, J., Sen, S. & Aslan, F., 2002. The first camel from the Upper Miocene of Turkey and the dispersal of camels into the Old World. *Comptes Rendus Palevol*, **1**: 117–22.

Vandermeesch, B., 1981. *Les hommes fossils de Qafzeh (Israel)*. CNRS.

van Dover, C.L., 2000. *The Ecology of Deep-Sea Hydrothermal Vents*. Princeton University Press.

van Gorsel, J.T., 1988. Biostratigraphy in Indonesia: methods, pitfalls and new directions. *Proceedings of the Indonesian Petroleum Association 17th Annual Convention, Jakarta, Indonesia*: 275–300.

van Hinte, J.E. (ed.), 1966. *Proceedings of the Second West African Micropalaeontological Colloquium, Ibadan, 1965*. E.J. Brill.

van Kolfschoten, T. & Laban, C., 1995. Pleistocene terrestrial mammals from the North Sea. *Mededelingen Rijks Geologische Dienst*, **52**: 135–52.

van Kranendonk, M.J., Pjilippot, P., Lepot, K. *et al.*, 2008. Geological setting of earth's oldest fossils in the ca. 3.5 Ga Dresser formation, Pilbara craton, Western Australia. *Precambrian Research*, **167**: 93–124.

Vannier, J., 2009. L'explosion cambrienne ou l'emergence des ecosystems modernes. *Comptes Rendus Palevol*, **8**: 133–54.

Varnam, A.H. & Evans, M.G., 2000. *Environmental Microbiology*. Manson Publishing.

Vasey, G.M., 1994. Classification of Carboniferous non-marine bivalves: systematic versus stratigraphy. *Journal of the Geological Society, London*, **151**: 1023–9.

Vecoli, M., 2000. Palaeoenvironmental interpretation of microphytoplankton diversity trends in the Cambrian-Ordovician of the northern Sahara platform. *Palaeogeography, Palaeoclimatology, Palaeoecology*, **160**: 329–47.

Vecoli, M., Clement, G. & Meyer-Berthaud, B. (eds.), 2010. *The Terrestrialization Process: Modelling Complex Interactions at the*

Biosphere–Geosphere Interface. The Geological Society, London (Special Publication No. 339).

Vecoli, M. & le Herisse, A., 2004. Biostratigraphy, taxonomic diversity and patterns of morphological evolution of Ordovician acritarchs (organic-walled microphytoplankton) from the northern Gondwana margin in relation to palaeoclimatic and palaeogeographic changes. *Earth-Science Reviews*, **67**: 267–311.

Vecoli, M., Paris, F. & Videt, B., 2007. Enigmatic, spore-like organic-walled microfossils from Middle–Late Cambrian sediments in Algeria: terrestrial or aquatic origin? *Abstracts, The Palaeontological Association 51st Annual Meeting, Uppsala*: 59.

Vecoli, M., Tongiorgi, M., Abdesselam-Roughi, F.-F. *et al.*, 1999. Palynostratigraphy of Upper Cambrian-Upper Ordovician intracratonic clastic sequences, north Africa. *Bollettino della Societa Paleontologica Italiana*, **38**(2/3): 331–41.

Veevers, J.J. & Powell, C.McA. (eds.), 1994. *Permian–Triassic Pangean Basins and Foldbelts along the Panthalassan margin of Gondwanaland*. The Geological Society of America.

Veevers, J.J. & Saaed, A., 2009. Permian–Jurassic Mahanadi and Pranhita–Godavari rifts of Gondwana India: provenance from regional paleoslope and U–Pb/Hf analysis of detrital zircons. *Gondwana Research*, **16**: 633–54.

Veevers, J.J. & Tewari, R.C., 1995. *Gondwana Master Basin of India between Tethys and the Interior of the Gondwanaland Province of Pangea*. Geological Society of America (Memoir No. 187).

Vega, F.J., Nyborg, T.G., del Carmen Perriliat, M. *et al.*, 2006. *Studies on Mexican Paleontology*. Springer.

Venkatachala, B.S. & Sinha, R.N., 1986. Stratigraphy, age and palaeoecology of Upper Gondwana equivalents of the Krishna–Godavari basin, India. *The Palaeobotanist*, **35**(1): 22–31.

Vennin, E., van Buchem, F.S.P., Joseph, P. *et al.*, 2003. A 3D outcrop analogue model for Ypresian nummulitic carbonate reservoirs: Jebel Ousselat, northern Tunisia. *Petroleum Geoscience*, **9**: 145–61.

Verati, C., Rapaille, C., Feraud, G. *et al.*, 2007. 40Ar/39Ar ages and duration of the Central Atlantic Magmatic Province volcanism in Morocco and Portugal and its relation to the Triassic–Jurassic boundary. *Palaeogeography, Palaeoclimatology, Palaeoecology*, **244**: 308–25.

Verdenius, J.G., van Hinte, J.E. & Fortuin, A.R. (eds.), 1983. *Proceedings of the First International Workshop on Arenaceous Foraminifera*. IKU.

Verzi, D.H. & Quintana, C.A., 2005. The caviomorph rodents from the San Andres formation, east-central Argentina, and global Late Pliocene climate change. *Palaeogeography, Palaeoclimatology, Palaeoecology*, **219**: 303–20.

Vetter, E.W. & Dayton, P.K., 1998. Macrofaunal communities within and adjacent to a detritus-rich submarine canyon system. *Deep-Sea Research II*, **45**: 25–54.

Vially, R. (ed.), 1992. *Bacterial Gas*. Edition Technip.

Viana, C.F., Gama, E.G. Jr, Simoes, I.A. *et al.*, 1971. Revisao estratigrafia da bacia Reconcavo/Tucano. *Boletim Tecnico de Petrobras*, **14**(3/4): 157–92.

Vickers-Rich, P. & Komarower, P. (eds.), 2007. *The Rise and Fall of the Ediacaran Biota*. The Geological Society, London (Special Publication No. 286).

Vickers-Rich, P. & Rich, T.H., 1993. *Wildlife of Gondwana*. Reed (Heinemann).

Vijaya, 1999. Palynological dating of the Neocomian-Aptian succession in the Indian peninsula. *Cretaceous Research*, **20**: 597–608.

2009. Palynofloral changes in the Upper Paleozoic and Mesozoic of the Deocha-Pachami area, Birbhum coal-field, West Benal, India. *Science in China, Series D, Earth Sciences*, **52**(12): 1931–52.

Vijaya & Bhattacharji, T.K., 2002. An Early Cretaceous age for the Rajmahal Traps, Panagarh area, West Bengal: palynological evidence. *Cretaceous Research*, **23**: 789–805.

Vijaya & Kumar, S., 2002. Palynostratigraphy of the Spiti shale (Oxfordian–Berriasian) of Kumaon Tethys Himalaya, Malla Johar area, India. *Review of Palaeobotany and Palynology*, **122**: 143–53.

Vining, B.A. & Pickering, S.C. (eds.), 2010. *Petroleum Geology: from Mature Basins to New Frontiers (Proceedings of the 7th Petroleum Geology Conference)*. The Geological Society, London.

Visscher, H., Looy, C.V., Collinson, M.E. *et al.*, 2004. Enironmental mutagenesis during the end-Permian ecological crisis. *Proceedings of the National Academy of Sciences, USA*, **101**: 12952–6.

Voldman, G.G., Albanesi, G.L. & Ramos, V.A., 2010. Conodont geothermometry of the lower Paleozoic from the Precordillera (Cuyania terrane), northwestern Argentina. *Journal of South American Earth Sciences*, **29**: 278–88.

Volkheimer, W., Rauhut, O.W.M., Quattrocchio, M.E. & Martinez, M.A., 2008. Jurassic climates in Argentina, a review. *Revista de la Asociacion Geologica Argentina*, **63**(4): 549–56.

von Hillebrandt, A., Krystyn, L., Kuerschner, W.M. *et al.*, 2007. A candidate GSSP for the base of the Jurassic in the northern Calcareous Alps (Kuhjoch section, Karwendel mountains, Tyrol, Austria). *International Subcommission on Jurassic Stratigraphy Newsletter*, **34**: 2–20.

Vorobeva, N.G., Sergeev, V.N. & Knoll, A.H., 2009. Neoproterozoic microfossils from the margin of the East European Platform and the search for a biostratigraphic model of lower Ediacaran rocks. *Precambrian Research*, **173**: 163–9.

Voros, A., 2005. The smooth brachiopods of the Mediterranean Jurassic: refugees or invaders? *Palaeogeography, Palaeoclimatology, Palaeoecology*, **223**: 222–42.

Vrba, E.S., Denton, G.H., Partridge, T.C. & Burckle, L.H. (eds.), 1995. *Paleoclimate and Evolution (with Emphasis on Human Origins)*. Yale University Press.

Vrba, E.S. & Schaller, G.B. (eds.), 2000. *Antelopes, Deer and Relatives: Fossil Record, Systematics and Conservation*. Yale University Press.

Wacey, D., 2009. *Early Life on Earth: A Practical Guide*. Springer.

Wacey, D., Kilburn, M.R., McLoughlin, N. *et al.*, 2008. Use of NanoSIMS in the search for early life: ambient inclusions trails in a c. 3400 Ma sandstone. *Journal of the Geological Society, London*, **165**: 43–53.

Waddington, C. (ed.), 2007. *Mesolithic Settlement in the North Sea Basin. A Case Study from Howick, North-East England*. Oxbow Books.

Wakefield, M.I., Cook, R.J., Jackson, H. & Thompson, P., 2001. Interpreting biostratigraphical data using fuzzy logic: the identification of regional mudstones within the *Fleming* field, UK North Sea. *Journal of Petroleum Geology*, **24**(4): 417–40.

Wakefield, M. & Monteil, E., 2002. Biosequence stratigraphical and palaeoenvironmental findings from the Cretaceous through Tertiary succession, central Indus basin, Pakistan. *Journal of Micropalaeontology*, **21**: 115–30.

Walker, M., 2005. *Quaternary Dating Methods*. John Wiley & Sons.

Walker, M., Johnsen, S., Rasmussen, S.O. *et al.*, in press (2010). A proposal for the Global Stratotype Section and Point (GSSP) and Global Standard Stratigraphic Age (GSSA) for the base of the Holocene series/epoch (Quaternary system/period). *Journal of Quaternary Science*.

Wallace, A.R., 1858. On the tendency of species to form varieties, and on the perpetuation of varieties and species by natural means of selection. *Proceedings of the Linnean Society, Zoology*, **3**(9): 53–62.

Walsh, S.L., Gradstein, F.M. & Ogg, J.G., 2004. History, philosophy and application of the Global Stratotype Section and Point (GSSP). *Lethaia*, **37**: 201–18.

Wandrey, C.J., 2004. Bombay geologic province Eocene to Miocene composite total petroleum system, India. *United States Geological Survey Bulletin*, **2208**-F.

Wandrey, C.J., Law, B.E. & Shah, H.A., 2004. Sembar Goru/Ghazij composite total petroleum system, Indus and Sulaiman-Kirthar geologic provinces, Pakistan and India. *United States Geological Survey Bulletin*, **2208**-C.

Wang, L., Lu, H.Y., Wu, N.Q. *et al.*, 2006. Palynological evidence for Late Miocene–Pliocene vegetation evolution recorded in the red clay sequence of the central Chinese Loess Plateau and implication for palaeoenvironmental change. *Palaeogeography, Palaeoclimatology, Palaeoecology*, **241**: 118–28.

Wang, Z. & Tedford, R.H., 2008. *Dogs: Their Fossil Relatives and Evolutionary History*. Columbia University Press.

Ward, I., Larcombe, P. & Lillie, M., 2006. The dating of Doggerland – post-glacial geochronology of the southern North Sea. *Environmental Archaeology*, **11**(2): 207–18.

Ward, P.D., 2004. *Gorgon: Paleontology, Obsession and the Greatest Catastrophe in Earth History*. Viking Press.
  2006. *Out of Thin Air: Dinosaurs, Birds, and Earth's Ancient Atmosphere*. Joseph Henry Press.
  2007. *Under a Green Sky*. HarperCollins.

Warraich, M.Y. & Nishi, H., 2003. Eocene planktic foraminiferal biostratigraphy of the Sulaiman Range, Indus basin, Pakistan. *Journal of Foraminiferal Research*, **33**(3): 219–36.

Warraich, M.Y., Ogasawara, K. & Nishi, H., 2000. Late Paleocene to Early Eocene planktic foraminiferal biostratigraphy of the Dungan formation, Sulaiman Range, central Pakistan. *Paleontological Research*, **4**(4): 275–301.

Warwick, P.D. (ed.), 2005. *Coal Systems Analysis*. Geological Society of America (Special Publication No. 387).

Warwick, P.D., Johnson, E.A. & Khan, I.H., 1998. Collision-induced tectonism along the northwestern margin of the Indian subcontinent as recorded in the Upper Paleocene to Middle Eocene strata of central Pakistan (Kirthar and Sulaiman Ranges). *Palaeogeography, Palaeoclimatology, Palaeoecology*, **142**: 201–16.

Waters, C.N., Browne, M.A.E., Dean, M.T., *et al.*, 2007. Lithostratigraphical framework for Carboniferous successions of Great Britain (onshore). *British Geological Survey Research Report*, **RR/07/01**.

Waters, C.N., Waters, R.A., Barclay, W.J. & Davies, J.R., 2009. A lithostratigraphical framework for the Carboniferous successions of southern Great Britain (onshore). *British Geological Survey Research Report*, **RR/09/01**.

Watkins, J. & Drake, C.L. (eds.), 1982. *Studies in Continental Margin Geology*. American Association of Petroleum Geology (Memoir No. 34).

Watkinson, M.P., Hart, M.B. & Joshi, A., 2007. Cretaceous tectonostratigraphy and the development of the Cauvery basin, southeast India. *Petroleum Geoscience*, **13**: 181–91.

Webb, E.A. & Longstaffe, F.J., 2010. Limitations on the climatic and ecological signals provided by the del^{13}C values of phytoliths from a C4 North American prairie grass. *Geochimica et Cosmochimica Acta*, **74**: 3–41–50.

Webb, G.E., 1998. Earliest known Carboniferous shallow-water reefs, Gudman formation (Tn1b), Queensland, Australia: implications for Late Devonian reef collapse and recovery. *Geology*, **26**(10): 951–4.

Webb, P.W., 1984. Body form, locomotion and foraging in aquatic vertebrates. *American Zoologist*, **24**: 107–20.

Webb, S., 2006. *The First Boat People*. Cambridge University Press.

Webby, B.D., Paris, F., Droser, M.L. & Percival, I.C., 2004. *The Great Ordovician Biodiversification Event*. Columbia University Press.

Weedon, G.P., 2003. *Time-Series Analysis and Cyclostratigraphy*. Cambridge University Press.

Weijers, J.W.H, Schouten, S., Spaargaren, O.C. & Sissinghe Damste, J.S., 2006. Occurrence and distribution of tetraether membrane lipids in soils: implications for the use of the TEX86 proxy and the BIT index. *Organic Geochemistry*, **37**: 1680–93.

Weimer, P. & Link, M.H., 1991. *Seismic Facies and Sedimentary Processes of Submarine Fans and Turbidite Systems*. Springer-Verlag.

Weiner, S., 2010. *Microarchaeology: Beyond the Visible Archaeological Record*. Cambridge University Press.

Welcomme, J.-L., Benammi, M., Crochet, J.-Y. *et al.*, 2001. Himalayan forelands: Palaeontological evidence for Oligocene detrital deposits in the Bugti Hils (Balochistan, Pakistanm). *Geological Magazine*, **138**(4): 397–405.

Wellman, C.H., 2005. Dispersed spores as evidence of the origin and early evolution of land plants. *Abstracts, The Micropalaeontological Society Annual General Meeting Presentations, 2005*.

Wellman, C.H., Osterloff, P.L. & Mohiuddin, U., 2003. Fragments of the earliest land plants. *Nature*, **425**: 282–5.

Wells, N.A., 1986. Biofabrics as dynamic indicators in nummulite accumulations – discussion. *Journal of Sedimentary Petrology*, **56**: 318–20.

Wendorf, F.W. & Schild, R. (eds.), 1976. *Prehistory of the Nile Valley*. Academic Press.

Wengrow, D., 2006. *The Archaeology of Early Egypt*. Cambridge University Press.

Werdelin, L. & Sanders, W.J. (eds.), 2010. *Cenozoic Mammals of Africa*. University of California Press.

Wescott, W.A., Krebs, W.N., Sikora, P.J., Boucher, P.J. & Stein, J.A., 1998. Modern applications of biostratigraphy in exploration and production. *The Leading Edge*, September 1998: 1204–10.

Wesselingh, F.P. & Macsotay, O., 2006. *Pachydon hettneri* (Anderson, 1928) as an indicator for Caribbean–Amazonian lowland connections during the Early–Middle Miocene. *Journal of South American Earth Sciences*, **21**: 49–53.

West, R.G., 1980. Pleistocene forest history of East Anglia. *New Phytologist*, **85**: 571–622.

Whateley, M.K.G. & Spears, D.A. (eds.), 1995. *European Coal Geology*. The Geological Society, London (Special Publication No. 82).

Whatley, R.C., 1991. The platycopid signal: a means of detecting kenoxic events using Ostracoda. *Journal of Micropalaeontology*, **10**: 181–3.

1995. Ostracoda and oceanic palaeo-oxygen levels. *Mitteilungen aus dem Hamburgischen Zoologischen Museum und Institut*, **92**: 337–53.

Whatley, R.C. & Maybury, C. (eds.), 1990. *Ostracoda and Global Events*. Chapman & Hall.

Whatley, R.C., Arias, C.F. & Cornas-Rengifo, M.J., 1994. The use of Ostracoda to detect kenoxic events: a case history from the Spanish Toarcian. *Geobios, Memoire Speciale*, **17**(2): 733–41.

Whatley, R.C., Pyne, R.S. & Wilkinson, I.P., 2003. Ostracoda and palaeooxygen levels, with particular reference to the Upper Cretaceous of East Anglia. *Palaeogeography, Palaeoclimatology, Palaeoecology*, **294**: 355–86.

Wheeley, J.R., Smith, P. & Boomer, I., 2009. Conodonts as palaeothermometers for ancient oceans? *Abstracts, Palaeontological Association Annual Meeting, Birmingham*: 36.

White, M.E., 1986. *The Greening of Gondwana*. Reed.

1990. *The Nature of Hidden Worlds – Animals and Plants in Prehistoric Australia and New Zealand*. Reed.

White, T.D., Ambrose, S.H., Suwa, G. et al., 2009a. Macrovertebrate paleontology and the Pliocene habitat of *Ardipithecus ramidus*. *Science*, **326** (5949): 67, 87–93.

White, T.D., Asfaw, B., Beyene, Y. et al., 2009b. *Ardipithecus ramidus* and the paleobiology of early hominids. *Science*, **326** (5949): 64, 75–86.

Whiteside, J.H., Olsen, P.E., Kent, D.V. et al., 2007. Synchrony between the Central Atlantic Magmatic Province and the Triassic-Jurassic mass extinction event? *Palaeogeography, Palaeoclimatology, Palaeoecology*, **244**: 345–67.

Whittaker, J.E. & Hart, M.B. (eds.), 2010. *Micropalaeontology, Sedimentary Environments and Stratigraphy: A Tribute to Dennis Curry (1912–2001)*. The Micropalaeontological Society, London (Special Publication No. 4).

Whittaker, J.E., Horne, D.J. & Jones, R.W., 2003. Micropalaeontology in the service of archaeololgy: advances in Quaternary biostratigraphy and palaeoenvironmental analysis using foraminifera and ostracods. *Abstracts, The Micropalaeontological Society Annual General Meeting Presentations, 2003*.

Whittaker, J.E., Jones, R.W. & Banner, F.T., 1998. *Key Mesozoic Benthic Foraminifera of the Middle East*. The Natural History Museum, London.

Wichura, H., Bousquet, R., Oberhansli, R. et al., 2010. Evidence for Middle Miocene uplift of the East Africa Plateau. *Geology*, **35**(6): 543–6.

Wienberg, C., Hebbeln, D., Fink, H.G. et al., 2009. Scleractinian cold-water corals in the Gulf of Cadiz – first clues about their spatial and temporal distribution. *Deep-Sea Research I*, **56**: 1873–93.

Wignall, P.B., 1990. Benthic palaeoecology of the Late Jurassic Kimmeridge Clay of England. *Special Papers in Palaeontology*, **43**: 1–74.

1994. *Black Shales*. Oxford University Press.

Wignall, P.B., Newton, R.J. & Little, C.T.S., 2005. The timing of palaeoenvironmental change and cause-and-effect relationships during the Early Jurassic mass extinction in Europe. *American Journal of Science*, **305**: 1014–32.

Wignall, P.B., Sun, Y.D., Bond, D.P.G. et al., 2009a. Precise coincidence of explosive volcanism, mass extinction and carbon isotope fluctuations in the Middle Permian of China. *Science*, **324**: 1179–82.

Wignall, P.B., Vedrine, S., Bond, D.P.G. et al., 2009b. Facies analysis and sea-level change at the Guadalupian-Lopingian Global Stratotype (Laibin, south China), and its bearing on the end-Guadalupian mass extinction. *Journal of the Geological Society, London*, **166**: 655–66.

Wilgus, C., Hastings, B.S., Kendal, C.G.StC. et al. (eds.), 1988. *Sea-Level Changes: An Integrated Approach*. American Association of Petroleum Geologists (Special Publication No. 42).

Wilkinson, I.P., Tasker, A., Gouldwell, A. et al., 2010. Micropalaeontology reveals the source of building materials for a defensive earthwork (English Civil War?) at Wallingford Castle, Oxfordshire. *Journal of Micropalaeontology*, **29**: 87–92.

Wilkinson, I.P., Williams, M., Young, J.R. et al., 2008. The application of microfossils in assessing the provenance of chalk used in the manufacture of Roman mosaics at Silchester. *Journal of Archaeological Science*, **35**: 2415–22.

Wilkinson, K. & Stevens, C., 2003. *Environmental Archaeology: Approaches, Techniques and Applications*. Tempus.

Williams, H. F. L., 1999. Foraminiferal distributions in tidal marshes bordering the Strait of Juan de Fuca: implications for paleoseismicity studies. *Journal of Foraminiferal Research*, **29**(3): 196–208.

Williams, D. B. & Sarjeant, W. A. S., 1967. Organic-walled microfossils as depth and shoreline indicators. *Marine Geology*, **5** (5/6): 389–412.

Williams, M., Haywood, A. M., Gregory, F. J. & Schmidt, D. N. (eds.), 2007a. *Deep-Time Perspectives on Climate Change*. The Geological Society, London (The Micropalaeontological Society Special Publication).

Williams, M., Leng, M. J., Stephenson, M. H. *et al.*, 2006. Evidence that Early Carboniferous ostracods colonised coastal flood plain brackish water environments. *Palaeogeography, Palaeoclimatology, Palaeoecology*, **230**: 299–318.

Williams, M., Siveter, D. J., Popov, L. E. & Vannier, J. M. C., 2007b. Biogeography and affinities of the bradoriid arthropods: cosmopolitan microbenthos of the Cambrian seas. *Palaeogeography, Palaeoclimatology, Palaeoecology*, **248**: 202–32.

Williams, M. D., 1958. Stratigraphy of the Lower Indus basin, West Pakistan. *Proceedings, 5th World Petroleum Congress*: 377–94.

Williamson, K. & Stevens, C., 2001. *Environmental Archaeology*. Tempus.

Willoughby, P. R., 2007. *The Evolution of Modern Humans in Africa*. Altamira Press.

Wilson, B., 2003. Foraminifera and paleodepths in a section of the Early to Middle Miocene Brasso formation, central Trinidad. *Caribbean Journal of Science*, **39**(2): 209–14.

2004. Benthonic foraminiferal paleoecology across a transgressive-regressive cycle in the Brasso formation (Early-Middle Miocene) of central Trinidad. *Caribbean Journal of Science*, **40**(1): 126–38.

2005. Planktonic foraminiferal biostratigraphy and paleoecology of the Brasso formation (Middle Miocene) at St Fabien quarry, Trinidad, West Indies. *Caribbean Journal of Science*, **41**(4): 797–803.

2006. Trouble in paradise? A comparison of 1953 and 2005 benthonic foraminiferal seafloor assemblages at the Ibis field, offshore eastern Trinidad, West Indies. *Journal of Micropalaeontology*, **25**: 157–64.

Winchester-Seeto, T., Foster, C. & O'Leary, T., 2000. The environmental response of Middle Ordovician large organic-walled microfossils from the Goldwyer and Nita formations, Canning basin, Western Australia. *Review of Palaeobotany and Palynology*, **113**: 197–212.

Wing, S. L., Gingerich, P. D., Schmitz, B. & Thomas, E. (eds.), 2003. *Causes and Consequences of Globally Warm Climates in the Early Paleogene*. Geological Society of America (Special Paper No. 369).

Wnuk, C., 1996. The development of floristic provinciality during the Middle and Late Paleozoic. *Review of Palaeobotany and Palynology*, **90**(1–2): 79–98.

WoldeGabriel, G., Ambrose, S. H., Baroni, D. *et al.*, 2009. The geological, isotopic, botanical, inverterate and lower vertebrate surroundings of *Ardipithecus ramidus*. *Science*, **326** (5949): 65, 65e1–e5.

Wong, K., 2006. Lucy's baby. *Scientific American*, **295**(6): 56–63.

Wood, C., 2003. *Environmental Impact Assessment: A Comparative Review*, 2nd edition. Pearson/Prentice Hall.

Wood, L., 2000. Chronostratigraphy and tectonostratigraphy of the Columbus basin, offshore eastern Trinidad. *American Association of Petroleum Geologists Bulletin*, **84**(12): 1905–28.

Woodside, P. R., 1981. Petroleum geology of Trinidad and Tobago. *Oil and Gas Journal*, 28 September 1981: 314–87.

Worsley, D. & Aga, O. J., 1986. *The Geological History of Svalbard*. Statoil.

Wright, T., 2001. *Bolivinoides* saves Ipswich from flooding disaster! *Newsletter of Micropalaeontology*, **65**: 17.

Wynn, J. G., Alemseged, Z., Bobe, R. *et al.*, 2006. Geological and palaeontological context of a Pliocene juvenile early hominin at Dikika, Ethiopia. *Nature*, **443**: 332–6.

Xiangjun Sun & Pinxian Wang, 2005. How old is the Asian monsoon system? Palaeobotanical records from China. *Palaeogeography, Palaeoclimatology, Palaeoecology*, **222**(3–4): 181–222.

Xiao-Yan Song, Spicer, R. A., Jian Yang *et al.*, 2010. Pollen evidence for an Eocene to Miocene elevation of central Southern Tibet pre-dating the rise of the high Himalaya. *Palaeogeography, Palaeoclimatology, Palaeoecology*, **297**: 159–68.

Xiao-Yu Jang, 2008. A male cone of *Pseudofrenelopsis dalatiensis* with *in situ* pollen grains from the Lower Cretaceous of northeast China. *Geobios*, **41**(5): 689–98.

Xiqiu Han, Suess, E., Yongyang Huang *et al.*, 2008. Jiulong methane reef: microbial mediation of seep carbonates in the South China Sea. *Marine Geology*, **249**: 243–56.

Xuefeng Yu, Weijian Zhou, Xiaoqing Liu *et al.*, 2010. Peat records of human impacts on the atmosphere in northwest China during the late Neolithic and Bronze Ages. *Palaeogeography, Palaeoclimatology, Palaeoecology*, **286**: 17–22.

Yalden, D., 1999. *The History of British Mammals*. Poyser Natural History.

Yanagisawa, Y. & Akiba, F., 1998. Refined Neogene diatom biostratigraphy for the northwest Pacific around Japan, with an introduction of code numbers for selected diatom biohorizons. *Journal of the Geological Society of Japan*, **104**: 395–414.

Yang Wang & Tao Deng, 2005. A 25 m.y. isotopic record of paleodiet and environmental change from fossil mammals and paleosols from the NE margin of the Tibetan Plateau. *Palaeogeography, Palaeoclimatology, Palaeoecology*, **236**: 322–38.

Yang-Logan, L. C., Tveit, R., Bailey, H. W. & Gallagher, L. T., 1996. Biosteering – a biostratigraphic application to horizontal drilling in the Eldfisk field, Norwegian North Sea. *Transactions of the 1995 AAPG Mid-Continent Section Meeting.*

Yasuhara, M., Yoshikawa, S. & Nanayama, F., 2005. Reconstruction of the Holocene seismic history of a seabed fault using relative sea-level curves reconstructed by ostracode assemblages: case study on the Median Tectonic Line in Iyo-nada Bay, western Japan. *Palaeogeography, Palaeoclimatology, Palaeoecology*, **222**: 285–312.

Yongbo Peng, Huiming Bao & Xunlai Yuan, 2009. New morphological observations for Paleoproterozoic acritarchs from the Chuanlinggou formation, north China. *Precambrian Research*, **168**: 223–32.

Yongqiang Zong, Kemp, A. C., Fengling Yu *et al.*, 2010. Diatoms from the Pearl River estuary, China and their suitability as water salinity indicators in coastal environments. *Marine Micropaleontology*, **75**: 38–49.

Young, E., 2007. Pharoahs from the Stone Age. *New Scientist*, 13 January, 2007: 34–8.

Young, M. T., Rayfield, E. J., Barrett, P. M. *et al.*, 2007. Elucidating the feeding mechanics of *Diplodocus longus* using the finite-element method. *Abstracts, The Palaeontological Association 51st Annual Meeting, Uppsala*: **101**.

Yu, K.-F., Zhao, J.-X., Wei, G.-J. *et al.*, 2005. Del 18O, Sr/Ca and Mg/Ca records of *Porites lutea* corals from Leizhou peninsula, northern South China Sea, and their applicability as paleoclimatic indicators. *Palaeogeography, Palaeoclimatology, Palaeoecology*, **218**: 57–73.

Yuan Xunlai, 2002. *Doushantuo Fossils: Life on the Eve of Animal Radiation*. University of Science and Technology of China Press. In Chinese with English summary.

Zaborski, P. M. & Morris, N. J., 1999. The Late Cretaceous ammonite genus *Libycoceras* in the Iullemeden basin (west Africa) and its palaeogeographical significance. *Cretaceous Research*, **20**: 63–79.

Zaigham, N. A. & Mallick, K. A., 2000. Prospect of hydrocarbon associated with fossil rift structures of the southern Indus basin, Pakistan. *American Association of Petroleum Geologists Bulletin*, **84**(11): 1833–48.

Zalasiewicz, J., Taylor, L., Rushton, A. W. A. *et al.*, 2009. Graptolites in British stratigraphy. *Geological Magazine*, **146**(6): 785–850.

Zalasiewicz, J., Williams, M., Miller, M. *et al.*, 2007. Early Silurian (Llandovery) graptolites from central Saudi Arabia: first documented record of Telychian faunas from the Arabian peninsula. *GeoArabia*, **12**(4): 15–36.

Zamora, S., 2010. Middle Cambrian echinoderms from north Spain show echinoderms diversified earlier in Gondwana. *Geology*, **38**(6): 507–10.

Zanchetta, G., Leone, G., Fallick, A. E. & Nonadonna, F. P., 2005. Oxygen isotope composition of living land snail shells: data from Italy. *Palaeogeography, Palaeoclimatology, Palaeoecology*, **223**: 20–33.

Zapata, E., Padron, V., Madrid, I. *et al.*, 2003. Biostratigraphic, sedimentologic and chemostratigraphic study of the La Luna formation (late Turonian-Campanian) in the San Miguel and Las Hernandez sections, western Venezuela. *Palaios*, **18**: 367–77.

Zdrakov, A. & Kortenski, J., 2004. Maceral composition and depositional environment of the coals from Beli Breg basin, Bulgaria. *Review of the Bulgarian Geological Society*, **65**(1–3): 157–66.

Zhang, S., Barnes, C. R. & Jowett, D. M. S., 2006. The paradox of the global standard Late Ordovician-Early Silurian sea level curve: evidence from conodont community analysis from both Canadian Arctic and Appalachian margins. *Palaeogeography, Palaeoclimatology, Palaeoecology*, **236**(3–4): 246–71.

Zhang Zhongshi, Huijun Wang, Zhengtang Guo & Dabang Jiang, 2007. What triggers the transition of palaeoenvironmental patterns in China, the Tibetan Plateau uplift or the Paratethys Sea retreat? *Palaeogeography, Palaeoclimatology, Palaeoecology*, **245**: 317–31.

Zhong-Qiang Chen, Jinnan Tong, Zhuo-Ting Liao & Jing Chen, 2010. Structural changes of marine communities over the Permian-Triassic transition: ecologically assessing the end-Permian mass extinction and its aftermath. *Global and Planetary Change*, **73**: 123–40.

Zhong-Qiang Chen, Kaiho, K. & George, A. D., 2005a. Survival strategies of brachiopod faunas from the end-Permian mass extinction. *Palaeogeography, Palaeoclimatology, Palaeoecology*, **224**: 232–69.

2005b. Early Triassic recovery of the brachiopod faunas from the end-Permian mass extinction: a global review. *Palaeogeography, Palaeoclimatology, Palaeoecology*, **224**: 270–90.

Zhuravlev, A. Y. & Riding, R. (eds.), 2001. *The Ecology of the Cambrian Radiation*. Columbia University Press.

Ziegler, A. M., Eshel, G., McAllister Rees, P. *et al.*, 2003. Tracing the tropics across land and sea: Permian to present. *Lethaia*, **36**: 227–54.

Ziegler, K., Turner, P. & Daines, S. E. (eds.), 1997. *Petroleum Geology of the Southern North Sea: Future Potential*. The Geological Society, London (Special Publication No. 123).

Ziegler, M. A., 2001. Late Permian to Holocene paleofacies evolution of the Arabian plate and its hydrocarbon occurrences. *GeoArabia*, **6**(3): 445–503.

Zutshi, P. L. & Panwar, M. S., 1997. *Geology of Petroliferous Basins of India*. Dehra Dun Oil and Natural Gas Corporation.

Index

Bold type indicates figures.

Printed in the United States
By Bookmasters